The Lynx and the Telescope

Scientific and Learned Cultures and Their Institutions

Editor

M. Feingold (*California Institute of Technology*)

VOLUME 21

The titles published in this series are listed at *brill.com/slci*

The Lynx and the Telescope

The Parallel Worlds of Cesi and Galileo

By

Paolo Galluzzi

Translated by

Peter Mason

BRILL

LEIDEN | BOSTON

The translation of this work has been funded by SEPS (SEGRETARIATO EUROPEO PER LE PUBBLICAZIONI SCIENTIFICHE), Via Val d'Aposa 7, 40123 Bologna, Italy, seps@seps.it, www.seps.it

This book was originally published in Italian as *Libertà di filosofare in naturalibus: I mondi paralleli di Cesi e Galileo* (Storia dell'Accademia dei Lincei, Studi 4). Rome: Scienze e Lettere, Editore Commerciale, 2014.

Cover illustrations: Front: Stroganov bridge, Odessa, by Gennadiy Verschagin. Courtesy of the painter. Back: The author, photographed by Vassilis Colonas.

The Library of Congress Cataloging-in-Publication Data is available online at http://catalog.loc.gov

Typeface for the Latin, Greek, and Cyrillic scripts: "Brill". See and download: brill.com/brill-typeface.

ISSN 2352-1325
ISBN 978-90-04-34231-6 (hardback)
ISBN 978-90-04-34232-3 (e-book)

This book is printed on acid-free paper and produced in a sustainable manner.

Contents

List of Figures

Preface

In the virtually endless bibliography on Galileo, Cesi and the First Academy of the Lincei the contributions that propose a detailed analysis of the relations between the Tuscan scientist and the Roman Prince, comparing their respective cultural projects, the visions from which they drew inspiration and the methods that they practised, can be counted on the fingers of one hand.

Even more surprisingly, we still lack an exhaustive intellectual biography of Federico Cesi, a critical reconstruction that throws light on the motivations that guided his brief but intense existence, on his points of reference, on how he experienced the encounter with Galileo and reacted to the celestial novelties, to his mechanico-corpuscular analysis of physical phenomena and to his thesis of the mathematical structure of the Book of Nature. In a word, how did Cesi perceive the relation between Galileo's innovative project and the research agenda that he had drawn up as an adolescent and that had led him to found the Accademia dei Lincei?

Shifting our point of view to Galileo's perception of the motivations and intellectual commitment of his generous and affectionate Patron, the panorama appears even more depressing. One would search in vain for scholarly contributions which focus on the Tuscan scientist's evaluation of the priorities of Cesi and the judgment that he formed of the way in which the Prince and his collaborators interpreted the role of supporters and promoters of his new conception of the universe and of the relations between man and nature.

In the course of what can be defined as a new season of studies, opened thirty years ago by the contributions of Giuseppe Olmi, the cliché of the 'Galilean Academy', or rather of its 'conversion' to the new science after the Tuscan scientist's association, has appeared increasingly problematic.

The longstanding success of the image of the Lincei as a 'Galilean Academy' has prevented a correct interpretation of the way – neither passive, nor subaltern – in which Cesi and the founding members experienced Galileo's clamorous arrival on the scene and the conflicts in which he became embroiled. Similarly, Galileo's perception of the programme and goals of the Academy and of the activities and studies of its Patron have long appeared to scholars as topics of marginal interest.

The present work aims at filling this striking lacuna in the historiography on the First Academy. I set myself the task of viewing Cesi and his closest collaborators through the eyes of Galileo, and Galileo through the eyes of Cesi. Focusing on the documents that illustrate the dynamics of a relationship that developed over the course of twenty years of continuous physical and epistolary contact, a significantly different picture emerges.

The subtitle of this work expresses my provisional conclusions. Due to the considerable differences that characterise their visions and their reform projects, it is neither possible nor useful to consider the personalities of Galileo and Cesi as perfectly aligned. Their relationship can be more adequately represented by the image of two parallel lines that, while directed at a common target, never meet. Both men set out with a resolute determination to demolish the traditional system of knowledge, not only in the sphere of natural philosophy. However, the instruments and methods with which they confronted that challenge and the results that they hoped to achieve appear in many respects far from converging.

The sincere and loyal collaboration established between Cesi and Galileo was based on their full awareness of what might be defined, to use an oxymoron, as divergent convergence. They needed each other, even if they both knew that they would be able to proceed together along only a part of the journey on which they had embarked.

Motivated by the vision conceived many years earlier under the influence of authors (Della Porta, Heckius, Telesio, Persio, Campanella, Ficino and Paracelsus) whom it would be difficult to associate with the Tuscan scientist, Cesi strove to establish an organic link between the adventure he had undertaken and the new methods and worlds revealed by Galileo. Though not without elements of ambiguity, Federico convinced himself that their trajectories were substantially convergent: the Prince and the Academy would complete the work begun by Galileo, extending his rigorous researches on the structure of the universe to the no less important project of casting light on phenomena on the surface of the Earth, in its bowels and in its atmosphere. In this way Cesi hoped to demonstrate that the same vital principles operate in the heavens explored by Galileo thanks to the telescope and in the animal, plant and even mineral kingdoms on which he focused his gaze.

This was the image that Cesi and his closest collaborators, above all Stelluti and Faber, publicly disseminated to underline the direct continuity and full integration of the work and aims of the Academy with the scientific endeavours of Galileo.

As for the Tuscan scientist, one receives the impression that, if he had ever tried to do so, he soon abandoned the attempt to establish thematic and methodological convergences with the programmes that were dear to the Princeps of the Lincei.

The closest point of convergence between Cesi and Galileo was reached in their shared conviction of the fluid, unitary and corruptible nature of the heavens. However, even in their interpretation of this principle noteworthy differences in their personalities and expectations, as well as in their overall vision of nature and its operations, clearly emerge. Also the documents that bear witness to the generous and impassionate support that Cesi offered to Galileo in his dramatic clashes with the Church authorities betray their different views on the significance of those battles, on the objectives to be attained, and how to estimate and, above all, to accept the risks that that enterprise entailed.

My research on the relations between Galileo and Cesi, which began many years ago, led to a first result in the lecture that I delivered at the meeting organized by the Accademia dei Lincei on 16 December 2011 to celebrate the 400th anniversary of the induction of Galileo.[1] Research conducted for the preparation of that presentation, which was limited to a reconstruction of the first encounter between Galileo and Cesi in Rome in the winter of 1611, convinced me that it would be worthwhile to widen the perspective following the evolution of their personal and intellectual relations, as well as of the circles with which they were connected, down to Federico's death in 1630.

During the preparation of this work, I have contracted profound debts of gratitude towards many friends and colleagues. I discussed the conclusions to which my research was leading on several occasions with Tullio Gregory, who provided continuous encouragement and a determinant support to the publication of this work in the prestigious Academy's editorial series *Storia dell'Accademia Nazionale dei Lincei*.[2] Thanks to his stimulus and the efforts of the then Vice President (now President) of the Lincei, Alberto Quadrio Curzio, the Institution founded by Cesi has drawn up an ambitious programme of publications of the sources for the history of the Academy. Massimo Bucciantini, Michele Camerota, Giuseppe Olmi and Patrizia Ruffo read this work before it went into print and offered valuable suggestions. I am indebted to Marco Guardo for enabling me to consult texts and documents that would otherwise have been extremely difficult to access.

As the reader will see, I have examined a not inconsiderable quantity of sources bearing on Galileo, Cesi and the history of the First Accademy of the Lincei. In carrying out this vast programme I benefitted enormously from the on-line consultation of the manuscripts of the Archivio Linceo, published on the website of the Museo Galileo in Florence,[3] as well as from the wealth of primary sources and critical literature held by the library of the Museo Galileo and from the competent assistance of the library staff.

1 Galluzzi 2014b.

2 Galluzzi 2014a.

3 http://www.museogalileo.it/istituto/biblioteca-digitale-tematica/lincei/iarchivio.html (accessed on 10 July 2017).

The English translation of this work has been made possible by a generous contribution from SEPS, a meritorious institution that supports the international circulation of the most qualified Italian cultural production, as well as the translation into Italian of important works by authors from other countries.

I am grateful to Peter Mason for having tackled and brought to completion a particularly complex translation.

I express my gratitude to Mordechai Feingold for accepting this work in the prestigious series of Koninklijke Brill N.V. that he directs.

Florence, July 2017

Abbreviations

CL	G. Gabrieli (ed.), *Il carteggio linceo*, Rome 1996 [reprint of the original edition published in *Atti della Reale Accademia Nazionale dei Lincei*, a. CCCXXXV, 1938, Serie Sesta. Memorie della Classe di Scienze Morali, Storiche e Filologiche, VII, Rome 1938].
Contributi	G. Gabrieli, *Contributi alla storia della Accademia dei Lincei*, Rome 1989, 2 vols. [the two volumes contain the innumerable contributions of Giuseppe Gabrieli on Federico Cesi, the history of the First Academy and its fellows].
DBI	*Dizionario biografico degli italiani*, Rome 1960–.
Gesta	M. Guardo, R. Orioli (eds.), *Cronache e statuti della prima Accademia dei Lincei*, Rome 2014.
Indicatio	*Indicatio philosophicorum operum quae Federicus Caesius Lynceus Princeps I Fed: F. Princ etc.: sibi condixit*, in Contributi I, pp. 30–58.
Lynceographum	F. Cesi, *Lynceographum, quo norma studiosae vitae Lynceorum philosophorum exponitur*, A. Nicolò (ed.), Rome 2001.
OG	*Le opere di Galileo Galilei: Edizione Nazionale sotto gli auspicii di Sua Maestà il Re d'Italia*, A. Favaro (ed.), Florence, Tip. di G. Barbèra, 1890–1909, 20 vols. (reprint: Florence 1929–1939).
TM	*Rerum Medicarum Novae Hispaniae Thesaurus, seu Plantarum Animalium Mineralium Mexicanorum Historia, ex Francisci Hernandez Novi Orbis Medici Primarij relationibus in ipsa Mexicana Urbe conscriptis, a Nardo Antonio Reccho* [...] *collecta ac in ordinem digesta, a Ioanne Terrentio* [...] *notis illustrata. Nunc primum in naturalium rerum studiosorum gratiam lucubrationibus Lynceorum publici iuris facta*, Romae, ex Typographeio Vitalis Mascardi, 1649 [facsimile edition, Rome 1992].
Zibaldone	See Indicatio.

Note to the Reader

Federico Cesi adopted the title of Marquis of Monticelli from early adolescence. After the papal breve of Paul V of 28 January 1613, he acquired that of Prince of St Angelo and St Polo. Towards the end of his brief life, in June 1630, he also became Duke of Acquasparta (see Contributi 1989, I, p. 118). For the sake of simplicity and to avoid confusion, Cesi is always indicated as Prince in the present work, even when the reference is to events preceding the granting of that title.

The correspondence between Cesi and the Linceans, on the one hand, and Galileo, on the other, was published in the National Edition of the *Opere di Galileo*, edited by Antonio Favaro, as well as in the *Carteggio Linceo*, edited by Giuseppe Gabrieli. To avoid the extremely frequent double reference for the many epistolary documents cited, the following criteria have been adopted:

- for the letters of Galileo, those sent to him by the Linceans and other correspondents, and those between correspondents outside the circle of the Linceans that contain direct or indirect mention of Galileo, the reference is to the National Edition of the *Opere* di Galileo (OG);
- for the letters between Linceans and fellow Linceans, between Linceans and their correspondents, and between persons outside the circle of the Linceans that contain information related to the Lincean Academy and to its direct or indirect relations with Galileo, the reference is to the *Carteggio Linceo* (CL).

Latin texts are always translated into English. The original Latin text is provided in the footnotes only for those passages of Cesi's *Indicatio* (*Zibaldone*), *Apiarium* and *Tabulae Phytosophicae* which present a particularly complex syntactical construct and/or a peculiar terminology that render their interpretation problematic.

CHAPTER 1

"The Secret of the Eyeglass"

A 'Piece of Nonsense'?[1]

I have seen the secret of the eyeglass and it's a piece of nonsense, as well as being taken from Book 9 of my *De Refractione*;[2] and I will describe it so if you want to build it you will have fun.

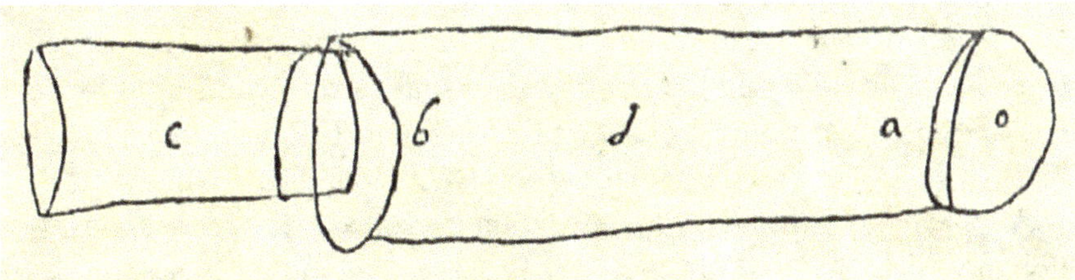

It is a solid silver tube measuring a palm from *a* to *d*, three inches in diameter, with a convex lens at one end (*a*). There is another tube 4 inches long that fits into the first, with a concave [lens] at the end (*b*), like the first. If you look through the first one alone, distant things will appear close by; but because the vision is not made inside the tube, they appear obscure and indistinct. If you insert the other tube with the concave lens, this produces the opposite effect: you see things clearly and distinctly. It goes in and out like a trumpet to adjust to the eyesight of the beholder, since it varies from one person to another.[3]

1 Favaro, Gabrieli and all the scholars who have published or cited the letter of Della Porta to Cesi have transcribed as 'coglionaria' the term used in that letter (Ms. Archivio Linceo XII, cc. 326–332v). A close reading of the manuscript shows that Della Porta wrote not 'coglionaria' (Freedberg 2002, p. 101 renders it as 'a load of balls') but 'castronaria', that is, a piece of nonsense as silly as a castrated ram. Obviously the meaning does not change significantly. In his translation of an essay by A. van Helden citing Della Porta's letter, Giorgio Strano was the first to draw attention to the error of Favaro and Gabrieli (Strano 2008, pp. 39 and 132).

2 As Favaro pointed out (OG X, p. 252, n. 1), neither Book IX nor Book VIII (*De specillis*) of *De refractione* (Della Porta 1593) contains any reference to the telescope.

3 CL, pp. 114–5.

 | DOI 10.1163/9789004342323_002

This detailed description of the functioning of the telescope, including the instructions for its manufacture, accompanied by a diagram for clarification, is contained in a letter sent by Giovan Battista Della Porta in Naples to Federico Cesi in Rome on 28 August 1609.

A year had passed since the arrival of the first reports of the sensational performances of the new device, made by craftsmen in the Netherlands who each claimed to have been the first to invent it and to be entitled to the rights to its exploitation. News of the invention had spread among the main cities of Europe, where exemplars of the curious object were put on sale in increasing numbers.[4]

Italy was no exception. A few weeks before Della Porta's letter to Cesi, on 21 July 1609, the learned Servite Paolo Sarpi had written to Francesco Castrino in Paris that 'that telescope that makes distant things seem near' had appeared in Venice.[5] Although struck by the new invention, Sarpi was not convinced of its practical utility for military purposes, in spite of the claims of its promoters. Another letter from Sarpi shows that news of the telescope had already reached Venice around the beginning of November of the previous year.[6]

Sarpi had not revealed the identity of the person who introduced the invention to Venice. Neither did Giovanni Bartoli, secretary to the Tuscan Resident to the Serenissima, in his letter to the Secretary of State of the Grand Duchy of Tuscany, Belisario Vinta, dated 22 August (that is, a week before Della Porta's letter to Cesi):

> Here somebody has offered to the Senate to reveal the secret of a telescope or eyeglass or some other instrument capable of observing objects at a distance of 25 to 30 miles so clearly that they seem to be on the spot. Many have seen it and tried it from the bell tower of San Marco. But it is said that this secret is common knowledge by now in France and elsewhere and that it can be bought at little expense. Many say they have had it and seen it [...].[7]

It is clear from the testimony of Sarpi and Bartoli that the invention had been brought to Venice from abroad and that the attempts of its promoters to sell

4 On the circulation of telescopes in the main cities of Europe and the debates that it aroused, see Bucciantini-Camerota-Giudice 2015. See also Van Helden-Dupré-Van Gent-Zuidervaart 2014.

5 Ibid., pp. 33–7 and 53–7; see also Biagioli 2010.

6 On 9 December 1608 Sarpi wrote to Castrino that he had received news of the new eyeglass a month earlier; cf. Sarpi 1931, p. 15, Biagioli 2010, pp. 203–30, and Bucciantini-Camerota-Giudice 2015, pp. 33ff.

7 OG X, p. 250.

exclusive rights to the government of the Venetian Republic were doomed to fail since the instrument could be purchased inexpensively in France and other countries. It is worth noting that the scepticism of the two men regarding the usefulness of the telescope is in complete agreement with the definition of the instrument as a 'piece of nonsense' that Della Porta was to employ a few days later.

Bartoli's letter tells us that the privileged observatory for the public demonstration of the powers of the instrument, on whose name there was no consensus (it was variously referred to as *tromba*, *trombone*, *cannone*, *visorio*, etc.), was the bell tower of San Marco. As elsewhere, the choice of the tallest accessible location was dictated by the need to show how much in advance the telescope enabled observers to spot ships on the lagoon that were approaching the city. It is not surprising that it was above all the instrument's military potential that was emphasised to enhance its value in the eyes of potential buyers.

The failure of the unknown promoter of the telescope to impress the Venetian Senate predicted in Bartoli's letter is in striking contrast with another document, penned only one day before by Galileo, which confirmed many of the details contained in the Tuscan diplomat's letter but which presented a completely different scenario:

> I was summoned by the Serenissima Signoria, to which and to the whole of the Senate I demonstrated it [the telescope], to the utmost bewonderment of all; and there were very many gentlemen and senators who, in spite of their years, scaled the staircases of the tallest bell towers in Venice more than once to descry the sails and vessels so far away that, although they were travelling with full sail, more than two hours passed before they could be seen without the aid of my telescope.[8]

This animated depiction of the life and customs of the Venetian élite refers to the public demonstration of a telescope with a magnifying power of 10,[9] organised in the presence of the supreme authority of the Venetian Republic on the bell tower of San Marco on 21 August, that is, one day before Bartoli's sceptical report to Florence. On 24 August, Galileo officially presented the instrument to the Doge Donato,[10] and a few hours later received confirmation of life tenure at the University of Padua with a remarkable increase in salary.[11]

8 Ibid., p. 253 (letter to Benedetto Landucci, 29 August 1610).

9 In Galileo's words, 'the effect of this instrument is to represent an object 50 miles away as large and clear as if it were 5 miles away', ibid.

10 Ibid., pp. 253–4.

11 Ibid., p. 254.

It remains puzzling why the secretary to the Resident, whose principal function was to transmit reliable information as soon as possible, had passed on to Florence such an outdated report, moreover one concerning a citizen of the Grand Duchy of Tuscany.

But to return to Naples and Rome, how did Cesi and Della Porta learn of the 'secret of the eyeglass'? Since the two men exchanged letters in the same days that saw the triumph of Galileo's telescope in Venice, the possibility that Della Porta's letter was written in response to those events must be excluded. It is regrettable that the letter of Cesi to which the Neapolitan replied has been lost, as it would have provided valuable information on the source from which news of the appearance of the telescope spread in Rome. We must therefore try to identify that source only on the basis of clues.

From boyhood Federico Cesi was on friendly terms with Cardinal Bellarmine, a prominent figure in the Curia and a fundamental reference point for the Society of Jesus.[12] The founder of the Lincei also frequented the Collegio Romano, where the German mathematician Christopher Clavius and a group of fellow Jesuits had distinguished themselves in recent decades for the quality of their studies and publications, above all in astronomy and mathematics, as well as for their skill in the construction of precision instruments.[13] As we shall see, from the spring of 1610 onwards, the scientific authority of the Fathers of the Collegio Romano would play a decisive role in the controversial process that led to recognition of the existence of the celestial discoveries made by Galileo with the aid of the telescope. Little disposed to acknowledge the merits of others – at least, this was how they were regarded by many contemporaries and among the Linceans[14] – the Jesuit Fathers strongly emphasised the fact that they had built their own telescopes before the publication of the *Sidereus Nuncius* (i.e. before March 1610).[15] If they were not boasting, they had thus fabricated their own instruments before knowing about Galileo and independently of him. In that case, the Fathers of the Collegio Romano would have been entitled to claim to have been the first, along with

12 Cesi probably made Bellarmine's acquaintance in Capua, where the Jesuit was Bishop, during his return journey to Rome after his long stay in Naples in 1604 (cf. Odescalchi 1806, p. 78).

13 On the scientific activity of the Fathers of the Collegio Romano, see Baldini U. 1995 and 2003; see also Lattis 1994.

14 On Stelluti's harsh judgements see below, p. 138, n. 104. Fabio Colonna also criticised the inclination of the Jesuits to appropriate discoveries made by others (see his letter to Stelluti of 10 October 1628, CL, p. 1188).

15 OG XI, pp. 33–4. Letter of Christoph Grienberger to Galileo, 22 January 1611.

the Tuscan scientist, to have realised that the new instrument not only had military potential, but that its most sensational novelty lay in the possibility it offered to explore the sky.

Other documents, to be discussed later, indicate that the principal contribution to the early construction of telescopes with which the associates of the so-called 'Academy of Father Clavius' in the Collegio Romano conducted the first celestial observations came from Father Giovanni Paolo Lembo, who was particularly skilled in mechanical operations.[16]

The rapidity with which the Fathers had news of the telescope and their early dedication to the construction of this new instrument suggest that the Roman Jesuits may have been the source from which Cesi and Della Porta first learned of the 'secret of the eyeglass' at a date that precludes the reverberation of the events in Venice.

At the centre of an unparalleled information network, the Fathers of the Collegio Romano became rapidly acquainted with the appearance of the new instrument in numerous cities of Europe. They would receive news of the novelties revealed by Galileo in the *Sidereus Nuncius* with even greater celerity. In fact, on 12 March 1610, one day before the *Nuncius* had even been published, Markus Welser wrote to Clavius from Augsburg that he had heard from Padua 'as a certain and established fact' that Galileo had observed novelties in the sky thanks to an instrument that he called *visorio*:

> he has discovered [...] four new planets, new to our knowledge and never seen before [...], with very many fixed stars that have never been known or seen before [...]. Welser was sceptical, but deferred to the authority of Clavius: I well know that doubt is the beginning of wisdom; for this reason I suspend judgement, but beg your Reverence to freely give me your opinion in confidence [...].

As to the channel by which Della Porta received information about those novelties, we may hazard an even more precise identification. The source from which he learnt of the new instrument might have been the above mentioned Jesuit Giovanni Paolo Lembo, a native of Benevento who had joined the Society of Jesus in Naples in 1600. A letter written a few months later to Galileo by another Father of the Collegio Romano, Christoph Grienberger, provides evidence that Lembo had made some telescopes for the use of his fellow Jesuits before hearing about those made by Galileo: 'before he [Father Lembo]

16 Ibid., p. 33; cf. note 19 and p. 14.

had had any knowledge of your lenses, he built these instruments without ever having seen them, but simply by dint of conjecture [...]'.[17]

That the Jesuit was Della Porta's source of information about the new invention before the latter had received Cesi's letter is suggested by the declaration that the Neapolitan later made to Federico concerning the frequent contact and reciprocal esteem between himself and Father Lembo. As Della Porta told Cesi, although he had been unable to observe the celestial novelties revealed in the *Sidereus Nuncius*[18] because of the modest performance of the instruments at his disposal, he had no doubt about their objective existence due to the assurances received from Father Lembo, in whom he placed the greatest confidence:

> The Jesuit Father Paolo Lembo, highly skilled in mathematics and mechanics [...], reported that his observations of their motions [of the satellites of Jupiter] concurred with those of Galileo, which I have no difficulty in believing.[19]

---

17 OG XI, p. 33.

18 Early on Della Porta received news of the celestial discoveries disclosed by Galileo in the *Sidereus Nuncius* from the Neapolitan Marquis Giovan Battista Manso, who had received them from Paolo Beni, an ex-Jesuit and colleague of Galileo in the University of Padua. In his undated reply to Beni, which we can assign to the second half of March 1610, Manso wrote that he had 'reread your letter many times, and shared it with Signor Porta'. He added that 'the invention of the eyeglasses' by Galileo 'has provoked considerable jealousy on the part of our Signor Porta, who had thought some time ago that [telescopic magnification] could be extended infinitely (I mean as far as vision can extend, once obstacles have been removed, by proportioning the distance between the concave and convex lenses)', OG X, p. 292.

19 CL, p. 309. Gabrieli assigned this undated letter to 1612. Although Della Porta was on excellent terms with Father Lembo, his relationship with the Society of Jesus was far from idyllic. The Fathers not only viewed his activities in astrology, alchemy and natural magic with suspicion, but they also regarded him as a pole of attraction that interfered with their ambition of recruiting the sons of the Neapolitan nobility in the Jesuit schools (Gatto 1994, pp. 33–7). Della Porta's path crossed that of Robert Bellarmine for a brief period when the latter was Provincial of the Society of Jesus in Naples from the autumn of 1594 to the end of 1596. The later testimony of Father Santagata, reveals not only the fierce competition between the Jesuits and Della Porta, but also the reasons for it and the role played by Bellarmine: 'The most useful result of the private conversations of Bellarmine was the bringing into discredit of the new Academy of dubious reputation created by Gio. Battista Della Porta. This good philosopher and better physiognomist had built up a sizable following and created a school called 'of Secrets'. Nobody was admitted who had not merited entry with some new discovery beneficial to the health of the body or to the progress of the arts. Since not all those seeking admission had the means to do so, it resulted in very serious errors: some studied condemned books, others became

Let us leave the exchange of letters between Naples and Rome to follow the brilliant success that Galileo was enjoying in the Republic of Venice thanks to the generous resolutions of the Senate of the Serenissima. It should be borne in mind that the appearance of the telescope on Galileo's horizon came absolutely out of the blue. In the spring of 1609 he was neither engaged in studying optical phenomena and laws nor in polishing lenses. Describing his current researches in a letter of 11 February 1609 to Antonio de' Medici, an influential member of the Florentine dynasty with whom he was on friendly terms, he reported that he was immersed in mathematical demonstrations and experiments bearing on his treatise on mechanics, focussing in particular on the resistance of materials and the parabolic trajectory of projectiles.[20]

Between the spring and summer of 1609, fundamental sections of the project to reform traditional mechanics and the science of motion that Galileo had conceived right from the start of his scientific career were taking shape. That project was to reach completion with the publication of the *Two new sciences*[21] almost thirty years later. That his priority in those months was in the study of mechanics and motion is confirmed by the letter sent to him on 4 April from the learned mathematician Luca Valerio. Valerio expressed his approval of a theorem on the centre of gravity of solids (unfortunately lost) that Galileo

followers of women poisoners, others squandered their wealth in chemical experiments and destroyed precious items. Many of those admitted hoped to emulate the glory of their Master, who for some divinations regarding future events was held to be little less than a prophet, and therefore other abuses proliferated that it is not difficult to imagine. When Bellarmine was asked about this School, he disapproved of it in every aspect, and his opinion helped to reduce its reputation and popularity. After the Jesuit's disapproval, no persons of note frequented it except Gio. Battista Marino, though he too withdrew when he saw the Master summoned for impious conduct' (ibid., pp. 34–5). Della Porta's religious conduct appears elusive and contradictory. As Romano Gatto has emphasised, an anonymous manuscript in praise of Della Porta, housed in the Bibliothèque Nationale in Paris, presents a completely different picture from that outlined by Father Santagata: 'He then lived a very Christian life and generously gave of his wealth to the poor [...]. He was for 30 years a brother of the Congregation of Jesuit Fathers in the Casa Professa, where he set a very good example. After his death, he left a thousand ducats to the Monte dei poveri vergognosi of this Congregation' (ibid., p. 33, n. 56). The doubts about Della Porta's devotion worried Cesi, who, in the eulogy that he delivered to the Academy after the death of the Neapolitan Lincean, tried to dissipate the aura of heresy that surrounded him: 'He led a very virtuous life – he wrote – contrary to the opinion that the world had formed of him' (Contributi I, p. 677).

20 OG X, pp. 228–9.

21 Galileo's *Discorsi e dimostrazioni matematiche intorno a due nuove scienze* (Galilei 1638) were translated into English by Stillman Drake (Galilei 1974).

had appended to a previous letter.[22] We further learn from Valerio, who in the meantime had received another letter from Galileo (also lost), that in late May the Tuscan scientist was working on the completion of 'the most excellent work on the natural motion of heavy bodies and on the motion of projectiles, a subject that, as you rightly consider, nobody has seriously approached before'.[23]

Another letter from Valerio, dated 18 July,[24] on the eve of the explosion of public interest in the telescope (Sarpi's letter to Castrino mentioned above is dated 21 July), confirms that Galileo was concentrating all his energies on the definition of the fundamental laws of his new mechanics.[25] Valerio was replying to a letter (also lost) of 5 June[26] in which Galileo had informed him of the two principles on which the edifice of his new science of motion was based.

The immediately successive letter in the Galilean correspondence is that of 22 August in which Giovanni Bartoli wrote to the Tuscan court that the foreign telescope had appeared in Venice.[27] This document is followed by Galileo's letter of 24 August to the Doge Donato offering the Venetian Republic exclusive rights to his invention.[28]

Thus there can be no doubt that the appearance of the telescope in Venice took Galileo by surprise. It induced him to change his own research programme. In the weeks between the end of June and the middle of August he abandoned the science of motion. The priority now was to improve the modest performance of the instrument that had come from France or Flanders and to grasp how to exploit it to increase his own fortunes and to finally achieve his long-cherished ambition of returning to Tuscany.[29]

---

22 Ibid., pp. 239–41.

23 Ibid., pp. 244–5.

24 Ibid., pp. 248–9.

25 Galluzzi 1979, esp. pp. 303–7. Valerio's letter is a fundamental document that has been the object of numerous studies and interpretations for its crucial importance for understanding the process by which Galileo came to define his new conception of the motion of heavy bodies.

26 The date of Galileo's lost letter is indicated in Valerio's reply (ibid., p. 248).

27 OG X, p. 250.

28 Ibid., pp. 250–1.

29 Although he declared that he could not refuse the offer of such a generous salary, Galileo's reaction to the decision of the Senate of the Serenissima to give him tenure for life in the University of Padua was lukewarm. Informing, on 29 August 1609, Benedetto Landucci of the extraordinary recognition that he had received, he concluded bitterly that the new scenario would prevent him from returning to Tuscany: in spite of the 'useful and honoured occasion', he declared, 'it removes my hope of returning home'. And he concluded his own account of the tumultuous speed at which events had passed by reiterating his

After the letter of 22 August, Bartoli continued to keep the Secretary of State in Florence informed about the course of events. In one of his dispatches he mentions Galileo for the first time. In this, as well as in subsequent dispatches, the secretary to the Tuscan Resident took a very negative view of Galileo's behaviour, insinuating that he was amassing remarkable benefits by falsely claiming to have invented the telescope.[30]

In spite of Bartoli's efforts to present Galileo in a bad light, extremely encouraging signals were arriving from Florence. In the months between the moment when the Tuscan scientist first conceived the idea of using the telescope for astronomical observations – a true reinvention of the instrument – and the publication of the *Sidereus Nuncius*, which revealed sensational discoveries that were changing the face of the heavens, the conditions were created for Galileo's triumphal return to Tuscany.

Published in Venice on 13 March 1610 and dedicated to the Grand Duke Cosimo II, the *Sidereus Nuncius* paid an extraordinary homage to the Tuscan dynasty.[31] At the conclusion of some tight bargaining with the Secretary of State, Galileo had christened 'Medicean planets' the most spectacular of his celestial discoveries, the four satellites of Jupiter. The agreement with the Florentine Secretary of State for Galileo's return to Tuscany was rapidly drawn up to the scientist's full satisfaction. The time to pack his bags had finally arrived.

---

regret at seeing the prospect of the return to Florence that he so much longed for go up in smoke: 'So I find myself tied up here for life, and will have to be content with enjoying my home country on a few occasions during the holiday months' (ibid., pp. 253–4). Galileo had expressed the same regret in a letter of February 1609 to a Florentine correspondent in which he stated that he was ready to return home at a simple sign from the Grand Duke. He was tired of the obligation to serve a thousand patrons by which he was bound in Padua. He wanted to dedicate all his energies to composing the great works that he had conceived: 'Having by now worked for twenty years, the best of my life, in dispensing by the minute, as one says, to every request that little talent that had been granted me by God and by my efforts […]' (ibid., p. 232; letter to 'S. Vesp' from Padua, February 1609). The expression 'twenty years, the best of my life' has been traditionally interpreted as a declaration of regret at the prospect of concluding his lengthy residence in the Republic of Venice. In actual fact, Galileo simply intended emphasising that he had spent in Padua the most productive years, those of his maturity. Indeed, it would appear strange for him to have shown regret at having to leave his duties in the University of Padua in a letter expressing his keen desire to return home. This, however, does not call into question the fact that he had enjoyed the lengthy sojourn in the Serenissima.

30 OG XI, pp. 255 and 257.

31 Galilei 1610; OG III[1], pp. 53–96.

Let us return to the correspondence between Rome and Naples. Some months had passed since Della Porta's letter to Cesi on the 'secret' of the telescope. In a successive letter to Cesi (probably of late spring-summer 1610) the first mention of the name of Galileo in the correspondence of the young Lincean Academy occurs. Having received from Cesi a summary of the sensational novelties contained in the *Sidereus Nuncius*, Della Porta expressed both admiration and regret:

> I regret to say that the invention of the eyeglass in a tube was mine, but the Paduan Professor Galileo adapted it, and with it he has found four other planets in the sky, and thousands of fixed stars, and just as many, never seen before, in the Milky Way, and great things on the globe of the Moon, which fill the world with astonishment.[32]

The pride of the aged natural magician had suffered a severe blow, but he reacted with moderation. His acceptance without hesitation of the novelties of the *Sidereus Nuncius* shows that he had grasped their importance. The telescope that he had been the first to devise had undergone a metamorphosis in the hands of Galileo: it was now anything but a 'piece of nonsense'. On the other hand, Della Porta consoled himself: those novelties, discovered by an instrument that he felt he had invented himself, dealt a mortal blow to the Aristotelian natural philosophy that he had always vigorously opposed.

We lack direct evidence of Cesi's reaction to those discoveries, but a letter of 15 September 1610 to his brother Giambattista from Francesco Stelluti, one of the four founders of the Academy, who would never deviate a hair from the course set by Cesi, suggests that initially Federico's perplexities outweighed the certainties. Stelluti passed a very harsh verdict on the *Sidereus Nuncius*, echoing Della Porta's regret and his proud claim to have been the true inventor of the instrument:

> By now I am sure that you have already seen Galileo's book, that is, his *Sidereus Nuncius*, and the great things he says. But now Kepler, a student of Tycho, has already written against it, and an exemplar of his book has reached Father Clavius from Venice [Kepler's *Conversation with Galileo's Sidereal Messenger*][33] saying that Galileo claims the invention of that instrument although it is now thirty years ago that Giovanni Battista Della

32 CL, pp. 148–9.

33 Kepler 1965.

> Porta wrote about it in his *Magia Naturale* and he also mentioned it in his book *De refractione*. So poor Galileo will be shamed; but in the meantime the Grand Duke has given him 800 piastre, and the Signoria of Venice has increased his salary.[34]

It is difficult to believe that the source of this information was anyone other than Federico Cesi, and it is even more improbable that the Prince took a different view.[35] Moreover, Stelluti's letter explicitly referred to the Collegio Romano as the principal centre for the reception and dissemination of information, accompanied by expressions of doubt, about the new discoveries. The active presence of Cesi between the lines of Stelluti's letter is also confirmed by the credit assigned to the authoritative views of Kepler as well as of his mentor Tycho Brahe: both authors whom Cesi had read, and would continue to read, attentively.

Winning Over the Minds at the Collegio Romano

In the closed chambers of the Lincean Academy in Rome as in Naples, Venice, Bologna and Florence, the appearance of the telescope and the publication of the *Sidereus Nuncius* provoked the same negative response, instigated by the fear that the celestial novelties might lead to the collapse of an authoritative and reassuring system of beliefs. Galileo's opponents looked above all to the Collegio Romano to demonstrate that the house of cards erected by the Tuscan scientist would be very easy to topple. At least down to the autumn of 1610, everything seemed to indicate that the Jesuits were unlikely to give their assent to the discoveries. As Galileo's trusted friend Lodovico Cigoli wrote to him in October of that year, seven months after the publication of the *Sidereus Nuncius*:

---

34 CL, pp. 152–3.

35 From what source had Stelluti and his mentor Cesi drawn the information that led them to regard Galileo as an impostor who would soon be unmasked by Kepler, given that they had not yet read the latter's *Dissertatio*? One of their sources may have been Horky's *Peregrinatio* (Horky 1610), which the author sent to numerous persons in Venice and Rome. Neither can we exclude the possibility that the prediction of the imminent collapse of the *Sidereus Nuncius* and of its author was based on information sent by Magini to the Fathers of the Collegio Romano. Cf. Bucciantini-Camerota-Giudice 2015, pp. 99ff.

> These followers of Clavius [...] do not believe anything; and among the others Clavius, head of them all, said to a friend of mine about the four stars that they made him laugh. It would take a telescope capable of creating them and then of showing them. Galileo can think what he likes, but he will hold to his opinion.[36]

Immediately after the publication of the *Sidereus Nuncius*, a request for information was sent from the close vicinity of the Collegio to a trusted correspondent in Venice, where the Jesuits could no longer count on direct informants. Following the expulsion of the Jesuits by the Serenissima in 1606, it was strictly forbidden to allow documents or letters from members of the Society into the territory of the Republic or to write to them.[37] The person who tried to obtain information about the discoveries announced in the *Sidereus Nuncius* was Johann Schreck (Terrentius), a figure extremely close to both Cesi and Clavius. Schreck was to join the Linceans in 1611 only to resign later on when he entered the Jesuit order.[38] Schreck consulted a colleague who was resident in Venice at the time, the Neapolitan mathematician Camillo Gloriosi, who had studied in the Jesuit college in his home town and was drawn to the algebra of François Viète. Possibly for reasons of professional jealousy, Gloriosi disliked Galileo.[39] The Neapolitan mathematician's opening words in the letter sent to Schreck on 29 May 1610 indicate that he was replying to a confidential request for information made, in all probability, by and on behalf of the Clavius circle and perhaps even by Cesi himself: 'You wish to know what I think about the *Nuncius Sidereus* of Galileo'.[40] Drawing the same negative portrayal of the Tuscan scientist as other witnesses of those months, Gloriosi declared that the celestial phenomena that Galileo claimed to have discovered were not new, nor was he the inventor of the telescope: 'He makes many announcements that are not new and fails to give recognition to their true discoverer'.

Down to the autumn of 1610, that is, many months after the publication of Galileo's book, Federico Cesi, Francesco Stelluti and others who were soon to join the Lincean Academy (such as Della Porta and Schreck) all displayed

---

36 OG X, p. 442, letter dated 1 October 1610.

37 The fundamental study on the relations between Sarpi and Galileo and on the climate created in Venice by the Interdict, above all in the circle of Sarpi, is still that of Gaetano Cozzi (1978, esp. pp. 135–234); see also Bucciantini-Camerota-Giudice 2015, pp. 33–52.

38 On Schreck, see Contributi II, pp. 1011–51.

39 On Gloriosi, see Favaro 1983 (I, pp. 323–72).

40 OG X, p. 363.

evident signs of hesitation in the face of Galileo's celestial discoveries, in spite of the fact that they were all severely critical of traditional knowledge. While the publication of the *Sidereus Nuncius* had caused an uproar, full adhesions from the circles of natural philosophers and astronomers to the novelties it announced were extremely limited. Even Kepler's substantial endorsement in his *Dissertatio* (published before the end of April) was seen by many as a severe critique of Galileo's assertions.[41] Manifestations of disbelief had also sometimes passed the limits of a firm but civil opposition, as in the case of the violent *Peregrinatio* of the Bohemian Martinus Horky.[42]

If Galileo could not fail to be concerned by this climate, it was above all the deafening silence of the Collegio Romano and the reports of the scepticism of the Jesuits, on which his enemies were counting, that appeared threatening to him. Galileo knew that the Jesuits were by no means indifferent to the announcement of the new discoveries and that the Fathers at the Collegio Romano were actively engaged in observations of the celestial vault. The Jesuits' prolonged silence was also due to the fact that the *Sidereus Nuncius* made bold claims based on data that Clavius and his colleagues could not verify because they lacked adequate telescopes. Kepler had unquestioningly given his assent to Galileo; how could the Jesuits remain silent in the face of such astonishing novelties?

The exchange of information and opinions that passed through the Jesuit correspondence network between the spring-summer of 1610 and February 1611 enables us to reconstruct the slow and controversial process by which the Fathers became convinced of the objectivity of the images shown by the telescope and therefore of the effective existence of the celestial novelties.

---

41 Kepler 1610; OG III[1], pp. 99–126. The interpretation of Kepler's *Conversation* as a rebuttal of the *Nuncius* was widely shared. On 28 May 1610, Georg Fugger, the Imperial ambassador in Venice, congratulated Kepler for having unmasked Galileo: 'I have read your learned *Dissertatio* on Galileo's *Nuncius*, from which, if he wishes, he will easily realise that his mask has been pulled off [...]' (OG X, p. 361). On Fugger, who was in contact with Magini and Camillo Gloriosi, see Bucciantini-Camerota-Giudice 2015, pp. 104–7. According to Michael Maestlin, Kepler had torn from Galileo not the mask, but the feathers of the raven from the Aesopian fable ('*Galilaeum deplumasti*': OG X, p. 428; letter to Kepler of 7 September 1610; cf. Camerota 2004, pp. 175–6 and 601). This biased interpretation of Kepler's opinion was based above all on the affirmations with which the astronomer had stressed that Galileo had not discovered, but simply confirmed an interpretation – that of the irregular surface of the Moon – which had been advanced in the ancient world, in particular by Plutarch in his *De facie in orbe Lunae* (see Fabbri 2012.)

42 Horky 1610; OG III[1], pp. 129–45.

The Jesuits tried to make their own telescopes from the first news of the invention. Thanks to these efforts, primarily on the part of Fathers Lembo and Odo van Maelcote, they managed to build telescopes before the publication of the *Sidereus Nuncius*. These instruments enabled them to observe the celestial phenomena that Galileo was to describe in the first section of his book (the Milky Way, Orion and the Pleiades, the uneven surface of the Moon). They did not, however, permit a sighting of the satellites of Jupiter, the discovery considered by Galileo as the most important. It is extremely likely that Cesi and other persons close to the Jesuits (such as Schreck) took part in some of the sessions of observations in the Collegio Romano. At any rate, they were certainly informed about the progress of the Jesuit efforts to build performing instruments.

A significant aid to bridging the gap between the instruments of the Collegio Romano and those of Galileo came from Venice. Galileo was kept regularly informed about the progress of the Jesuits' observations by Antonio Santini, a native of Lucca who was operating as a merchant in Venice and who had studied under Galileo in Padua.[43] At the same time, Santini kept the Fathers of the Collegio Romano up to date on the celestial discoveries of Galileo and furnished them with lenses and telescopes that were more powerful than the ones that they had managed to construct.

Venice continued to play a fundamental role not only in the dissemination of information about the telescope and the new discoveries made in the heavens, but also in the production of more powerful instruments. The activity of Antonio Santini deserves special attention in this respect. Between the late spring and the autumn of 1610, Santini acted in Venice in the name of and on behalf of Galileo as an assiduous dispatcher of telescopes to Clavius and the Collegio Romano, as well as to personalities of strategic importance for the authentication of the new discoveries. The latter included the renowned

43 There is no exhaustive biography of this remarkable native of Lucca. After embarking on a career as a merchant, he devoted himself to the study and dissemination of the algebra of Viète and became involved by Galileo, with whom he maintained relations of friendship throughout his life, in the production and distribution of quality telescopes. On the few biographical data known, on the analysis of the correspondence that Santini maintained (not only with Galileo, but also with the Jesuits in Rome, particularly Clavius, as well as Magini, Ghetaldi, Gian Antonio Rocca, Baliani, Gloriosi and even Gassendi), and on a detailed list of his printed and manuscript works, see Arrighi 1965; cf. also the biography outlined by Ugo Baldini and Pier Daniele Napolitani in Clavius 1992, I, pp. 92–4, and Reeves 2011, pp. 43–4. On the interest of Italian mathematicians in the algebra of Viète and the influence of the algebra of Clavius (Clavius 1609), see Baldini U. 1995 and 2003.

astronomer Giovanni Antonio Magini, as well as the ambassador of the Emperor in Venice, Georg Fugger. The latter, who was highly sceptical of the existence of the new celestial phenomena, was in direct contact with Kepler, Welser and the German Jesuit circles.

Given Galileo's reluctance to allow performing instruments to fall into the hands of potential rivals, it may seem strange that he did not manage these relations in person, but permitted Santini to distribute telescopes. Santini's correspondence during those months indicates that the merchant from Lucca, who was working with and for Galileo in the last year of the latter's residence in the Republic of Venice, was capable of constructing telescopes that were of superior quality to those in circulation. If we put Santini's correspondence with the Jesuits in Rome and with Magini between the second half of 1610 and the first months of 1611 side by side with his intense correspondence with Galileo, it becomes evident that his distribution of telescopes between Venice, Florence, Bologna and Rome was approved by Galileo and responded to a precise strategy. Santini did not intend to enter into competition with Galileo, nor to show that he could match his talent in constructing telescopes.[44]

Let us start with the case of Magini. In the eyes of Galileo, Magini had not only firmly opposed the celestial discoveries at first, but was also guilty of protecting and keeping as his close collaborator Martinus Horky – the same Horky who only a few weeks after the publication of the *Sidereus Nuncius* had violently attacked Galileo not only for his false discoveries, but also for his moral conduct.[45]

Acceptance of the new phenomena by a growing number of authoritative exponents of astronomical research, and in particular Kepler's endorsement in the *Dissertatio*, gradually caused Magini to change his attitude. The Paduan astronomer realised that it was opportune to make amends with Galileo. It was clear to him that the most controversial topics of astronomical debate were by now closely connected with the observation and interpretation of celestial phenomena discovered with the aid of the telescope. If he was to continue to play an important role in those debates, he needed proper instruments for observation. Unable to construct them himself, and given that relations between him and Galileo were cold, he asked Santini to intervene on his behalf to quell the Tuscan scientist's resentment. At the same time he asked Santini to send

44 Mario Biagioli expressed a different opinion: that in sending lenses to Magini and the high-quality telescope to the Collegio Romano, Santini acted in total independence of Galileo (Biagioli 2006, esp. pp. 36–7 and 90–3).

45 Cf. Horky 1610.

him lenses of sufficient quality to observe the heavens in detail. It was certainly with Galileo's approval – the Tuscan scientist accepted Magini's apologies in view of the importance of the authoritative Paduan astronomer's acceptance of the celestial novelties – that Santini began sending him high-quality lenses from Venice.[46]

The attitude of Galileo's pupil towards the Jesuits of the Collegio Romano was different. During the long months between the presentation of the telescope to the Senate of the Republic of Venice in August 1609 and his move to Florence in September 1610, Galileo, as a public employee (Professor of Mathematics at the University of Padua), as well as being notoriously close to the Servite Friar Paolo Sarpi and his circle, could not correspond directly with members of the Society of Jesus. For Galileo, however, the approval of the new celestial discoveries by the Collegio Romano was crucial, and to obtain it he felt it was essential to maintain an open dialogue with Rome. Unable to do so in person, he entrusted this task to his faithful assistant who, as a citizen of Lucca engaged in commercial activity, could take advantage of extraterritorial privileges. This explains why Antonio Santini acted as the link between the two poles: he passed messages from Galileo on to Clavius and the Roman Jesuits and vice versa.

This enabled Galileo to transmit carefully doctored messages to Rome, sometimes in code (as in the case of the discovery of the phases of Venus and of the curious appearance of Saturn), on the progress of his observations with the telescope. At the same time Santini furnished him with information regarding the slow progress of the Jesuits in their celestial investigations: no doubt welcome news for the Tuscan scientist. Galileo hoped that this indirect channel would enable him to put sufficient pressure on the Roman Fathers and prompt them to side with him. He was very concerned about the news from the Eternal City regarding the scepticism or even outright opposition of the Jesuits. It was essential to Galileo that they should manage to observe the new phenomena with their own eyes. For that purpose, he was even prepared to break a rule that he had rigorously followed until then: preventing high-quality telescopes from ending up in the hands of dangerous competitors.

Although at first sight Santini's activities might appear as an arbitrary move that could have undermined the privileged position Galileo had acquired, it is more than reasonable to assume that he was authorised by the Tuscan scientist to send one of the most powerful telescopes to Clavius in Rome between the

46 OG X, p. 437, letter of Magini to Galileo, 28 September 1610, in which he mentions a '*canone* or *tromba* that Signor Santini has sent to me as a gift'. That instrument enabled Magini to distinctly observe the satellites of Jupiter.

end of October and early November 1610. That Santini had explicitly asked Galileo for permission to send the instrument to Rome is confirmed by his letter of 9 October 1610 in which he informed the Tuscan scientist that the Jesuits remained sceptical with regard to the new celestial discoveries, especially the satellites of Jupiter, because they were unable to see them. Santini voiced his suspicions: 'I wonder whether some of these major figures, I mean those held in high repute, are not being obstinate to pressure you to send them an instrument'.[47] If this was the aim of Clavius and his companions, they were soon satisfied. The victor, however, was Galileo. The high-quality telescope that Santini sent to Rome had miraculously revealed to the Jesuits those satellites of Jupiter that they would never have expected to see. On 4 December Santini wrote triumphantly to Galileo: 'Father Clavius writes to me from Rome that they have observed Jupiter'.[48] But the sting was in the tail: the Jesuit had added 'we are not yet sure whether they are planets or not'.

On 17 December Clavius wrote directly to Galileo (who, now that he was back in Florence, could freely correspond with the Jesuits) that thanks to the telescope sent by Santini, the Fathers had been able to observe the satellites of Jupiter and also the curious appearance of Saturn, after Santini (certainly, once again, with the prior approval of Galileo) had solved the anagram used by the Tuscan scientist to circulate the announcement of the triple aspect of the planet.[49] The role of Santini in serving the cause of Galileo, overcoming the barriers erected between Rome and the Venetian Republic as a result of the Interdict, finds further confirmation in the fact that it ceased once the Tuscan scientist was back in Florence.

Galileo's latest discovery, made on the eve of 1611, concerned the phases of Venus, perhaps the most important among his celestial novelties for its cosmological implications. On 30 December 1610 Galileo wrote directly to Clavius that, having observed Venus with his telescope for three months, he had discovered that it displayed phases like those of the Moon during its cycle. He drew a conclusion from this that must have landed in Rome like a thunderbolt:

> It is now clear [...] how Venus (and Mercury will no doubt do the same) moves around the Sun, the indubitable centre of the revolutions of all the planets [...].[50]

---

47 Ibid., p. 445.

48 Ibid., p. 480.

49 Ibid., p. 495; cf. Reeves-Van Helden 2007.

50 Ibid., p. 500.

Although the Jesuit Fathers knew that the celestial discoveries announced by Galileo were not the product of his fertile imagination, on hearing that Venus was sickle-shaped they hesitated before expressing their opinion. Yet as they devoted themselves to lengthy and painstaking scrutiny of the planet's movement, they gradually acquired the conviction that on this occasion too Galileo was right. In his long letter to Galileo of 22 January 1611, the Jesuit Christoph Grienberger declared that every trace of uncertainty had been dissipated. Venus too could be included on the impressive list of Galilean celestial discoveries. We learn from Grienberger that the observations that removed the last doubts of the Fathers had been conducted a few days before, on 17 January, and had been based on a close comparative inspection of the Moon and Venus. The orbit of the planet around the Sun appeared as an incontestable fact to all the Fathers.[51]

There remains a final point regarding Santini's role to be clarified. Why was the telescope sent to Rome on the orders, or at any rate with the prior approval, of Galileo constructed not by the Tuscan scientist himself but by Santini? Galileo was compelled to interrupt his own activities as instrument maker in the summer of 1610 at a delicate stage in the process of recognition of the objective functioning of the telescope and therefore of the effective existence of the new celestial phenomena. He had fitted out a laboratory for grinding lenses not only in Padua but also in Venice, where he counted on the assistance of Santini. Absorbed as he was in his systematic observations of the sky, defending himself from the attacks of his opponents, negotiating the transfer to Tuscany and making preparations for a large-scale move (which took place in the first half of September) after almost twenty years in the Venetian Republic, he temporarily delegated the task of producing quality lenses to Santini. After his return to Florence, he needed time to reinstall his optical workshop. This is confirmed by his reply to a request for a telescope from Giuliano de' Medici, Ambassador of the Grand Duchy in Prague. Galileo responded that he could not satisfy his request for the time being as he had been compelled to leave his lathes and other lens-grinding equipment in Padua.[52] He was therefore obliged to ask Santini to continue sending from Venice the lenses that he wanted to reach the hands of illustrious patrons and colleagues. He called once more on the assistance of Santini to meet the

51 OG XI, pp. 33–55.

52 OG X, pp. 440–1: 'My home is not yet in order [...]. I have not been able to fit out my equipment for lens-grinding, some of which is fixed to the wall [in my house in Padua] and cannot be transported'.

demands of an unidentified ambassador (perhaps again Giuliano de' Medici) in Florence.[53]

Galileo soon completed fitting out his new Florentine laboratory to become a fully autonomous producer again. At this point Santini abandoned the scene in his role as telescope maker. He left Venice soon afterwards, but did not disappear entirely from the Galilean orbit. A few years later he was to reappear in Rome in the entourage of Federico Cesi and the Lincean Academy.[54]

By the end of 1610 an important stage in Galileo's battle appeared to have drawn to a close: the recognition of the effective existence of the celestial phenomena revealed by the telescope. The authority of the Collegio Romano silenced those who regarded the images seen through the telescope with scepticism or disbelief. However, the explicit confirmation of the existence of the new celestial discoveries contained in Grienberger's letter did not reflect the view of all the Jesuits. Some of them were not prepared to give way in the face of the evidence. On 3 February 1611, a few days after Grienberger's letter to Galileo, another Jesuit, eye-witness to the observations made with a telescope in the Collegio Romano, wrote to a member of the same order in Munich that the satellites of Jupiter did not exist; they were merely 'a hallucination or visual deception'.[55]

Now that the doubts regarding the new celestial phenomena had been dissipated, it was time to face another challenge, the one that was most important to Galileo: what were the implications of the celestial discoveries for the traditional conception of the universe? The Fathers in the Collegio Romano had not made any pronouncements on this issue, nor had Galileo asked them to, so the best course seemed to them to be to smooth over the matter. After all, Clavius and his circle did not consider themselves natural philosophers but

53 Ibid., pp. 464–5.

54 Cf. OG XII, pp. 83–4, letter of Santini to Galileo, 11 July 1614.

55 The sender was the Jesuit Paul Guldin, who informed his fellow Jesuit Johann Lanz that the mathematicians of the Collegio Romano, headed by Clavius, were sceptical about the existence of the satellites of Jupiter. However, on 29 January, Father Clavius had told Welser that he and all the Fathers of the Collegio were by now fully convinced of the truth of the telescopic observations reported by Galileo: 'Regarding what you write to me on 7 January [Welser had asked whether it was true that the Fathers had verified the celestial discoveries: cf. OG XII, p. 14], I was for a long time in doubt and did not believe in those four Medicean planets, thinking that they were a hallucination produced by the telescope. But my mind is now made up as here in the Collegio we have seen those planets several times, both in the morning and in the evening' (OG XX, pp. 600–1; Camerota 2004, p. 201).

mathematicians and astronomers, whose task it was to 'save the phenomena'. Nor had their superiors invited them to express their views on the cosmological implications of the celestial discoveries. The highest authority of the Collegio Romano, the Rector Cardinal Robert Bellarmine, had asked them only to confirm or deny the existence of those new phenomena.[56]

From the beginning of 1611 the hierarchies of the Society of Jesus decided to face up to Galileo's discoveries. They celebrated the divine greatness and lauded the man who had invented the telescope and revealed new celestial phenomena. At the same time, they devoted all their efforts to stressing that the Fathers had arrived independently at the same conclusions as Galileo. It was important to suggest that the scientific competition between the Tuscan scientist and the Collegio Romano had ended in a tie. On the other hand, Bellarmine's intervention had made it crystal clear that the determination of the true structure of the universe was not a matter to be left to astronomers and mathematicians. It was rather the competence of natural philosophers, although the last word was that of theologians. Like his learned companions, Clavius abode strictly by the Cardinal's decision. In the addenda introduced in the latest edition of his commentary on the thirteenth-century *Tractatus de Sphaera* by Johannes de Sacrobosco he listed the celestial discoveries that Galileo (who was praised unconditionally) had made with the aid of the telescope. Clavius spurred on the astronomers to devise appropriate solutions to revise the traditional system of the world after the shockwave produced by the *Sidereus Nuncius*.[57] The call was to 'save the phenomena', not to explain either their causes or their effective nature. This was the preserve of astronomers. Cardinal Bellarmine would take care of the other questions.

Second Class Telescopes

It is impossible to precisely establish when Federico Cesi read the *Sidereus Nuncius* or handled a telescope. As the letter sent to him by Della Porta in August 1609 seems to indicate, he had probably been able to look through a Flemish instrument or a replica made on the spot as early as the summer of 1609. As for the *Sidereus Nuncius*, published in Venice on 13 March 1610, Cesi certainly

56 OG XI, p. 87. Bellarmine's request for the opinion of the mathematicians of the Collegio Romano is dated 19 April 1611. The Cardinal's concerns bore exclusively on the confirmation of the effective existence of the celestial phenomena discovered with the aid of the telescope.

57 Clavius 1611–1612.

had direct knowledge of it before 15 September 1610, when Francesco Stelluti informed his brother, who had not yet been able to read it, of the incredible news announced by that book, relying on information most probably furnished him by the Prince. If 15 September is the *terminus ante quem*, the letter from Venice of 29 May 1610 in which Camillo Gloriosi wrote to Schreck about the contents of the *Sidereus Nuncius* can be taken as the *terminus post quem*.[58] The Princeps of the Lincean Academy probably read Galileo's book between June and August 1610 after his second visit to Della Porta in Naples in May. Back in Rome, Cesi informed the Neapolitan of the sensational celestial discoveries disclosed in the *Nuncius*.[59] This confirms that Della Porta's undated letter to Cesi in which he commented on the information he had received from Federico belongs to the summer of 1610.

From another undated letter of Della Porta, this time sent to an unnamed correspondent, we learn that, on the occasion of Cesi's visit to Naples in May 1610, the Neapolitan had shown him a telescope (*specillum*) he had made that could distinguish a man at a distance of three thousand paces. This is an important piece of evidence because it shows Cesi's burning curiosity about the telescope at the time of his visit to Naples and his hope that the author of the *Magia Naturalis* would reveal its secret to him.[60]

The oldest document attesting the presence in Rome of a Galilean telescope, accompanied by a copy of the *Sidereus Nuncius*, is the letter of thanks for the extremely welcome gifts sent to Galileo on 28 April 1610 by Cardinal Francesco Maria del Monte, who was closely linked with the Medici.[61] Other documents indicate that Galileo sent telescopes to Rome in the immediately

58 OG X, p. 363.

59 CL, pp. 148–9, letter in which Della Porta replied to a lost letter of Cesi. There is no evidence that an exemplar of the *Sidereus Nuncius* was in Cesi's library. Schettini Piazza has argued that the Prince owned the exemplar with five drawings of the surface of the Moon in Galileo's hand (Schettini Piazza 2011, pp. 23–6) which appeared on the antiquarian market in 2005 and which has since been shown to be a forgery (see Wilding 2015, especially pp. 17–22). No copy of the *Sidereus Nuncius* is recorded in either of the two inventories of Cesi's library preserved in the Archives of the Accademia dei Lincei (see Biagetti 2008).

60 Gabrieli showed (CL, pp. 308–10) that Della Porta's letter was addressed to Johannes Faber and can be dated to 1612. This letter was first published by Odescalchi (Odescalchi 1806, pp. 92–3). It was mentioned also by Favaro (OG XIX, pp. 611–2), who did not identify the addressee and assigned it to 1613. Faber probably wanted first-hand information for use in the composition of the short history of the invention of the telescope that was later inserted in the *Thesaurus Mexicanus* (cf. below, pp. 144–6).

61 OG X, pp. 343–4, letter of 28 April 1610.

successive weeks. Among the recipients of these cherished gifts was Cardinal Scipione Borghese, who had requested an exemplar of the instrument from Florence on several occasions, even though since August 1609 he had been in possession of a telescope sent him from Flanders.[62] In the first half of June one of the instruments made by Galileo reached the Cardinal,[63] who was not slow to demonstrate his heartfelt gratitude to the Florentine scientist.[64] Son of Ortensia Borghese and nephew of Pope Paul V, Scipione was on cordial terms with the Cesi family, including the young Federico. A few years later he was a witness to Federico's second marriage to the Florentine noblewoman Isabella Salviati.[65] Thus the authoritative Cardinal might well have given the young Prince his first opportunity to explore the heavens with an instrument made by Galileo.

At the request of the Grand Duke, Galileo dispatched numerous other telescopes to Cardinals and high-ranking representatives of the Roman patriciate between the summer and autumn of 1610. Cardinal Borghese did not mention the *Sidereus Nuncius* in his letter of thanks, nor did Cardinal Montalto, another fortunate recipient of a Galilean telescope, receive a copy of the book.[66]

Galileo behaved differently towards competent colleagues. The letter from Clavius to Galileo of 17 December 1609 confirms that for a long time Galileo had deliberately avoided letting the Fathers of the Collegio Romano get their hands on a quality instrument,[67] although he was quick to send them an exemplar of the *Sidereus Nuncius*. His attitude towards Luca Valerio was the same: he sent him a copy of the book, but failed to provide him with an instrument that would have permitted him to verify with his own eyes the extraordinary discoveries announced in the *Nuncius*.[68]

Galileo had developed a precise strategy of dissemination. Baldino Gherardi, a Tuscan living in Padua who was closely linked with Paolo Gualdo, travelled especially to Rome to present precious gifts to prelates, patricians and the most competent mathematicians and natural philosophers on

62 Ibid., p. 255, letter to Paolo Gualdo, 31 August 1610, in which Lorenzo Pignoria writes: 'Signor Galileo has pocketed a thousand florins for life, it is said with the help of a telescope similar to the one that was sent from Flanders to Cardinal Borghese'.

63 Ibid., p. 375.

64 Ibid., p. 382, letter from Scipione Borghese to Galileo, 26 June 1611.

65 Contributi I, p. 589.

66 OG X, p. 409.

67 Ibid., pp. 384–5.

68 Ibid., pp. 362 (Valerio to Galileo, 29 May 1610) and 434 (Valerio to Galileo, 24 September 1610).

Galileo's behalf.[69] Gherardi had received precise instructions on how he was to distribute the *Nuncius*, the telescope, or both: the leading prelates and most influential members of the nobility were to receive telescopes alone, while both book and instrument were to be presented only to those persons who were on particularly friendly terms with the Florentine dynasty, such as Cardinal Del Monte. The mathematicians and natural philosophers, such as Luca Valerio, on the other hand, were to receive the *Nuncius* but not the instrument.

It is not difficult to grasp the rationale of this strategy. Galileo was still intensely engaged in his exploration of the celestial vault and did not want to run the risk that, thanks to his most performing telescopes, competent colleagues might anticipate him in discovering other sensational novelties.

Galileo had naturally taken into account the possibility that the telescopes sent to illustrious Cardinals might fall into the hands of expert competitors and he was perfectly aware that the most dangerous among this category were the Jesuits of the Collegio Romano. This is why he only sent instruments of modest quality to Rome, which permitted observation of the uneven surface of the Moon, the composition of the Milky Way, and the constellations densely populated with stars, but not of the satellites of Jupiter.

Galileo's fears were justified. A few days after the arrival of the first exemplars of his telescopes in Rome, they were passed from the hands of the Cardinals, who were curious about the new discoveries but had not read the *Nuncius*, into those of the Fathers of the Collegio Romano, who were among the first readers of the book published in March 1610. We have explicit evidence of the intense circulation within the city of the telescopes sent to Rome by Galileo and their passing into the hands of Clavius and his companions in the letter from Grienberger to Galileo of 22 January 1611. The German Jesuit, a highly skilled mathematician and astronomer, stated that the Fathers had obtained one of the instruments 'that you sent to various people in Rome'. It may have been the one sent to Cardinal Borghese. From Grienberger's words his disappointment at Galileo's failure to send a telescope directly to the Collegio Romano clearly emerges. The Jesuits compared the telescope of Galileo that they had managed to obtain with the instruments that they had fabricated themselves and declared that the latter were as good if not better.[70] Grienberger's letter confirms

69 It was Gherardi who delivered the telescope to Cardinal Del Monte (ibid., p. 343) and the *Sidereus Nuncius* to Luca Valerio (p. 434).

70 OG XI, p. 34.

that with the telescopes sent by Galileo to Rome it was impossible to clearly distinguish the satellites of Jupiter.

The Jesuit recognised that the turning-point in their campaign of celestial observation came with the arrival of the telescope that Antonio Santini sent to the Collegio from Venice. That instrument was immeasurably superior to those that the Fathers had been able to test until then.[71] With its capacity to magnify 'thousands of times' it was in a different class.

Galileo did not confine this strategy only to his colleagues and patrons in Rome. Even though he was eager for Kepler to side with him, he carefully avoided the risk of giving him the opportunity to make independent observations. Although Kepler had immediately subscribed to the celestial discoveries, for months the Imperial astronomer was unable to verify them himself because Galileo ignored his request for one of the instruments that he was liberally distributing to the powerful German Electors, Archdukes and Princes. Like the Jesuits of the Collegio Romano, Kepler had to borrow a telescope – the one that Galileo had given to Giuliano de' Medici, the Tuscan Resident in Prague – in order to scrutinise the novelties in the heavens.[72] A year had passed since the publication of the *Sidereus Nuncius*.

However, like the exemplars sent to Rome, the telescope presented by Galileo to Giuliano de' Medici had severe limitations: it enabled observation of the novelties announced in his *Nuncius*, while the satellites of Jupiter could only be distinguished with difficulty. Galileo's most recent discoveries, those made between the summer and autumn of 1610 (the triple aspect of Saturn, soon followed by the phases of Venus) thus remained for Kepler a closed book, obliging him to take Galileo's assertions on trust once more.

The shortcomings of the Galilean instruments sent to Rome and Germany down to the late autumn of 1610 are confirmed by a letter to Clavius from the Elector of Cologne of 24 March 1611. The Elector had learned from Clavius that the Jesuits had managed to observe Saturn and Venus thanks to the instrument 'sent them from Venice', that is, the one dispatched to Rome by Antonio Santini. The Elector expressed to Clavius his disappointment at 'having at my disposal one sent me by Signor Galileo, which is not sufficient to achieve the same result'.[73]

71 Ibid.

72 Ibid., p. 78; Kepler to Galileo, 28 March 1611.

73 Ibid., p.73.

CHAPTER 2

Parallel Convergences?

The Encounter

So far we have paid little attention to Cesi's reaction to these momentous events. Like Tommaso Campanella, the young Prince will certainly have thought that the exceptional conjuncture of Jupiter, Saturn and Mercury in the summer of 1603 when the Lincean Academy was founded had prophetically anticipated the intellectual revolution that the telescope and the new celestial discoveries seemed to herald.[1] Again like Campanella, he must have interpreted that extremely rare astronomical event as a sign of the predestined convergence of the paths that the Linceans and Galileo had been following separately until then. Cesi's hopes were to meet with more success than those of the Calabrian Friar, whose repeated offers of alliance were never taken into the slightest consideration by Galileo.

Like so many others, after reading the *Nuncius* Cesi endeavoured to explore the sky through the lenses of a telescope. He does not appear among the fortunate recipients of the instruments sent to Rome by Galileo. However, with the assistance of the faithful Stelluti, he managed to make one himself, possibly following instructions received from Della Porta. Federico might previously have obtained an instrument from Flanders. Nor can we rule out the possibility that, given his excellent relations with Cardinal Borghese, he held the telescope that Galileo had sent to the Cardinal in his hands.[2] At all events, Cesi certainly closely followed the animated debates on the celestial novelties at the Collegio Romano and the Jesuits' attempts to construct instruments that would enable them to verify the discoveries disclosed in the *Nuncius*.

1 In the extraordinary letter that Campanella wrote to Galileo from prison in Naples on 13 January 1611, he declared that he had foretold the sensational novelty of the celestial discoveries as early as 1603, reflecting on the radical renewal of knowledge and morals generated by the 'great conjunction' of Mercury in the sign of Sagittarius: 'You should also know that in the astrological prediction that I made in view of the great conjunction of 1603 – he wrote to the Tuscan scientist – I predicted among many other things that a new science of the celestial phenomena would be founded in this century [...]' (OG XI, p. 26; Campanella 2010, pp. 193–4). On the relation between the 'great conjunction' of Campanella and the foundation of the Lincean Academy, see De Angelis 2012b; cf. also below, p. 50.

2 Cf. above, p. 22.

 | DOI 10.1163/9789004342323_003

Apart from the scanty clues provided by the letter of Francesco Stelluti to his brother and by Della Porta's reply to the lost letter of Cesi, we have to rely on conjecture to reconstruct the Prince's reaction to the sensational news from Padua and Venice and to the novelties spread by the *Nuncius*. Documentation on the debates provoked in Rome by the celestial discoveries between autumn 1610 and Galileo's arrival in the Eternal City at the end of March 1611 is extremely limited. It is confined to Clavius' tardy response on 17 December[3] to the letter sent to him by Galileo on 15 September[4] (when he had just returned to Florence from Padua), and the Tuscan scientist's reply to the Jesuit of 30 December.[5]

Galileo's letter of December 1611 is particularly important because it sought to sound out how the Jesuits would react to the revolutionary cosmological implications of the celestial novelties to which he explicitly alluded: 'this system of the planets – Galileo wrote to Clavius – is certainly different from the way it is commonly conceived to be'.[6] He had thrown down the gauntlet. The Fathers of the Collegio Romano could no longer hide behind the fragile screen of declarations of admiration for the new celestial phenomena. It was now clear that Galileo, once in Rome, would pose crucial questions. This explains why, in his letter of 22 January 1611, the Jesuit Christoph Grienberger gave Galileo a clear signal of the unwillingness of the Fathers of the Collegio to question the centrality and immobility of the Earth. It was one thing to recognise the existence of new phenomena and to entrust to mathematicians and astronomers the task of saving them. It was quite another to call into question conceptions, such as the centrality and immobility of the Earth, on which the entire edifice of traditional knowledge rested.[7]

Even if we cannot establish exactly when Federico Cesi was able to leaf through the pages of the *Nuncius* and peer through the lens of a telescope, it is certain that he was extremely impressed by the sensational novelties

3 OG X, pp. 484–5, 431–2 and 499–502 respectively.

4 OG X, pp. 431–2.

5 Ibid., pp. 499–502.

6 Ibid., p. 500.

7 While giving his endorsement of the celestial novelties, Grienberger was unwilling to follow Galileo in the Copernican interpretation of the phenomena observed in the sky with the aid of the telescope. In his letter he stressed more than once that the Earth is immobile and at the centre of the world and that the Sun, Moon and other planets revolve around it: 'I noticed that phenomenon on Earth [the illumination of the mountain peaks at dawn, while the valleys are still shrouded in darkness], because the Sun gradually rises above the horizon *in its daily motion* [...]'; further on, writing about the Moon, he added: 'It follows the Sun *in its daily motion* [...]' (OG XI, p. 32; emphasis added).

announced in that work before Galileo arrived in Rome. Those novelties undermined the authority of fundamental aspects of traditional conceptions that had remained unchallenged for centuries. For these reasons, after his initial hesitation, Cesi saw in Galileo a formidable ally in the battle for a radical reform of received knowledge to which he and his Academy were dedicated. It is therefore hardly surprising to find the Prince among the most active and passionate protagonists of the communication campaign launched by Galileo among the prelates and nobility after his arrival in Rome on 29 March 1611.

That long-awaited journey had been in the forefront of Galileo's mind ever since his return to Tuscany. But the need to unpack his belongings and to fit out his home properly for the study and observation of the celestial phenomena, the delicate task of re-constructing the laboratory for lens-grinding that he had left behind in Padua and Venice, as well as problems with his health had forced him to postpone the trip for many months. Another reason also suggested deferring his departure: Galileo wanted to arrive on the Roman scene after the Fathers of the Collegio Romano had been able to observe and confirm the novelties discovered with the aid of the telescope. As we have seen, he did not receive encouraging responses on this score until the end of January.

The aim of his mission was clear: he intended to obtain full endorsement of the celestial novelties from the most influential scientific authorities in Papal Rome, above all from the Jesuits whom he knew to be hesitant. Moreover, as official Mathematician and Philosopher of the Grand Duke of Tuscany, he had a mission to carry out for his patron. After Galileo's dedication of the satellites of Jupiter to the Florentine dynasty, Cosimo II wanted to spread the news of the ascension of the Medici among the stars in the major centre of influence and power of the Christian world. In one of the numerous letters of recommendation to influential persons in Rome, the Grand Duke wrote to Cardinal Del Monte that Galileo's journey was of special importance 'for him [Galileo] as a son of Florence, for the public good, and for the glory of our age'.[8] Precisely because it was a delicate mission that called for careful preparation, the Grand Duke hesitated for a long time before giving his impatient Philosopher and Mathematician permission to leave for the Eternal City.[9] Galileo's visit to Rome

8 Ibid., pp. 60–1, letter of 27 February 1611. The Grand Duke recommended that the Cardinal 'guide and introduce him wherever necessary […]' in order that 'he should easily manage to give a good impression of his own motivations and to clarify the truth'.

9 On 12 January Belisario Vinta invited Galileo, on behalf of the Grand Duke, to check his impatience to go to Rome: 'As for your journey to Rome, I believe that His Highness is very much in favour, but has not yet made a final decision […]' (ibid., p. 20).

was of such an official nature that Cosimo urged his ambassador to treat Galileo as a person 'of whom we are very fond', to offer him hospitality in his residence (Palazzo Firenze), and to meet all expenses incurred during his stay.[10]

In the accounts of an agenda densely packed with meetings, presentations and debates in which Galileo featured as a brilliant protagonist during those memorable Roman weeks, we finally see Federico Cesi emerge from the shadows. It took only a few days for the Prince to abandon his initial hesitation and to fall under the spell of the personality of Galileo. His meeting with the Tuscan scientist came as a revelation. Cesi promptly put himself at his service and became his principal promoter. On 25 April, little more than three weeks after Galileo's arrival in Rome, on the pressing invitation of the Prince, he was inscribed in the Lincean Academy and received the ring with an emerald engraved with a crowned lynx from the hands of his patron.[11] Cesi was almost always in the front row during the animated meetings and debates triggered by the presence of Galileo in Rome.

We are granted a lively description of the extraordinary excitement generated by Galileo's energetic defence of the objectivity of images seen through the telescope, of the newly discovered celestial phenomena, and of their cosmological implications, in a letter that the Prince sent to Francesco Stelluti in Fabriano, where the latter was detained on family business, on 13 April 1611, five days after Galileo's induction into the Academy. Cesi was bitterly disappointed that his friend was not in Rome at a moment that opened up such promising scenarios to the young Academy and its projects to renew traditional knowledge: 'If there was ever a time for you to be in Rome, it is now. Every clear evening we see the new things in the sky, truly an appropriate exercise for the Linceans'.[12] These words reveal the motivations that led the Prince to invite Galileo to join the Academy and to promote his sensational celestial discoveries. The perplexities of the summer of 1610 had been replaced by feverish expectations, nourished by the awareness of living an epochal moment. Federico's enthusiastic tone was similar to that with which Campanella had welcomed the celestial discoveries that he was convinced announced an imminent and radical renewal of knowledge and morals, thanks to the *Nuncius*

10 Ibid., p. 60, letter from the Grand Duke on 27 February 1611 to the Tuscan ambassador in Rome, Giovanni Niccolini. Cosimo II emphasised that Galileo was coming to deal with an affair to which he [Cosimo II] attached the highest importance: 'the affair that he will deal with is very dear to us both for the benefit of learned persons as well as for his glory'.

11 Cf. Contributi I, p. 514.

12 CL, p. 158.

and its author, whom the Calabrian Friar likened to a new but even bolder Christopher Columbus, as well as to another great Florentine, Amerigo Vespucci, who had revealed a new continent.[13]

The appearance of Galileo on the Roman scene marked the turning-point that the Linceans had been waiting for amid the infinite difficulties of the first eight years of the institution. This favourable conjuncture led them to speed up the process to provide the Academy with proper statutes. Cesi wrote to Stelluti that they could no longer postpone 'establishing our principles and our academic constitution [...] in order that not we alone may proceed in the fervour of the sciences [...]'. The phrase 'fervour of the sciences' drew on the vocabulary of magic and alchemy of Della Porta – Cesi's principal point of reference before Galileo burst on the scene – to express the sensation of witnessing a unique moment and experience. It was as though with Galileo the messenger of the Lord, the man of Providence, had appeared who would act as a catalyst for the plans of the Linceans. Cesi was already planning to send Galileo to Naples, where his presence 'would be beneficial to our common cause'.[14] The Prince regarded Galileo as an extraordinary ambassador who, in his vision, would forge links with Della Porta, Vice Principal of the young and fragile Liceo in Naples.

Mutant Lynxes in the Academy's Menagerie

To understand the reasons that led Cesi, after his initial hesitation, to recognise Galileo not as the bearer of a vision that would impose adjustments in the strategies and objectives that the Linceans had been striving to achieve, but as an extraordinary resource to implement the very principles on which the Academy had been founded, we have to go back to the motivations that had inspired Federico in 1603, his original priorities, his models and the identity of his first associates.

Until a few decades ago, there was consensus among historians in considering the Academy as the most evident expression of the mental transformation brought about by the Galilean revolution. Inspired by the flattering prospect of assigning to Cesi's institution the prestigious title of the oldest scientific academy of the Modern Age, scholars overlooked its difficult and discontinuous early years and neglected the evidence that, after the induction of Galileo, the

13 OG XI, p. 24, letter to Galileo, 13 January 1611.

14 CL, pp. 159–60.

FIGURE 1 *The lynx facing left in the frontispiece of the eight books of the* Phytognomonica *by Giovan Battista Della Porta* (DELLA PORTA 1588)

Linceans had continued to pursue the same objectives that had led the young Federico to embark on his adventure. Research conducted in the last thirty years on the history of the First Academy has made it abundantly clear that its initial phase was marked by mentalities, models and aims that were far removed from the priorities of Galileo's agenda as well as from the methods he practised in the investigation of natural phenomena.

The choice of the lynx as the Academy's emblem was probably suggested by Della Porta. The image of a lynx within a rectangular frame accompanied by the motto 'aspicit et inspicit' (he looks at and into) can be found at the centre of the frontispiece to the 1588 edition of the eight books of the Neapolitan's *Phytognomonica,* and of the edition of his *Magia Naturalis* published in the following year (fig. 1).[15] Both the image and the motto seemed to Cesi the perfect icon to communicate the programmes and expectations of the first Lincean nucleus: the radical renewal of the study of nature from which a profound reform of moral conduct would also ensue.

Gilberto de Angelis has questioned that Della Porta was the source of Cesi's decision to adopt the lynx as the Academy's emblem. According to De Angelis,

15 Della Porta 1588 and 1599.

Federico drew inspiration above all from Ficino, Campanella and Bruno. The Prince would have chosen the lynx not simply for the keen eyesight with which it was traditionally attributed, but primarily for its nature as a solar animal (*animal lucis*).[16] Not by chance – as De Angelis stressed – Johannes Faber will suggest that 'lynx' derives from 'lux'.[17]

De Angelis' interpretation seems to be in contrast with the authoritative testimonies of Faber and Stelluti, who declared on several occasions, as did Cesi himself, that the lynx had been chosen for its capacity to see further and better (*plus et melius*), seeing at a great distance with the eyes of its forehead and penetrating the essence of things with the eyes of its mind. The Linceans evoked the classical myth of the Argonaut Lyncaeus, who was credited with such extraordinary vision that his eyes were able not only to clearly distinguish objects at a great distance but also to penetrate their internal structure.[18]

That in choosing the emblem Cesi and his associates had in mind the myth of Lyncaeus is confirmed by the *Gesta Lynceorum*, in which the seat of the Academy, drawing inspiration from Plato's λυκαίον, is defined as a *locus lyceus* and the spirit that animated its founders as perfectly expressed by the lynx, 'the animal endowed with the most powerful eyesight'. The lynx represented the pursuit of 'the science of nature and the destruction of all vices and vanities'.[19] Besides the mention of Plato as a fundamental point of reference, this passage highlights the twofold objective of the founders: to combat the degeneration of natural philosophy and to promote a profound moral reform.

The lynx of the Academy was a lynx *ex libris* in that neither Cesi nor his fellow Linceans had ever seen the feline in flesh and blood. This is confirmed by Faber in a passage of the *Thesaurus Mexicanus* written around 1624, in which he states that, apart from a few dried skins in the store of a Roman merchant, he had only been able to see coloured drawings of the lynx.[20] A few years later, Francesco Stelluti inserted a woodcut of a lynx in his commentary on the *Satires* of Persius Flaccus (fig. 2). His image of the feline was based on direct observation of two live exemplars, a male and a female, that had been given to

16 De Angelis 2012b, pp. 13–34.

17 TM, p. 526. On the basis of a subtle etymological analysis of the Latin word *lynx*, Faber concluded that it came from λυκαίον (wolf, but in Greek mythology, the mountain of Arcadia on which some gods, including Zeus, lived) and stressed that many authors derived that word from *lux* (light) (cf. below, p. 470).

18 On the myth of Lyncaeus, its exploitation by Cesi and the Linceans, and its extension to the dimension of microscopic vision, see Lüthy 1996.

19 Gesta, c. 7r.

20 TM, pp. 526–7.

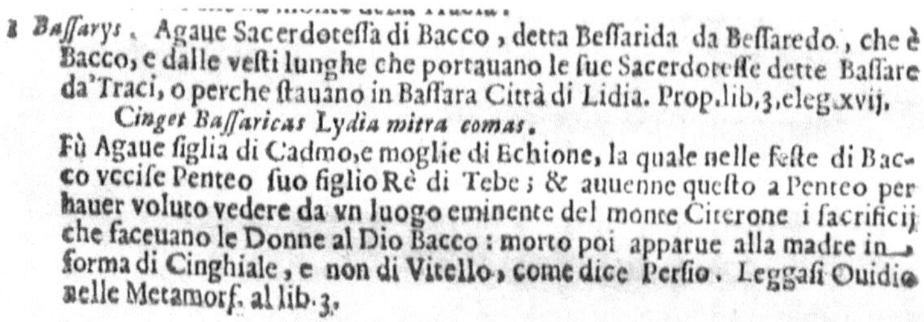

1 *Baſſarys*. Agaue Sacerdoteſſa di Bacco, detta Beſſarida da Beſſaredo, che è Bacco, e dalle veſti lunghe che portauano le ſue Sacerdoteſſe dette Baſſare da'Traci, o perche ſtauano in Baſſara Città di Lidia. Prop.lib.3,eleg.xvij.

Cinget Baſſaricas Lydia mitra comas.

Fù Agaue ſiglia di Cadmo, e moglie di Echione, la quale nelle feſte di Bacco vcciſe Penteo ſuo figlio Rè di Tebe; & auuenne queſto a Penteo per hauer voluto vedere da vn luogo eminente del monte Citerone i ſacrificij che faceuano le Donne al Dio Bacco: morto poi apparue alla madre in forma di Cinghiale, e non di Vitello, come dice Perſio. Leggaſi Ouidio nelle Metamorf. al lib. 3.

FIGURE 2 *The lynx facing left in the* Persio *of Stelluti* (PERSIUS FLACCUS 1630, p. 41)

Cesi by Cardinal Francesco Barberini.[21] Since an image of the lynx identical to that of Stelluti's *Persio* graces the frontispiece of the *Praescriptiones Lynceæ Academiæ* (fig. 3) published in Terni in 1624,[22] this date can be taken as a *terminus ante quem* for the first representation of the animal from life in the Lincean circle.

In his commentary on Persius, Francesco Stelluti gave a detailed explanation of the reasons that had led the four founders to choose the lynx as the emblem of the Academy. It is noteworthy that at this late date, on the eve of the dissolution of the Academy, the motivations on which the Academy had been founded remained unchanged in spite of the prolonged exposure to the ideas and programmes of Galileo:

21 Persius Flaccus 1630, pp. 34–6.

22 Cesi 1624.

PRÆSCRIPTIONES
LYNCEÆ
ACADEMIÆ
CURANTE
JOANNE FABRO
LYNCEO BAMBERGENSI
Simpliciario Pontificio, Academiæ
Cancellario, prælo ſubjectæ.

INTERAMNÆ.
In Typographeio Thomæ Guerrerii. M. DC. XXIV.
SUPERIORUM PERMISSU.

FIGURE 3 *The lynx facing left in the frontispiece of the* Praescriptiones Lynceae (CESI 1624)

> [Cesi] chose this [the lynx] as the emblem of our Academy and as a stimulus and a continuous spur to remind us of the acuteness of vision, not of the corporeal eyes, but of those of the mind, necessary for the contemplation of nature that we practise; and it being all the more necessary in those matters to penetrate the inside of things, to know their cause and the operations of nature, which works internally, as one says in a beautiful simile that the lynx does with its glance, seeing not only what is outside, but also what hides within.[23]

It must be underlined that the lynx originally assumed as symbol by the founders of the Academy had a very different appearance and symbolic meaning from that portrayed from life which was later to replace it. It was the image of the lynx ripping apart the triple-headed Cerberus barring the entrance to Hades, accompanied by the motto 'Sagacius ista' (endowed [the lynx] with exceptionally penetrating senses)[24] (fig. 4) – an eloquent allusion to the inclination of the Academy to study the secrets of nature and of the supernatural. The idea of adopting that image was probably the product of the vivid imagination of Johannes Heckius (Van Heeck), who compiled the section of the *Gesta Lynceorum* in which the lynx that tears apart Cerberus is described. However, this suggestive logo, pregnant with references also to the moral mission of the Academy, soon disappeared from the Lincean horizon. At an unknown date, a rapid stroke of the pen in the manuscript of the *Gesta* transformed the metaphorical image of the feline into the naturalistic lynx, which has remained the symbol of the Academy through its complex vicissitudes to the present day.[25]

---

23 Persius Flaccus 1630, p. 37, cited in Lüthy 1996, p. 8.

24 Cf. Odescalchi 1806, p. 14.

25 Guardo-Orioli 2014, p. 67, n. 58. It escaped the notice of the unidentified author of this deletion that the same description of a lynx ripping a three-headed Cerberus to pieces appears a few pages earlier in the same manuscript (3v). This passage also provides a detailed explanation of the reasons behind the choice of that image, expressing as it did the desire of the founding members to immerse themselves in the study not only of natural phenomena but also of supernatural ones: 'Therefore they adopted the image of the three-headed Cerberus overpowered and torn apart by a lynx as the common emblem of the Academy, demonstrating that the lowest sublunary worlds can easily be subjected to dissection by all our senses, but not so the higher one. That is why the lynx gazes on the stars and celestial bodies, fixing all the keenness of its gaze on them, as indicated by the inscription *sagacius ista*, i.e., it embraces both the natural and the supernatural reality, as well as all that is knowable'.

FIGURE 4 *Odescalchi's graphic reconstruction of the first Lincean symbol: the feline tears the three-headed Cerberus to pieces* (ODESCALCHI 1806)

The lynx dismembering Cerberus has survived in a document embossed with the first seal of the Academy (*sigillum lyncealitatis*), the signet ring encrusted with an emerald that symbolised the obligations assumed at the moment of induction. It shows the lynx attacking Cerberus with the words 'sagacius ista' above and the word 'Lyncaei' below. While sinking its claws into the three-headed monster, the lynx turns its gaze to the sky illuminated by three six-pointed stars and the rays of the Sun (fig. 5), probably an allusion to the conjunction of the three planets under whose influence the Academy had been founded.[26]

As Gilberto De Angelis has demonstrated, the *sigillum lyncealitatis* was the product of lengthy reflection.[27] First of all, it followed the five precepts for the 'perfect emblem' formulated by Paolo Giovio in the *Dialogo dell'imprese militari e amorose*.[28] The choice of the emerald was likewise the fruit of careful meditation, inspired by the vast literature on the symbolic value of precious stones.

26 Cf. De Angelis 2012c, pp. 33–8 (fig. 8).

27 Ibid., pp. 37–49.

28 Ibid., pp. 33–4.

FIGURE 5
The Lincean seal, Ms. Vat. Lat. 9684, Biblioteca Apostolica Vaticana, c. 66v

Cesi and his first associates had certainly had in mind Pliny's comparison of emeralds with the sharp eyes of cats and panthers in his *Naturalis Historia.*[29] Based on Pliny's authority, the lynx, traditionally celebrated for the acuteness of its vision, must have seemed the most suitable creature to associate with that luminous stone. The emerald was not only an expressive symbol. It also transmitted active impulses to whoever possessed it and fixed his gaze on it. The idea of the symbol as spur (*incitatio*) was often evoked by Cesi. That choice reflected Federico's close reading of Ficino, who, in his *De Vita*, had stressed the efficacy of images and their capacity to stimulate man to action. It is not fortuitous that Cesi defined the signet ring as an incentive to chastity and hope in the *Lynceographum.*[30]

The significance of the Lincean ring was not limited to its symbolic value and its power to encourage moral actions. It was also an expression of the personality and expectations of its wearer, as Lodovico Dolce had theorised in his treatise *Delle Gemme*, a text present in Cesi's library.[31]

29 Ibid., pp. 37–9.

30 'The emerald encrusted in the ring engraved with the lynx expresses both the flourishing hopes and the positive influence of the stone' (Cesi 2001, p. 193). See De Angelis 2012c, p. 40.

31 Ibid, p. 41; Biagetti 2008, p. 291, item 1728: 'Lod.ci Dolce delle gemme Ven.a 1565'.

From Dolce, or from Marbod's treatise on stones (another book present in Federico's library)[32] the Prince might have also drawn inspiration in defining the emerald as an effective stimulus to chastity, a fundamental quality of the Lincean spirit, emphasised not only in the *Lynceographum* but also in all the foundational documents of the First Academy, as well as in the *Apiarium*. As Dolce had written:

> this stone [the emerald] is suitable for those who love chastity: that is why it does not suffer a virgin to be violated, but breaks. It restrains the movement of lasciviousness [...].[33]

In Dolce's treatise Cesi could also find elements that suggested a direct association between the emerald and the lynx: 'It [the emerald] restores tired vision and renders it more efficacious'.[34]

The conviction of the propitiatory value of the emerald also emerges from the letters with which Cesi accompanied the dispatch of the signet ring to new academicians. The Prince gave precise instructions on its use, significance, and on the reasons that had led him to choose that particular stone. As he wrote to Galileo, the ring should be considered as the sign and seal that 'every Lincean should wear on his finger all the time, also using it to seal letters of philosophical content or that bear on Lincean matters'. He invited Galileo to consider 'that the stone and its size were not proportionate to merit but expressed the fraternal spirit, and that the stone had been chosen for its natural qualities and properties and its significations [...]'.[35]

The choice of the lynx and the emerald and the hidden meaning attributed to them reveal the motivations of the Linceans in the early years. Cesi attached great importance to the display of the emblem of the lynx in the coats of arms and devices of the members of the Academy. In the compendium of the Lincean constitution that Cesi sent to Galileo on 2 June 1612, the Prince took pains to stress the importance of that practice: the Linceans 'should bear the lynx in an oval or circular frame in the ornament at the top of the shield of the coat of arms like another small device within that enlarged one'.[36]

---

32 Ibid., p. 293, item 1742.

33 De Angelis 2012c, p. 44.

34 Ibid.

35 OG XI, p. 286, letter of 22 March 1612.

36 Ibid., p. 312. Fabio Colonna inserted a rampant lynx opposite a siren in the family device published in the section of the *Thesaurus Mexicanus* that he edited (fig. 6).

FIGURE 6 *Family device of Fabio Colonna* (THESAURUS MEXICANUS, p. 847)

Each of the four founder members added to the collective one of the lynx, a personal emblem, accompanied by a motto that symbolically expressed his leanings and personality. The name *Tardigradus* (slow-paced) assumed by Stelluti betrayed his melancholy character, the result of the influence of Saturn, the slowest of the planets, which appeared at the centre of his emblem with the motto 'quo serius eo citius' (the slower, the swifter). Heckius opted for *Illuminatus* (Enlightened), with the image of the Sun in a trigon and the motto 'a patre luminum' (from the Father of lux). De Filiis chose *Eclipsatus* (Eclipsed), with the Moon eclipsed by the Earth and the motto 'spero lucem' (I hope for light), to underline his hope of receiving illumination from his fellow Linceans. Cesi, finally, adopted *Coelivagus* (Wanderer of the heavens) and chose the eagle bathed by the Sun's rays bearing the Earth in its talons with the motto 'utrumque', to indicate his knowledge of both the celestial and the terrestrial worlds.[37]

It is not by chance that the names chosen by the four founders all alluded to the stars, and in particular to the essential function of the Sun and of the physical and intellectual light that emanates from it. However, these academic names were not long-lived: they were employed only in the initial stages and

37 Gesta, c. 7r.

only by the four founders. When the Academy re-awakened in 1609–1610, after a prolonged hibernation, none of the new members took on an academic name.

Cesi's decision to place the eagle at the centre of his own emblem expanded the range of the symbolic menagerie of the Academy, which was already considerably variegated at that time. If the bee was to dominate its later symbolism, right from the start it included the rampant lion, attested in the marquis crest of the Cesi family that appeared in the frontispieces of Della Porta's works published at Federico's expense.[38] It has passed unnoticed that in those frontispieces Federico revisited his original marquis crest. The simple original crest of the Marquis of Monticelli still graced the frontispiece of Heckius' *De nova stella disputatio*, published at the Prince's expense in 1605.[39] It shows a tree rising from a rocky mountain (Monticelli) (fig. 7). The same crest can be found in the wooden ceiling of one of the salons of the Cesi palace in Acquasparta and, in a more elaborate version, in the centre of the vaulted ceiling of the Camerino degli specchi (the Chamber of Mirrors).[40] In this latter version the rocky mountain surmounted by the tree is inserted in a frame surrounded by the heraldic devices of the *gens caesia*. In the frontispieces of the two works of Della Porta published at Federico's expense two rampant animals on either side of the family crest appeared for the first time: a spotted lynx on the left and a lion on the right (fig. 8). It was the visual announcement of the imminent induction of Della Porta and of the adoption of the Neapolitan naturalist's vision in the programme of the Academy.[41]

Not only in the lynx, but also in its pairing with the rampant lion, do we perceive Cesi's fascination for Della Porta, who had portrayed a rampant lion in the frontispiece of his *Caelestis physiognomoniae libri sex* of 1603 (fig. 9). Later the lion was to feature prominently in the frontispiece of the edition of the *Satires* of Persius Flaccus translated and annotated by Francesco Stelluti (fig. 10).

This systematic deployment of the allusive language of images betrays the influence of Giovan Battista Della Porta, who had continuously referred to the symbolic significance of animals and to their relations with human features and inclinations. In the frontispieces of Della Porta's edition of Ptolemy's *Almagest*, his vernacular edition of the *Magia Naturale*, published in Naples in 1611, and the *Tre libri de' spirituali*, published in the same city five years

38 Della Porta 1610a and 1610c.

39 Heckius 1605.

40 Sapori-Vinti-Conti 1992, pp. 26 and 31.

41 Della Porta 1603. The modified crest in the frontispieces of the two works by Della Porta dedicated to Cesi was noted by Gabrieli (Contributi II, p. 1667), but he did not comment on it.

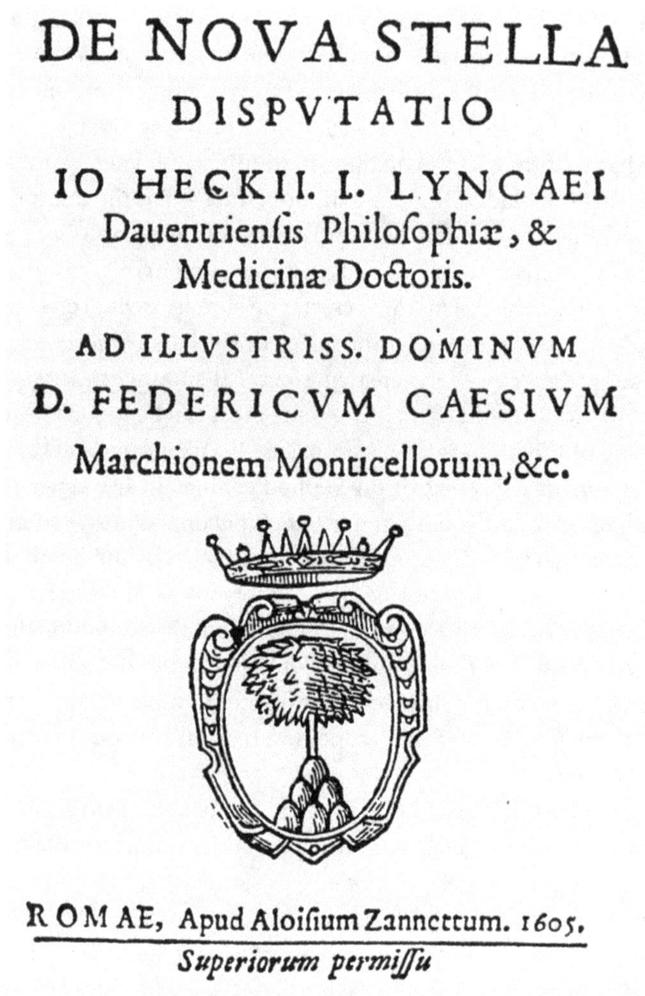

DE NOVA STELLA
DISPVTATIO
IO HECKII. I. LYNCAEI
Dauentrienſis Philoſophiæ, &
Medicinæ Doctoris.
AD ILLVSTRISS. DOMINVM
D. FEDERICVM CAESIVM
Marchionem Monticellorum, &c.

ROMAE, Apud Aloiſium Zannettum. 1605.
Superiorum permiſſu

FIGURE 7 *The device of the Marquis of Monticelli in the* De nova stella disputatio *of Heckius* (HECKIUS 1605)

earlier, the reader's eye was caught by the image of the phoenix (fig. 11).[42] The resurrection of the mythical creature was accompanied by the motto 'ex me ipso per saecula novor cum simus ab uno' (I resurrect from myself for ever, as we all derive from the same principle). Neither did Della Porta pass over the winged horse Pegasus: its image in flight dominates the frontispiece of the 1592 edition of his *Villae* (fig. 12).[43] It was probably under the influence

42 Ptolemy 1605; Della Porta 1611 and 1606.

43 Della Porta 1592.

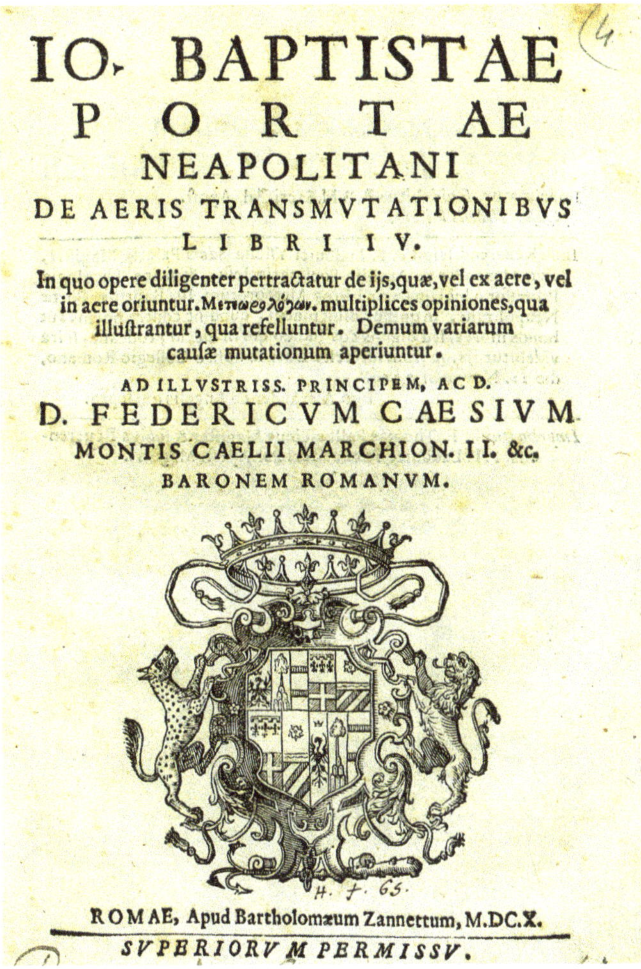

IO. BAPTISTAE
PORTAE
NEAPOLITANI
DE AERIS TRANSMVTATIONIBVS
LIBRI IV.
In quo opere diligenter pertractatur de ijs,quæ,vel ex aere, vel in aere oriuntur. Μετεωρολόγων. multiplices opiniones,qua illuſtrantur, qua refelluntur. Demum variarum cauſæ mutationum aperiuntur.
AD ILLVSTRISS. PRINCIPEM, AC D.
D. FEDERICVM CAESIVM
MONTIS CAELII MARCHION. II. &c.
BARONEM ROMANVM.
ROMAE, Apud Bartholomæum Zannettum, M.DC.X.
SVPERIORVM PERMISSV.

FIGURE 8 *The 'revised' device of the Marquis Cesi in the frontispiece of the* De aeris transmutationibus libri IV (DELLA PORTA 1610a)

of that model that Stelluti chose the title of his interminable epithalamium, *Il Pegaso*, published in 1617 to celebrate the marriage of Federico Cesi to Isabella Salviati.[44]

Besides the use of their personal emblems, the four founders also agreed to adopt a code in their correspondence. Devised by Cesi with the collaboration of Stelluti, it was inspired by the *De zypheris* of Della Porta.[45] The Lincean code was finally deciphered by Domenico Morosini, at the instigation of

44 Stelluti 1617.

45 Della Porta 1602.

COELESTIS
PHYSIOGNOMONIAE
LIBRI SEX.
IOAN. BAPTISTAE PORTAE
NEAPOLITANI.
Vnde quis facile ex humani vultus extima inspectione, poterit ex coniectura futura praesagire.
In quibus etiam Astrologia refellitur, & inanis, & imaginaria demonstratur.
NEAPOLI, Ex Typographia Io. Baptistae Subtilis. M. DC III.

FIGURE 9 *The rampant lion in the frontispiece of the six books of the* Coelestis Physiognomoniae (DELLA PORTA 1603)

Francesco Cancellieri, in 1822.[46] His brilliant solution is preserved in the Biblioteca Apostolica Vaticana. The code consisted of converting the five vowels of the alphabet into the astrological aspects of the planets, while the consonants were replaced by the symbols of the planets and the signs of the zodiac (except for Pisces).[47] The fact that between 1625 and 1630 Francesco Stelluti was engaged in projecting a mechanical device to facilitate the writing of

46 Ms. Vat. Lat. 9684, c. 143.

47 Cf. Morosini 1865, pp. 24–7; see also Partini 1986 and Alessandrini 1986, p. 107.

FIGURE 10 *Frontispiece of Stelluti's annotated edition of* Persio (PERSIUS FLACCUS 1630). *The lion in the left-hand niche is kept on the leash by Moral Philosophy*

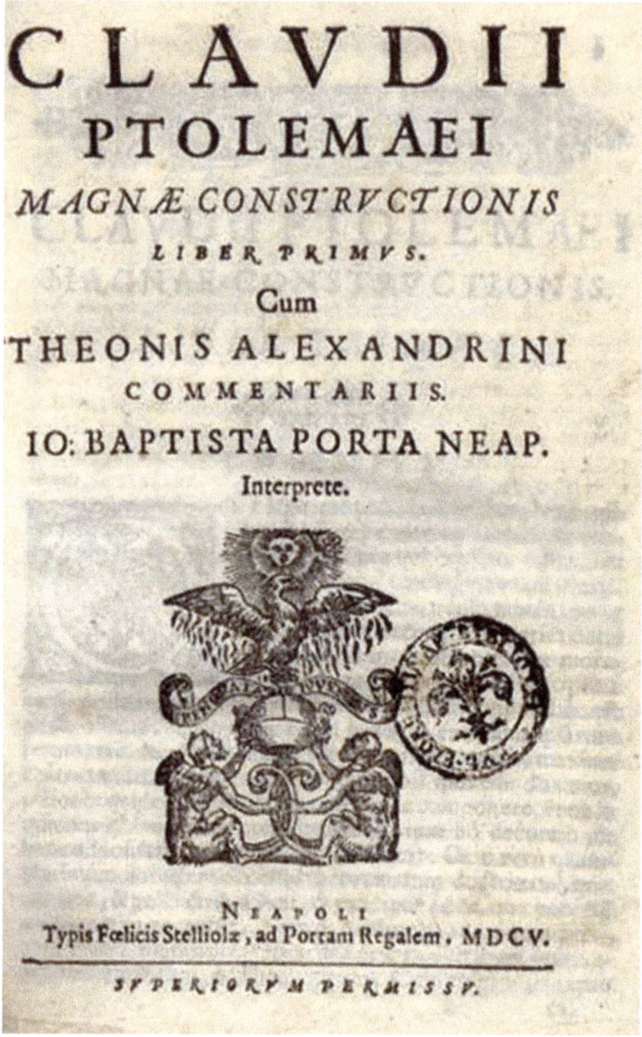

CLAVDII
PTOLEMAEI
MAGNÆ CONSTRVCTIONIS
LIBER PRIMVS.
Cum
THEONIS ALEXANDRINI
COMMENTARIIS.
IO: BAPTISTA PORTA NEAP.
Interprete.

NEAPOLI
Typis Fœlicis Stelliolæ, ad Portam Regalem. MDCV.

SVPERIORVM PERMISSV.

FIGURE 11 *The phoenix in the frontispiece of Della Porta's edition of Ptolemy's* Almagest (PTOLEMY 1605)

texts in code suggests that many years after the foundation of the Academy the need to guarantee the secrecy of communications between the members was still felt.[48]

48 Tiziana Gazzini proposed a reconstruction of the device, that she defined as a 'cryptographic machine', illustrated by Stelluti in Ms. Vat. Lat. 9685 (cc. 80v-81r), Biblioteca Apostolica Vaticana (Gazzini 1986b, pp. 302–13). This alphanumerical device would have

VILLAE
IO. BAPTISTAE
PORTAE, NEAPO-
LITANI,
LIBRI XII:

1, Domus
2, Syluacædua
3, Sylua glandaria
4, Cultus & insitio
5, Pomarium
6, Oliuetum
7, Vinea
8, Arbustum
9, Hortus coronarius
10, Hortus olitorius
11, Seges
12, Pratum.

IN quibus maiori ex parte, cùm verus plantarum cultus, certaq; insitionis ars, & prioribus seculis non visis producendi fructus viamonstrantur: tum ad frugum, vini ac fructuum multiplicationem experimenta propemodum infinita exhibentur.

Adiecto INVENTARIO quamcopiosissimo.

M D XCII.

FRANCOFVRTI
Apud Andreæ Wecheli heredes, Claudium Marnium, & Ioannem Aubrium.
Cum S. Cæsareæ Maiestatis privilegio.

FIGURE 12 *Pegasus in the frontispiece of Della Porta's* Villae (DELLA PORTA 1592)

encoded every word by means of simple operations on the instrument. In another publication (Gazzini 1986a), she illustrated a modified working model of the device, though she declared herself unable to decode the symbols used by Stelluti. Della Porta had already thought up similar devices to facilitate writing in code (cf. Della Porta 1953a, pp. 92, 98 and 103, with illustrations of the cryptographic machines). Gazzini also studied a different encoding device illustrated in the same manuscript (c. 82r), which she defined as an 'alphabetic cross' because of its shape. She hypothesised that Stelluti may have drawn inspiration for his cryptographic devices from Bacon's *De dignitate et augmentis scientiarum*.

Initiation Ceremonies

There should be no need to point out that the celebration of the capacity to penetrate to the essence of things with the eyes of the mind was far removed from Galileo's recourse to experience and mathematical reasoning to reveal the laws of nature. His telescope, and later his microscope, provided a quantitative improvement of vision, not a qualitative one like that attributed by the founders of the Academy to the eye of the lynx. The clear distinction between the performance of the telescope and that of the gaze of the feline would not prevent Heckius from declaring that Galileo had 'added his adventurous lenses to the eyes of the lynx'.[49] This image of the telescope as an instrument that rendered the eye of the lynx more powerful was to enjoy considerable fortune in the Academy: Faber was to define the instrument as 'the new eye of the lynx'.[50]

It has long been maintained that the foundation of the Academy in 1603 should be considered as a false start which did not produce any further developments, while the effective commencement of academic activities – a true re-foundation – took place between 1609 and 1610. This interpretation of the Academy's beginnings has contributed to drawing a veil over its first years and over Cesi's original project which, even if it failed to materialise because of the dispersal of the nucleus of founders, remained firmly implanted in the Prince's mind. This is demonstrated by the energy devoted by Federico to drawing up the statutes of the institution during those years of enforced isolation, as well as by the evidence preserved in the correspondence of the Academy until 1609, as well as in the proceedings of its early meetings.

The latter indicate the strong interest of the founders in astronomy and, above all, astrology. Work on the construction of a large astrolabe was begun immediately after the founding meeting of the Academy on 17 August 1603.[51] To avoid depending on outside assistance the founders undertook the task collectively: ' We all completed the plane projection of the sky with its many divisions in our spare time'. It was mainly Heckius and Stelluti who worked on the construction of the astrolabe. In this respect it is worth noting that Stelluti is indicated as 'calculator and predictor' of celestial movements in the early documents.[52] We learn from the *Gesta* that construction of the astrolabe took

49 Cf. below, p. 147.

50 TM, p. 462: 'tam mirabilis telescopio suo, uti novo Lyncis oculo in caelo primus animadverti [...]'.

51 Contributi I, p. 509. The date is attested in the Gesta (c. 7v).

52 CL, p. 473.

less than two months and was completed on 12 October. The *Gesta* offer precious information on the founders' motivations for devoting their efforts to the fabrication of the instrument:

> On 12 October of this year the planisphere, or rather the model of all the celestial orbs and motions, was finally completed. All the members were delighted that the immense intricacies of all the books and tables could now be easily grasped by rotating that instrument.[53]

Other clues regarding the construction and functioning of the instrument can be gleaned from the proceedings of the first meetings:

> The Planisphere or Great Astrolabe [...] displays not only the astronomical and astrological phenomena of the heavens but also the sublunary ones that depend on them, drawing on both traditional and more recent knowledge. It was therefore completed and kept with all its accessories.[54]

The passages in the *Gesta* that refer to the astrolabe indicate that it was considered a fundamental instrument for an extremely delicate operation to which the first Linceans intended to devote themselves: to accord Christian religion with the philosophy of Plato.[55] It is evident that Cesi and his companions assigned to the alliance between the Catholic religion and the Aristotelian conception of the universe responsibility for the dramatic degeneration of knowledge and ethics of contemporary times. And they trusted that that degeneration could be countered by replacing Aristotle's cosmology with the vision of the relations between man and the universe outlined by Plato in the *Timaeus*.

Not surprisingly, Cesi was determined to keep these intentions secret by extending the use of the Lincean code to the inscriptions on the instrument:

> In the meantime, while the celestial planisphere was being elaborated, to avoid communicating the art and skill of its construction and conception and to prevent the work itself from being revealed even to drones, Caelivagus first recommended the use of a coded notation and to conceal the

53 Gesta, c. 7v.

54 Contributi I, pp. 509–10.

55 Gesta, c. 6v: 'Placebant praecipue lectori [Heckius] et auditoribus cum Christiana religione Platonicae sapientiae mira concordatio.'

> parts and the whole by means of a symbolic language. The other members were convinced, and it was immediately decided that the signs of the planets would be in hieroglyphs, and those of all the stars in the alphabet to render them extremely difficult to identify and to prevent what was inscribed from being read. So that hardly anyone would be able to properly distinguish whether the computation of the Planets [orbits] was drawn from the difference of the heavens and orbs or independently.[56]

In view of the importance assigned to the construction of the astrolabe in the Lincean agenda, it is not surprising that one of the first concerns of the four associates was to draw up the horoscope of the newborn institution. A few weeks after its foundation, on 25 September 1603, while work was going ahead feverishly on the construction of the astrolabe, the founders organised what Massimo Luigi Bianchi has called 'a magical ceremony conducted according to the canons of Renaissance hermeticism'.[57] Bianchi suggested that the four Linceans drew inspiration from Francesco Cattani da Diacceto and Tommaso Campanella. The description of the ceremony in the *Gesta*, in the handwriting of Cesi, is preceded by a brief text in Lincean code on the propitious position of Mercury followed by the horoscope of the Academy, based on the positions of Mercury, Jupiter and Saturn on 25 September. The Linceans aimed above all at capturing the favourable influence of Mercury. To render the propitiatory rite more effective, the three metals and plants corresponding to the three planets were displayed. Cesi and his companions also recorded 'Pythagorean mysteries' on sheets of paper and silk as they carried out operations of numerical magic. Finally, they resorted to the positive radiation of a magnet: 'A magnet was put in place to receive its radiant heat, so that under this influence the scientific studies should be enabled, assisted and facilitated'.[58] It was after this propitiatory ceremony that the four founding members adopted the lynx as the emblem of the Academy and chose their individual devices.

The extant minutes reveal the heterogeneous and peculiar nature of the topics discussed in the first meetings: reflections on *transnaturalis philosophia*; the enunciation and discussion of Platonic theses; experimentation with miraculous remedies; readings in elementary geometry and logic; the world

56 Ibid., c. 7v. In the final passages of this somewhat obscure Latin text an allusion to the fluid nature of the heavens can be perceived. For the structure and functioning of the instrument, see Giorgio Strano's remarks in Galluzzi 2014, pp. 539–44.

57 Bianchi M.L. 1991, pp. 53–6.

58 Gesta, c. 3r-v.

before the Flood; and so on. Besides these sublime scenarios for reflection, the minutes record questions of a strictly practical nature, such as the project expounded by Stelluti for an extendable ladder.[59] Other evidence confirms that the founders assigned a magical significance to numbers: they considered the structure of the Academy with its four members perfect because when one spoke, the listeners formed the perfect number of three.[60] Internal discussions adopted the model of the *quaestiones* practised in the universities. During the first session after the foundation, on 19 August, Heckius defended Platonic conclusions that were called into question with great erudition immediately afterwards by Cesi.[61]

The accounts of the first steps of the institution in the *Gesta Lynceorum* perfectly illustrate the expectations of the first Linceans and the ceremonial adopted in their meetings. On Christmas Eve, 1603, benefitting from a lull in the opposition of Federico's parents, the four founders met in the Cesi palace. The Academy's session was opened by hanging the image of a lynx – we do not know whether it was a drawing, a painting or a sculpture – from a beam in the roof as a stimulus to the study of truth and the fraternal union of the academicians:[62] a symbolic act of both physical and spiritual elevation.[63]

The following day, Christmas Day, while Federico dined with his parents, his three associates met on the same premises. Heckius improvised a solemn speech to reinforce the sacrosanct bond of the Lincean union,[64] which was enthusiastically welcomed by those present. Federico joined them later on, donned a cassock, and mounted the pulpit to renew the exhortation to cement the bonds of friendship. Subsequently he invited his companions to approach him one by one. He hung a gold necklace around their necks with a pendant in the form of a lynx and explained the meaning of that ceremony in the following words: 'Receive [...] this symbol of fraternity that is shared with me; let

59 Contributi I, p. 510: 'Miram scalae cuiusdam ascensoriae fabricationem, quae tum brevis tum longa fieri posset'.

60 Ibid., p. 508. On the interest of the founding members of the Academy in numerical magic and, more in general, on the interpretation of mathematics as a 'code' that expressed the hidden significance of things, see Eugenio Garin's remarks (1986, esp. pp. 19–23). The *Liber secundus* of Della Porta's *Taumatologia*, a work to which Cesi attributed great importance, is dedicated to the explanation of the magical properties of numbers (Della Porta 2013b, pp. 25–74).

61 Contributi I, pp. 506–7.

62 Gesta, c. 9v; cf. Odescalchi 1806, p. 24.

63 Cf. De Angelis 2012b, pp. 15 and 38–40.

64 Gesta, cc. 9v-10r.

it be a sign of virtue and fraternity, let it be the reward for present and future efforts'.[65] Then they prayed. It was in the course of that ceremony that Federico, abandoning the initial function of 'Greater Counsellor', assumed the prerogatives and functions of Princeps of the Academy.

The choice of Christmas 1603 for the solemn ceremony of inauguration of the Academy and of the election of Federico as its Princeps was inspired by the 'great conjunction' (Jupiter and Saturn in the ninth house of Sagittarius with Mercury in ascendant) that Campanella had interpreted as the announcement of an imminent major upheaval.[66] It is worth recalling that, after reading the *Nuncius*, Campanella had written enthusiastically to Galileo that his sensational celestial revelations heralded the dawning of the new era that he had predicted eight years earlier.[67] Campanella had explained in his *Prognosticum* why he believed that extremely rare astronomical event to be the harbinger of a new phase in the history of mankind: the last time that that phenomenon had been witnessed had coincided with the birth of Christ.[68] It is also worth stressing that Campanella sent his letter to Galileo via Antonio Persio.[69] As we shall soon see, the mention of Persio and the role assigned him by Campanella are revealing.

The ceremony of Christmas 1603 underlines the solemnity that Federico and his associates intended to confer on Academic functions. It is not surprising that the procedure conceived for induction evoked the ritual of investiture as a knight or of ordination into the priesthood. The desire to provide the Academy with a ceremonial had manifested itself right from the start, at any rate considerably before Della Porta sent insistent demands to the Prince calling for the compilation and formal adoption of a 'book of ceremonies'.[70] Eloquent testimony of the importance attributed to the rituals of academic procedures can be found in the letter that the Linceans jointly wrote, probably

65 Ibid., c. 10v; cf. Odescalchi 1806, p. 29.

66 De Angelis 2012b.

67 Cf. above, p. 25 and n.1.

68 De Angelis 2012b, p. 12.

69 Ibid., pp. 6–7, OG XI, p. 26, letter of 13 January 1611, republished in Campanella 2010, p. 189.

70 CL, pp. 225–6, letter of Della Porta to Cesi, 31 May 1612, in which the Neapolitan dictated the rules for formal investiture as a Lincean that he wanted to see recorded in a 'libretto delle cerimonie'. Adopting this strategy, he concluded: 'With time it will become more important to be a Lincean than to have received the Cross of these Knights [of Malta]'.

in 1606, to Emperor Rudolf II.[71] They stated that, during academic sessions, they wore an open silk gown extending from the neck to the feet, unbound at the waist, with half-length sleeves, and evenly trimmed with lynx fur from the neck to the chest.[72] It is difficult to imagine that they really thought that the Emperor would be interested in such a detailed description of their academic attire! It was evidently information that they considered important to underline the nobility and dignity of the enterprise and to give it status among the powers that be. It is not by chance that in the same letter they informed Rudolf of their intention to pursue the most noble sciences, those that are not disputable, and that they attributed the highest moral value to that mission. They furthermore announced their intention to spread the Lincean initiative in 'various parts of the world', as well as their desire to dedicate themselves to the study of stones, minerals, plants and all productions of nature.[73]

The documents of the first years of the Academy clearly reveal the influence of the models which inspired the production of naturalistic encyclopaedias from the second half of the sixteenth century. Cesi and his companions shared the ambition to compile an entirely new inventory of nature in the awareness that the traditional encyclopaedia had been superseded by the flood of new data emerging from the geographical discoveries. This stimulated the formation of 'museums' or 'theatres' dedicated to the illustration of the recently discovered productions of nature, to highlight their properties and to introduce new schemes of classification. Federico Cesi and his companions shared the same motivations as intellectuals like Ulisse Aldrovandi and sovereigns like Francesco I Medici, the Prince of the Studiolo. The goals pursued by the Lincean Academy were thus the development of a sixteenth-century programme. For Cesi and his fellow Linceans, it was clear from the start that they were embarking on a very important enterprise: that of liberating consciences and nature itself from the tyranny of Aristotle and his followers. The lynx-eyed academicians aimed not only to collect and reorganise data; they wanted to give knowledge a new significance in order to counter the cultural, moral and religious degeneration of their age. This required great commitment, the

71 Ibid., pp. 1267–8. It is not certain that this remarkable document was ever dispatched to the Emperor.

72 'Toga serica utantur a iugulo ad pedes aperta atque descincta, largioribus manicis in mediocrem gyrum detinentibus, quam lyncea pelles ambiat a collo ad medium pectoris pertinens altitudini aequabili', ibid., p. 1268. In a medal of the Academy Cesi appears wearing the toga with half-length sleeves, cf. ibid., p. 113, fig. 10.

73 Ibid., p. 1267.

awareness that they were fighting for a noble cause and, consequently, also strong religious motivations.

In the letters written between 1603 and 1605 to famous European naturalists such as Caspar Bauhin, Charles de l'Écluse (Carolus Clusius), Mathias de l'Obel and Thomas Mermann, the four Linceans repeatedly referred to themselves as 'very keen-eyed researchers into the occult sciences and disciples of Paracelsus'.[74] The main purpose of these letters was to request seeds and opinions on their medicinal properties. Their strong interest in botany, their extreme curiosity about secrets, their confidence in remedies produced by chemical operations and their expectations of a radical moral and religious reform fully justify their defining themselves as followers of Paracelsus.

The correspondence between the members of the initial nucleus after their forced separation displays the same type of curiosity and interests: seeds, secrets, medicinal remedies, stones with singular properties, exotic animals, rare books. The letters of Heckius, the most eccentric and unsettled of the four, are particularly revealing. As evidence of his ascendancy over Cesi and the other Linceans, Heckius is referred to as 'Master' in the earliest documents. After being forced to abandon Italy, he sent letters to Federico from England, Scotland, Germany and Prague where he had finally settled. Heckius informed Federico of the promising results achieved as travelling ambassador of the Academy. He had even received the favours of Emperor Rudolf II whom he described as a 'desperate lover of the Lyncean programme'.[75] Besides the Dutch Johannes Heckius, we shall soon find among the associates of Cesi's Academy several Germans, all extraordinarily receptive to the Paracelsian tradition, such as Johannes Faber, Theophilus Müller and Johannes Schreck (Terrentius), author of a Paracelsian *Compendium*.[76]

The other characteristic aspect of the early experience of the Academy is the interest in the secrets of nature, and in particular in natural magic. It is not surprising that Della Porta was the intellectual most sought after by Cesi.

74 Ibid., p. 30, letter from Heckius to Thomas Mermann, 17 February 1604. Similar expressions recur in their joint letter of 20 March 1604 (ibid., p. 31) to Adone Campello ('Greetings, most learned sir, from the Linceans, lovers of science who investigate the spagirical disciplines and are well informed about your fame), as well as in the joint letter addressed to Rudolf II discussed above (ibid., pp. 1267–8).

75 Ibid., p. 47, letter from Prague to the Linceans, 19 December 1604.

76 The *Compendium* is preserved among the Lincean manuscripts (formerly Albani) in Montpellier: 'Compendium eorum quae a philos. Paracelso suis in scriptis dispersa sunt'. Cf. Contributi I, p. 1029 and Alessandrini 1978, pp. 106–11.

Federico visited him in Naples in 1604 securing his enthusiastic adhesion to the Lincean project. The Prince projected the opening of a branch of the Linceans in Naples and set about acquiring for the Academy the museum, library and above all the enormous mass of secrets of the Neapolitan.[77] The terms in which he informed Stelluti about his meetings with Della Porta betray a keen enthusiasm for the 'most beautiful secrets' that the Neapolitan had told him about and which he had promised to reveal.[78]

Gardens, libraries, the exchange of seeds and rare naturalia, secrets, alchemical transmutations, interpretations of the influences of the stars and the capturing of their favourable powers, the expectations of a radical renewal to rescue mankind from the drama of original sin – these were the priorities in the agenda of the Lincean Academy in its early years.

The insistence in the programmatic declarations of the initial period on their decision to assume the teachings of Paracelsus and his followers as their models reveals the academicians' willingness to engage in a project of radical reform of knowledge. This was intended not only to put the methods and aims of the study of nature on a new footing, but also to restore dignity to morals

---

77 Cesi went back to Naples with Stelluti in the spring of 1610 to start negotiations to acquire the rare natural productions and above all the 'secrets' collected by Della Porta in the course of his long life for the branch of the Academy that he intended to set up in Naples. Later on, Federico sent Stelluti back to Naples to formalise the deal. The Prince gave his faithful Lincean fellow precise instructions on how to bring the negotiations to a successful conclusion (CL, pp. 342–52; mid-April 1613). The idea of setting up a Lincean branch in Naples, of making Della Porta its head, and of furnishing it with the Neapolitan's material and intellectual legacy must have emerged in the Prince's mind after his 1604 trip to Naples, in the course of which he and Della Porta discovered that they shared the same motivations and goals. This is shown by the dedication to Cesi of the edition of *De distillatione* of 1608 (Della Porta 1608). The dedication itself is dated four years earlier, and must therefore be considered as the immediate token of appreciation of the visit that Cesi had made to Naples in that year. During that visit Federico met Giovan Battista on several occasions manifesting a reverent admiration for him. The 1608 edition of Della Porta's *De distillatione* displays a splendid frontispiece in which the Neapolitan is framed by objects alluding to the variegated range of his interests as a researcher and collector of rarities and natural secrets. Cesi must have seen that image as the perfect representation of the interests and aims of his Academy.

78 'I have held discussions with Sig. Gio. Battista Porta and Sig. Ferrante Imperato, who are all very dear friends of mine and of the Linceans, and really are wonders of nature [...]; I have learnt a great deal in conversation with them, and I have spent a large part of the time in Naples with these two to my great benefit' (CL, p. 41, letter to Francesco Stelluti, 17 July 1604).

and religious feeling. It is natural to connect those expectations to the ideals of the Rosicrucians who, starting from explicitly Paracelsian positions, strove to introduce a radical renewal of knowledge. It may not be coincidental that Cesi wrote his *Discourse on the natural desire for knowledge* (*Discorso sul natural desiderio di sapere*) only months after the publication in Germany of the *Fama fraternitatis*, the manifesto of the ideals and inspiring principles of the Rosicrucians. The timing is telling inasmuch as the call in Federico's *Discourse* to pursue knowledge as a good in itself and not as a way to acquire wealth and privileges is particularly close to the definition of the disinterested nature of knowledge in the *Fama* and other Rosicrucian texts.[79]

The Linceans closely followed the German developments of Paracelsianism, paying particular attention to the Rosicrucian movement. It is sufficient to recall that Schreck – the Lincean who had left the Academy to join the Society of Jesus – had been entrusted by the Academy with the task of purchasing 'all the books about that Rosicrucian society' at the Frankfurt book fair.[80] So it is not surprising to find the *Fama fraternitatis* in the personal library of Prince Cesi.[81]

The main impulse to assume the explicitly Paracelsian orientations attested in the documents of the first years of the Academy certainly came from Heckius. The information that he sent to Cesi after his forced departure from Rome in 1605 reveals his strong interests in chemical philosophy and its most extraordinary practical applications. At the beginning of his journey to the heart of Europe, Heckius had stopped over in Florence, where, thanks above all to Antonio de' Medici, interest in chemical philosophy was intense. While in Florence, he visited the Biblioteca Palatina, consulted its rich collection of chemical texts and met with the Grand Duke's chemist.[82] Although Heckius does not supply the name of this person, who worked in the Casino di San Marco with Antonio de' Medici, it is reasonable to suppose that it was Benedetto Punta, physician of the Grand Duchess Christine of Lorraine and of Antonio de' Medici.[83]

---

79 On the presence of Rosicrucian themes in the production of the first Linceans and in the writings of Cesi, see Galluzzi 1982, esp. pp. 47ff. On the influence of Paracelsian motifs on the programmes and activities of the founders of the Academy, see Saverio Ricci 1992.

80 CL, pp. 599–600, letter to Faber, 10 June 1617. Cf. Galluzzi 1982, p. 89.

81 Cf. Biagetti 2008, p. 404, item no. 2686: 'Fama fraternitatis rosae Crucis in tod. 1616'.

82 Gesta, c. 23v. Schreck informed Faber of the meeting with Prince Antonio de' Medici in a letter from Florence of 24 September 1615 (CL, p. 516); cf. also Galluzzi 1982, p. 61.

83 On Punta see below, pp. 59–60.

Although Faber, Terrentius and Heckius were closer to Paracelsianism than Cesi, on several occasions Federico was to display a marked interest not only in the theoretical foundations of chemical philosophy, but also in the chemical transmutations of natural substances. He was hostile, however, to the unscrupulous commercial exploitation of alchemical transmutations, a widespread practice at the time. In the *Lynceographum* he vigorously denounced the improper use of 'that part of Chemistry that bears on the transmutations, production and modifications of metals for profit'.[84] On the contrary, he attributed great value to the disinterested study of chemistry, which he considered a useful means to throw light on the operations of nature. Provided it was conceived as an instrument to reveal natural processes that satisfied man's innate thirst for knowledge, alchemy was not only allowed but actively encouraged by the Lincean rule.

The expectation of a radical reform of knowledge and morality also animated Cesi's exhausting efforts to define statutory regulations that would fully express the ideals of the Academy and lay down precise rules of conduct. The composition of the *Lynceographum*, begun in 1605, reflects the aims to which the activities of those early years were directed. The *Lynceographum* contains an extraordinary quantity of prescriptions, similar to the rules of a monastic order and elaborated down to the most minute details, regarding not only the procedures to be followed by the Academy, but also the model of life and the moral ideals to which its members were expected to conform. Cesi began to circulate among the Linceans the first drafts of the *Lynceographum* and its abridged version (*Ristretto delle costituzioni lincee*) in 1612.[85] At that point the institution was characterised by two poles, as a result of the induction of its most distinguished members, Della Porta and Galileo. Cesi took great pains to maintain an equilibrium between these two protective deities. An eloquent example of this effort is offered by the letters accompanying the dispatch of the *Ristretto* to the two members.[86] This text was sent to Della Porta through the intermediary of Stelluti, who took it to Naples in the spring of 1613. Stelluti was asked by Federico to inform the Neapolitan that the *Ristretto* was an

84 Lynceographum, p. 69. It is not clear to what extent Cesi had adhered to chemical philosophy. Although there are documents that attest his hostility towards chemical remedies, his appreciation of the theoretical principles of Paracelsus and of the variegated range of his successors appears evident.

85 On the *Ristretto* see Marco Guardo (2005) [rev. ed. in Guardo-Orioli 2014, pp. 125–43].

86 Guardo 2005, pp. 512–7.

enterprise 'founded on his reputation'.[87] Cesi had dispatched the *Ristretto* to Galileo a year earlier, inviting the Tuscan scientist, whom he defined as the leading light of the Lincean enterprise, to pass his comments on to him:

> Please review it and tell me what you think should be changed or corrected, as I wish this enterprise, rendered so glorious by you, to be based in particular on your reputation.[88]

The great amount of energy expended over a considerable period of time in precisely defining the rules of conduct governing the life of the Academy, as well as in their continuous refinement, display an almost schizophrenic character, especially if one bears in mind that those rules rarely matched the real life and functioning of the institution or the behaviour of its members. Moreover, despite Cesi's repeated announcements, the *Lynceographum* was never published. It was only many years after its compilation that a few general indications on the aims of the Academy and on its internal regulations were issued. But not even these concise prescriptions, very different, also as to their content, from the long-winded and peremptory precepts of the *Lynceographum*, were actually put into practice.[89]

The founders of the Academy also failed to maintain the commitment, solemnly proclaimed in the joint letter to Emperor Rudolf II and stressed in the *Lynceographum*, to keep any interference with religion and politics from the agenda of the institution.[90] Yet Cesi attributed such importance to this rule that he recommended that the Linceans who wanted to write historical works should confine themselves exclusively to protagonists and themes of Antiquity, programmatically avoiding contemporary issues. The distance from the

---

87 CL, p. 345.

88 OG XI, p. 312, letter of 2 June 1612. On Galileo's lack of interest in the *Lynceographum*, the *Ristretto*, and later the *Praescriptiones Lynceae*, see Guardo 2005, who analysed in detail the relations between the first two documents and between the *Ristretto* and the *Praescriptiones*. The *Praescriptiones* were published in 1624 (Cesi 1624).

89 Cesi 1624.

90 CL, p. 1267: 'They decreed that all controversies in civil and divine matters should be banished from their writings and debates [...]. Prelates and priests are closer to God, and are therefore to be honoured by all men with veneration, respect and observance. We desire this to be placed before the eyes of the Linceans, that they should always hold them in the greatest honour [...]. We wish the Linceans to be faithful to Princes in whose regions there are Lycea, or where Linceans live, and reverent, and ready to serve and assist them [...]' (Lynceographum, pp. 66–7).

Renaissance models of the 'civil' historiography of a Guicciardini or a Machiavelli could not be greater.[91] Eloquent proof of the continuous breaking of this rule is provided by the activity of Johannes Faber, the man around whom all the initiatives of the Linceans would revolve after his induction in 1612. Faber was actively engaged in the conversion to Catholicism of Germans who had joined the Protestant reform movement. His talent in this delicate mission is confirmed by the letter in which one of his German correspondents, the fellow Lincean Markus Welser, acknowledged Faber's unrivalled skill in converting heretics.[92] Another German, Johannes Schreck (Terrentius), who was an active member of the Academy during his brief period of association, was equally gifted in proselytism.[93] Faber also excelled in the most delicate political mediations; he acted on behalf of third parties to get them permission to read prohibited books; moreover he operated as an agent of the Fugger family in

91 'Linceans must avoid historical themes bordering on political matters because many are harmed by that, even without any responsibility on the part of the author, as well as writing about the deeds of Princes, but with their approval they will be allowed to sing the praises of Lords, Regions and Nations where there are Lycea, and to deal with events pertaining to antiquity' (ibid., p. 69). On this point see Olmi 1981, p. 194.

92 An unpublished, undated and unsigned letter to Faber, attributed by Irene Fosi to Welser, describes the activity of converting German Protestants conducted by Faber: 'I wish the German gentlemen who come to Italy and return to Germany to leave satisfied with this province, and most of all with Rome and the territory belonging to the Church [...]'. Welser went on to lavish unconditional praise on Faber's extraordinary capacity in this delicate task: 'It is most foolish of me to overlook speaking of you, who with your marvellous knowledge and singular prudence have earned the title of one of the Seven Sages of Italy' (Fosi 2005, pp. 194–5).

93 A letter from Ignatius Hanniel in Rostock to Joseph Justus Scaliger contains a description of the activities of Schreck in converting Protestants in Germany in 1607, when he was not yet a Jesuit and no longer a Lincean. Ottaviani has drawn attention to this document as evidence for the dedication to converting lost souls on the part of Schreck, Faber and Schopp, three Germans closely tied to the Linceans (Ottaviani 2005, pp. 101ff). The substantial German colony of the Academy, jealously protected by the Prince (in this connection it is worth recalling Cesi's sharp rebuke of Galileo, who had denigrated the entire virtuous German nation in his polemic with Scheiner: OG XI, p. 423) was marked by this proselytising activity, which used the art of memory, the practice of a medicine with clear Paracelsian connotations, and recourse to the sensational effects of natural magic as instruments to win over consciences. According to the portrait delineated by Hanniel, these were the weapons that Schreck used in his well-attended meetings with the young Protestants of Rostock: 'A few days ago we received news of Terrentius [Schreck] who professed the arts of memory and other more secret ones and used exotic languages among a large public of young men, and eventually secretly spread the Catholic religion [...]' (Ottaviani 2005, p. 102).

Augsburg on economic and political matters. Nor was he an admirable example of the disinterest in money matters that the Linceans were expected to show, accustomed as he was to receiving payment – considerable sums at times – for the services that German visitors and nobility requested from him in Rome, such as the dissolution of marriages, the granting of ecclesiastical privileges, and assistance in civil trials or those instigated by the Inquisition.[94]

Another solemnly proclaimed obligation – chastity – that also found expression in the symbolic value of the Lincean emerald, was not respected, even though we read in the *Gesta* that those who contravened this rule would be suspended for three days from the activities of the Academy.[95] It is worth recalling that among the reasons that prompted Cesi's parents to intervene to put a stop to his academic initiative were their concerns about the dissolute life that the adolescent Federico was held to be leading together with his companions. As Odescalchi emphasised, the detractors of Federico and of the Academy reported to his father 'that under cover of study, his son was leading an extremely degenerate life and, with the collusion of the members of the Academy, kept up a shameful correspondence with women of ill repute, and had even conceived an impure passion for his chambermaid [...]'.[96] That young Federico was not particularly attracted by chastity nor by the rules of monastic life is confirmed by the letter he sent to Stelluti from Naples in 1604 after fleeing there subsequent to the dissolution of the Academy imposed by his parents:

> Naples paradise of delights, full of amenities and pleasures, enjoyable in its loveliness, most beautiful, gentle and charming, the abode of fertile Ceres, abundant Neptune, and most pleasing and courteous Venus [...][97]

Other founding members displayed no greater respect for the Lincean rule of chastity either. In April 1605 Cesi and Heckius wrote a severe admonition to Francesco Stelluti, who had moved to Parma. The joint signature gave that message the status of an official injunction to maintain a conduct in line with

94 On the personality of Faber and his role as a point of reference for German Catholics visiting Rome as well as a professional converter, see Fosi 2005, Brevaglieri 2009 and De Renzi 1996, 2007 and 2008.

95 'Christian modesty set above all else, chastity professed above all else, and the penalty for the offender will be that of being denied the company of the associates and society of the Academy for three days, during which the penitent is in exile; afterwards, repented, he may request readmission' (Gesta, c. 8v). Cf. also Odescalchi 1806, p. 20.

96 Ibid., pp. 21–2, drawing on Gesta, c. 8v.

97 CL, p. 41, letter of 17 July 1604.

the principles of the Academy after Stelluti seemed to have forgotten the solemn pledge to morally irreprehensible conduct and above all chastity. He was accused of shameful sexual practices ('*venereas vanitates*') and the frequentation of women of ill repute ('*foeminarum indignae prophanaeque servitutes*'). Cesi and Heckius peremptorily ordered him to return to the ranks, to resist every libertine temptation, and to dedicate all his energies to the study and acquisition of knowledge.[98]

It has to be stressed that the motivations and principles that characterised the initial stage of the Academy did not disappear after the induction of Galileo. Although after 1611 Cesi and the Linceans placed more emphasis on the direct observation of nature, on experimental practice and on mathematical speculations, these remained declarations of principle rather than models of research actually implemented.

Galileo, on the other hand, never showed interest in any attempt to reform the original identity of the Academy. While making it clear that he did not want anything to do with him, he tolerated the presence of members such as Giovan Battista Della Porta, of whom we can be certain that he shared Sagredo's denigratory opinion: 'I consider that among learned persons he [Della Porta] occupies the same place that bells have among musical instruments'.[99]

Galileo explicitly showed his dislike of Della Porta to Cesi on only one occasion. In May 1613, the Prince asked Galileo for information on a Tuscan physician who had gone to Naples 'to pay a visit to Porta'.[100] To Federico's knowledge, that physician, Benedetto Punta, had gone to Naples 'on the orders of the Grand Duke and of Don Antonio [de' Medici]'.[101] The reason for Cesi's move emerges in a successive letter to Galileo: Della Porta had told him about Punta, who had paid him a visit asking him, in the name of the rulers of Tuscany, to reveal some of his secrets. From Cesi's letter we learn that Punta's requests had been favourably received by Della Porta. The latter had developed

98 Ibid., p. 72, letter of 29 April 1605.

99 OG XI, p. 398, letter to Galileo, 22 September 1612.

100 Ibid., p. 508. Cf. Galluzzi 1982, pp. 59–62.

101 Ibid., p. 528, letter from Cesi to Galileo, 29 June 1613: 'I have had news from Porta about the chemical physician sent to Naples to deal with Sig. Porta [...]'. Cf. Galluzzi 1982, p. 59. The concise biography of Della Porta that Cesi composed after his death contains an interesting reference to Benedetto Punta: the physician of Christine, Grand Duchess of Tuscany, had been sent to Naples on the orders of the Grand Duke himself. Punta presented Della Porta with a gold crown as a gift: 'The Grand Duke of Tuscany also sent his physician [...] with a gold crown to get them to share some secrets together' (Contributi I, p. 677).

a strong admiration for the Tuscan physician to the point of proposing that Cesi invite him to join the Lincean Academy. Federico turned to Galileo for information and advice.[102]

Della Porta had given Punta a copy of the *Index* of the secrets of his *Taumatologia*,[103] in which he exalted the beneficial effects of spagirical medicine, frequently mentioning Paracelsus and his German epigones in glowing terms for having perfected, based on experience, chemical philosophy after its uncertain beginnings under the sign of Ramon Llull. Cesi defined Punta as a 'chemical physician' and a 'spagirian'. He told Galileo that Francesco Stelluti had been present at the 'discussion of this spagirian with Porta', who set 'great store by the profound knowledge of rare medicaments and natural things that he found in him'.[104]

Galileo's lost reply to Cesi's request for information would have helped us to fill in the sketchy biography of Punta. However, from a letter from Cesi to Galileo on 19 July we learn that the information received from the Tuscan scientist led Federico to drop, without hesitation, Della Porta's proposal of Punta's induction in the Academy: 'I will neither do nor think any more about the chemist whom I had mentioned to you at the instigation of Porta'. Cesi invited Galileo to pardon the infatuations of Della Porta with improbable figures like Punta as a consequence of his advanced years.[105]

Although Punta's mission to Naples had been completely successful since he had taken back to Florence many of Della Porta's secrets, he could not show off the Lincean emerald. Galileo's negative opinion did nothing however to diminish the appreciation of him by Della Porta and the other Linceans. A year later Della Porta begged Galileo 'if he met with Signor Benedetto Punta, most worthy physician of her Most Serene Excellency the Grand Duchess, to remind him of my affection'.[106] Moreover Johannes Schreck, who stopped in Florence in 1615 on his way to Germany after joining the Society of Jesus, wrote to Faber that he had visited 'our Dr Punta' as though he were a fellow Lincean. Most probably the meeting took place in the Casino di San Marco in the presence of Antonio de' Medici.[107]

102 OG XI, p. 509.

103 See Alessandrini 1978, pp. 94–6 and Della Porta 2013b.

104 OG XI, p. 529, letter of Cesi to Galileo, 29 June 1613. Cf. Galluzzi 1982, pp. 59–60.

105 Ibid., p. 538: 'As for Porta and his obsession with sending secrets and with many other things, it is necessary that you bear in mind that he is an octogenarian [...]'.

106 OG XII, p. 102, letter of 26 September 1614.

107 CL, p. 516, letter of 24 September 1615.

CHAPTER 3

Fluid Heavens

Wavering Certainties

It is now time to return to the bustle in Rome in the spring of 1611 and the encounter between Cesi and Galileo. Until this moment, cosmology had not represented a priority in the Prince's research programme. His letter to Stelluti of 30 April 1611 provides the first evidence of the new horizons opened up by the meeting with Galileo, the discussions of the new celestial phenomena and of their implications for the definition of the true constitution of the universe. The meeting had left a strong impression, raising doubts that led Federico to suspend judgement:

> The philosophers debate that the sky is fluid and no different from the air, or, in accordance with the ancient opinion of the Pythagoreans and the new observation of today, the spheres are disposed in this form [see fig. 13]. But there is no little difficulty in accepting that the Earth is the centre of all the spheres.[1]

This text reveals both the enthusiasm and the hesitation of Cesi. He refers not only to his own personal reactions, but also to the positions of others in Rome on how to conceive the true system of the universe in the light of the new celestial phenomena. From Cesi's summary of those discussions two possible scenarios emerged, both running counter to tradition: the admission of the fluid nature of the heavens, with the consequent demolition of the solid spheres, and the revival of the ancient Pythagorean conception.

The problematic nature of this text is accentuated by the diagram of the system of the world that Cesi sketched in the letter (fig. 13). The Earth appears at the centre of the orbits of the Moon, Sun, Jupiter and Saturn, while Mercury and Venus accomplish their revolutions around the Sun. The diagram also shows the four satellites around Jupiter (Cesi wrote 'Galilei' beside the outermost of these) while Saturn is represented as composed of three bodies, in accordance with Galileo's observations of the planet. It should be noted that the diagram does not take into account the doubt expressed by Cesi regarding the possibility of maintaining the Earth immobile at the centre of the universe.

1 CL, pp. 158–9.

 | DOI 10.1163/9789004342316_004

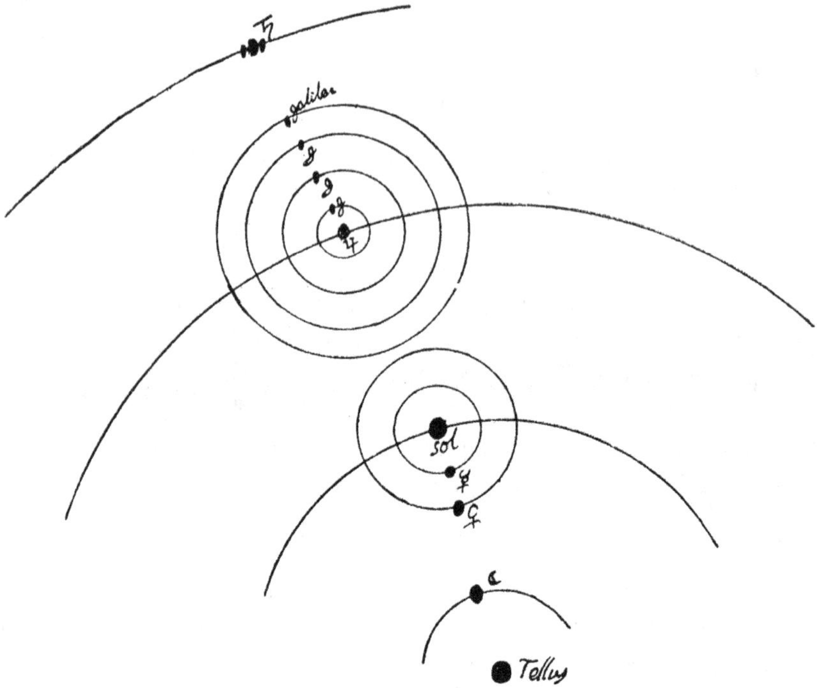

FIGURE 13 *Cesi's semi-geocentric system of the world*

The diagram sketched by Federico raises a series of questions that are not easy to solve. Before tackling them, however, it is necessary to go back in time to the conception of the universe to which the Prince adhered before his reading of the *Nuncius* and the possibility of discussing the system of the world with Galileo. Cesi had certainly been aware for some time of the cosmological conception of Tycho Brahe, whose conviction that the planets do not revolve fixed in crystal spheres, but move through the 'liquid' sky he shared. His knowledge of Brahe and his acceptance of the thesis of the fluidity of the heavens emerge from documents that date to the very beginnings of the Academy. One of its founders, Johannes Heckius who, though Dutch, had not joined the Protest ant Reformation, had written a short treatise on the *nova* of 1604 during the exile imposed on him by the hostility of Federico's parents.[2] The *De nova stella disputatio* of Johannes Heckius, published in 1605, was the very first publication of the Academy, even if the lynx did not yet appear on its frontispiece. Heckius vigorously

2 Heckius 1605. On the *Disputatio* of Heckius, see Saverio Ricci 1988b and Finocchiaro G. 2003.

opposed the thesis of the fluidity of the celestial spaces advanced by Tycho in order to maintain that comets moved beyond the sphere of the Moon. The Danish astronomer's revolutionary conception meant abandoning dogmas that had been shared for centuries and made it necessary to rethink the traditional structure of the universe. To rebut Brahe's perilous theses, Heckius maintained that the *nova* was not a real celestial body that had suddenly appeared among those already known. Like all the others, this star had existed ever since the Creation. According to Heckius, to explain its unexpected and temporary appearance, it was enough to hypothesise that, because of the variable density of the crystalline spheres in continuous rotation, a 'window' had been opened for a limited period through which the brilliance of the star had become visible from the Earth. Contrary to the heretical opinion of Brahe, Heckius reaffirmed that the heavens were solid and incorruptible.

As Saverio Ricci has shown, Cesi intervened in Heckius' manuscript before it went to print.[3] The Prince's emendations seem inspired by a twofold preoccupation. In the first place, Federico toned down the violence with which his fellow Lincean had raged against Tycho Brahe, who, blinded by his perfidious heresy, aimed at subverting the ancient doctrine of the immutability of the heavens, going so far as to question the supreme authority of Holy Scripture. Secondly, Cesi watered down his fellow Lincean's resolute denunciations of logical and philosophical absurdity and of incongruence for which Heckius castigated not only Tycho, but all the critics of Aristotelian cosmology.

According to Ricci, Cesi's emendations of the manuscript of *De nova stella disputatio* indicate his disagreement with the positions of Heckius, motivated by his own favourable reception of Tycho's system and, in particular, his thesis of the fluidity of the heavens. This would account for Cesi's embarrassment in reading the *Disputatio* and for his emendations to moderate its aggressive and resolute tone.[4]

If this were the state of mind of the young Federico, the question arises as to why he decided to go ahead with printing Heckius' text at his own expense. As the first publication of the Academy, the *Disputatio* was inevitably destined to characterise its identity and research programme. It is important to note that the edition of Heckius' treatise opened with a dedication to the Princeps of the Academy and bore the device of the Cesi family (cf. fig. 7). Finally, in the letter in which

3 The manuscript of the *Disputatio* with Cesi's corrections is preserved in the library of the Accademia dei Lincei in Rome (Ms. Archivio Linceo 11).

4 Ricci 1988b, pp. 115ff.

he acknowledged receipt of the manuscript from his Dutch friend, Federico had warmly expressed his appreciation of the work:

> I was so happy to receive your text on the new star, since everything that comes from my brother Illuminatus is so wonderfully congruent with my own inclination, and since all the natural philosophers and mathematicians are hallucinating on this subject. Only Maginus [Giovanni Antonio Magini] has denounced as absurd the opinion that its matter [that of the new star] is made up of condensed vapours. The others, particularly the Peripatetics, seem to be totally at sea and are now denying the celestial doctrines of Aristotle.[5]

Less than two weeks later, Cesi reiterated his positive verdict, this time also in the name of the other Linceans: 'Your opinion on the star is now in press and is warmly welcomed by all'.[6]

In the light of these documents, neither Ricci's interpretation of Cesi's intervention in the manuscript of the *De nova stella disputatio* as a sign of disapproval of Heckius' virulent attacks on Tycho, nor his interpretation of the letters to Heckius in which Cesi praised the work of his fellow Lincean, appear convincing:

> The fact – Ricci remarked – that the Academy published the hypothesis advanced by Heckius, who adopted a midway position between the Aristotelians and their opponents […], is a sign of the difficult situation in which the Linceans found themselves with regard to traditional knowledge and the ecclesiastical authority, as well as the plurality of positions and the complexity of maintaining a balance between them within the Academy.[7]

It is likely that Cesi's interventions aimed not at mitigating the violence of the attack on Tycho's interpretation of the *nova* due to his admiration for the Danish astronomer, but rather to avoiding the impression that the Academy was publicly adopting a position on questions bearing on religion and politics

5 CL, pp. 56–7, letter of 19 March 1605. Note Cesi's reference to Magini's refutation of the interpretation of the *nova* as made up of condensed vapours. Galileo figured among the supporters of that interpretation (see below, pp. 249–50).

6 Ibid., p. 59, letter of 1 April 1605.

7 Ricci 1988b, p. 118.

that he had decided, from the outset, to keep out of its programme. In Cesi's view, Heckius' denunciation of the 'hallucinations' of Tycho as resulting from his adhesion to a heretical confession and his insistence on the irresponsible audacity with which the Danish astronomer had distanced himself from the affirmations of Holy Scripture, were a contravention of that rule. Although it is extremely probable that Cesi was already convinced at this date of the untenability of the Aristotelian conception and that he recognised in the work of Tycho an important contribution to its demolition, it would, in my view, be going too far to make the interventions in the manuscript of Heckius depend on these concerns alone.

We lack explicit evidence to reconstruct the stages through which Cesi came to the conclusion that the heavens are unitary and fluid and, therefore, permeable. Nor do we have indications of the discussions and/or readings that drove him in this direction. What is certain is that regarding the homogeneity of the heavens Bernardino Telesio could have offered him declarations as clear and authoritative as those of Tycho. The young Federico might have received analogous stimuli also from other authors with whom he was certainly familiar, such as Patrizi and Bruno, who had resolutely subscribed to the fluid and unitary nature of the celestial spaces. Like Telesio they had come up against not only the opposition of the Aristotelians, but also the resolute hostility of the Church. It was probably the awareness of the problems encountered by those illustrious contemporaries, as well as by others among his points of reference – such as Persio, Della Porta and Campanella –that led Federico to the conviction that it was necessary to keep the Academy far from that perilous terrain.[8]

Tycho or Telesio?

The studies dedicated to Federico Cesi, which have increased considerably during the last thirty years, have curiously neglected to focus on his early education. Great importance has been attached to the role of Heckius and Della

8 Seen through the eyes of both Federico's parents and the guardians of orthodoxy, the Academy was bound to appear as a circle to be kept under careful surveillance. Even leaving aside the peculiar religious vocation of Heckius and his strong inclination towards astrology, including prediction of future events, many of the most illustrious members of the Academy had already had dealings with the Tribunal of the Inquisition or were to do so: Della Porta, Persio, Stelliola, Galileo, not to mention the bad reputation in the eyes of the guardians of orthodoxy of authors whom many of the Linceans admired: Paracelsus, Bruno, Telesio, Campanella, Patrizi.

Porta as points of reference for the young Federico. However, other stimuli and suggestions came from authors who had nothing to do with chemical philosophy and natural magic. Among these, an important role was certainly played by the Abbot Antonio Persio.

Cesi must have known the Abbot from adolescence, as from 1596 Persio was employed as personal theologian to Federico's uncle, Cardinal Bartolomeo Cesi.[9] Federico's familiarity with the Abbot is confirmed by a number of sources. To mention just one, Antonio Persio was to be one of the privileged few who, as guests of Federico, attended the demonstration of the spectacular celestial discoveries by Galileo on the Janiculum Hill in Rome on that fateful evening of 11 April 1611.[10]

A few months later, in January 1612, Antonio Persio passed away, before he had time to celebrate the induction into the Academy promised by Cesi. The Prince nevertheless kept his word in electing him as a posthumous member with the unanimous approval of the academicians, including Galileo. The Tuscan scientist manifested his interest in Persio's manuscripts now in the hands of Cesi and encouraged the Prince to publish them. To Federico's promise to send him the index of the works left by the Abbot,[11] the Tuscan scientist replied that 'he was eagerly awaiting the things of Signor Persio to see them and to hear what the Peripatetic will say about them'.[12] Once in possession of Persio's manuscripts, Cesi was determined to publish them, convinced as he was of their importance. However, he was able to publish only their index and this not without difficulties, due to the opposition of the censors explicitly alluded to in the Lincean correspondence.[13]

Persio was a resolute opponent of Aristotle and his followers. A pupil of Telesio, after the death of the Master he had continued with the struggle to free natural philosophy from the yoke of Aristotle. The Abbot was an unusual character. Although the personal theologian of illustrious Cardinals (he had served Cardinal Enrico Caetani before entering into the service of Cardinal

---

9 Cf. Contributi I, p. 867.

10 Cf. below, pp. 84–5 and n. 67.

11 OG XI, pp. 302–3, letter of Cesi to Galileo, 26 May 1612: 'I was thinking of sending you a printed index of the subjects dealt with by Persio [...]'. Cesi referred to the edition of the index of the unpublished works of Persio that he was to publish at his own expense (Persio 1613).

12 Ibid., p. 301, letter to Cesi, 26 May 1612.

13 Cf. OG XI, p. 303 (letter of Cesi to Galileo, 26 May 1612): 'The censors are keeping it [the index of the unpublished works of Persio] for being largely contrary to Aristotle. This shows how powerful is his tyranny'. See also Contributi I, pp. 871 and 885.

Bartolomeo Cesi) and a censor of the Congregation of the Index, he openly followed the teachings of a philosopher whose works had been suspended by the Tribunal of the Inquisition pending expurgation of doctrines dangerous to the faith. Moreover, Persio made no secret of his personal sympathy and solidarity with Campanella, another author of whom the ecclesiastical authorities took a dim view. And that was not all. He engaged in intense discussions with Francesco Patrizi, whose works had received a treatment analogous to that of Telesio at the hands of the Inquisition.[14]

From the months following the death of the Abbot in January 1612 down to late 1621, the Lincean correspondence contains many testimonies to the determination with which Cesi attempted to publish the two manuscripts that the deceased had bequeathed to the Academy: the *De ratione recte philosophandi*, which has been lost, and the *De natura ignis*, which is among the manuscripts of the first Academy kept in the library of the Accademia Nazionale dei Lincei in Rome.[15] Federico, who acted as executor of Persio's will, tried, in vain, to acquire also the Abbot's library.[16]

Cesi entrusted the task of preparing Persio's manuscripts for publication to his fellow Lincean Johannes Faber, who displayed evident sympathy with the Telesian legacy. During a trip to Naples in 1608 to plead that Campanella be freed from imprisonment, Faber often met the followers of Telesio who frequented the Museum of Ferrante Imperato.[17]

There is no explicit evidence that the publication of the works of Persio was held up by difficulties in obtaining authorisation from the ecclesiastical authorities, as continuous references to the internal revision of the work for

14 There is no exhaustive biography of Antonio Persio. Information on the life and works of the Abbot is provided by Gabrieli (Contributi I, pp. 865–7); see also Artese 1987 and Persio 1999.

15 Ms. Archivio Linceo 7; see Contributi I, pp. 883–5.

16 Ibid. Persio's library was dispersed.

17 Faber was a Lecturer in simples and Superintendent of the Vatican garden under Pope Paul V. His official motivation for going to Naples in 1608 was to obtain plants for the Vatican garden. In reality, one of the reasons, if not the main one, for his trip was to negotiate with the Viceroy for the release of Tommaso Campanella from prison. The plan had been prepared with his friend and fellow countryman Caspar Schopp with the authoritative encouragement of Duke Maximilian of Bavaria and Archduke Ferdinand of Austria. The operation failed, but Faber did obtain a mitigation of Campanella's conditions in prison. On that occasion Faber formed a long-lasting friendship with Ferrante Imperato, whose Museum was frequented by admirers and followers of Telesio. On Faber's journey to Naples, see Berti 1881, Amabile 1887 (pp. 82ff.) and, above all, Gabrieli, Contributi I, pp. 335–98, esp. 387–9.

publication in the correspondence between Cesi and Faber seem to suggest. Nor does the manuscript of *De ignis natura* show any signs of revision.[18] Cesi might have intended to moderate the tone of the many passages in which Persio proposed conceptions such as that of the soul of the world that were regarded with great suspicion. Persio had in the past run up against the sensitivity of the censors. This may have been the reason why he planned to have his last works printed in Germany.

Cesi's keen admiration for the aged theologian is also evident from the attention that he devoted to collecting information about his life, scientific production and character with a view to composing an official biography of the posthumous Lincean.[19]

If one bears in mind the active presence of Persio in the family setting in which Federico grew up, it is natural to suppose that the Abbot played an important part in his education. Clear echoes of Persio's influence can be found in many of the Prince's orientations and in numerous aspects of his production. Federico must have appreciated the resolute manner in which the Abbot had assumed the role of defender of freedom of thought in philosophy, not only in matters pertaining to the natural world. Moreover he sympathized with Persio's efforts to demonstrate the fragility of the foundations of traditional natural philosophy.

Like Cesi and the other first Linceans, Persio did not confine himself to highlighting the contradictions in the physics of Aristotle, but drew on his theological skills to forcibly denounce the deviations from the genuine interpretation of the Christian mission that resulted from a reading of Holy Scripture through Aristotelian eyes.[20]

Both in the works that Persio had published during his lifetime and in those that had passed into his hands after the Abbot's death, Federico could find other elements of great interest for the project that he had conceived. As we shall see in due course, traces of his reading of Persio and of his Master Telesio will become evident above all in the production of the last fifteen years of the Prince's short life.[21] For the moment it is sufficient to underline the extreme

18 Ibid., p. 848. Besides the fear of not overcoming the hostility of the highly suspicious censors, it is also possible that the project did not materialise because of the considerable expense involved, bearing in mind that the manuscript of *De natura ignis* alone occupies three large volumes.

19 Ibid., pp. 872ff.

20 On the arguments against Aristotle by Telesio and Persio inspired by religious motivations, see Redondi 2004, p. 135.

21 Cf. below, esp. Chaps. 10 and 12.

likelihood that Persio's personality had a strong impact on the young Cesi. Among those who drew the curiosity and admiration of Cesi, Persio was not the only one to have subscribed to Telesio's claim to the primacy of sense experience. As shown by the Dominican friar's influence on the Academy's inaugural ceremony on Christmas Eve 1603, Tommaso Campanella, a follower of Telesio and a close friend of Persio, was also one of the Prince's early points of reference. Cesi could have met Campanella during his visits to Naples in 1604 or in 1610.

The relation between Campanella and Cesi was to remain constant over the years,[22] as emerges also from the Dominican friar's letters to Galileo. We learn from one of these that, after the Tuscan scientist's admission to the Academy, Cesi had asked him to draw up his horoscope. Galileo declined the invitation on the grounds that he did not believe in the influence of the stars. After the Prince had informed Campanella of Galileo's refusal, the latter took up his pen to protest vigorously to the Tuscan scientist. Probably aware that Galileo had derived material benefits from drawing up horoscopes for wealthy individuals during his years in Padua, Campanella asked him why, on the pretext that he did not believe in astrology, he had refused to draw up Cesi's horoscope, while he had not hesitated to produce one for Cosimo II de' Medici: 'So you – he concluded – made fun of him. God forbid'.[23]

---

22 Cesi too intervened in a personal capacity to plead for Campanella's release from imprisonment in Naples. Federico, as well as other Linceans, admired the Dominican friar; cf. Contributi I, pp. 385–98; cf. also Guerrini 2007a, pp. 119ff.

23 OG XII, p. 32, letter of 8 March 1614; see also Campanella 2010, pp. 208–11. Campanella is referring to a passage in the dedication of the *Nuncius* to Cosimo II in which Galileo declared that the Grand Duke's sublime destiny had been revealed from birth by an astral conjuncture that witnessed the ascendancy of Jupiter (no random choice if one bears in mind that the satellites of Jupiter were dedicated to the Medici dynasty): 'It was Jupiter, I say, who at Your Highness's birth, having already passed through the murky vapours of the horizon, and occupying the midheaven and illuminating the eastern angle from his royal house looked down upon Your most fortunate birth from that sublime throne and poured out all his splendour and grandeur into the most pure air, so that with its first breath Your tender little body and Your soul, already decorated by God with noble ornaments, could drink in this universal power and authority' (Galileo 1969, pp. 31–2). Although Cesi failed in his attempt to obtain a horoscope from Galileo, he was to have more success later on, thanks to the good offices of Faber and Colonna, with the well-known Neapolitan physician and astrologer Mario Schipani, nominated for admission to the Academy (cf. Gabrieli 1987, I, pp. 533–4: the proposal of his induction was made by Cesi at the session held on 26 January 1615, attended by Galileo; it was the same meeting in which Federico read the *Discourse on the natural desire for knowledge*). In particular, the Prince wanted

Since the presence of Telesian motives in Cesi's writings, assimilated by impulse and through the mediation of Persio, has not until now been taken into consideration,[24] it will be worthwhile to focus on them, and above all on the use to which Cesi put them in his own ambitious reform project of natural philosophy and morals.[25]

Let us begin with cosmology, one of the areas in which the young Federico showed early signs of a rift with the Aristotelian tradition. It was not only his familiarity with the world system of Tycho Brahe that led the Prince to abandon the solid spheres of the planets in favour of the fluid and homogeneous nature of the heavens. The first impulse to reconsider the fundamental question of the nature of the celestial spaces probably came from a reading of the writings of Telesio and from his early contacts with Persio, the most refined interpreter of his works and their most passionate and effective promoter.

Telesio had neither adhered to the Copernican vision nor abandoned the conviction that the planets were fixed in spheres that carried them around the Earth immobile at the centre of the world.[26] However, he had criticised the division of the universe into two distinct zones: the supralunary and the sublunary. He argued that the celestial spaces were composed of a homogeneous substance in different states of condensation. The sublunary world was therefore not different from the rest of the heavens; there were no fractures in the universe. The only clear distinction that Telesio maintained was that between the hot and bright celestial spaces and the cold and dark body of the

---

to know from Schipani whether his wife's continuous miscarriages and thus inability to produce a male heir were the result of baneful astral influences (CL, pp. 1098, 1101, 1108 and 1117). The horoscope cast by Schipani is preserved in Ms. Archivio Linceo 2 (c. 103) of the Library of the Accademia Nazionale dei Lincei. It indicates Federico's date of birth (which he personally gave to Schipani) as 1 March 1585 (cf. Partini 1986, pp. 66–7). However, that date refers to Cesi's baptism which took place three days after his birth on the 26 February (cf. Contributi I, pp. 21–5).

24 To the best of my knowledge, the only indications in this sense are contained in the suggestion of Gilberto De Angelis to consider the Abbot as the *trait d'union* between Cesi and Campanella (De Angelis 2012b, pp. 15–21); in the proposal of Barbara Caredda, who argued that the influence of Telesio can be perceived in Cesi's *Coelispicium* and *Apiarium* (Caredda 2008, pp. 166–72); and in some remarks of Guerrini on the presence in the Prince's writings of elements that reveal the influence of Telesio and Persio (Guerrini 2006, p. 82).

25 There is a vast literature on Telesio's conception of nature. I confine myself to referring to the exhaustive biography by De Franco 1995, and the outline of his life and works in Bondì 1997.

26 Ibid., pp. 31ff.

Earth. Heat was diffused in the heavenly spaces by the Sun through light, while cold and rest were the essential qualities of the Earth. Although fluid, Telesio's heavens were variously permeable. Only light could traverse their entire extension. The spheres that bore along the planets fixed in them, composed of a more condensed celestial matter, prevented less refined substances from penetrating them.

Telesio opposed cold, the principle of rest, to heat 'father' of light and therefore of motion. There was no need, in his opinion, to resort to a Prime Mover, given that motion was an intrinsic property of heat, the substance of which the heavens were made. Likewise, given that the substance of our planet is cold, Telesio concluded that the Earth is immobile at the centre of the universe.[27]

While affirming that the celestial spaces are made of fluid matter, Telesio's position on the corruptibility or incorruptibility of the celestial bodies appeared elusive. He felt that dilemma to be pointless. Aristotle had stated that heat was not radiated by the Sun: there was neither heat nor cold in the immutable spaces and bodies of the heavens. For the Greek philosopher the heat felt in the sphere of the elements was generated by the friction produced by the rotation of the solid celestial spheres. Telesio pointed out a contradiction in that interpretation: was not friction a form of alteration, while Aristotle had banished alteration from the supralunary heavens? To escape from this inconsistency, Telesio maintained, it is enough to abandon the Aristotelian conception and to suppose that the Sun produces heat by its intrinsic virtue. In Telesio's view, heat and light constitute the essential substance of the heavens. Their inexhaustible source is the Sun, from which they are radiated through the entire universe without encountering obstacles because the heavens are fluid and permeable.

This essential outline is sufficient to indicate that Telesio proposed a cosmological vision that contained important novelties vis-à-vis the Aristotelian *De Caelo*, above all due to the demolition of the dualistic conception of the universe and to the affirmation that the fundamental principles governing its operations were heat and cold. Notwithstanding his original contribution, Telesio's cosmological conception appears extremely modest when compared with the lively contemporary debates on the system of the world taking place amongst the most renowned astronomers of Europe. The fact that he never mentioned their names leads one to suspect that he was not familiar either with the cardinal astronomical works of those decades, starting with the

27 Cf. Telesio 1965–1974, 1976 (*Libri I-II*), pp. 30–45.

publication of Copernicus' *De revolutionibus* in 1543, or with the innovative contributions of Tycho.

Telesio's main objective was not that of proposing a new architecture of the cosmic machine. He was primarily interested in introducing a radical revision of the fundamental concepts on which Aristotle's *Physica* was based, to be replaced by principles – heat, cold, light, matter and motion – that could be verified by the senses.

It is difficult to imagine that the young Cesi did not read Telesio's *De rerum natura*, which was in his personal library as well as in that of Johannes Faber.[28] It is equally difficult to suppose that the theologian of his Cardinal uncle Bartolomeo had not warmly recommended him to read that book. It is thus extremely plausible that Cesi's early adhesion to the thesis of the unitary nature of the heavens was stimulated by that reading and by his conversations with Persio. As the blatant hostility to the conception of the fluidity of the heavens displayed in Heckius' *De nova stella disputatio* seems to indicate, in 1605 Tycho had not yet been perceived as a point of reference, at least not by the founders of the Academy as a group. Besides, although Tycho, unlike Telesio, demolished the celestial spheres, he still maintained a division between the substance of the supralunary sky and the more imperfect and corruptible matter of the spaces below the lunar orb.

Let us now return to the *nova* of 1604 and Cesi's attitude towards Heckius' *De nova stella*. During the months following the appearance in 1604 of a new star in the skies of Europe heated debates flared up among astronomers. It was probably during those months that Persio introduced Cesi to the interpretation of the nature of the comets proposed by Telesio in *De cometis et lacteo circulo* that the Abbot had inserted in the collection of treatises by his teacher published in Venice in 1590.[29] For Telesio, comets were the products of exhalations and vapours from the Earth. Thus far he did not depart from the Aristotelian interpretation to which, however, he introduced two corrections that radically unsettled it. First, contrary to what Aristotle maintained, comets were not formed below the sphere of the Moon. The exhalations from the Earth rose

28 Cf. Biagetti 2008, p. 450 (item no. 189): 'Toletius [sic] de Rerum Naturae [sic] Neapoli 1570', and again on p. 135 (item no. 472): 'Bernardinij Telesij de re[rum] nat[ur]a Libri IX, Neapoli 1587'. Cesi also owned the compendium of the philosophy of Telesio by Sertorio Quattromanni (Quattromanni 1589), as well as the edition of the *Opuscula* of Telesio (Telesio 1590) published by Antonio Persio; cf. Biagetti 2008, pp. 247 and 148. For Faber's copy, see Contributi II, p. 1225.

29 Telesio 1590 (*sub indice*). *De cometis* was republished by De Franco (Telesio 1981).

to much higher levels, penetrating the spaces beyond the orb of the Moon. This implied that the thesis of a structural distinction between the sublunary and the supralunary regions had to be rejected.

The second correction was even more radical. Telesio refuted Aristotle's interpretation of comets as real bodies set alight by the friction between the higher parts of the sublunary sky and the solid sphere of the Moon. In his opinion, comets were illusory images produced by the reflection of the solar rays on the Earth's vapours. Those exhalations remained in the upper regions of the sky for a limited duration of time before falling back to Earth.[30] There was thus a constant exchange of matter and influences between the sublunary sky and the rest of the universe. Once again we encounter a principle that will play an important role in all of Cesi's scientific production.

It should be remembered that the interpretation of comets as phenomena generated by the reflection of the solar rays on terrestrial exhalations was widespread in the final decades of the sixteenth and the early ones of the seventeenth century. Besides Telesio, this interpretation was also maintained, though with sometimes significant modifications, by Cardano, Pena, Rothmann and others.[31] It also exerted a strong fascination on the young Galileo, who advanced it in the three crowded lectures that he gave on the new star of 1604 in Padua at the end of the same year. Unfortunately the text of those lectures is not extant, and the few fragments that remain do not suffice to establish whether Galileo was aware of how close Telesio's interpretation was to his own.[32]

Fifteen years later, Galileo was to propose the same explanation for the comets that he had given for the *nova* of 1604. It is therefore not surprising that when the Jesuit Orazio Grassi denounced the interpretation of the comets proposed in the *Discourse on the comets* of Galileo-Guiducci, he emphasised that it betrayed the influence of reprehensible writers such as Telesio and Cardano, who had both been rightly condemned by the Church for their perilous doctrines:

> Here Galileo seems to be somewhat indebted to Cardano and Telesio [on the interpretation of the nature of comets], although having elaborated

30 The comet is 'light of the Sun reflected by terrestrial vapours; comets are generated by the Sun's rays reflected by condensed exhalations' (Telesio 1981, pp. 2–7).

31 On Telesio's astronomical conceptions, see Lerner 1990. Lerner drew attention to the relation between Telesio's writings on comets and the contemporary debate on their nature (pp. 104–5).

32 OG II, pp. 267–84. Cf. Camerota 2004, pp. 116–24.

> a sterile and unhappy philosophy, they were blessed by no offspring and left to posterity books rather than disciples.[33]

Galileo replied to these malign insinuations in the *Assayer*, accusing Grassi of having gratuitously cast aspersions on the reputation of those authors.[34] The Jesuit reiterated his denunciation of the impious teachings on comets of Telesio and Cardano in a passage of the *Ratio*, his rejoinder to the *Assayer*,[35] to which Galileo added the withering comment: 'You make me a follower of Telesio and Cardano with the declaration, for whoever did not know it, that their philosophy and doctrine are equally condemned'. Given, he continued, that you want to condemn me as a Telesian for maintaining that 'the comet is a pure illusion', should one conclude that to affirm that thesis is 'to deny Divine Providence or God Himself'?[36] Obviously Galileo's statement that he had not read the works of Telesio or Cardano ('I, for my part, have never studied such authors') should not be taken literally, given that their books were on the Index of prohibited works.

Father Grassi was not the first to insinuate that Galileo belonged to the same category of suspect authors as Telesio and Cardano. A few weeks after the publication of the *Sidereus Nuncius*, Ludovico delle Colombe had stigmatised Galileo's arrogance in daring to criticise Aristotle and his followers. In the same months in which influential scholars all over Europe were celebrating the celestial novelties, comparing Galileo to the great explorers Columbus and Vespucci who had boldly crossed the ocean to discover a new world, Delle Colombe reversed the terms of the comparison: vanity had diverted Galileo from the 'ocean of truth' and led him to irresponsibly challenge the Pillars of Hercules and the *non plus ultra* represented by the philosophy of Aristotle. According to Delle Colombe Galileo's audacity rivalled not only that of some of the most execrable ancient philosophers, but also of contemporary authors like Telesio and his followers, who had deluded themselves that they could destroy the authority of the Stagirite.[37]

---

33 Drake-O'Malley 1960, p. 74 (OG VI, p. 118).

34 'Does he [Sarsi] not realise with how little piety he strips and robs them [Cardano and Telesio] of their reputation to cover a tiny blemish on that of his Master?' (OG VI, p. 236).

35 'What kind of piety is that, Galileo, to defend those [Cardano and Telesio] [...] whose philosophy, condemned in many parts as hostile to Catholicism, deserves to be buried in oblivion?' (ibid., p. 397).

36 Ibid., p. 398.

37 OG III[1], p. 253.

The presence of Telesio in the animated debates on celestial physics in the Roman scenario linked with the Linceans is confirmed by the testimony of Francesco Ingoli, who declared that the frequent meetings in Cesi's palace in 1613 were also attended by followers of Telesio's natural philosophy.[38]

Associating Galileo with Telesio may appear strange. Yet the Tuscan scientist was depicted as an ally of Telesio not only by the most strenuous defenders of Aristotle, but also by persons who had enthusiastically welcomed the celestial novelties, such as Tommaso Campanella. In explaining to Galileo his reasons for inclining towards the Copernican system, Campanella wrote that he saw in it confirmation of the principle that the substance of the heavens was heat, as Telesio had argued: 'Thus the principles of our erudite Telesio – he wrote to the Tuscan scientist – stand firm'.[39]

If Persio introduced Cesi to the Telesian conception of the unitary nature of the heavens in the early stages of his education, and if we suppose that Galileo knew Telesio's text on comets when he delivered his lectures on the *nova* in 1604, then we may speak of a first unconscious convergence between the two men, many years before their paths were destined to cross.

There is another feature to take into account in reconstructing the relations between the young Federico and the elderly Abbot Persio. Some of the theses discussed by Telesio in his *De rerum natura* were the object of a series of critical remarks by Francesco Patrizi. Although starting from different positions, Patrizi too was engaged in the battle to reform the natural philosophy of Aristotle.[40] Patrizi questioned Telesio's thesis that the planets were fixed in orbs of denser hot material and were borne along by them 'like knots in a wooden plank' around the Earth at the centre of the Universe at different speeds and on variously inclined axes of revolution.[41] According to Patrizi it was simpler and more logical to attribute to the planets an intrinsic principle of motion, as suggested by authoritative exponents of ancient cosmological and religious traditions. Nor did Patrizi share the Telesian conception of the identity between the matter of the heavens and the matter of the planets. He also called into question the spherical shape of the universe maintained by Telesio.[42] The most logical solution in Patrizi's eyes was to suppose that the stars, like animals,

38 On the meetings in Cesi's palace in 1613 see below, pp. 160–5.

39 OG XI, p. 22, letter of 13 January 1611.

40 The documents of the discussion of cosmological questions between Telesio and Patrizi were published by Francesco Fiorentino (Fiorentino 1872–74, pp. 358–414).

41 On the 'knots in a wooden plank' see Lerner 1980 and 1990 (esp. pp. 84–9) and Dollo 1995. Cf. also Rossi 1977b, pp. 109–47.

42 Lerner 1990, p. 90.

were endowed with movement of their own. He stressed how this conclusion followed logically from the very Telesian conception of an *anima* or vital spirit permeating all of nature. Why not extend that spirit to the celestial bodies and abandon the clumsy and superfluous spheres? That spirit would also resolve the contradictions into which those who, following Aristotle, postulated the existence of a *Primum Mobile,* fell.

Telesio replied to Patrizi's objections without retreating an inch. Persio also intervened to defend his Master in a polite crossing of swords with Patrizi. Persio knew Patrizi personally and the two men held one another in high esteem, in spite of their diverse conceptions of the system of nature and the structure of the universe. Persio defended the centrality and immobility of the Earth as his teacher Telesio had done, while Patrizi, although with some distinctions, followed the Copernican hypothesis.[43]

Cesi's familiarity with Persio enabled him to acquire precious information about the lively debate centred on different models for interpreting nature. Those models shared not only the unitary conception of the heavens and the presence of a principle of movement, identified as light and heat, diffused through the cosmos, but above all the need to proceed to a radical reform of the natural philosophy of Aristotle, nourished by hopes of renewal also in the art of medicine, as well as in ethics and religion.

What is striking about the debates between Telesio, Persio and Patrizi, which were still resonant when the young Federico began his studies, is above all the convergence, and at times hybridisation, between traditions and conceptions that we tend today to consider alternative to one another. The dialogue between Telesio and Persio, on the one hand, and Patrizi, on the other,[44] makes patently clear the inadequacy of all analyses of the philosophical, scientific and religious culture between the end of the sixteenth and the beginning of the seventeenth centuries in terms of rigidly delineated and isolated categories such as Aristotelianism, Platonism, Hermeticism and natural magic, Corpuscularism, Stoicism, Empiricism, Sensism, Rationalism, Spiritualism, Mechanicism, Vitalism, and so on. It is almost always impossible to fit the positions of the authors who animated the debate of those decades into those schemes. While resolutely proclaiming the primacy of the senses, Telesio and Persio did not hesitate to use elements drawn from the Eleatic tradition, to assume positions

43 On the discussions between Persio and Patrizi, see Garin 1961, De Franco 1995 and Artese (Telesio 1991), as well as Lerner's essay on the cosmological questions (Lerner 1990).

44 Cf. Rossi 1977b, Puliafito 1992 and Bondì 1997.

that echoed the ideas of Ficino, to flirt with the supporters of living nature while seeking, through new ways of reading the works of Aristotle, to overturn the image of the Greek philosopher petrified by the Scholastic tradition. They sought to replace that image with that of an author whose texts were open to doubt, to the possibility of arriving at new certainties by reasoning without prejudice about principles, scrupulously observing nature, questioning it by means of ingenious experiments, or subjecting it to measurements and quantitative analyses. Nor did they resist the temptation to interpret Holy Scripture in a way that would confirm the positions they maintained, or at least to show how distorting and perilous was the alliance between Aristotelian natural philosophy and Scholastic theology.

Within that scenario the originality of the individual authors ended up by coinciding with the capacity to propose new syntheses by variously combining elements from diverse traditions. Nor should it be forgotten that, if Telesio, Patrizi and many other protagonists of those years concentrated their efforts on demolishing the dogmatic block formed by the Aristotelian philosophy of nature and Post-Tridentine theology, fundamental elements of Aristotelian physics were still present, although mixed with other ingredients, within those bold constructions. Likewise, Aristotelian logic continued to play an important part in new horizons.

It was certainly a stimulating panorama for a young man like Federico, strongly inclined to fight for the renewal not only of natural philosophy but also of religion and ethics. This is why it is important to focus on the points of reference from which the adventure of the founder of the Academy set out, following Garin's invitation to pay attention to Cesi's precocious reading of Patrizi, Telesio and Persio.[45]

In the absence of documents that would provide conclusive answers to questions about Federico's course of study, we are bound to rely on conjecture. The young man was certainly fascinated by Heckius; not for nothing was the latter elected *Magister* and *Provisor* of the most exceptional and arcane experiments during the session of the Academy on 12 October 1603.[46] Moreover, it is certain that the Prince's journey to Naples in 1604 was undertaken not only to escape from the oppressive family atmosphere after the resolute intervention

45 Garin analysed the blend of cultural ingredients, considered irreducibly diverse today, that characterised the vision of the First Academy of the Linceans and its protagonists (1986, esp. pp. 16–17).

46 Gesta, c. 4r: 'In addition he [Heckius] was officially appointed Master of the Linceans in astral science, Proposer of experiments on the hidden marvels of nature and on medicine, and wise Superintendent of secrets'.

of his father against the academic initiative, but also to make contact with Giovan Battista Della Porta, a fundamental point of reference for Federico's own research projects.

The role of Antonio Persio in the shaping of the intellectual horizon of Cesi during his tender years, as well as the motivations that led him for many years after the death of the Abbot to endeavour to have his works published, deserve further attention. A close analysis, which has not yet been conducted, of Persio's manuscript treatise *De natura ignis*, would probably throw light on a relation about which we know too little. It would also help us to understand why, while convinced that the light and heat concentrated in the Sun were the vital forces of the universe, the Abbot refused to accept Galileo's advocacy of the Copernican structure of the universe and the determinant role assigned in it to the Sun. If the aged Abbot did not want to relinquish his own conviction regarding the central position of an immobile Earth, he was nevertheless open to the new prospects revealed by the Tuscan scientist and the contribution they offered to the demolition of the tyranny of Aristotle. As Cesi wrote to Galileo immediately after the death of the Abbot, Persio 'was entirely on your side apart from on the opinion of Copernicus'.[47]

The Revival of Martianus Capella

Having reconstructed the horizons on which the Prince's gaze was fixed before the earth-shaking revelations of the *Sidereus Nuncius* and the suggestions arising from his encounter with Galileo, we can return to the letter from Cesi to Stelluti of 30 April 1611. The evidence for Cesi's early adherence to the thesis of the fluidity of the heavens helps to understand the enthusiasm with which he took part in the lively discussions on the structure of the universe during Galileo's visit to Rome. Cesi, it will be recalled, had informed Stelluti that the participants included some 'Pythagoreans',[48] among whom there is no doubt that Galileo played the role of protagonist. The solution proposed by the 'Pythagoreans' as alternative to the systems of Ptolemy and Tycho had sown the seed of doubt in the Prince's mind. As he wrote to Stelluti, 'there is no little difficulty in accepting that the Earth is at the centre'.[49] However, the diagram of the system of the world (fig. 13) enclosed with the letter, corresponding – as Federico took

47 OG XI, p. 285, letter of 22 March 1612.

48 CL, p. 159, letter of Cesi to Francesco Stelluti, 30 April 1611.

49 Ibid.

pains to point out – to the 'new observation of today', indicates that Galileo's arguments had not yet convinced him to join the ranks of the Copernicans: the diagram shows the Earth firmly in place at the centre of the universe.

Let us examine that remarkable diagram of the system of the world in more detail. It represents neither the universe of Copernicus (the Sun orbits the Earth at the centre) nor that of Tycho (only Mercury and Venus revolve around the Sun, while the paths of Jupiter and Saturn are geocentric). The absence of the orbit of Mars is also surprising and prevents us from knowing whether Federico took it to revolve around the Sun or around the Earth. Was it a simple slip? Or had Cesi deliberately removed the stumbling block? For it had been the inevitable impact between the sphere of Mars and that of the Sun in Tycho's semi-heliocentric system that had led the Danish astronomer to banish the solid spheres from the skies.

Upon closer inspection, the diagram in Cesi's letter seems to represent the system of the universe conceived in Late Antiquity by the fifth-century Roman writer Martianus Capella, whose scholastic manual[50] enjoyed widespread popularity throughout the Middle Ages. There is no documentary evidence for the presence of followers of the ancient system of Capella in the lively Roman debates of those weeks. Nor can we rule out the possibility that Cesi, an amateur as to observational astronomy, may have confused it with the solution proposed by Tycho. However, the absence of Mars casts doubts on the possibility that Cesi had the Danish astronomer's system of the world in mind. It is worth underlining that in the system of Capella only Mercury and Venus revolved around the Sun. The circumterrestrial revolution of the sphere of Mars was external to that of the Sun, thus excluding any risk of collision between the orbs of the two celestial bodies.

Until now Cesi's diagram of the structure of the universe has not been the object of in-depth analysis. Yet it deserves serious attention because it represents the oldest explicit pronouncement by the Prince on the system of the world. It should not be forgotten that Cesi's primary aim in writing to Stelluti was to inform him of the debates in Rome and of the orientations that had emerged from Galileo's numerous meetings with the lay and ecclesiastical intelligentsia of the Eternal City. We cannot therefore rule out the possibility that his diagram of the celestial constitution echoed arguments or proposals advanced by some of the Fathers of the Collegio Romano.

The revival of the system of Capella – or, at least, of a combination of the systems of Tycho and Ptolemy – would have admirably contributed to 'save'

50 Minneius Felix Martianus Capella, *De nuptiis Mercurii et Philologiae.*

the new celestial phenomena, avoiding the threat to traditional cosmology posed by the phases of Venus. Moreover, it would have offered a reassuring image of the universe. In contrast to the revolutionary scenarios – not only in the field of astronomy – opened up by the Copernican system and, to a lesser degree, by that conceived by Tycho, Capella's solution only required renunciation of the geocentric orbits of Mercury and Venus. The Earth continued to occupy its central position; the planets remained fixed in solid spheres that bore them along; the conception of a structural distinction between the heavenly spaces beyond the sphere of the Moon and those surrounding the Earth was maintained; finally, in the sphere of the elements, the natural philosophy of Aristotle upheld its authority.[51]

Several clues point to the Collegio Romano as the source of inspiration for the diagram that Federico sent to Stelluti. Having verified the existence of the celestial novelties, the Fathers of the Collegio might have regarded Capella's system as an excellent solution with which to confront Galileo's challenge to the traditional geocentric system. Besides, the Latin author from Late Antiquity had had a strong impact on the Christian culture of the Middle Ages and was therefore perfectly acceptable, unlike Tycho, who, before being recognised as an authoritative point of reference, was to be bitterly opposed by the Jesuits (as well as by Heckius) for belonging to the pernicious sect of Protestant heretics. It is telling that Father Clavius had steadily declined Tycho's request to correspond with him. It should also be emphasised that, in his letter to Clavius, Tycho had complained of being the victim of plagiarism by Ursus, who had conceived a system practically identical to that of Capella.[52] As recent studies have highlighted, Clavius was perfectly informed about the many

51 On the system of Capella see Eastwood 1982. For its diffusion in the Middle Ages see Eastwood 1993.

52 Clavius 1992, III[1], pp. 96–103, letter of 5 January 1600 sent from Benátky. In vain Tycho had tried to make direct contact with Clavius to convince the Jesuit and his colleagues in the Collegio Romano to adhere to his geo-heliocentric system and to accept the penetrable nature of the heavens, which formed an essential part of his conception. Clavius ignored Tycho's advances and never showed any willingness to accept the thesis of the fluidity of the heavens. Until his death, he continued to support the complex geocentric system elaborated by Magini. Moreover, the German Jesuit never mentioned the name of Tycho in any of the numerous editions of his popular commentary on the *Tractatus de Sphaera* by Johannes de Sacrobosco. When the fellow Jesuit Johann Ziegler asked him for his opinion of the system of Tycho, Clavius' reply in 1606 left him dumbfounded: 'I cannot give you my opinion on Tycho because I have not seen his work' (Kepler 1937-, XV, p. 293). On the attitude of the Jesuits towards Tycho down to the official adoption of his system, see Lerner 1995.

semi-Tychonic systems, which often differed from one another in only minor details, devised in particular by German astronomers.[53] They sought to find an acceptable compromise between the data of ever more precise astronomical measurements that demonstrated the incongruences of the Ptolemaic system, on the one hand, and the need to prevent the collapse of traditional natural philosophy and cosmology, on the other.

What is certain is that some mathematicians and natural philosophers of the Society of Jesus did contribute to the ephemeral revival of the ancient system of Capella as an alternative to that of Tycho, at least down to the final years of the second decade of the seventeenth century, when the Jesuit hierarchies decided to embrace the system of the Danish astronomer, overcoming any hesitations generated by the heretical nature of the author. It was above all Fathers Scheiner and Biancani who claimed the Capellan origin of the system of Tycho in order to endorse the geo-heliocentric conception without embarrassment.

Neither Scheiner nor Biancani was present in Rome in the spring of 1611. However, evidence that the idea of exhuming the ancient Capellan hypothesis was on the agenda of Clavius and the Collegio Romano is offered by the subtle operation conducted by Scheiner in the *Disquisitiones Mathematicae*, published in 1614 under the name of his pupil Locher.[54] The work confirms that the idea of reviving that ancient cosmological system was prompted by the need to encapsulate the potentially destabilising celestial novelties within the traditional architecture of knowledge. Some of the pages of Scheiner's work, offering a bird's-eye view of the recent cosmological conceptions (from Copernicus to Clavius passing through the celestial discoveries thanks to the telescope), help us to understand which suggestions may have led Cesi to draw his peculiar diagram of the system of the world. In Scheiner's concise history of contemporary astronomy, Galileo is cited with the greatest respect and without calling into question his role in transforming the telescope from a toy into an extraordinary research instrument. However, for Scheiner Galileo was not the inventor of the telescope. The Jesuit defined it as 'tubus opticus, nuper a Germania inventus',[55] thus adding a new candidate to the fierce competition for the title of inventor of the instrument. For Scheiner, the most advanced solution in the cosmological discussions between the middle of the sixteenth and the first decade of the seventeenth centuries had been devised by Magini

53 See Schofield 1981, Gingerich-Westman 1988, Lattis 1994 and Lerner 1995.

54 Scheiner 1614.

55 Ibid., p. 46.

and Clavius. While duly taking *De revolutionibus* into account, they had formulated the hypothesis of a geocentric universe formed by eleven interlocking spheres in continuous movement. That solution, which many regarded as definitive because it perfectly accounted for all observable celestial phenomena, had been unhinged by the telescope and the celestial discoveries. Magini and Clavius were thus forced to go back to work. This is the point at which Scheiner introduced Martianus Capella:

> They both [Clavius and Magini] changed their opinions. It seems that Magini is working at a new system, in which the movements of the heavens are saved by slowing them down; something similar can be found in Martianus Capella. As to Clavius, he sang his swansong at the end of his life with the final edition of his works.[56]

In the metaphor of Scheiner, Clavius performed his 'swansong'[57] in his posthumously published commentary on *de Sphaera,*[58] in which he invited the astronomers to devise a system of the world consistent with the new celestial phenomena discovered by Galileo: 'since matters are so – the Jesuit wrote – let the astronomers see how the celestial spheres are to be arranged in order to save these new phenomena'.[59] Scheiner expressed his confidence that if Clavius had not been overtaken by death in January 1612, he would 'no doubt have produced a new convenient system'.[60] He concluded that the most illustrious exponent of the Collegio Romano, together with Magini, with whom Clavius had always been in perfect agreement, would probably have paid extremely close attention to the geo-heliocentric system of Martianus Capella.

Immediately afterwards, Scheiner accused Tycho of having taken his ideas from the Latin author without acknowledging his source:

> As for this system [the geo-heliocentric system], although he [Tycho Brahe], hiding the truth, claims to have invented it himself, yet it seems that he drew it from Martianus Capella, who in Book 8 of his Philologiae [...] openly states: 'Although Venus and Mercury rise and set [seen from the Earth] every day, their orbits do not embrace the Earth at all,

56 Ibid., p. 50.

57 Cf. Lerner 1995.

58 Clavius 1612–1613.

59 Scheiner 1614, p. 51.

60 Ibid.

> but revolve around the Sun in a larger orbit. So the centre of their circuits lies in the Sun'.[61]

The operation of substituting Capella for the Danish astronomer, thus removing the embarrassing shadow of the heretic Tycho, was performed unscrupulously by Scheiner, who even pretended to ignore what the text of Capella that he had cited stated in clear terms. For Capella, only Mercury and Venus moved along heliocentric orbits (as indicated by Cesi in his diagram), while for Tycho all known five planets at the time followed heliocentric orbits. Performing a sleight of hand, Scheiner declared that:

> In Capella's system, the Sun is the regulator and centre of the five planets. Hence it is probable that Tycho constructed his system from him [Capella] with a few minor additions.[62]

Once the geo-heliocentric system had been rid of the taint of heresy, Scheiner could insinuate that, if Clavius had not passed away, he would probably have reflected on the opportunities offered by the system of Capella to confront novelties that threatened to destroy ancient certainties: 'I do not know whether the system of Martianus is the one whose investigation Clavius recommended'. But what is certain, Scheiner continued, is that the Latin author's system

> finds favour with many astronomers today because it is congruent with the celestial phenomena. Moreover, it is opposite to the Copernican one, being closest to the Ptolemaic system; it reduces the superfluity of celestial revolutions and easily explains everything with fewer heavens.[63]

Scheiner's text provides authoritative confirmation of that which, in the most recent studies on the revision of traditional astronomical conceptions after the publication of the *De revolutionibus*, has been defined as a 'revival of interest in the Capellan system',[64] whose most prominent exponents were members of the Society of Jesus.

Scheiner's testimony is also important because it comes from an author who kept up an intensive exchange of information with the Fathers of the

61 Ibid.
62 Ibid., p. 53.
63 Ibid.
64 Schofield 1981, p. 38.

Collegio Romano. This evidence reinforces the suspicion that the diagram in Cesi's letter to Francesco Stelluti of 30 April 1610 reflected attitudes taking shape in the minds of some of the Fathers of the Collegio, to whom the option of geo-heliocentrism seemed to offer the only way to incorporate without traumatic consequences the celestial novelties within the traditional framework of knowledge.

As we recall, 'there is no little difficulty in accepting that the Earth is at the centre'[65] was the expression with which Cesi had concluded his account to Stelluti of the exciting celestial novelties. His words show that at the time of writing, Galileo's arguments had shaken the Prince's previously held certainties, but had not led him to abandon the centrality and immobility of the Earth.

The letter to Stelluti is a crucial document because it enables us to follow almost live the Prince's reflection on the constitution of the universe, on the reciprocal positions and on the motions of the planets after meeting Galileo and witnessing his impassioned promotion of the Copernican system. This document is all the more important because in subsequent years Cesi would rarely confront questions about the system of the world, and then only with great hesitation.

Coup de théâtre

When Cesi wrote his letter to Stelluti, Galileo had been in Rome for exactly a month, frenetically engaged in the promotion of the telescope and the celestial discoveries, driven by the desire to win converts to the Copernican system. During these weeks Cesi appears constantly at his side organising appropriate occasions for Galileo to meet and try to convince the most influential members of the clergy, nobility and cultural élite.[66]

It is worth considering more closely at least a couple of the many cultural and social events in which Cesi and Galileo were protagonists. Let us start with the one organised by Cesi himself on 14 April in the setting of the garden of Monsignor Malvasia on the Janiculum Hill. Cesi took pains to invite a select delegation of natural philosophers and mathematicians to meet Galileo and experience the performance of his telescope in person. We know the identity of almost all of them: Antonio Persio, whose presence at the event

65 CL, p. 159.

66 On Galileo's activities in Rome in the spring of 1611, see Bucciantini-Camerota-Giudice 2015, pp. 202–26.

is at this point hardly surprising;[67] Francesco Pifferi, Professor of Mathematics at the Sienese Studio and expert in the construction of mathematical instruments; Giulio Cesare Lagalla, Professor of Medicine and Philosophy at the University in Rome;[68] Girolamo Sirtori, who a few years later would provide a detailed account of that memorable evening in his treatise on the telescope;[69] Johannes Faber, who had joined the Academy of the Linceans a few days after Galileo and was soon to become its Chancellor; Giovanni Demisiani, a scholar and philologist of Greek origin, admitted to the Academy not long afterwards; Johannes Schreck (Terrentius), who also soon became a Lincean; and Paolo Monaldeschi, a relative of Cesi. At least eight other persons were present whose identity cannot be established.[70] As far as is known, none of those who witnessed that evening belonged to the regular clergy. The event was organised as a lecture with demonstrations by Galileo aimed specifically at Prince Cesi and his associates and a select group of mathematicians and natural philosophers linked to that circle.

The evening was a repetition – in a different context – of the meeting organised in Bologna by Antonio Magini a year earlier, on 25 April 1610. On that occasion Galileo was offered the opportunity of demonstrating the wonders that the telescope made it possible to observe in the sky to Magini's guests assembled at the home of the nobleman Massimo Caprara.[71] Organised a few weeks after the publication of the *Sidereus Nuncius*, the Bologna event was hardly a success for Galileo. The opinion clearly prevailed of those who declared that they were unable to observe the celestial novelties announced by

---

67 The memorable evening on the Janiculum was also mentioned in the draft of the biography of Persio outlined, at least in part, by Federico Cesi after the death of the Abbot. It confirms the information contained in other accounts of the event adding a note of colour: during that 'philosophical meeting' Persio had 'disputed all the time with his customary vigour, in spite of being a septuagenarian' (Contributi I, p. 877).

68 On Lagalla, see Preti 2004.

69 We learn from Sirtori's account that to satisfy their curiosity, Galileo removed the two lenses from the tube and showed them to Cesi's guests (Sirtori 1618, p. 27); cf. also Rosen 1947, pp. 52–3.

70 Our main sources for that evening, besides Cesi and Sirtori, are Lagalla 1612 (OG III[1], p. 330), Faber (TM, p. 473), and the *Avviso* for the Court of Urbino of 16 April 1610 (Orbaan 1920, p. 283). On the dinner and demonstration of 14 April, see Redondi 1982 and 1984 and Bucciantini-Camerota-Giudice 2015, pp. 215–6. On the course of events of that evening and the discussions between the participants, see Redondi 1984 (pp. 11ff.), Camerota 2004 (pp. 205–6) and Bucciantini-Camerota-Giudice 2015 (p. 215).

71 Bucciantini-Camerota-Giudice 2015, pp. 89–102. Both Martinus Horky and Magini had taken care to secure the widest possible diffusion of the news of the unsuccessful evening.

the Tuscan scientist, or who appealed to unquestionable arguments of principle to resolutely deny that the confused phenomena revealed by the telescope corresponded to bodies that really existed in the heavens.

Discussions on the evening on the Janiculum took a different course. Galileo's main aim was to get Cesi's guests to observe his latest discoveries: the strange appearance of Saturn and the phases of Venus. Galileo was particularly concerned to show a sickle Venus: a phenomenon absolutely incompatible with Ptolemy's system of the world. To use his words, the phases of the planet formed a 'powerful argument for the Pythagorean and Copernican constitution'.[72] He attributed no less importance to the singular appearance of Saturn: the telescope revealed curious swellings on either side of the planet, which Galileo took to be satellites like those that he had observed circling Jupiter. This was new important evidence that, together with the celestial novelties described in the *Sidereus Nuncius,* certainly already known to Cesi's guests, strengthened the Copernican vision of the universe.

As they waited for darkness to fall, Galileo invited them to direct his telescope at terrestrial targets. The testimonies of some of those present reveal their astonishment at the fantastic performance of the instrument. It was powerful enough to enable them to read the time on the quadrant of a tower clock three miles away. Once night set in, Galileo was finally able to show them the novelties that his telescope had revealed in the heavenly vault. As in Bologna, some of Cesi's guests expressed their reservations, arguing that the fact that the telescope provided a faithful magnified image of buildings and architectural details in the distance was no proof of the real existence of the new celestial phenomena.

Reviving an argument based on the authority of Aristotle that had been used by the Florentine Francesco Sizi in his *Dianoia,*[73] some of Cesi's guests pointed out the substantial difference between the celestial and the terrestrial worlds. As there was no correspondence between the corruptible terrestrial bodies and those immaculate ones that illuminate the sky,[74] it was not legitimate to assume that the objective functioning of the telescope in terrestrial operations could be extended to its astronomical use. On the Earth's surface, Galileo's detractors argued, the image of objects progressively loses definition with the increase in distance: a few miles are sufficient for the eye to be no longer able to perceive an object clearly. How was it possible to trust telescopic

72 OG X, p. 483, letter of Galileo to Giuliano de' Medici, 11 December 1610.

73 Sizi 1611 (OG III[1], pp. 205–50).

74 Ibid., pp. 14–16; see Redondi 1982 and 1984, p. 181.

images of celestial bodies that were situated at distances immeasurably more remote? Besides, they continued, the telescopic vision of terrestrial objects takes place in the medium of air, which does not produce significant distortions by refraction because of its subtlety. But when the telescope is focused on the sky, the visual rays are subjected to bizarre deviations as they pass through the dense crystalline structure of the heavenly spheres. The images generated are consequently illusory. The emphasis on refraction was obviously based on the conviction of the substantial difference between the supralunary and sublunary worlds.

In spite of the hesitations of some of those present, the evening on the Janiculum was a success. They all managed to observe the celestial novelties. Consequently, the debate focused on their effective existence, that is, on the reliability of the astronomical images produced by the telescope. It was a heated and lengthy debate: one of the documents indicates that it came to an end at a late hour.[75] Cesi's guests invited Galileo to produce conclusive demonstrations that the bodies observed through the lenses of the telescope but invisible to the naked eye really existed and that the images produced by the instrument were faithful. It was at this point in the discussions that the thorny issue of the nature of light arose.

Based on the authority of Aristotle, Giulio Cesare Lagalla emphasised that light is a quality manifested by a luminous body in a transparent medium: an absolutely stable quality that can only be perceived by the naked eye.[76] It was therefore impossible that the telescope could make visible bodies that were invisible without the aid of the instrument. The phenomena revealed in the sky by the telescope had therefore to be considered visual deceptions, deformations produced by the lenses due to refraction.

We know from Lagalla that a few days later a further opportunity to view the celestial novelties through the lenses of the telescope was offered to some of those who had been present on the Janiculum in the presence of Galileo.[77]

75 *Avviso* of Urbino: 'Although they remained until one o'clock in the morning, they could reach no consensus' (Orbaan 1920, p. 283).

76 OG III^2, pp. 928–9, n. 11.

77 Accounts of these events take it for granted that it was during the meeting on the Janiculum that Galileo staged the *coup de théâtre* of the demonstration of the remarkable Bologna stone (see below, pp. 93–4). However, close reading of the extant documentation shows that the evening organised by Cesi in the garden of Malvasia was confined to scrutiny of the sky with the telescope. After meeting with obstinate resistance to the objectivity of the telescopic vision manifested that evening, Galileo decided to focus on that problem on a later occasion, basing his argument on the singular phenomenon of the glowing stone.

This second meeting was attended by Lagalla, Cesi, Demisiani and the Roman naturalist Clementi. The weather that evening was variable, obliging them to interrupt the observations time and again. The question of the reliability of the celestial images produced by the instrument was raised again during those intervals, with the consequent resumption of the animated discussions about the nature of light, on which no agreement had been reached during the night on the Janiculum.[78]

The problem at the centre of the observations and discussions during this second meeting, above all for its cosmological implications, was represented by the phases of Venus: a phenomenon that could not be seen by the naked eye because of the bright scintillation and radiation surrounding the planet. Only when viewed through the telescope did the phases of Venus become apparent. Some of those present called attention to the contradictory behaviour of the instrument: it removed the luminous radiation of the planets but not that of the stars. They drew arguments from this apparent contradiction to stress once again the illusory nature of the telescopic images of the celestial bodies.

These arguments made it quite clear to Galileo that, to remove scepticism about the reliability of the telescopic images, it was necessary to demolish the Aristotelian conception of light. He had been aware of this obstacle for some time, and during the previous months he had concentrated his attention on this problem. While in Rome, he had not missed any opportunity to emphasise that, although not visible to the naked eye, the real existence of the celestial bodies revealed by the telescope could not be denied. In a letter to Monsignor Dini of 21 May 1611, a few weeks after the event on the Janiculum, Galileo had denounced the presumption of those who wish 'to make our eyes the measure of the expansion of light, such that where luminous objects are not visible to

---

It is to this second occasion that Lagalla refers in his *De luce et lumine disputatio* (OG III^2, pp. 927–9). Lagalla's account indicates that the occasion on which Galileo exhibited the Bologna stone was not the same as the one on which he showed the celestial novelties on the Janiculum. Various details make this clear. In the first place, Lagalla had already given a precise description of the participants and the course of events in the demonstration on the Janiculum in that same work (ibid., p. 330). There was therefore no reason to return to the topic in a different chapter of the volume. Besides, the persons present at the demonstration of the stone listed in Lagalla's *Disputatio* are not the same as those who took part in the evening on the Janiculum: the second meeting was attended by only four persons, viz. Cesi (Lagalla does not mention him, but it is difficult to imagine that the Prince was not there), Demisiani, Giovanni Clementi and Galileo (ibid., p. 928).

78 'When clouds prevented us from observing the constellations, we discussed various matters [...]. The nature of light also came up for discussion [...]'. (ibid., p. 927).

us, we should affirm that they do not exist'.[79] The Tuscan scientist's message was clear: what effectively exists cannot be limited to what can be seen with the naked eye. The telescope was an eloquent example of this principle, as it provided an absolutely faithful image of real celestial bodies and phenomena that remain hidden from direct vision:

> […] as visible images are nothing but figured light [*luce figurata*] where those images reach light arrives too; now, if the images of the four Medicean planets were to vanish and to be completely lost before reaching the Earth, no amount of Murano glass would be sufficient to render them visible, because what is nothing cannot be multiplied, and both dilation and augmentation presuppose the existence of something to dilate and augment.[80]

The telescope did not alter, nor did it make illusory images appear as real bodies due to refraction. It simply magnified celestial bodies by bringing their images closer to our eyes. Visible images, Galileo declared, 'are nothing but figured light'. That means that light is a corporeal substance, albeit a very subtle one, and not a quality manifested by a luminous body in a transparent medium, as Aristotle and his followers claimed. The telescope caused the feeble, because distant, luminous images of the celestial bodies to converge, without deforming them in the slightest, magnifying them until they became visible. The operation

> of the telescope is nothing but a closing in on the images of visible objects, bringing them closer to us according to the tenth, twentieth, thirtieth or other smaller or larger proportion compared to their true and real distance, representing the same objects to us as we would see them at such small distances.[81]

There was still a problem raised by the followers of traditional natural philosophy that had to be tackled: Why did the lenses obliterate the scintillating corona observed around the planets with the naked eye, while they did not remove the luminous radiation of the stars? As he had written to Giuliano de' Medici in February 1611, Galileo was convinced that the radiation of the stars and the

79 Ibid., XI, p. 115, letter of 21 May 1611.

80 Ibid.

81 Ibid., p. 195, letter of Galileo to Grienberger, 1 September 1611.

tremulous flickering of light that we see when we gaze at them were produced by the vibrations of the intrinsic 'splendour' of their own substance.[82] This vision anticipated the explicit formulation of the theory of the oscillatory nature of light in the letter that his trusted disciple and collaborator Benedetto Castelli was to send a few months later to Galileo.[83] According to Castelli and Galileo the stars appeared surrounded by their luminous radiation even when observed through the telescope because their luminosity, emitted through vibrations travelling at great speed, was so intense that the lenses were unable to remove it.

An explanation still had to be found for why the planets, which lacked luminosity of their own, displayed a luminous radiation to the naked eye that disappeared when observed through a telescope. To account for this apparent anomaly, Galileo declared that the scintillation of the planets is 'greater or lesser depending on their distance from the source (the Sun) by which they are illuminated.'[84] This was why, even when observed through the telescope, Mercury displayed radiation because of its great luminosity due to its proximity to the Sun, while the disks of Jupiter and Saturn, the planets farthest from the Sun, appeared as clear disks without radiation. As for Venus, since it occupied an intermediate position, the telescope removed the radiation perceived by the naked eye almost completely, thereby enabling observation of the periodic variations in its aspect and size.

Galileo replied to Lagalla's reservations and objections that evening by proposing a revolutionary conception of light. Light was not a quality of luminous bodies in a transparent medium, as Aristotle had argued, but a special substance that transmitted, without deforming them and with astonishing rapidity, luminous images of objects to our organs of sight through vibratory movements. Galileo based on this innovative conception of the nature of light his claim to the absolute objectivity of the telescopic images of celestial bodies, rejecting objections centred on distortions produced by the instrument.[85] At the same time, given that light transports the 'figured images' through the celestial spaces in exactly the same way as in the sublunary region of the elements, the Tuscan scientist vigorously postulated the thesis of the homogeneity of terrestrial and celestial matter.

---

82 Ibid., pp. 61ff, undated letter.

83 Ibid., letter of 8 May 1612. Cf. Galluzzi 2011, pp. 60–3.

84 Ibid., p. 195.

85 On Galileo's use of the revolutionary conception of light as a substance to demonstrate the objectivity of the enlarged images of celestial bodies produced by the telescope, see Redondi 1982.

The long letter that Galileo sent to the Jesuit Grienberger on 1 September 1611, three months after the conclusion of his stay in Rome, helps to understand his strategy.[86] Galileo responded to the Jesuit's perplexity regarding his interpretation of the telescope's selective removal of the luminous radiation of celestial bodies. He did so by proposing an experiment that he had actually devised to demolish the objections of another Jesuit, Mario Bettini, who had rejected Galileo's thesis of the presence of valleys and mountains on the Moon.[87] According to Father Bettini, if there were mountains on the Moon, the lunar disk observed with the telescope would appear jagged, while on the contrary, it looked perfectly smooth. In his letter to Grienberger, Galileo rebutted Bettini's argument by declaring that the mountain tops on the lunar disk could not be perceived through the lenses of the telescope because of the strong contrast between the extremely bright luminosity of the disk of our satellite and the darkness of its background. For Galileo it was a phenomenon analogous to that which took place when observing Venus, though in reverse: because of the planet's strong radiation, the naked eye seemed to perceive it as a perfect disk, while through the lenses of the telescope it presented a sickle shape. The same causes operated in both cases, depending on the strict laws that govern the nature of light and its perception by our organ of vision.

The experiment illustrated to Grienberger consisted of making two fine curved incisions (to simulate the curvature of the Moon), one with a jagged edge and the other with a smooth edge, in a metal plate and placing a light source behind it. When observed at close quarters with the naked eye, the smooth incision displayed a uniform luminosity, while the jagged one (representing the uneven surface of the Moon) produced a clear effect of radiation. However, when the observer moved 100 to 150 paces away from the plate, both incisions showed radiation. Observed at the same distance through a telescope, one appeared jagged and the other smooth, exactly as when at close quarters with the naked eye. Finally, if the observer moved 1,000 to 1,500 paces away, the profiles of both incisions appeared completely smooth through the

86 OG XI, pp. 178–203; cf. Dupré 2003, in particular pp. 390–1.

87 *De lunarium montium altitudine problema mathematicum ter habitum Mantuae* [...] *coram Serenissimo Duce, et in cubiculo coram Illustriss. Cardinali Gonzaga*. On the so-called 'problem of Mantua', see Camerota 2005, p. 211. Father Biancani's objections echoed the doubts raised in Kepler's *Dissertatio cum Nuntio Sidereo*. Kepler had asked Galileo why when observed through the telescope the lunar disk appeared perfectly round and not irregular and jagged, as it would have been logical to suppose if very high mountains were present on our satellite (OG III[1], p. 115); see Kepler 1965, pp. 28–9.

lenses of the telescope, 'nor can the roughness and irregularities of the jagged incision – Galileo stated – be seen any longer'.[88]

The fact that, when observed from a considerable distance through the telescope, the outline of the jagged incision appeared as smooth as the other not only confirmed the existence of mountains on the lunar disk, even though their peaks could not be distinguished; it also demonstrated that the anomalies produced by telescopic vision were only apparent.

The letter to Grienberger echoed the polemics on the reliability of the telescopic images that had animated the spring nights in Rome on the Janiculum and elsewhere. The fact that the aspect of the celestial bodies observed with the telescope proved to be entirely faithful to their actual configuration – as in the case of Venus – or presented an altered image – as in the case of the smooth lunar disk or the radiant stars – did not mean that the telescope deceived. All those phenomena depended on the rigorous laws that govern the perception of telescopic images in various conditions of illumination. The telescope failed neither in the sky nor on Earth.

The experiment described in his letter to Grienberger is an eloquent example of a method of argumentation that Galileo was to use systematically in his *Letters on Sunspots*, the *Assayer* and the *Dialogue concerning the two chief world systems*. It consisted of extending to bodies beyond the reach of our senses – such as the celestial bodies – the results obtained by experiments conducted on phenomena that can be directly controlled. It goes without saying that this method was based, once again, on the premise of the physical homogeneity of the entire universe.

Galileo was perfectly aware that he would run up against resolute opposition during the meetings that the Prince organised in Rome. His experience the previous year in Bologna and his close reading of Sizi's *Dianoia* had enabled him to foresee with almost absolute certainty the arguments that would be advanced against the use of the telescope to scan the sky. He was thus prepared to face the most obstinate reservations, those based on the Aristotelian conception of light as a quality of luminous bodies that is instantly diffused in a transparent medium.

To demolish the fortress in which his opponents had taken refuge, Galileo advanced the alternative conception of light as a substance. To affirm that light is a body and substance was to unleash a frontal attack on the very foundations of Aristotelian physics. He knew perfectly well that such a bold sally would scandalise the resolute followers of Aristotle and reinforce their determination

88 OG XI, pp. 198–9.

to deny the existence of the celestial novelties revealed by the telescope. He had therefore prepared a surprise move. The most suitable moment to pull the rabbit out of the hat would be the evening of the second meeting promoted by Lagalla, when the heated discussions had reached such a dead end that Galileo wondered whether he would be able to win the day in the face of the opponents' objections. He invited the discussants to suspend the debate until the following morning after a few hours rest. To overcome their hesitations, he announced that at sunrise he would show them an astonishing phenomenon from which precious indications could be drawn to conclusively demonstrate the nature of light.

Thanks to the testimony of Lagalla, we can reconstruct in detail the extraordinary mise-en-scène that Galileo staged to produce the utmost amazement:

> In the crepuscular light of the early morning, before the Sun's rays shone brightly, he [Galileo] took a wooden box and showed us some small stones in it. In order for us to see whether they had any light, he first showed them by the light of a candle, and afterwards in complete darkness. When we all declared that they did not have any light, he exposed the box outside the window, not yet to the primary light of the Sun's rays, but to the secondary, feeble and uncertain light of the breaking of dawn. After a short time he closed the window again and showed us the stones glowing brightly like burning coals in the total darkness; the light had been produced or retained without manifesting any heat. Even more remarkably, we watched this light slowly fade and eventually become extinguished, from which we deduced many diverging arguments.[89]

The significance that Galileo attributed to the experiment is clear: the Sun's light had been absorbed by the stones, which retained it displaying for a while an evident luminosity. Here was the proof that light is not a simple quality but a substance. It also explained why, like all other substances, the luminous images of the most remote celestial bodies were faithfully magnified by the telescope.

Although it failed to convince all those present (Lagalla did not change his opinion), the Galilean interpretation of the extraordinary experiment made a very strong impression. The Bologna stone, or *spongia solis* as it was called by its discoverer,[90] seemed to open the way to a process of radical reform of

89 OG III^2, p. 928.

90 Cf. Redondi 1992 and 1994; see also Gómez 1991.

natural philosophy. At the same time, it confirmed the results of the telescopic exploration of the celestial spaces which demolished the picture of the cosmos that had prevailed until then.

We have no direct evidence of Cesi's reaction when the box was opened in the dark room, and may wonder whether the Prince was convinced by the direct connection established by Galileo between the absolute fidelity of the telescopic representation of the celestial bodies and the remarkable phenomenon of the stones that absorb, retain and emit light. What is certain is that he remained profoundly impressed by the experiment. For months and even years after the event there are frequent references in the correspondence of both Cesi and Faber to Galileo's *coup de théâtre* and to the curiosity provoked on a European scale by the luminous stones (calcined barium sulphide). Ferrante Imperato, informed by Faber of the extraordinary properties of the stone, immediately thought of consulting Della Porta.[91] Imperato hoped that the visit of Galileo to Naples announced by Cesi would give him the possibility of discussing many things with the Tuscan scientist, 'and in particular the stones that absorb and retain light'.[92] Faber also made sure that the news of the Bologna stone circulated in the German world. He sent some of the stones to Welser, though the latter did not display any particular enthusiasm.[93] Welser did, however, ask an unnamed friend and philosopher for his opinion, who delivered his verdict with perfectly Aristotelian language and concepts.[94] A few years later, on behalf of Cesi, Faber repeatedly begged Schreck, who was in Munich at the time, for his opinion on that remarkable phenomenon. The ex-Lincean, who had difficulty in calcining the stone to get it to emit luminosity,[95] finally

---

91 CL, p.163, letter of 10 June 1611.

92 Ibid., p. 165, letter to Faber, 7 July 1611.

93 Ibid., p. 172, letter of 26 August 1611: 'I do not have the time to discuss the stones that glow in the dark with you'. He went on to state that he had not observed anything extraordinary about them: 'I thank you for the stone you sent, but I lack the spirit to extract a miracle from it'. It must be stressed that Welser was and would remain sceptical about the astonishing secrets of nature on which he received continuous information from the German Linceans: 'I died of laughter at that gentleman who wants to shut up the human voice in a tube. What a fine invention! This invention will surely overshadow the long tube of Galileo'. He expressed the same scepticism regarding 'snow that passes through glass' and 'the tarantula'.

94 OG XI, p. 140, letter to Faber, 15 July 1611. Given the familiarity of Welser with Scheiner, the German Jesuit might be identified with the anonymous philosopher.

95 CL, p. 566, letter from Schreck in Milan to Faber in Rome, 2 May 1616. Schreck had stopped in Bologna (where he knew that the luminous stone had been prepared for the first time) for more precise information on the method to calcine the *spongia solis*.

came up with a singular interpretation of the causes of the emission of light.[96] However, he drew the line at Faber's continued requests for him to put his interpretation down on paper: 'You should not expect anything from me regarding the discourse on that light, for not only is it very difficult, but it requires an alchemical-optical anatomy of the stone itself and many other things'.[97] Via Schreck, Scheiner was also infected by curiosity regarding the Bologna stone.[98]

Cesi was no less active in spreading news of the phenomenon and distributing pieces of the stone as well as information on the interpretation of the causes of its luminosity. He repeatedly turned to Galileo for fresh supplies.[99] Conversations with Cesi on the Bologna stones led Grienberger to invite one of his pupils to ask Galileo whether they 'lost light at the point where they were touched or rubbed'.[100] Almost two years after the demonstration of the remarkable effects of the Bologna stone in Rome, Cesi asked Galileo to send him new samples for the Prince of Bamberg.[101] Later still, Federico informed the Tuscan Lincean that during a lecture to the Collegio Romano, a Jesuit who had violently contested the existence of mountains on the Moon and of spots on the Sun had 'taken the luminous stones as his personal device'.[102]

The Prince himself continued to ponder the nature of light and the Bologna stones. He mentioned them in a section of the *Zibaldone*, compiled between 1613 and 1616, as well as, later on, in the *Tabulae Phytosophicae*.[103] In the *Zibaldone*, the phenomenon in question was mentioned not, as one would expect, in the *Physica Mathesis*, but in the *Taumatombria*, the section in which he analyzed the most marvellous meteorological phenomena. This indicates that, in the eyes of the Prince, the properties of the stone appeared extraordinary

---

96 In a letter of 9 September 1617, Schreck informed Faber that, in his opinion, the light penetrated the minuscule pores of the stone and was consequently forced to go through an infinite series of reflections, thus remaining imprisoned in the stone for a long time (ibid., p. 614).

97 Ibid., p. 605, letter to Faber, 23 July 1617. It transpires from this letter that Faber and Cesi planned to include a text by Schreck on the Bologna stone in the *Thesaurus Mexicanus*.

98 'I recently visited father Schemer [sic] in Ingolstadt, who asked me whether I knew how to prepare the luminous stone. I replied that only calcination was needed' (ibid., p. 614, letter to Faber, 19 September 1617).

99 'I have received the box with the luminous stones and thank you in every way; it was very welcome to me' (OG XI, p. 515, letter of 30 May 1613).

100 Ibid., p. 513, letter from Giovanni Bardi to Galileo, 24 May 1613.

101 Ibid., p. 481, letter to Galileo, 15 February 1613.

102 OG XII, p. 12, letter of 3 January 1614.

103 Contributi I, p. 40: 'De [...] arte in lapide lucifero'; cf. below, p. 128.

for reasons and goals different from those of Galileo. The latter had used the stones as evidence in favour of his claims that light is substance. He was not interested in the properties of the stones in themselves. After the show on the Janiculum, he replaced them in the box and only returned to them on two occasions in the course of his later career,[104] although he continued to send fragments to Rome to satisfy the constant requests of the Prince and other influential persons in the Eternal City. When he visited Rome again in 1616, he brought fresh supplies of the luminous stone with him, but not enough to satisfy the expectations of nobles and Cardinals, who exploited their personal relations with the Tuscan court to obtain them. In February 1616, while Galileo was in Rome, Francesco Maria del Monte wrote to Antonio de' Medici to request the extraordinary stone together with an explanation of the process of calcination by which it became luminous:

> Galileo has brought here a solar stone that, when placed in daylight and subsequently removed to a dark place, glows like a burning coal. I understand that these stones are found in the territory of Tuscany, and are prepared by calcination. I therefore beg your Excellency to enlighten me in this respect, which I will hold among my other infinite obligations to you.[105]

After waiting in vain for two months, the Cardinal repeated his plea: 'I shall be awaiting receipt of the luminous stone together with the recipe and method of calcination from your Excellency'.[106]

It is worth pointing out that in the vast Lincean circulation of the luminous stones we never find any hint at the direct relation that Galileo had established between this curious phenomenon and the demonstration of the objectivity of

104 These were in a letter to Cesare Marsili on 29 August 1626, and in a polemic with Liceti at the end of his career. In the former (OG XIII, p. 339), Galileo requested new supplies of stones that 'when calcined, receive and retain the light for a little while'. Marsili promised to send him some and explained that they came from Mt. Paderno not far from Bologna. He also mentioned a curious effect produced by the radioactive stones: 'I have seen a water or ash made from that stone that causes hair to fall out' (ibid., p. 340, letter of Marsili to Galileo, 2 September 1626). For the dispute with Liceti, see Liceti 1640 (OG VIII, pp. 481–6); cf. also Galluzzi 2011, pp. 121ff.

105 Francesco Maria del Monte's letter to Antonio de' Medici, 5 February 1616.

106 Letter to Antonio de' Medici, 2 April 1616. The Cardinal's great interest in the stone is not surprising: he was passionately intrigued by alchemy and organised meetings in his residence in the Palazzo Madama attended by the leading representatives of chemical medicine in Rome, including Johannes Faber. Cf. Clericuzio 2009, pp. 281–302 (esp. p. 287); see also OG, Appendice, 2015, II, p. 160, n. 2.

the telescopic images of celestial bodies. The references to the theme of light and the properties of the Bologna stone in the research programmes and publication projects of Cesi and his closest collaborators suggest that it had stimulated them to engage in completely different kinds of speculations which were of fundamental importance for the agenda of the Prince and of his Academy.

The selective filter through which Cesi viewed the stone indicates that, even when confronted with the revolutionary novelties proposed by Galileo, his sensibility was still that which had led him to found the Academy. It is thus not surprising that he took care to provide Della Porta with a sample of the *spongia solis*. It was an efficacious way of reciprocating the numerous secrets that the Neapolitan had revealed to him during their many years of friendship and bait for his insatiable curiosity about the most extraordinary natural phenomena. Cesi hoped to receive from Della Porta illuminating gleanings on the nature of those mysterious stones and of the light that they absorbed and emitted. Finally, he believed that the Bologna stones would have found their ideal location in the collection of *arcana naturae* of Della Porta's museum that the Prince was making every effort to acquire for the young and short-lived Neapolitan branch of his Academy.

Storm Clouds Gather

Let us now follow the delicate negotiations in which Galileo was involved in Rome in the spring of 1611, amiably assisted and constantly accompanied by his patron, Prince Cesi. During those weeks, Federico and his closest collaborators, especially Faber, Schreck and Demisiani, often spent nights following the movements of Galileo, armed with his telescope, on the hills of the Eternal City. His innumerable demonstrations provoked the most unpredictable reactions. A letter from Schreck to Faber written more than ten years later informs us about yet another social event organised to demonstrate the performance of the telescope and the celestial novelties. It did not take place on the Janiculum this time, but on top of Trinità dei Monti, in the presence, among others, of Faber and most probably Cesi. Schreck recalled the strange attitude on that occasion of an unnamed philosopher who, imitating the resolute gesture of Cesare Cremonini in Padua, refused to observe the sky 'through Galileo's tube'.[107]

107 'Pietro Castelli seems to outdo that philosopher who on the hill of Santa Trinità refused to observe the stars through the lenses of the telescope to avoid having to confess that he had seen them' (letter of 22 April 1622). I am unable to identify the unnamed philosopher.

As the time for him to return to Tuscany approached, Galileo felt the need to draw up a balance of the results achieved during the long weeks spent in Rome. He was fully aware of not having overcome the resistance of the Aristotelians, fortified as they were by the thesis of the unreliability of telescopic vision and of the irreducible diversity of the celestial world from the terrestrial one. Nevertheless, Galileo was comforted by the fact that he had obtained recognition of the telescopic discoveries from the Collegio Romano, the institution that he was most eager to win over to his side. This was certainly the main objective of his mission, and one that he could proudly claim to have attained.

His conviction had certainly been reinforced by the solemn celebration of the telescope, of the celestial novelties and of their discoverer organised on 13 May by the Jesuits with the active collaboration of Cesi in the Collegio Romano. The conference room was packed with influential representatives of the Society, prelates and the most distinguished noble families, as well as students of the Collegio. Accounts of the event underline the presence of at least three Cardinals. Prince Cesi, significantly defined as the 'promoter' of Galileo, was seated in the front row.[108] It was a way of emphasising his fundamental role in organising that ceremony in which the new leading light of the Academy was guest of honour. The tenor of an anonymous chronicle of the event leaves no doubt that Galileo was given an extremely flattering reception:

> A Latin oration was read together with other compositions in praise of Signor Galileo Galilei, Mathematician of the Grand Duke, exalting and praising to the stars his discovery of new planets, that were unknown to the ancient philosophers, facilitated by the magnifying powers of the telescope invented by the Neapolitan Della Porta. After this public demonstration, Galileo will return to Florence highly pleased and as it were crowned by the universal consent of this university.[109]

Gabrieli was the first to publish this letter (CL, pp. 764–5), which is kept in the Archivio degli Orfani in the Library of the Accademia Nazionale dei Lincei, but transcribed it in an incorrect and incomplete manner. For a faithful transcription of this document see OG, Appendice, 2015, II, pp. 206–7.

108 Orbaan 1920, p. 284. On the ceremony organised in the Collegio Romano in honour of Galileo, see Pantin 2005 and Bucciantini-Camerota-Giudice 2015, pp. 206–8.

109 Orbaan 1920, p. 284. Note the genealogy of the invention of the telescope dear to Cesi, on which cf. below, pp. 142–52; see also Pantin 2005 and Reeves-Van Helden 2007, esp. p. 138.

The oration in praise of Galileo, the *Sidereus Nuncius Collegii Romani*,[110] was delivered by the Belgian Jesuit Odo van Maelcote, who devoted a significant part of his speech to confirming without reservations the existence of the phenomena observed in the sky thanks to the telescope.[111] However, he avoided dwelling on the cosmological implications of those new discoveries. He explained that he had simply donned the guise of a messenger of the celestial novelties, a messenger who had arrived late with respect to the Galilean *Nuncius* because he limped (*claudus tabellarius*). This was a way of referring to the condition of inferiority due to the delay before the Jesuit Fathers had been able to focus a high-performance telescope on the heavens.[112]

Van Maelcote's eulogy also mentioned the Galilean discovery – once again without in any way hinting at its implications – that posed the greatest threat to the traditional conception of the structure of the cosmos: the phases of Venus.[113] Many among his audience will have been aware that this phenomenon overturned the Ptolemaic system.

The account that a Jesuit present at that ceremony, Grégoire de Saint-Vincent, a pupil of Van Maelcote, sent two months later to his fellow Jesuit Van der Straeten, Rector of the College in Bruges, confirms the fulsome praise bestowed on Galileo by the Collegio in Rome.[114] Alluding to Clavius' initial reservations, he declared that shadows of doubt remained only with regard to the interpretation of the uneven surface of the Moon.[115] However when, almost half a century later, Grégoire de Saint-Vincent again evoked that memorable Roman ceremony, he recalled an extremely interesting detail; when

110 OG III[1], pp. 291–8.

111 In the official verdict on the celestial novelties delivered by the Jesuits of the Collegio Romano to Bellarmine on 24 April 1611 at his explicit request, Clavius had expressed his own doubts regarding the presence of mountains and valleys on the surface of the Moon: 'The great irregularities of the Moon cannot be denied; but it seems more probable to Father Clavius that the surface is not irregular, but that the lunar body is not uniformly dense but has denser and less dense parts' (OG XI, p. 93). However, Clavius soon had to fall in line with the opinion of the other Jesuit Fathers, who had confirmed the irregularity of the surface of the Moon. This is shown by the fact that no mention was made of the reservations of the distinguished elderly mathematician in the *Nuncius* of the Collegio Romano, read out three weeks after the account sent to Bellarmine.

112 OG III[1], p. 293.

113 Ibid., p. 297.

114 OG XI, pp. 162–3, letter from Rome, 23 July 1611: 'He received the unconditional approval of the many learned and noble men who were present'.

115 However, Father Grégoire declared that he was convinced that 'it is impossible to save the spots on the Moon as a consequence of the variable density of its matter', ibid., p. 163.

Van Maelcote mentioned that Venus, viewed through the telescope, presented phases similar to those of the Moon, a murmur of disapproval on the part of the natural philosophers, was clearly heard: 'We illustrated – Father Grégoire wrote – the new phenomena […] to the whole university in our Collegio. And we demonstrated that Venus clearly revolves around the Sun, *not without murmuring on the part of the philosophers*'.[116] Those murmurings were a clear alarm signal that did not escape Galileo and which was bound to produce strong apprehension also in the mind of Prince Cesi.

Although Galileo's stay in Rome in the spring of 1611 is traditionally depicted as a triumphal success, clear symptoms of hesitation and embarrassment emerged in the face of the celestial novelties. It would not be long before they would turn into resolute opposition.[117]

This alarm signal was immediately followed by another that was considerably more threatening. Only four days after the solemn celebration in the Collegio Romano, the Holy Office called for an investigation to verify whether the name of Galileo appeared in the documents of the proceedings against the Aristotelian Cesare Cremonini a well-known friend and colleague of Galileo in Padua.[118] This decision was taken in the presence of Bellarmine, who was probably one of the three Cardinals attending the ceremony at the Collegio Romano.[119] It is difficult not to see a connection between that celebration and the decision to instigate an inquisitorial procedure. The decision of the Holy Office in those same days seems to mark the start of a campaign to cast a slur on the reputation that Galileo was gaining. Many viewed that success with concern.

In spite of the proverbial secrecy of the procedures of the Holy Office, some news of the opening of the inquisitorial procedure must have leaked out. A few years later, in 1616, the Tuscan Resident in Rome, Piero Guicciardini, who

116 Father Grégoire's letter to Christiaan Huygens, 4 October 1659 (Huygens 1889, II, p. 480; emphasis added). Cf. Camerota 2004, pp. 209–11 and Bucciantini-Camerota-Giudice 2015, p. 207 and n. 24.

117 Cf. ibid, pp. 216–9.

118 'May 17th 1611 […] It was determined to verify whether Galileo, Professor of Philosophy and Mathematics, is mentioned in the Inquisitorial procedure against Dr. Cesare Cremonini' (OG XIX, p. 275). Cf. Bucciantini-Camerota-Giudice 2015, pp. 217–8.

119 During those weeks Bellarmine also held informal meetings with Galileo in which the two discussed the Copernican conception. This can be deduced from a letter from Piero Dini to Galileo on 7 March 1615, in which the conversation between the Cardinal and Galileo on questions *de systemate mundi* is mentioned: 'He [Bellarmine] assured me that he had not heard it mentioned [the Copernican hypothesis] since your private conversations' in Rome in 1611 (OG XII, p. 150).

had succeeded Giovanni Niccolini while Galileo was still in the Eternal City, reported to the Secretary of State of the Grand Duke of Tuscany that, on the occasion of the Tuscan scientist's visit in 1611, not only did his opinions displease many, but:

> his doctrine and some other things caused misgivings on the part of the consultors and *Cardinals of the Holy Office*, and among others *Bellarmine* told me that great respect was due to everything connected with his Highness, but that if he [Galileo] should remain here too long, they would be obliged to ask him to justify his doctrines.[120]

Cesi himself must have been aware of the inquisitorial proceedings and of the scrutiny of Galileo' position. Soon after the return of the Tuscan scientist to Florence, Federico invited him to proceed with the utmost caution and not to imagine that his victory in the first battle would guarantee him success in the war that he had declared on the supporters of the traditional ideas: it was necessary to win the approval not only of right-minded and free-thinking persons, but also of unfair ones, such as the Peripatetics, who were 'enemies of truth'. As he would continue to do with modest success in the following years, Cesi invited Galileo not to be impatient: 'I hope [...] that your haste will not give rise to any altercation or controversy'.[121]

The Galilean discoveries had hurt the pride of the Jesuits in the Collegio Romano who were jealous of their reputation as the most advanced outpost of scientific knowledge in Christendom. These feelings are explicit in the letter of Grégoire de Saint-Vincent to Christiaan Huygens. Father Grégoire confessed the dismay caused by the celestial novelties among the ranks of his fellow Jesuits, who would never have believed it possible that such new phenomena as the strange appearance of Saturn could be discovered by anyone but a member of the Academy of Father Clavius.[122]

The ambiguous character of the celebratory welcome received by Galileo in the Collegio Romano is further confirmed by the fact that also Christoph Grienberger, the most receptive among the Fathers to the celestial discoveries

---

120 OG XII, p. 207, letter of 5 December 1615 to Curzio Picchena. Emphasis added to words written in code due to their confidential nature.

121 OG XI, p. 210, letter of 17 September 1611.

122 Huygens 1889, p. 490, letter of 4 October 1659. Father Grégoire expressed the same embarrassment at the fact that the Jesuits had not been the first to discover the phases of Venus: 'I would never have believed that the strange appearance of this planet could be discovered by others than we Jesuits who defined ourselves as the Academy of Father Clavius'.

and their cosmological implications, revealed his embarrassment at Galileo's priority in perfecting the telescope and using it as a revolutionary astronomical instrument. In his letter of 22 January 1611, Grienberger stressed that, before hearing of Galileo's telescope, his fellow Jesuit Lembo had made an instrument with which the Fathers of the Collegio had observed the satellites of Jupiter.[123] Grienberger's account did not correspond to the actual course of events: as we have seen, it was only after receiving a high-quality telescope from Venice on the initiative of Santini that the Fathers of the Collegio Romano had observed the moons of Jupiter. It mattered to Grienberger to emphasise the complete independence of the Jesuits in making their own telescope, in observing the new phenomena, and even in compiling the periodic tables of the satellites of Jupiter, undertaken – as he wrote to Galileo – 'before we learned your method of recording distances'.[124] Statements like these were bound to wound the pride of the Tuscan scientist.

A few months later other alarming symptoms reached Galileo's ears. In his opening lecture of the annual courses in the Collegio Romano, the unidentified Jesuit speaker praised the recent celestial discoveries without ever citing Galileo.[125] Furthermore, only a few days after the triumphal celebration at the Collegio Romano, a Jesuit in Parma, Father Mario Bettini, had denied the existence of valleys and mountains on the surface of the Moon. Following Galileo's vigorous protests, Grienberger attempted to minimize inviting him to consider the letter of apology – full of fulsome praise – that Giuseppe Biancani, another Father in the Jesuit College of Parma and a former pupil of the Tuscan scientist, had sent him.[126] Although he thanked Grienberger for his reassurance, Galileo was not mollified. As he wrote to Grienberger, he suspected that Father Biancani's words in praise of him were a false contrivance:

> Because no one is so simple that he does not understand how praise can be meant ironically or to flatter [...], but not reproaches and insults, which always come from the heart.[127]

123 OG XI, p. 33.

124 Ibid., p. 34. In his *Sidereus Nuncius Collegii Romani* Van Maelcote had also claimed the Fathers' independence in the celestial discoveries, omitting to mention the fundamental role played by the telescope sent to the Jesuits of the Collegio by Santini.

125 OG XII, p. 12, letter of Cesi to Galileo, 3 January 1614.

126 OG XI, p. 126, letter of Biancani to Grienberger, 14 June 1611. See Camerota 2004, pp. 213–4.

127 Ibid., p. 179, letter of Galileo to Grienberger, 1 September 1611.

CHAPTER 4

Building a Friendship

New Spectacles in the Heavens

Any attempt at reconstructing the intellectual biography of Federico Cesi between 1603 and his first meeting with Galileo runs up against a lack of documents that throw light on the reflections and research priorities of the young Prince. In fact, apart from the energy spent on drawing up the rules of the institution, the support offered to the publication of Della Porta's secrets and the efforts to maintain the cohesion of the Lincean nucleus after the dispersal of the founding members, Cesi's intellectual identity only emerges as the reflection of the activities of those individuals whom he admired and trusted to implement a plan which, at the end of the first decade of the seventeenth century, was bound to appear precarious to him.

His relation with Della Porta was clearly one of intellectual subordination, which is hardly surprising in view of the tender age of the Prince and the great authority of the elderly Neapolitan master. Their correspondence bears no trace of Federico's personal positions, or of any independent evaluation of the often outlandish analyses and proposals of his correspondent. Della Porta was to remain an oracle of truth for Cesi and an infinite source of those secrets that he found so fascinating, encouraging him to invest considerable sums in having them printed.

In 1610 Federico gradually began to free himself from the watchful eyes of his parents, who had opposed his academic dreams, above all for their not unfounded suspicions of plagiarism on the part of Heckius. Although burdened with heavy debts, the family fortune passed largely under the control of Federico a few days before Galileo arrived in Rome. The Prince could finally resume his long-cherished academic project, as he hastened to inform Francesco Stelluti ('now my property is mine') inviting him to return to Rome to 'establish our principles and our constitutions'.[1]

That letter, marked by the enthusiasm and excitement provoked by the celestial novelties announced by Galileo, offers the first known piece of evidence for the research programme envisaged by the young Prince while showing that

1 CL, p. 159, letter to Francesco Stelluti, 30 April 1611.

 | DOI 10.1163/9789004342323_005

his guiding light was still Della Porta. Putting the confidential parts of the letter in code, he informed Stelluti that he had prepared rooms in his Roman residence 'for you and me, for the studies and sciences for which I am preparing and mainly those of Della Porta and mirrors'.[2]

The Prince therefore had a double objective in mind: to complete the compilation of the *Lynceographum* and to dedicate himself to the study of natural philosophy following the suggestions and models of Della Porta, who by that time had become a member of the Academy.[3] The reference to mirrors indicates that, under the expert guidance of Della Porta, Federico intended improving his understanding of the laws of optics, in order to master the principles that governed the functioning of the telescope.

The lively discussions with followers of Aristotle on hydrostatics, in which the Grand Duke of Tuscany had invited Galileo to participate immediately after his return from Rome, obliged him to put aside observations of the sky and to postpone writing the text on the constitution of the universe that he had promised in his letter to Belisario Vinta.[4]

As for Prince Cesi, he was mainly occupied with the revival of the Academy and recruiting new associates to fill its meagre ranks. These concerns explain why Cesi's first letter to Galileo in Florence was not dispatched until the end of July 1611.[5] The few letters exchanged with Galileo between the autumn of that year and the beginning of 1612 offer no clues as to whether Cesi was engaged in the programme of research mentioned to Stelluti. However, it is evident that he had assumed by now the role of scrupulous observer of the reactions of the *literati* and clerics of the Eternal City to the novelties announced by Galileo. At the same time, he was making frequent naturalistic excursions to satisfy his interest in botany, which had been and would remain one of his research priorities.

Galileo sent Cesi the discourse on the *Bodies that stay atop water or move in it*[6] as well as his letters to Gallanzone Gallanzoni and to Grienberger (both dealing with the controversial question of mountains on the Moon) with the request to circulate them.[7]

2 Ibid., p. 160.

3 Della Porta was inducted into the Lincean Academy on 6 July 1610 (Contributi I, p. 513). He was the first new member since the founding of the Academy. The Neapolitan filled the vacancy created by the death of Anastasio De Filiis in 1608.

4 OG X, pp. 348–53, letter of 7 May 1610.

5 Ibid., p. 158, letter of 23 July 1611.

6 Galilei 1612 (OG IV, pp. 59–114) and 1981.

7 OG X, pp. 141–55 (letter to Gallanzone Gallanzoni, 16 July 1611) and pp. 179–203 (letter to Christoph Grienberger, 1 September 1611).

From February 1612 the exchange of letters between Florence and Rome became progressively more intense, though both now and in the future Federico was to prove the more assiduous correspondent of the two.

A letter from the future Lincean Markus Welser to his close friend Johannes Faber on 6 January 1613 brought Galileo back to the forefront of the research programme of the Roman Academy, which was now more visible and public. Welser anticipated the arrival in Rome of a book dedicated to him, without mentioning the name of its author, that he was certain would be of interest to Cesi and his circle.[8] The work was *Tres epistolae de maculis solaribus* (*Three letters about sunspots*), published a few days earlier in Augsburg.[9] Faber hastened to pass the news on to Galileo and Federico, who had left Rome for Acquasparta in Umbria to take care of his family affairs, to go on excursions in the surrounding territory, rich in botanical and other natural resources, and to complete the drawing up of the statutes of the Academy and the definition of his personal programme of study. On 4 February Cesi wrote to Galileo that the author of the work on the sunspots had adopted the pseudonym of Apelles hiding behind the painting ('Apelles latens post tabulam').[10]

Galileo received the news with apprehension. He had been observing that singular phenomenon for more than a year and reflecting on its nature and causes. During his stay in Rome he had shown the sunspots to numerous influential persons, including Cesi and Faber. Besides, he had mentioned his solar observations in the discourse on the *Bodies that stay atop water or move in it*.[11] The presence of dark spots on the Sun were the latest of his celestial discoveries. He had made them the stunning concluding act of the theatrical revelation of the celestial spectacles staged in Rome in the spring of 1611.

---

8 CL, pp. 186–7. Welser had been favourably impressed by the work: 'I made fun of it at first, but confess that it has left me convinced. I am keen to hear what Signor Galileo will say about it' (p. 187).

9 Scheiner 1612a (OG V, pp. 21–70; for an English translation of the text see Galilei-Scheiner 2010, pp. 58–85). In September of the same year Scheiner published a second work, the *Accuratior disquisitio* (Scheiner 1612b; for the English translation of the Jesuit's text see Galilei-Scheiner 2010, pp. 182–230) in which he gave an account of new observations that in his opinion confirmed that the spots were small planets orbiting the Sun.

10 OG XI, p. 271. The allusion is to the famous episode in which the great Greek painter hid behind his paintings to eavesdrop on the comments of viewers.

11 Galilei 1612 (OG IV, p. 64) and 1981 (pp. 19–20). 'I add to these things [the list of celestial discoveries made with the telescope] the observation of some dark spots that are seen on the Sun's surface'.

Evidence of Galileo's concern triggered by the publication of Apelles is offered by his letter to Alfonso Antonini on 20 February 1638. The Tuscan scientist recalled the demonstrations of the spots on the surface of the Sun 'at the time when I was in Rome, where I showed them several times to many important prelates in the gardens of the Quirinale'.[12] The Lincean Angelo De Filiis would confirm that claim in the preface to Galileo's *Letters on sunspots*. He wrote that, during his stay in Rome in 1611, Galileo had announced his discovery of the 'spotted Sun' not only verbally, but also by demonstrating 'the effect itself' to numerous eye-witnesses 'in the Quirinale garden'.[13] Finally, in his biography of Galileo, Vincenzo Viviani was to state that in the context of a series of public demonstrations in which Galileo 'showed the new spectacles of the sky' disclosed by the telescope he revealed the sunspots 'in the Quirinale Garden' in the presence of Cardinals, representatives of the clergy, patricians and natural philosophers.[14]

Upon receiving news of the publication of Apelles, Galileo will instinctively have recalled the Prince's recommendation in his letter of 20 August 1611 to make haste in publishing a revised edition of the *Sidereus Nuncius*, including in it those celestial novelties (the triple-bodied Saturn, the phases of Venus, the sunspots) discovered after the publication of the book in March 1610, before others could appropriate them.[15] It had become a matter of urgency to develop an effective strategy to avoid being robbed of a priority to which he attached special value. The 'spotted Sun' confirmed first of all that the heavens were corruptible. Moreover, thanks to his assiduous observations of the spots, he was confident of being able to demonstrate that the Sun rotates around its own axis, as Copernicus had supposed.[16] That demonstration assumed the greatest importance in the eyes of Galileo as it provided convincing evidence of the heliocentric structure of the universe. This is confirmed by his reply to

12 OG XVIII, p. 297. See Galluzzi 2009.

13 Galilei-Scheiner 2010, p. 376 (OG V, pp. 81–2).

14 OG XIX, p. 612.

15 'You have not yet written anything on the horned Venus and the triple Saturn. Please do so as soon as possible to prevent your children from finding some brazen father who is dying to adopt them' (OG XI, p. 175).

16 Ibid., V, pp. 219–20. The idea that the spots were an incontrovertible proof of the spinning of the Sun on its own axis had occurred to Galileo soon after he had observed the phenomenon. He explicitly mentioned it in the opening pages of his treatise on hydrostatics: 'which [the spots], changing their position in that [the Sun], offer a strong argument either that the Sun revolves, or perhaps that other stars, similar to Venus and Mercury, circulate around the Sun', Drake 1981, pp. 19–20 (OG IV, p. 64).

Apelles, in which he formulated the hypothesis that the motion of the planets, including that of the Earth, around the Sun depended on the rotation of the Sun on its axis.[17] That passage was not approved by the censors and had to be expunged from the printed text. However Galileo was to return to the idea of the spinning Sun as motor of the planets in the *Copernican letters*.

As soon as he got his hands on a copy of Scheiner's *Tres epistolae*, Galileo dropped everything else to concentrate on demolishing the interpretation put forward by Apelles (at this stage neither Cesi nor Galileo knew that the author was a Jesuit). According to Apelles, the spots observed on the Sun were not produced on its surface, but were shadows projected by small stars orbiting it. It was important for Galileo to refute this thesis not only because he considered it erroneous, but also because it seemed to have been devised on purpose to save the perfect and incorruptible nature of the Sun.[18]

Galileo sent a long letter to Cesi on 12 May 1612 in which he shared his conclusions on the nature of the spots. His tone was peremptory: 'I conclude and believe that I am able to demonstrate necessarily that they [the spots] are contiguous with the surface of the solar body, where they are generated and dissolved continuously like clouds around the Earth'. He added that 'the spots owe their circular motion to the Sun itself, which completes a rotation on its own axis within a lunar month'. He concluded triumphantly:

> This novelty [the sunspots] is likely, I suspect, to be the funeral – or rather the Last and Final Judgement – of pseudo-philosophy, as the marks have now been seen on the stars, the Moon and the Sun. I now expect to hear the Peripatetics spouting big things to maintain the immutability of the heavens. I do not know how it [the immutability of the heavens] can be

17 OG V, p. 140 (note to line 26): 'I would not venture to declare anything with certainty on that problem. I would only ask the speculative philosophers to consider that the fact that they [the spots] all fall within that strip of the solar globe corresponding to the part of the sky through which the planets move, and not anywhere else, is a sign that even the motions of the planets might depend on that effect'.

18 Like many of his fellow Jesuits, Scheiner had qualms about the corruptibility of the heavens, a thesis that was vigorously opposed by the hierarchy of the Society, whereas he did believe in the fluid nature of the celestial spaces. It is worth stressing that in 1614 Scheiner received a formal order to abandon the thesis of the fluidity of the heavens from the General of the Order, Claudio Acquaviva (see Camerota 2005). The Jesuit reaffirmed his hostility towards the corruptibility of the heavens in his posthumously published *Prodromus* (Scheiner 1651), in which he launched a frontal attack on the Copernican hypothesis based not only on scientific arguments but also on the authority of Holy Scripture.

> saved and hidden away now that the Sun manifestly demonstrates itself to us with sense experience. Whence I hope that the mountains on the Moon are about to become a joke and a mere whim compared with the violence of the clouds, vapours and smoke that are produced, moved and dissolved continuously on the very face of the Sun.[19]

Knowing that this interpretation would go down very well with the Prince, who was strongly inclined towards the conception of the fluidity and homogeneity of the heavens, Galileo took the liberty of proposing to publish his own interpretation of the sunspots under the aegis of the Lincean Academy. He was not mistaken. As soon as Cesi received the letter, he replied enthusiastically inviting Galileo to send him the text as soon as he could.[20]

As a condition of publication of his book under the emblem of the lynx, Galileo demanded the Prince's commitment not to claim priority for the celestial discoveries with high-sounding expressions nor to adopt an excessively polemical tone with respect to usurpers.[21] Cesi's reply fell in line with this demand: 'Neither I nor the other Linceans will write anything in these disputes that has not first passed through your hands'. Immediately afterwards however he gave a glimpse of those bellicose intentions that Galileo seemed to fear: 'The aim is to teach these obstinate adversaries a lesson and to celebrate your inventions and discoveries with all our might'.[22]

This reversal of roles comes as a surprise: Galileo recommends a low profile, while Cesi prefers a direct and resolute reply to Apelles. In the correspondence between Galileo and the Linceans it is usually the Prince and his colleagues who call for patience and caution while Galileo remains defiant.

A year later, when the *Letters on sunspots* were going to press, Galileo reiterated his desire to avoid aggressively claiming his priority and denouncing the usurpers. He told Cesi that he did not approve of the tone adopted by Angelo De Filiis in the draft of the preface. The Lincean had meticulously listed the many celestial discoveries of the Tuscan scientist vigorously asserting his priority over the numerous competitors who had sought to appropriate them. The letter in which Galileo asked Cesi to modify the original wording of the

---

19 OG XI, p. 296.

20 Ibid., p. 297, letter of 17 May 1612: 'Your opinion on the sunspots that shows the paralogisms of the pseudo-philosophers has exceedingly pleased me'.

21 I want – he wrote to Cesi – 'that you and the other Linceans show caution, in writing about my things, not to prejudice that esteem in which you have presented the world with so many of your most excellent other conditions' (ibid., p. 296, letter of 12 May 1612).

22 Ibid., p. 297.

preface, signed by De Filiis but certainly inspired by the Prince, has not survived. It is from Cesi's reply that we learn that Galileo had disapproved of the peremptory tone used to denounce the malign practices of the usurpers. Federico defended that choice and made it clear that he spoke on behalf of the entire Academy: 'The need we see to extirpate from these persons [...] the things spread by the envious and other adversaries who seek to defraud you of your achievements prevents us from sharing your view'.[23]

In a further attempt to overcome Galileo's resistance, the Prince turned to Lodovico Cigoli, a trusted friend of the Tuscan scientist, who urged Galileo to approve De Filiis' preface as it stood 'because we all wish it and consider it necessary for many reasons'. 'Give the *placet* and soon', he concluded, stressing that to persist with his opposition 'will displease us all'.[24]

Galileo did not give way to the pressure of his Roman friends. With the assistance of Cesi, De Filiis was obliged to revise his text. In the preface published in the *Letters on sunspots*, the emphasis on the priority of the discoveries was more restrained and the tone more moderate. The mention of Apelles as the usurper of Galileo's priority in the discovery of the sunspots contained in the initial version disappeared. So did that of Kepler, whom De Filiis had criticised for publishing private letters of Galileo without formal authorisation. The vigorous denunciation of the malpractice of those who study not to learn from the masters, but to steal from them and to cancel their memory, was replaced by an anodyne chronicle of the evidence for the priority of Galileo in discovering the sunspots.[25]

What prompted Galileo's firm opposition to the first version of the preface, provoking such great frustration among his Roman friends? He may have been annoyed by the emphatic style of De Filiis' text, as Cigoli was also to recognise,[26] or by the presence of a scientist of the stature of Kepler among the throng of his usurpers and malign slanderers. His attitude was probably also conditioned by the unrewarding efforts and unhealed wounds of his heated polemics, not only with the many fierce opponents to his writings on hydrostatics, but also with the considerable number of those who had denied his discoveries after the publication of the *Sidereus Nuncius*, or who had claimed to have made them themselves.

23 Ibid., p. 483, letter of Cesi to Galileo, 22 February 1613.

24 Ibid., p. 485, letter of 24 February 1613.

25 Favaro published the two versions of the preface (OG V, pp. 79–88).

26 '[Cesi] read me the preface to be printed in the book: I found the style rather clumsy [...]' (ibid., XI, p. 484).

Ten years later, Galileo publicly acknowledged that he had been mistaken in not following the pressing advice of his Roman friends. In the opening pages of the *Assayer*, in which he proudly defended himself from the denigrators and usurpers who had dogged him from the outset, he recalled the recommendations of Cesi and the other Linceans on the occasion of the publication of the *Letters on sunspots.* Galileo mentioned 'some gentlemen, my true and sincere friends' who had done their best to persuade him to confront the usurpers. It is impossible not to recognize in 'those gentlemen' Cesi and his Lincean friends who had urged Galileo to approve the resolute defence of his priority in the original version of De Filiis' preface. While tempted by their arguments, he continued, which had 'nearly destroyed my resolve to write no more' he had decided not to follow their counsel: 'yet my desire to live tranquilly and without such disputes prevailed, and thus, fixed in my purpose, I believed I had silenced in this way all the tongues which previously had shown such eagerness to contradict me'. He now frankly admitted his error:

> But it was in vain that I had reached this frame of mind, nor have I been able by remaining silent to avoid my persistent fate of having always to concern myself with those who write against me and quarrel with me.[27]

Let us leave, for the time being, the discussions on the nature of the sunspots and on the priority of the celestial discoveries. Those sunspots and Galileo's interpretation of them had led the Prince to temporarily abandon his quest for rare plants, curious minerals and secrets of nature and to fix his gaze on the heavens. The letter that Cesi sent to Florence on 20 June 1612, a few days after receipt of Galileo's proposal to publish his interpretation of the sunspots under the aegis of the lynx, provides the first evidence of the Prince's immersion in a clearly defined research project.[28] It was a project of particular importance given that he intended to exchange ideas with Galileo on the real structure of the universe. The letter reveals Federico's plan to devote himself to compiling a treatise on celestial matters, to which he later referred as the *Coelispicium* (in-depth analysis of heavenly phenomena). As Cesi wrote to Galileo, he was taking advantage of the fact that he had temporarily retired to his estate in San Polo dei Cavalieri, not far from Rome, to reflect on 'celestial and terrestrial matters'. Evidently the seed of doubt on the real configuration of the system

27 Drake-O'Malley 1960, p. 169 (OG VI, p. 218).

28 Ibid., XI, pp. 332–3.

of the world had been sown in his mind by Galileo in Rome. Although the 'spotted' Sun also seemed to confirm the Copernican vision so dear to his admired associate, the Prince still had reservations. The Copernican structure of the universe was difficult for him to accept, above all because it presented an excessively intricate and disorderly picture:

> I see that I would be very happy with the Copernican system if the epicycles and eccentrics were completely removed, as he [Copernicus] does in all the other parts except in the Earth and in the Moon because he seems to place the Earth in an epicycle or eccentric, and the Moon in the epicycle of an epicycle.[29]

Cesi also asked Galileo to confirm that Copernicus had maintained, as he thought, the solidity of the spheres 'which, according to Kepler, Tycho destroyed'.[30]

As in Francesco Stelluti's letter to his brother discussed earlier, the authority of Kepler was invoked to support a cosmological vision that differed from the Copernican one so strenuously defended by Galileo. At this date Cesi had certainly read Kepler's *Astronomia Nova*.[31] His letter clearly reflects Kepler's recourse to elliptical orbits and his confirmation of Tycho's revolutionary demolition of the solid celestial spheres.

Cesi's statements also echo the lively discussions on Kepler's conception of the universe within learned Roman circles. As Massimo Bucciantini has demonstrated, at the beginning of the 1610s Rome and Venice were the centres in which the presence of Kepler was most keenly felt.[32] It should be kept in mind that Remus Quietanus, who was on familiar terms with Kepler, spent a long period in Rome from 1611 on. Remus, who had followed Galileo's lectures in the University of Padua in 1610,[33] was in the service of the Sforza family with whom Cesi was connected: the Prince was to marry a daughter of the Sforza Duke only a few years later. In a letter sent from Rome to Kepler on 17 December 1611, Remus had expressed his reservations regarding Copernicus: 'Why should the Sun not move along an epicycle?'[34] As he illustrated with

29 Ibid., p. 332.
30 Ibid., p. 333.
31 Kepler 1609.
32 Cf. Bucciantini 2003, esp. pp. 206–7.
33 Ibid. (*sub indice*); see also Favaro 1983, I, pp. 407–10 and 446–7.
34 OG XV, p. 247.

a diagram, he was inclined to believe that the planets followed a spiral path, a hypothesis that coincided perfectly with observational data. Seen from the Earth, the planets do indeed seem to move along "knots" of a spiral shape. Like that of Cesi, Remus' vision presupposed the demolition of the solid spheres and the affirmation of the fluidity of the heavens.

Remus' reservations were similar to those that the Prince was to convey to Galileo a few months later. Both Remus and Federico were reluctant to abandon the Earth's central position in the universe. Remus' perplexity depended on the immense force that would be required to move our planet along a circumsolar orbit.[35]

It is therefore extremely likely that, in his letter to Galileo, Cesi expressed doubts that had emerged during his discussions with Remus. Not by chance, a few years later, the Prince was to encourage him to make direct contact with the Tuscan scientist.[36] Moreover, Remus would be nominated for enrolment in the Academy, although, in spite of Galileo's approval, that induction never materialised.[37]

Interest in the mathematical and cosmological works of Kepler was also very strong in the Collegio Romano. The Jesuit Odo van Maelcote had not yet read the *Astronomia Nova* when Cesi communicated to Galileo his inclinations towards Tycho and Kepler. The Jesuit Father, who was well informed about the scientific production of the Imperial astronomer, would obtain a copy of the book only a few months later. It made such a strong impression on him that he did not hesitate to write enthusiastically to the author, even though he was a heretic:

> I am very familiar with your works, the *Dioptrice*, the *De nive sexangula*, the *Dissertatio cum Nuncio Sidereo* and the others. Having recently obtained a copy of the work on the star of Mars [*Astronomia Nova*], I have had the pleasure of reading it again and again and admiring it. Good God! What sublime and wonderful ideas there are in that book. What a

35 Ibid.

36 OG XII, pp. 433–5, letter of Remus to Galileo, 12 January 1619: 'Our illustrious Prince Cesi has several times urged me to write to you'.

37 Contributi I, p. 541. Cesi proposed the induction of Remus into the Academy at a session held in May 1621 with the following motivation: 'He is the physician and mathematician of Archduke Leopold; he is especially skilled in astronomy and very accurate in astronomical calculations; he completely embraces the axioms of our Galileo; he is also highly distinguished in medicine and experiments'. See also CL, p. 736, letter of Faber to Galileo, 1 May 1621.

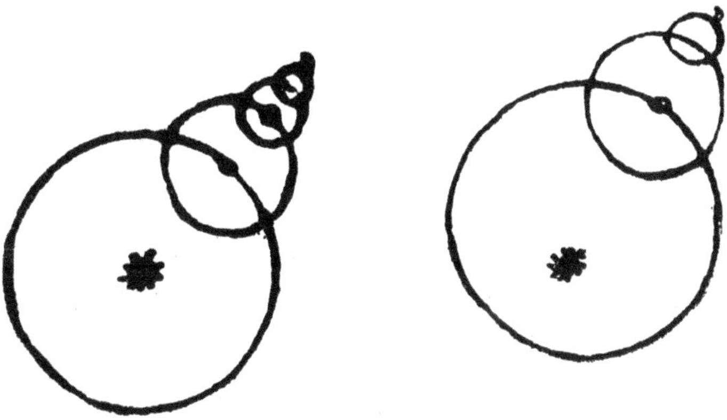

FIGURE 14 *The two diagrams of the Copernican system drawn by Cesi* (CL, p. 239)

> splendid service you have done, not only to Copernicus, but also to Tycho and Ptolemy and their followers![38]

Cesi's reservations regarding Copernicus were not determined only by his suspicion that the author of *De revolutionibus* maintained the solid orbs.[39] He was also hesitant with regard to the relative positions of the orbits of the Sun, Earth and Moon in the heliocentric system. As in the letter to Stelluti of 20 April 1611, Cesi expressed his dilemmas in visual form, sketching two diagrams [fig. 14] that presented alternative solutions. He asked Galileo 'whether we should consider the disposition of the movements in the Copernican system in accordance with the first or the second diagram, and whether with or without the solid spheres, or in some other way'.[40]

In the first diagram the Sun occupies the centre, while the Earth moves around it on a deferent and the Moon proceeds along an epicycle on the deferent of the Earth. This solution mirrored the Ptolemaic system, simply reversing the positions of the Earth and the Sun.

The second diagram presents a different arrangement: the Sun is eccentric with regard to the deferent of the Earth, while the Moon follows the Earth's motion along an epicycle.

38 OG XI, p. 445, letter to Kepler, 11 December 1612. Cf. Bucciantini 2003, p. 213.

39 'I do not see that Copernicus ever takes a clear position on the question of the solidity of the spheres', ibid., XI, p. 333.

40 Ibid.

In spite of his 'infinite obligations', Galileo replied without delay.[41] He was delighted to hear that the Prince was 'so occupied with the contemplation of the Copernican system and was inclined to prefer it to the Ptolemaic one, especially if the eccentrics and epicycles could be totally removed'.[42] His expression of satisfaction, if not of surprise, at the attention that Cesi was devoting to Copernicus confirms that Galileo knew that he had not succeeded in convincing Cesi of the truth of the heliocentric system during their intense discussions in Rome. This aspect has been neglected by studies on Cesi and his relations with Galileo. If Federico had been an unconditional admirer of the Tuscan scientist for some time, it does not mean that he had become a follower of Copernicus.

Before replying to Cesi's straightforward request for information on the position of Copernicus regarding the solid orbs of the planets, Galileo tackled the Prince's reservations on the eccentrics and epicycles because of the intricate and disorderly intertwining of spheres that they produced in the system conceived by the Polish astronomer. He abruptly swept aside the very principle on which the Prince's argument was based, that is the conviction that nature always works in ways that human reason perceives as simple, orderly and direct, programmatically avoiding any untidy or superfluous solution:

> We should not wish nature to accommodate itself to what seems better ordered and disposed to us; we should rather accommodate our own intellect to what nature has made, in the certainty that this is the best and only way. And since it has pleased her to make the stars move around different centres, we can be sure that such an arrangement is the most perfect and admirable and that any other would be lacking in all elegance, incongruous and puerile.[43]

To accept without difficulty the existence of motions along eccentrics and epicycles, it is enough to abandon the erroneous belief that nature operates according to our paradigms of order, simplicity and beauty. A similar, if not more serious, error is to pay heed to those (Galileo mentioned Lagalla as a telling example) who, setting out from the assumption that astronomy is a purely conjectural science, consider the motions along eccentrics and epicycles simply as hypotheses and fictions whose only purpose is to save the celestial phenomena.

41 Ibid., p. 344, letter to Cesi, 30 June 1612.

42 Ibid.

43 Ibid.

One may suspect that Galileo's brusque reaction was aimed at a target far more sensitive to Cesi than was Lagalla. Distancing himself from the Prince's anthropocentric conception of nature, Galileo aimed to strike at the heart of a person of a quite different scientific calibre, an author towards whom Federico had shown an evident inclination: Johann Kepler, who had developed the vision of a cosmos constructed in accordance with the rules of order, mathematical proportion, rationality and harmony based on the centrality of mankind in the plan of the divine Creation. It is telling that, after the discovery of the satellites of Jupiter, Kepler had wondered what influence they exerted on the inhabitants of the Earth.[44] Galileo was disappointed to discover that Cesi's wary approach to heliocentrism was by way of the obscure paths of the Imperial astronomer.

The existence of motions along eccentrics and epicycles had been conclusively demonstrated, Galileo continued, by the astronomical discoveries made with the telescope. Did not the the satellites of Jupiter, that revolves around the Sun, move in epicycles? Were not the orbits of Mercury and Venus around the Sun in the Tychonian system circles on circles? And how could the fact that Mars appears very close to the Earth at some times and very distant from

44 In his *Conversation with Galileo's Sidereal Messenger*, Kepler had wondered about the function assigned to the four satellites of Jupiter in the plan of the Creation, and in particular whether they were inhabited and exerted an influence on terrestrial creatures (ibid., III^1, pp. 120ff.: Kepler 1965, pp. 45ff.). Galileo's critique of the anthropocentric conception of Kepler emerges, although implicitly, also from the letter with which he replied, via Monsignor Dini, to the concern expressed by some astrologers of Perugia. They feared the complete overthrow of their profession as a result of the discovery of the four new celestial bodies whose influence had not so far been taken into account in astrological predictions. To defend themselves from that serious threat, they had supposed that the small scale of those new bodies rendered their influence negligible. That solution echoed the one advanced in his *Conversation* by Kepler (ibid., pp. 121–2), who had denied that the influence of the satellites of Jupiter could be perceived on the Earth because of their modest dimensions. If anything, it would have been perceived by the inhabitants that he believed were to be found on Jupiter. With this argument Kepler trusted to have removed the threat to the survival of astrology: 'For they [the celestial bodies] operate through the aspects, and an aspect is a disposition produced by the angle formed at the centre of the Earth or of the eye [...]. Now it is clear from your observations, Galileo, that these four planets are very tiny [...]. Hence [...] I am prepared to admit that, since their minute size offers no obstacle, they [Jupiter and its moons] coincide in the aspects for influencing terrestrial minds [...]. In this way. astrology maintains its standing' (Kepler 1965, p. 41: OG III^1, pp. 121–2).

it at others be explained if not by postulating the existence of eccentrics or epicycles?[45]

However, acknowledging the existence of eccentrics and epicycles did not mean admitting that the planets move along solid orbits: 'That nature requires solid orbs, eccentrics and epicycles to accomplish those motions I consider to be a simple work of our imagination, an absolutely unnecessary chimaera'.[46] On this point at least – that is on the fluid nature of the heavens – he satisfied Cesi's expectations.

Having imparted this stern lesson on the erroneous anthropocentric foundations of traditional cosmology (and, although implicitly, that of Kepler),

45 Galileo denounced once more the instinctive inclination to attribute to nature patterns of behaviour that seem logical and perfect to the human intellect in the final pages of his *Letters on sunspots*, where he held that the tenacious clinging to the incorruptibility of the heavens derived from fear of death: 'I suspect that our desire to evaluate everything according to our own meagre measure causes us to fall into strange fantasies and that our particular hatred of death makes us detest fragility […]' (Galileo-Scheiner 2010, p. 294: OG V, p. 235). On the distinction drawn by Galileo in various contexts between man and nature, and, in particular, on the indifference of nature vis-à-vis human expectations, as well as on his polemic against the widespread anthropocentric positions not only of his adversaries, but also of his friends and collaborators, see Torrini 1993, esp. pp. 34–6. For Galileo's insistence on the significance and implications of the inexorability of nature, which acts without paying the slightest heed to human expectations and mental paradigms, see Hamou 2008. Hamou provides an anthology of the numerous passages in which Galileo draws a sharp distinction between how humans expect nature to behave and the inexorable laws that govern its operations. On Galileo's anti-anthropocentrism see also Torrini 2002 and Stabile 2003. It should however be emphasised that to fully bring into focus Galileo's position on the relation between the way of thinking of man and the operations of nature, it is necessary to consider not only his numerous anti-anthropocentric pronouncements, but also the passages, often in the same works, in which he reveals an inclination to consider the principle of the harmonious, simple and orderly organisation of nature by no means unfounded. In at least one case, he went so far as to exclude, based on the rigorously rational foundations of its operations, the possibility that nature could have had recourse to the farrago of solid orbs, eccentrics and epicycles that Ptolemy devised to govern the motions of the planets. 'Now – Galileo-Guiducci wrote in the *Discourse on the comets* – this multiplicity of spheres, forever idly waiting in the hope that a comet should arrive, God knows when, to turn around in them for a brief period, through a small part of their circle, cannot reconcile with the extreme neatness which nature maintains in all her other works by retaining nothing which is superfluous or idle' (Drake-O'Malley 1960: OG VI, p. 87). On Galileo's ambiguous position with regard to the anthropocentric paradigm, see Galluzzi 1979.

46 OG XI, p. 345.

Galileo finally replied to the precise questions put to him by the Prince. With regard to his two diagrams, he declared that Copernicus had had recourse to the epicycles on epicycles as sketched in the first diagram as well as to the eccentric position of the Sun in the second. The Polish astronomer had not conceived of the orbs as solid spheres with the planets fixed in them, but as 'simple circles described by the revolutions of the stars'.[47] Galileo did not hesitate to superimpose his own convictions on the thought of Copernicus. Although he had not explicitly tackled the question of the solidity or fluidity of the heavens, the author of the *De revolutionibus* had used expressions from which it could be deduced that the solid spheres still continued to serve an important function in his system.[48]

Cesi had waited impatiently for Galileo's reply. He had given prior notice to Cigoli and Luca Valerio that, upon his return from a brief stay in Tivoli, they would meet with the other Roman Linceans to read and comment on it together.[49] The Prince's reply to the Tuscan scientist can therefore be considered the fruit of shared reflections with his associates, as had been the cosmological questions addressed to him. Cesi was not entirely convinced by Galileo's arguments. He recognised that the Copernican system offered 'great satisfactions', reducing 'the multiplicity of motions and spheres and their great and complex diversity'.[50] However, he would have been even more satisfied if all the eccentrics and epicycles had been eliminated. The human intellect is confounded when faced with such a farrago of spheres and revolutions, which appeared to be 'the abortion of some bizarre mind or a jumble of strange phantasms'. It 'willingly embraces the most pure and simple demonstrations, believing these to be the work of nature'.[51]

---

47 Ibid.

48 Copernicus 1543. Many passages of *Liber IV* in particular seem to suggest that the Polish astronomer considered that the planets moved fixed in solid spheres.

49 OG XI, pp. 365–7, letter of 21 July 1612. On this letter see Bucciantini 2003, pp. 207–12. The autograph letter lacks an indication of the addressee. Pelseneer 1932 (p. 201, n. 72) questioned that it was addressed to Galileo. At the request of Eric John Aiton, Stillman Drake dispelled Jean Pelseneer's reservations by showing that the document is a piece in the exchange of letters between Cesi and Galileo during summer 1612 on questions *de systemate mundi* (Aiton-Drake 1959, p. 152). It is also worth mentioning the earlier debate among authoritative Galileo scholars regarding the English translation by Erwin Panofsky of a passage from this same letter, which had seemed misleading to Edward Rosen (Rosen 1956, pp. 78–80). Drake defended Panofsky, showing that the translation was irreproachable (Drake 1957, pp. 65–6).

50 OG XI, p. 366.

51 Ibid.

So far Cesi's reply indicated his desire to cling to the anthropocentric paradigm so firmly rejected by Galileo. On one point, however, he declared himself in agreement with the Tuscan scientist: the planets were endowed with an intrinsic principle of motion and pursued their respective orbits in the fluid spaces of the heavens without being fixed in incorruptible crystal spheres. Granted the existence of souls that move the celestial bodies, Cesi had no difficulty in admitting that the planets proceeded along eccentric circles, epicycles or any other path. He acknowledged that Galileo was also right in inviting him to consider that 'nature has designed these [orbs] not according to the rules that we find beautiful, but in its own way'. Consequently, we should recognise and accept them as such considering them the most perfect solution.[52]

The Prince's cautious approval of Galileo's demolition of the anthropocentric arguments did not lead him however to adhere to the Copernican system, as the Tuscan scientist had hoped. If anything, it reinforced his inclination towards the solution proposed by Kepler in the *Astronomia Nova*. The Imperial astronomer had distanced himself from the perfectly circular nature of the orbits of the planets, which represented one of the most relevant anthropocentric principles of traditional astronomy. Moreover, he had resolutely demolished the solid spheres, maintaining 'that to bind the planets to the orbs was to attach them against their will to the mill-wheel and to compel them to follow a path from which they often escape'. It was based on Kepler, and perhaps on Patrizi, whom he read attentively, not on Copernicus that Cesi declared that the celestial bodies moved freely in the fluid spaces without being encrusted in solid spheres:

> And therefore I recognised with you many non-concentric motions neither around the Sun nor around the Earth, but some around the Earth, some around the Sun, and perhaps all, if the paths of the planets are elliptical, as Kepler asserts.[53]

Federico's position had not changed. He had perfectly understood that Galileo's defence of eccentrics and epicycles aimed to clear the field of Kepler's elliptical orbits, a solution that the Tuscan scientist did not like to see the Prince embrace. It did not escape Galileo's notice that the text in which Cesi affirmed the absurdity of tying the planets to the mill-wheel was quoted from a passage

52 OG XI, p. 366.

53 Ibid.

in the introduction to Kepler's *Astronomia Nova*.[54] To the Tuscan scientist's invitation to abandon the idea that nature operates according to principles that appear elegant, simple and rational to our intellect, the Prince responded proposing a solution that was bound to displease Galileo: should we not therefore subscribe to the Keplerian conception of the elliptical orbits of the planets?

Perhaps an awareness of the unbridgeable gap between their positions brought the dialogue between Cesi and Galileo on the structure of the universe and on the motions of the planets to a close. Galileo's reply to Cesi's second letter was evasive. He promised to send him his treatise on the system of the world which would answer his questions on the Copernican system.[55] Discussions between Galileo and Cesi on cosmological matters stopped here.

If still uncertain about the real structure of the universe, on one point the Prince manifested an unshakeable conviction: the celestial spaces are fluid, permeable and homogeneous. Any system of the world worthy of consideration had to assume this principle as a fundamental presupposition. This was why he had asked Galileo to clarify Copernicus' position regarding the existence of the solid spheres. For the same reason he had not disguised his own inclination towards Tycho's solution, which had the added value of avoiding the trauma of shifting the immobile Earth from its privileged central position, and, above all, towards that of Kepler, who, like Galileo, placed the Earth in motion around the Sun. Both Tycho and Kepler presupposed the penetrability of the heavens, and did not question the paradigm so dear to Cesi of the centrality of mankind in the plan of the Creation.

During the weeks in which he exchanged letters with Galileo on cosmological questions, Cesi had begun work on a treatise on celestial matters. As he had informed his correspondent, his questions regarding the Copernican conception arose from that endeavour, which he was developing 'in some of my notebooks to give a better foundation to my natural chaos'.[56] This is the first explicit reference to the *Coelispicium*, the text which was to occupy a privileged position in the projected *Theatrum naturale* (*Theatre of Nature*) and *Speculum Rationis*

54 Kepler 1609 (penultimate page of the Introduction). Explaining the tortuous path that had finally led him to discover that planets move along elliptical orbits, Kepler introduced the image, echoed by Cesi, of the planets tied to mill-wheels: 'under the spell of common opinion, I, under the fascination of the vulgar opinion, had tied them [the planets] as it were to mill-wheels'. Cf. Bucciantini 2003, pp. 208–11.

55 The content of Galileo's lost letter to the Prince can be reconstructed from Cesi's reply of 4 August 1612: 'I shall be pleased – Cesi wrote – to see what you promise me about the structure of the universe in your treatise' (OG XI, p. 370).

56 Ibid., p. 366.

(*Mirror of Reason*). Regrettably, the *Coelispicium* is lost. Nevertheless, as we shall see, the evidence that has survived clearly shows that Cesi would not have dealt with the problem of the real structure of the cosmos in that work. After his exchange of letters with Galileo in the summer of 1612, he would never make further pronouncements on questions *de systemate mundi*. He would, however, continue to fight for the recognition of the unitary and homogeneous nature of a fluid and corruptible universe. From this point of view, in his eyes, there was little difference between Telesio, Patrizi, Tycho, Kepler or Galileo: they were all perfectly aligned with this paradigm. On the contrary, the suspicion that Copernicus maintained the solid spheres prevented the Prince, in spite of pressure from Galileo, from adhering to the Polish astronomer's vision.

Books, Frontispieces, Theatres, Mirrors and Ladders

Cesi's letters to Galileo on the system of the world indicate that he was immersed in the study of celestial questions, guided above all by his eagerness to demonstrate that the heavens are fluid and homogeneous. That principle was the corner-stone of the ambitious project that he had been nurturing for some time.

The reference in the second of his letters to Galileo to the numerous notebooks in which he was giving form to what he called 'my natural chaos' is particularly significant. Federico's 'natural chaos' was the mass of extracts and notes deriving from his scrutiny of classical and contemporary texts. If the information stored in the notebooks was chaotic, Cesi's objective was to present it in an orderly and systematic way within an organic encyclopaedic treatise: the *Theatre of Nature* and the *Mirror of Reason*, conceived from the very start of his studies.

As we shall see, Cesi intended to devote a section of this treatise to expound the reasons for considering the heavens homogeneous and fluid and to highlight the consequences of that principle for the general interpretation of nature: an extremely important section to which he would devote many years of intense application. In his correspondence the Prince constantly defined that text as *Coelispicium*. On many occasions, he made it clear that he considered it not as an autonomous work, but as a tile in the ambitious mosaic of all knowledge on mankind and nature and their connections and filiations that he intended to represent in his *Theatre of Nature* and *Mirror of Reason*.

The only clues to the arrangement and themes of this lost (and possibly never completed) work can be found in an important document that testifies to Cesi's efforts to draw up the general structure and internal articulations of his encyclopaedic project. The document, traditionally known as a *Zibaldone* (commonplace book) because of its apparently chaotic organisation, bears a precise title that helps to bring into focus its nature and purpose: *Indicatio philosophicorum operum quae Federicus Caesius sibi condixit* (*Outline of the philosophical works that Federico Cesi plans to compile*).[57] With this title Federico meant to schematically account for his research project aimed at defining the fundamental categories of thought, the systematic organisation of nature, as well as the imperatives deriving from them for moral conduct. The character of the *Indicatio* as a research agenda was explained by Cesi in a gloss to the title in which he clarified that the articulations or chapters (he called them Tables or Frontispieces) of his encyclopaedic treatise

> are not only enumerated one by one, but also a brief account or rather designation of each one is given in the Frontispiece as an encouragement and constant stimulus to press on with the work undertaken.[58]

After this preamble, Cesi supplied other valuable specifications. The opening section of the *Indicatio,* entitled *Mirror of Reason*, would develop reflections of a more general character. He defined it as a universal synopsis (*typus synopticus universalis*), that is an observatory from which it would be possible to capture the entire articulation of nature and of human knowledge in a single glance. He further articulated the general structure of the *Theatre of Nature* in a series of Frontispieces: the sense organs with which man perceives nature; the instruments of reason; the orderly classification of knowledge, presuppositions and dogmas; and, finally, the moral benefits acquired through the dispassionate study of nature.

The schematic description is followed by a long *scholium* in which Cesi clarified the function of the *Mirror of Reason*. Its first goal was to emphasize the full correspondence between the correct way of thinking of man and the intrinsic articulation of nature, 'for we have supposed that the legitimate foundation of the workings of the intellect are the things themselves'.[59]

Following this principle, Cesi planned to establish a rigorous criterion of classification for all natural phenomena. Finally he would visualise the differences and divisions of all the sciences and disciplines in the form of a tree, from the most general categories to the ultimate components:

---

57 The *Zibaldone* was published by Gabrieli (Contributi I, pp. 27–77). For an analysis of its content, see Guerrini 2006, esp. pp. 21–87.

58 'Quae singillatim non solum enumerantur sed et brevis quaedam uniuscuisque ratio, sed potius designatio in simplex Frontispicium rudimento quodam exponitur ad propriam excitationem, et assiduo ad insumptos labores stimulum' (Contributi I, p. 30).

59 Ibid.

> We have established the articulations of everything from the general categories down to the ultimate particles of the ultimate branches, and we have divided up and described the classes in synoptic tables distinguishing them according to our own opinion, or as we found them in previous classifications.[60]

In the rest of this lengthy *scholium* Cesi provided further details on his plan for a general representation of mental categories and the organisation of nature. He would articulate the latter in a scheme showing the classes and families to which the various individuals belonged, their hierarchical order and the affinities or differences between them.

In the *Mirror of Reason* Federico intended to define the principles and methods underlying his ambitious project of integration of reason and nature, nature and the arts, nature and ethical values: a Universal Encyclopaedia, a mirror of mirrors, to be arranged in many closely linked Frontispieces.

The image of the Frontispieces directly echoed that of the Book of Nature. The Frontispieces would offer a panoramic view of that Book underlining its intrinsic unity, while the *Mirror of Reason* would make it possible to grasp the general articulation of the Frontispieces (in turn composed of Tables) and emphasize their connections.[61]

60 'Omnium membra a summitatum amplitudine ad ultimos usque ramulos ultimarum particularum constituimus, classesque prout ipsi dispeximus, prout in ea quae prius adnotata erant animadvertimus, partiti ex inventione sumus et synopticas inde tabulas descripsimus', ibid., p. 31.

61 'A very comprehensive synopsis [the *Mirror of Reason*] formed from the integration of practically all the tables' (ibid., p. 32). Cesi attached great importance to the synoptic visualisation of the matters under discussion in the Tables and Frontispieces. For this reason he took pains to give precise information in the *Indicatio* on how he would organize the printed edition of the *Theatre of Nature* and of the *Mirror of Reason*. The *Tabulae* would be structured on two levels: the first, *plenissimus*, consisting of an introductory text juxtaposed with 'the synoptic tables of all the parts [...] down to the smallest possible'; the second, *restrictior*, in which a single chart 'expounds all the faces and parts together at the same time at a glance, to furnish [...] the most concise compendium of the work itself and so that [...] this index will immediately fix them all fully illuminated in the mind' (Contributi I, p. 32). He went on to specify that these charts would be proposed in three formats: large, medium and small. Since the project was not completed, it is impossible to verify to what extent these prescriptions would have been respected. However, consideration of the Prince's works in the successive years suggests assigning the *Apiarium* to the category *plenissimus*, while the *Tabulae Phytosophicae* seem to afford an example of the *restrictior* format.

Having delineated the organisation of the *Mirror of Reason*, Cesi described the structure of the *Theatre of Nature*. Its initial Frontispiece showed the links and connections between the diverse areas of the world of nature. This would be followed by the table '*De scientiarum nexu, mutuisque respectibus*' (On the connections and mutual dependences of all sciences). This pair of tables was followed by the Frontispiece of the *Physica Mathesis*, which appears finely articulated in classificatory schemes deliberately contained within the space of a page to enable them to be taken in at a single glance.[62]

In this Frontispiece Cesi proposed an orderly classification of the aspects of nature that can be subjected to physical or mathematical enquiry. However, the dense enumeration of principles and elements (figure, composition, movement, light, fire, water, etc.) offers no clues as to any real intention of conducting precise quantitative analyses. Moreover, terms such as *ratio* or *proportio* are never found in the tabular prospectus of this Frontispiece, while the term *mathematica* occurs only in the category *de figura*, which Federico breaks down into *de figuris mathematicis/naturalibus*.[63] All the relations that Cesi establishes between the various categories of the phenomena under consideration appear to be of a qualitative nature. The title *Physica Mathesis* should therefore be understood as a tribute to the Galilean research model rather than as a methodological criterion that the Prince actually intended to follow. The same ambiguity between declarations of method and actual style of analysis will reappear in the *Tabulae Phytosophicae*.[64]

The Frontispiece of the *Physica Mathesis* evokes above all themes and motives of the Renaissance philosophy of nature, also in the terms used to underline the causes of interactions between natural substances (friendship, sympathy, opposition etc.). No less evident appears the presence of models and analytical procedures inspired by the Paracelsian tradition and chemical philosophy (*de solutione, de fluore, de solutorum in alia impetu, chimica naturalis*, etc.).

The graphic solution adopted by Cesi to present the categories of the Frontispiece of the *Physica Mathesis* deserves attention. He placed the most general categories on the left side of the page, highlighting their articulation in the right margin. For example, the general category *de lumine* on the left was broken down into *de luce calida, de luce frigida* and *de coloribus* on the right.[65] Recognising that this arrangement was anything but transparent, Federico

62 Ibid., pp. 33–5.

63 Ibid., p. 33.

64 See below, Chap. 12.

65 Ibid., p. 34.

stressed that clarifications and further comments would be provided in the next Frontispiece, where he would address celestial questions. This indicates that he did not consider the treatment of natural elements in the *Physica Mathesis* as a field of research separate from the study of celestial phenomena. He took pains to point out that there was no break in continuity between the Frontispieces of the *Physica Mathesis* and that dedicated to the *Coelispicium*. The latter was preceded by a connecting table containing illuminating declarations of principle: the cosmos is unitary; the same principles apply in the heavens as in the sublunary world; nature is organised as a finely graduated ladder without interruptions. Thus he had designed

> a table in which different columns for different entities converge in rungs from opposite sides, so that heat appears at the head of the column, cold at its base, and the intermediate steps closer to or further away from them are recorded.[66]

In this way the image of the Book of Nature with its Frontispieces was assimilated to the equally ancient and fortunate one of the Ladder of Nature where all creatures are connected without any break in continuity according to the functions that they perform in the general economy of nature. The table of the rungs of beings introduced the Frontispiece of the *Coelispicium*, explicitly drawing attention to its close link with that of the *Physica Mathesis*. Cesi entitled this section 'Caelispicium De caelo: De natura caelesti'.[67]

The structure of this Frontispiece shows the Prince's intention of proceeding to a general reform of celestial philosophy in order to shed light on the effective nature and substance of the heavens, the *caelum naturale* as he defined it. After listing the bodies and qualities present in the sky (stars, ether, air, heat, fire, light), Federico provided summary but clear information about the new image of the universe that would emerge from his reform of traditional cosmology. The sky would appear unitary, permeable like the air, immobile, and traversed by bodies that move in it freely.[68]

66 'Tabula in qua pro diversis conditionibus columnae diversae pariter producuntur ab extremis oppositis gradibus procurrentes, ut in eiusdem columnae capite calor, in basi frigus, intermedia in excursu gradus, prout ab illa magis recedentes, vel accedentes notantur', ibid., p. 36.

67 Ibid., p. 37.

68 Ibid.: 'Unum pervium alterabile idem, fere cum aere non a naturae opere segregatum, immobile motu lationis firmae et stabilis, sed in quo omnia moveantur. Omnia mersa sint et compraehensa corpora'.

He stressed how the themes dealt with in this Frontispiece formed the 'apex of natural philosophy'. Man could attain this peak by climbing all the rungs of the Ladder of Nature along two different but necessarily complementary paths: physical and mathematical inquiry (based on experiment, observation and detection of the causes of phenomena) and the interpretation of the oracles in Holy Scripture.[69]

He went on to provide an illuminating explanation: the aim of his research was not to outline the real system of the world but to establish the general concepts and principles indispensable for the study of the motions of the celestial bodies and for determining their reciprocal positions. Only by patiently following this course, one step at a time, would it be possible to remove paralogisms and to prepare the mind for the crucial question of the true structure of the universe:

> not by defining and establishing, but by encouraging, prompting and assisting philosophers so that, abandoning all other presuppositions, the only thing to be proven and declared is what motion and what place in the universal machine pertains to each body. For the proportion of their motions and their relative positions, as well as the nature itself of the celestial bodies and of their emanations, is abundantly clear.[70]

The differences between Cesi's vision and that of the Tuscan scientist are clearly perceptible. Like Galileo, Cesi wanted to found the new cosmology on the demolition of the paralogisms of the traditional one, but the task that he assigned the *Coelispicium* had nothing in common with the Tuscan scientist's agenda. Federico remained one step behind. Before advancing an organic interpretation of planetary motions, of the reciprocal positions of the celestial bodies and, above all, of which body occupies the central position in the system of the world, it was necessary to reply to fundamental questions such as whether the sky is fluid or solid, what is the nature of fire and light and how they are to be distinguished.

Cesi would never overstep the limit that he had set himself in the early outline of the *Coelispicium*. He aimed at demolishing the errors of tradition and bringing into focus the principles and rules required to set cosmological

69 Ibid.

70 Ibid.: 'Ipsum ad usque systema illud quidem non definiendo stabiliendoque, sed ad illud philosophos provocando, excitando, ducendoque reliquis omnibus praeparatis, ut id unicum dumtaxat comprobandum, contendendumque remaneat, quis cui corpori motus, quis respectu universalis machinae situs debeatur. Nam motuum ratio, nec non situum inter se, naturaque ipsa supernorum et illorum quae inde emanant abunde perspicitur'.

research on a solid foundation. This would be made possible by the programmatic integration of the methods of physical and mathematical inquiry with those of biblical hermeneutics. As we shall soon see, this 'concordist' programme was significantly different from Galileo's conception of the relations between the Book of Nature and the Book of Holy Scripture.

It is difficult to imagine that, while defining this original programme of research, Cesi was not aware of the distance that separated it from the vision and the expectations of Galileo. It seems likely that his continuous recommendations to the Tuscan scientist not to act in haste were intended not only to protect him from the dangers that he appeared to underestimate but also to incite him to work patiently with the Prince in elucidating the principles and concepts indispensable to face with any hope of success the challenge of designing a system of the world that faithfully reflected the 'natural sky'.

It is thus hardly surprising that the key questions in the programme outlined by Cesi were not those valued by Galileo concerning the motions of the celestial bodies and the central position of the Sun, but those bearing on the solid or fluid nature of the heavens, on the origin of *novae*, comets and sunspots, on ethereal heat and fire, on light, on air, on celestial waters, and on the formation of clouds.

The Prince's plan and its theoretical foundations emerge very clearly from the concluding passage of the section containing the outline of the Frontispiece of the *Coelispicium,* where he declared that there is a force (*vis residens*) that permeates the universe. This force operates in the celestial spaces exactly as it does on and below the surface of the Earth.[71] Investigation of the nature and mode of operation of this pervasive force afforded unexplored and sublime fields to natural philosophy, removing the scepticism of those who considered this exercise tedious and futile. Finally, he emphasised that acceptance of the fluid, homogeneous and unitary nature of the celestial spaces as well as of those that surround the Earth was an essential premise for the understanding of the 'natural sky':

> Certainly you exclude nature if you exclude its mutations. We do not recognise and consider the sky of others [philosophers] to be natural, but non-natural, i.e. solid, incorruptible and adamantine.[72]

---

71 'Inferne superneque', ibid., p. 38.

72 'Naturam certe ipsam excludis si mutationes excludis et praesertim alteratur. Non certe naturale caelum agnoscimus et contemplamur, aliorum caelum non naturale videtur, nempe, immateriatum, incorruptibile, adamantinum, ibid.

The Frontispiece of the *Coelispicium* is followed by the Table dedicated to one of the constitutive elements of the universe: air. Cesi believed that it was necessary to investigate and establish its nature prior to any serious reflection on the system of the world. He proposed examining air with the same method of passing from scrutiny of its more general properties to that of the particular phenomena in which this element plays a major role, including the beneficial effects of particular types of air on health. The same model was to be applied to the study of all natural phenomena (heat, fire, light, etc.). His choice of air as an exemplary model of inquiry is significant: in his vision of the unity and homogeneity of the heavens, air was the element in which the continuous exchange of substances and influences between the sky and the Earth occurred.

The *Coelispicium* occupies a strategic position in the sequence of the Frontispieces of the *Zibaldone*. It precedes the Frontispiece entitled *Taumatombria* (*On marvellous rains*), in which Cesi – as the title suggests – intended to analyse the natural productions generated thanks to substances that fall from the sky. At first glance, one may receive the impression of a radical change of scenario. This, however, is not the case: Cesi forcefully emphasizes the absence of a break in continuity between the two Frontispieces. If the *Coelispicium* concluded with the declaration of the existence of a force that pervades the universe, the *Taumatombria* set out just from the concept of the marvellous forces embedded in nature.[73] In the opening statement of this section of his *Theatre of Nature* Cesi defined the topics that he would analyse in the *Taumatombria*:

> On those substances that fall from the sky.
> A universal history, compilation and representation of all prodigies, followed by an explanation of their causes and the forces of nature. The sequence and classification start from the prodigious rains and other substances that fall from the sky.[74]

The analysis of these natural phenomena, he stressed, should be considered not only contiguous, but intrinsically related to that carried out in the *Coelispicium*. He had conceived the classification of meteorological phenomena

73 'De prodigiosa natura, de naturae mirandis, de admirandis naturae viribus', ibid.

74 'De iis quae e caelo labuntur praecisius. Prodigiorum omnium universalis historica collectio et repraesentatio, inde causarum et naturae viribus explicatio. Ordo autem et digestio ab imbribus admirandis aliisque e caelo cadentibus sumitur', ibid., pp. 39–40.

as a ladder descending rung by rung from the noblest celestial contemplations to the most minute terrestrial phenomena.[75] For Cesi the absurd Aristotelian treatment of celestial questions in his *De Coelo* distinct from that of meteorological phenomena derived from his erroneous dualistic conception of the universe. The Prince's profound conviction of the unity of the cosmos had led him to distance himself from the Greek philosopher's vision and to develop his investigation of celestial and meteorological questions in two directly connected sections of the *Theatre of Nature* employing the same methods and principles.

After this illuminating premise, Cesi concentrated his attention on one category of phenomena which in the *Taumatombria* played the same role of methodological testbed as air in the *Coelispicium*. He defined this category as *De succis oposcopia* (autoptic inquiry into natural juices that fall to Earth from the sky). His tabulation of these miraculous celestial rains would examine in detail, in order to reveal their causes and properties, dew, honey, manna, sugar, wax, a mythical substance called *laserpitium*, gums, resins, as well as all those earthly products generated by the combined action of celestial influences and substances that fell from the sky. This section of the *Taumatombria* also considered extraordinary luminous phenomena: *lux furtiva*, rainbows and parhelia, nocturnal and crepuscular light, as well as the Bologna stone and the method to calcine it ('de arte in lapide lucifero'). In the hands of the Prince, the stone had travelled from its close relationship with the telescope, the nature of light and the celestial discoveries in the palaces of the Roman nobility to its role in the analysis of prodigious rains.

The importance attached by Cesi to the interpretation of these phenomena is confirmed by the letter to Galileo on 30 January 1614. It was penned in reply to a lost letter from the Tuscan scientist, in which he had manifested the conviction that parhelia were not, as Aristotle had argued, products of the combustion of terrestrial exhalations and vapours in the higher reaches of the Earth's atmosphere, but optical illusions. Cesi did not hide his satisfaction at the authoritative verdict. It confirmed the arguments that he had raised against the fallacious opinions of Lagalla, the indefatigable advocate of the Aristotelian explanation of parhelia. Federico interpreted parhelia and comets as optical illusions produced by the reflection of the solar

75 'Quod in caelestibus admiratio maior eaque longe digniora notioraque, tum quod nobis a caelesti contemplatione Philosophiam exsorsis, inde ad globum descendendo, haec primum occurrunt quae quodammodo medias inter caelum terramque obtinere videri possunt', ibid, p. 40.

rays on terrestrial exhalations that rose beyond the sublunary sphere. In his opinion, parhelia offered fresh evidence that there was no solution of continuity between the topics discussed in the *Coelispicium* and those in the *Taumatombria*:

> I could not easily express the pleasure I felt in reading what you wrote to me about the nocturnal lights. I contradicted Lagalla on the parhelion, obtaining his agreement on denying the conflagrations and even beginning to convince him to accept the same explanation for the comets, on which he had previously sided with the Peripatetics. In my contemplations of prodigious phenomena I had already explained what I thought about these lights and red gleams: in opposition to the Peripatetics, I had argued that, as vapours sometimes rose above the sky of the Earth, they produced that appearance when light fell on them.[76]

Federico boldly claimed to have reached those conclusions himself, before receiving Galileo's opinion. Although he did not mention Telesio, it is highly likely that he had drawn inspiration from the interpretation that Telesio had given of those phenomena in his *De iis quae in aere fiunt et de terremotibus*, which Persio had inserted in the collection of his Master's writings published in Venice in 1590.[77]

We are now in a position to understand the chain of reasoning that guided Cesi in the organisation of the research programme outlined in the *Zibaldone*: from abstract reflection on the symmetrical correspondence between the categories of human intellect and the articulations of nature to the hierarchical and relational classification of physical elements; from here to the analysis of a number of celestial phenomena (programmatically avoiding the problem of the system of the world), to finally arrive, without any break in continuity, at the investigation not only of meteorological phenomena but also of those observed both above and beneath the surface of the Earth.[78] As other evidence confirms, fossils already excited the Prince's curiosity and imagination at the time of his compilation of the *Zibaldone*.

---

76 OG XII, p. 18.

77 Telesio 1590, pp. 14–27.

78 At the foot of the ladder that represented the extraordinary phenomena described in the *De succis oposcopia* Cesi had appended the continuous transformation taking place below the surface of the Earth: 'Admirandae Montium fossiones in aerem turbificae' (Contributi I, p. 40). See below, Chap. 10.

The Frontispiece on intermediate natures ('de mediis naturis') that came immediately after the *De succis oposcopia* casts further light on the Prince's design. Cesi called intermediate those creatures or phenomena characterized by properties and functions pertaining to different species, such as mammalian fish or fossil wood. As ambiguous or ambivalent natures, they occupied an intermediate position on the Ladder of Nature between mammals and oviparous creatures, or between plants and minerals. The words used to define this area of research indicate the particular importance that it held for him: 'A remarkable corner of the Theatre of Nature, all the more hidden, concealed and obscure because overlooked'.[79]

The Prince was engaged in compiling the *Indicatio* during the same months in which he was discussing with Galileo the world systems of Copernicus, Tycho and Kepler, the orbits of the planets, the fluidity of the heavens and the nature of sunspots. A few weeks after their exchange of opinions on cosmological matters, he informed Galileo that he had completed

> a work, which I have [...] called *Coelispicium*, which contains many celestial matters, as you will see, which I am also going through theologically, and I argue particularly against the solidity and hardness and multiplicity of spheres and the excessive number of movements.[80]

At the same time, as a direct development of the *Coelispicium*, he was attending to the compilation of the *Taumatombria*. This is confirmed by his letter of a few months later in which he asked Galileo to furnish him with a detailed description of a meteorite that had fallen in Florence and about which the Tuscan scientist had spoken to him in Rome.[81]

In spite of his repeated promises, Cesi never sent Galileo the *Coelispicium*. Nor do we find in his correspondence with Galileo any explicit mention of the projects of the *Mirror of Reason* and the *Theatre of Nature* to which he was devoting a great deal of energy. The *Mirror of Reason* and the *Theatre of Nature* were the linear development of the visions and expectations that had motivated him to create the Lincean Academy in 1603. It is not fortuitous that the personal emblem Federico had assumed in that year was the eagle clutching the Earth with its talons and gazing at the Sun with the motto *uterque*: a

79 'Angulus quidam mirabilis naturalis Theatri, magis reconditus et abditus, magis quoque praetermissus obscurior', ibid.

80 OG XI, p. 403, letter of 29 September 1612.

81 See below, p. 216.

symbolic image that perfectly expressed his conviction of the absence of any break in continuity between the heavens and the Earth.[82]

Cesi was aware that the models and goals pursued by him and his Academy, rooted as they were in Della Porta, Telesio, Paracelsianism and the Renaissance philosophy of nature, were distant from Galileo's vision. This is why he never let him read the *Coelispicium*. To communicate it as an autonomous text would have profoundly distorted its significance. On the other hand, he feared that revealing the central position occupied by the *Coelispicium* in the full organisation of the *Indicatio* would cause Galileo some embarrassment. Thus it is not surprising that the outline of Cesi's ambitious project was to remain unknown to the Tuscan scientist.

However strange it may appear, it was precisely in these years of intense and affectionate correspondence between the two men, both passionately convinced of the need for a radical reform of traditional knowledge, that the paths of Cesi and Galileo came to resemble two parallel lines proceeding in the same direction but never meeting.

Embarrassing Omissions

The *Letters on sunspots* is the first work of Galileo published under the aegis of the Linceans and the second official publication of the Academy, preceded only by Heckius' *De stella nova disputatio* published in 1605. In the *Letters* Galileo adopted a far more explicit position in favour of the heliocentric conception than he had done in the *Sidereus Nuncius* or in his work on hydrostatics of 1612. In the months between the arrival of the *Epistolae* by pseudo-Apelles in Galileo's hands (February 1612) and the publication of the *Letters on sunspots* (early April 1613), the correspondence between Cesi and Galileo reached its peak. Unfortunately, almost all the letters written by Galileo are lost. Besides, as Federico's polite but continuous complaints indicate, he failed to answer many of the letters sent to him by the Prince. Cesi and his associates in the Academy displayed admirable energy in first begging Galileo to complete the reply to Apelles, and then in sparing him the unpleasant and delicate task of facing the keen-eyed censors, with whom Federico negotiated to obtain the *imprimatur*.

The process of publication and the content of the *Letters on sunspots* have been the object of penetrating research in the last decades, the conclusions of

82 See above, p. 38.

which there is no point in repeating here.[83] It is of more interest to focus on the idea that Cesi formed of this text and on the reflections that it stimulated in his mind. Apart from innumerable manifestations of admiration, expressions of scorn for the Jesuit who wanted to deprive Galileo of priority in making the discovery, and declarations of solidarity against the attacks that his adversaries were launching against the Tuscan scientist, no explicit comment on this work is to be found in his correspondence. One is struck in particular by the absence on the part of the Prince of any reaction, expression of approval, reservation or personal evaluation of the evidently Copernican implications that Galileo attributed to the discovery and interpretation of the nature of the sunspots.

This is not to say that Cesi did not share those orientations, but rather that he did not regard them as the most important feature of that work. He found himself in complete agreement with other aspects of the text, above all with the clear position adopted in favour of the fluidity and corruptibility of the heavens, demonstrated almost experimentally by Galileo's observations of the sunspots.

The Prince dedicated himself to the task of convincing the censors to authorise those passages in the *Letters on sunspots* in which Galileo declared that the corruptibility of the heavens was confirmed by Holy Scripture. It was in the course of these animated discussions with the censors that Cesi came to realise that the defensive strategy of the Peripatetics and the ecclesiastical authorities would fall back on a firm denial above all of the corruptibility of the heavens: 'they will have less difficulty in conceding that the heavens are fluid than that they are corruptible and susceptible to change'.[84]

In his letter of 10 November 1612 he informed Galileo that, to overcome the resistance of the censors to approving the agreement between the fluidity and corruptibility of the heavens and Holy Scripture, he was engaged in the interpretation of those passages of the Sacred texts that seemed to be contrary to that conception. In doing so he was resuming analyses and reflections that he had developed in the *Coelispicium* .[85]

The pressure exerted by the Prince and his arguments were in vain. The keen eye of the censors also rested on the biblical verse ('the kingdom of heaven

83 Camerota 2004, pp. 238–59; Bucciantini 2003, esp. pp. 206–44; Reeves 2005, pp. 185–210. For the English translation, with introduction and notes by A. van Helden and E. Reeves, of the *Letters on sunspots, Tres epistulae* and the *Accuratior disquisitio* of Scheiner see Galilei-Scheiner 2010.

84 OG XI, p. 394, letter of Cesi to Galileo, 14 September 1612.

85 OG XI, p. 429.

suffereth violence, and the violent take it by force'),[86] that Welser had adopted as epigraph to the letter in which he had invited Galileo to state his views on Scheiner's interpretation of the sunspots. In refusing to grant permission to print the verse, the censors suppressed not only the bold affirmations on the corruptibility of the heavens formulated by Galileo, but also the innocent intellectual exercise of a German scholar who was very close to the Jesuits, a personal friend of Scheiner and extremely hesitant in the face of the celestial novelties. The biblical verse had to be replaced by an innocuous Horatian one chosen by Cesi.[87]

Galileo had spared no effort in trying to overcome the obstinate opposition of the censors, urging them to accept at least the alternative epigraph (lost) that he had conceived. Cesi wrote to him that he had no illusions about the possibility of overcoming such a clearly prejudiced hostility:

> I agree with you that the censors should reasonably have been satisfied with the toned down version [of the epigraph] that you sent me; but the only method to follow is that of trying to gradually remove the tyranny of the Peripatetics. I wrote *in iure*, adducing ten passages of Scripture and the same number from the Holy Fathers in confirmation of what you claim, that the corruptibility of the heavens is in agreement with Scripture and clearly indicated by it. To no avail [...]. We must be patient: they do not want anything from Scripture to be included there.[88]

Cesi's initial optimism had evaporated. According to his plan, the *Letters on sunspots* would have included also letters from the Jesuit Fathers Clavius and

86 'Regnum coelorum vim patitur et violenti rapiunt illud' (*Matthew* 11:12).

87 The 'offending' epigraph was to be replaced by a quotation from Horace that appeared less compromising to the defenders of celestial immutability: 'Virtue, opening heavens to those who do not deserve to die, makes her course by path untried' (*Carmina* 3.2: 21–2; Galilei-Scheiner 2010, p. 87). In an attempt to sweeten the bitter pill, Cesi had offered Galileo the alternative of a verse from Petrarch: 'Already the minds of men are making an attempt on the heavens, and the bravest are now conquering them', adding 'You are the first to scale them, etc.' (OG XI, p. 438). It is not known whether Cesi's proposal was rejected by the censors or by Galileo. When Faber informed Welser of the change that they had been forced to make to his epigraph, Welser expressed his full comprehension: 'changing the passage [...] requires no excuse on your part, but I am grateful for it: because what would be the point of leaving those words in if they might displease the Master of the Sacred Palace?' (OG XII, p. 483).

88 Ibid., letter of 30 November 1612.

Grienberger to certify, in all probability, that they had witnessed Galileo's demonstration of the sunspots in Rome in the spring of 1611, that is, before the observations of Scheiner.[89] However the two Fathers changed their minds and withdrew their initial assent. Furthermore Cesi had had the idea of adding two letters to Galileo's text, to be written by himself under a pseudonym and in the form of a dialogue in which the Peripatetic was won over by the arguments of the follower of the new ideas.[90] Federico's letters have not survived, but it is certain that he defended in them the fluid and corruptible nature of the heavens on the basis of the arguments developed in the *Coelispicium*. After reading Cesi's letters, Galileo adduced instrumental motivations to convince the Prince to relinquish his plan.[91] Although it is impossible to establish the reasons that led Galileo to decline Cesi's generous offer of support, he was probably embarrassed to see his own conception of the fluidity and corruptibility of the heavens 'contaminated' by the functions assigned by the Prince to that doctrine in his *Taumatombria*.

The correspondence of these months reveals symptoms of divergences of opinion between Galileo and Cesi. The Prince, supported by Welser (who motivated his request with the need for Scheiner to read the work) and Cigoli, insisted that the *Letters on sunspots* be published in Latin,[92] but Galileo turned a deaf ear.[93] For Cesi the vernacular represented an impoverished alternative

89 Ibid., p. 444, letter of Cesi to Galileo, 1 December 1612.

90 Ibid., p. 458, letter of Cesi to Galileo, 4 January 1613.

91 'I have to inform your Excellency that, after showing the two letters you sent me to various erudite friends, they have been either declared fictitious, or by the same author [Galileo], or by you, which has greatly surprised me. The same opinion was expressed here [in the villa 'alle Selve' of Filippo Salviati] by Sig.r Salviati, to whom I confessed everything in confidence [...]. He considered that, if they were printed, they could be used by my detractors to launch a terribly fierce attack, declaring that to palliate my falsehoods I was forced to try to deceive the world with fictions and frauds [...]. So [to Sig.r Salviati] it seemed better to give the content of these letters [...] a more direct form not giving a foothold to the malign' (ibid., pp. 468–9, letter of 25 January 1613). At Galileo's explicit invitation to unmask the author of the letters, Cesi retreated (ibid., p. 475, letter to Galileo, 1 February 1613). The letters that Federico had invited Clavius and Grienberger to include in the volume met with the same fate. Grienberger was the first to get cold feet ('Father Grienberger was not very satisfied for his letter to be printed', ibid.). Afterwards Clavius too abandoned Federico ('The letters of Clavius and others will be left out': ibid., p. 481, letter of Cesi to Galileo, 15 February 1613).

92 Ibid., p. 420, letter of 28 October 1612.

93 Ibid., p. 408, letter of Welser to Galileo, 5 October 1612; ibid., p. 418, letter of Cigoli to Galileo, 19 October 1612.

to the noble classical language. Galileo defended his position by stressing that Latin lacked adequate terms for new phenomena like the sunspots. His arguments in favour of a language in constant evolution, echoed those in defence of the vernacular of the Accademia della Crusca. Thus it is not surprising that Cesi directed his criticisms at the Florentine Academy:

> Indeed, if the Cruscanti valued our language above Latin, from which it is derived, I think they would be making a great mistake. Latin likes to adorn itself beautifully with Greek words. But what am I to say? In a similar way, our language takes all the scientific names and terms from Latin; and if it has not yet borrowed those that we are now trying to compose [sunspots], it is because it has not had the opportunity to do so.[94]

The Prince's vehement disapproval of the determination with which the Accademia della Crusca proclaimed the superiority of the vernacular will certainly not have pleased Galileo, who, as Cesi must have known, had been a member of that Academy since August 1605:

> I am also sorry for the difficulty that my having written in our Florentine tongue has caused Apelles – the Tuscan scientist wrote in the third of his *Letters on sunspots*. I did this for several reasons, one of which is that I do not wish to somehow neglect the richness and perfection of a language that is adequate for treating and explaining the concepts of all disciplines and therefore writing in this idiom rather than in any other is most appreciated by our Academies and by the entire city.[95]

Cesi had the support of Cigoli in yet another episode that reveals the different sensibilities of the Prince and the Tuscan scientist. Federico begged Galileo, also in the name of the Tuscan painter, to limit his attack to Scheiner, restraining himself from negatively judging the entire German nation, as he had evidently done in the manuscript of the *Letters on sunspots* sent to Rome: 'It is a nation – Cesi objected – that is very fond of learning and of *literati* [...] whom it celebrates; thus the Linceans in particular must treat them [the German scholars] as friends: they philosophize free from prejudices'.[96]

94 Ibid., p. 420.

95 Galilei-Scheiner 2010, p. 256 (OG V, pp. 189–90).

96 OG XI, p. 422.

It is not difficult to understand Cesi's irritation. The *Letters on sunspots* were dedicated to an influential German, Markus Welser, who had recently been made a Lincean.[97] The German component of the Academy had been and would remain significant, both in terms of numbers and as a point of reference for the definition of its programme. The Chancellor of the Academy, Johannes Faber, who was doing all he could (he had maintained contacts with both Welser and Scheiner) to assist the Tuscan scientist in publishing the *Letters*, was also a German.

In another instance Cesi delivered a severe rebuke to Galileo for his apparent disregard for academic procedures. Opening the copy of the book on floating bodies sent to him by Galileo, he came in for a shock: the frontispiece made no mention of its author's membership of the Lincean Academy in spite of the fact that, on Galileo's induction in April 1611, the Prince had laid great stress on the rule which bound all members to explicitly declare their academic affiliation in all published works.[98]

In his letter of 6 October 1612, in which he acknowledged receipt of two copies of the discourse on *Bodies that stay atop water and move in it,*[99] he informed the Tuscan scientist that he had suspended distribution of the work until the *Letters on sunspots* had been published. He was held back by the fear that circulating the text on floating bodies might create the impression that Galileo paid scant heed to the Academy:

> I will send it [the discourse on floating bodies] to Della Porta and the other Linceans after the solar letters are printed, in order that, seeing the name 'Lincean' there, they will know that it was not omitted from that text on purpose. This is why I have postponed the distribution of the book.[100]

---

97 Welser was inducted on 3 August 1612 (Contributi I, p. 520).

98 Galileo was not the only one to disregard this rule. Although they were funded by Cesi, the frontispieces of the two works of Della Porta published after his induction into the Academy (Della Porta 1610a and 1610c) lacked the emblem of the lynx. In those publications, however, the feline did appear in the modified emblem of Cesi (see above, pp. 39–42). Moreover, Della Porta was called by Demisiani *Neapolitanus Lynceus* in his dedicatory poem to the *Elementorum curvilineorum libri tres* (Della Porta 1610c).

99 Galilei 1612 and 1981.

100 OG XI, p. 409.

The embarrassing omission was compensated *ad abundantiam* in the printed edition of the *Letters*. The crowned lynx dominated the frontispiece.[101] The work was dedicated to the Lincean Filippo Salviati by the fellow Lincean Angelo De Filiis. The Academy was frequently evoked in the dedication and in the preface, although, as would also be the case in the *Assayer*, the Prince's name was not mentioned.[102] This reticence to put himself in the limelight is a recurrent character trait of Federico. During his lifetime his name appeared in only one of the works published by the Academy, the *Apiarium*, and even then without any particular emphasis. However the absence of any reference to the Prince in the *Letters on sunspots* is striking. It is difficult to attribute it to reasons of prudence like those motivated in the *Assayer* by the condemnation of the Copernican system in 1616. Although the celestial novelties had met with violent reactions before 1613, the ecclesiastical authorities had not yet formally rejected them. Nonetheless, Cesi declined Galileo's offer in a (lost) letter to dedicate the *Letters on sunspots* to him motivating his refusal with clumsy and contrived arguments: he claimed to find himself in a conflict of interests, as in his *Coelispicium*, the imminent publication of which he announced to Galileo, he upheld the same theses as those contained in the *Letters on sunspots*. He suggested that Galileo dedicate the work to his patron the Grand Duke instead:

> As for the dedication of your solar observations, I agree with you on the dignity and nobility of the work, which I value highly and know will be held in high esteem by all; and I would do what remains [i.e. accept the dedication] if I did not find myself in the situation of having almost

101 Moreover, Francesco Stelluti made an implicit reference to the feline in the celebratory dedicatory ode of the *Letters on sunspots*, in which Galileo was praised for having been the first to discover 'the pure cloak dotted with spots' of the Sun (OG V, p. 92). The term *cloak* recalls the pelt of an animal – like the spotted fur of the lynx – rather than the surface of the Sun. It was an effective way of underlining that the lynx of the Academy had prophetically announced the spotted Sun (cf. Galluzzi 2014).

102 Cesi's name was mentioned only in the letter of Welser to Galileo of 5 October 1612 (published at the opening of the *Terza Lettera* in the *Letters on sunspots*), in which the German Lincean referred to Federico's intention to print the solar letters: 'And Prince Federico Cesi will act in a manner worthy of his station and profession [...] in printing the two letters [...] as I understand he has resolved to do' (ibid., V, p. 184). However, replying to Welser in his *Terza Lettera* Galileo stated that he was unaware of the Prince's intention to publish his letters on the sunspots, which, he added, had not been written for publication (ibid., p. 188).

> completed my treatise, in which I fully demonstrate the importance of the celestial observations and novelties discovered by you and the obligation that all learned and studious persons must feel towards you [...]; as it is a work of yours, I thought from the first that it should be dedicated to the Grand Duke. Thus, since I must fully satisfy myself and be of service in my own work, I could not repeat it in a simply dedicatory letter.[103]

A curious argument indeed! Everyone knew that the Linceans De Filiis, Salviati and Welser, who featured prominently in the frontispiece and in the opening pages of the *Letters*, acted on behalf of and in the name of the Prince, as well as that it was he who had provided the funds necessary to print a work that had become highly expensive due to the numerous images of sunspots engraved by Matthias Greuter after drawings by Cigoli.

Cesi was reluctant to expose his name in public. It became him to act as a discreet patron rather than as a direct protagonist or author, especially when dealing with matters that were challenging the traditional system of knowledge. Patrons are usually pleased to accept, or even impose, a prominent role for themselves; Cesi preferred to fling the stone and hide the hand.

This prudent strategy was adopted to protect not only himself but also the Academy. He was perfectly able to discern the clouds gathering on the horizon. The *Letters on sunspots* picked a fight with a member of the Society of Jesus, an institution with which Cesi did not want his Academy to clash.[104] He knew that among the Jesuits there were authoritative Fathers, such as Christoph Grienberger, who were by no means hostile to the Galilean interpretation

---

103 Ibid., XI, p. 403.

104 The fact that the founders of the Academy demonstrated a strong competitive spirit with the Society of Jesus does not mean that they did not admire the compactness and determination with which the Fathers pursued their Order's goals. Cesi sought to maintain close ties with the Fathers, at least down to the death of Bellarmine in September 1621 and the explosion of the polemic on the comets. However, he was highly critical of the objectives of the Jesuits and of their methods to attain them. The first severe judgement on the Fathers of the Society is to be found in the sonnets composed by Francesco Stelluti during his exile in Parma (1605–8?), where the Jesuits had an important College. The sonnets show the profound disregard of the principal collaborator of Cesi for the followers of the rule of St Ignatius. Those modest poetic exercises contain serious condemnations of the wicked vices of the Fathers: hypocrisy, belligerence, sodomy, alchemy of the worst kind, lack of political scruples, unbelief, theft and more (cf. Contributi I, p. 918). Among the Linceans also Fabio Colonna made disparaging remarks on the Jesuit fathers (cf. above, p. 4, n. 14).

of the sunspots, but he was familiar with the rules of the Society, which would oblige all the Fathers to stand by a member of theirs under attack from Galileo.[105]

The letter in which Filippo Salviati, recently admitted to the Academy, proposed in Galileo's name the induction of two new members, Cosimo Ridolfi and Benedetto Castelli, provided yet another occasion for a rebuke from Cesi.[106] In advancing the candidature of Castelli, Salviati and Galileo showed that they had forgotten another rule of the Linceans, which excluded the affiliation of anyone who belonged to a regular religious order: and Castelli was a Benedictine monk.

---

105 The correspondence of Galileo and the Linceans contains many signs of the ambiguous attitude of the Jesuit Fathers to the novelties proposed by Galileo in cosmology and natural philosophy. These ambiguities often reflected their oscillation between the desire to participate as protagonists in the most advanced scientific discussions and the obligation to respect the rigorous rules of obedience imposed by the Order. Grienberger himself was often obliged to disguise his own ideas for reasons of expediency. This led the followers of Galileo who had contact with him in Rome to suspect him of duplicity. For example, after presenting Grienberger with two works against Galileo's treatise on floating bodies – the *Considerazioni* of the 'unknown Academician,' Arturo Pannocchieschi (Pannocchieschi 1612), and the *Operetta* of Giorgio Coresio (Coresio 1612) – Cigoli wrote to the Tuscan scientist: 'I was at Father Grienberger's, who told me that he had received the two small books but had not yet finished reading them. I think he said that to disguise his opinion, although he did manage to say that you are mostly right, but that he had only read it [Galileo's discourse on *Bodies that stay atop water*] hastily. It is enough: I see there a behaviour that is always suspect and not free' (OG XI, p. 369). A few months later, Faber wrote to Galileo that Grienberger was acting with calculated prudence in the matter of the polemic with Scheiner: 'Eight days ago Father Grienberger was in my house, and he told me he had not yet seen the latest work by Apelles [*Accuratior disquisitio*: Scheiner 1612b]; but in fact, although he knows that he [the author, that is Scheiner] is a Jesuit, he is in pretty close agreement with you [...]. But, as a son of holy obedience, he does not dare to give his opinion' (OG XI, p. 434). The suspicions of Cigoli and Faber proved unfounded. In the end Grienberger decided to say what he thought. He did it in a letter to Galileo on 1 February 1613, in which he stated that he approved without reservations of his book on hydrostatics. The Father was convinced that Galileo was already aware of his favourable opinion: 'You will already have heard from others on several occasions, I think, what I have thought and expressed' (ibid., p. 477). It must be stressed that Cesi had previously informed Galileo that Grienberger was the only Jesuit to assume a loyal attitude, having recognised without hesitation that the Tuscan scientist had been the first to discover the celestial novelties (ibid., p. 422).

106 Ibid., pp. 510–1, letter of Salviati to Cesi, 20 May 1613.

Well knowing who was behind the nomination, Cesi replied not to Salviati but directly to Galileo. His tone was polite but clear. While sincerely appreciating the qualities of the candidate, the rules of the Academy excluded the possibility of his being accepted among its ranks:

> The strong ties with which he [Castelli] already found himself bound for all time bars him from the equal and common functions of the Academy; hence we are unable to inscribe his name in the list of its dearest and most esteemed friends, as will be the case for any other persons who find themselves in a similar condition.[107]

Suspecting that Galileo had forgotten that rule, he drew attention to it once again in his postscript:

> From the first we have thought that members of regular religious orders could not become academicians because of the impediments of their rule […]. Even Terrentius returned the symbol [the Lincean signet ring] the day before he became a Jesuit.[108]

He reminded Galileo that he had personally explained the reasons for adopting this rule during their discussions in Rome. Moreover, when Cesi had informed him that Schreck had joined the Society of Jesus, Galileo had replied that he had grasped the inevitable implications of the German fellow's decision: 'The news of Terrentius has displeased me for the great loss to our company, but has pleased me for his holy resolution and the acquisition of another company […]'.[109]

Cesi promised to send Galileo the *Lynceographum* to remind him of the rules of the Lincean constitution. Although in the *Lynceographum* there is no statement that explicitly stipulates that affiliation to a regular religious order is incompatible with membership of the Academy, the obligation of total dedication to the Academy is frequently evoked, particularly in the *Pars prima*.[110] Moreover, the *Lynceographum* states that, in the event of the death or of entrance into a religious order of a member, all records of his membership (emerald ring, certificate of induction, correspondence with other members,

---

107 Ibid., p. 515, letter to Galileo, 30 May 1613.

108 Ibid., pp. 515–6.

109 Ibid., p. 247, letter of Galileo to Cesi, 19 December 1611.

110 Cf. Lynceographum, p. 40.

etc.) would be kept in the Lincean Archive.[111] Incompatibility is not mentioned either in the *Ristretto* of the Lincean constitution, sent to Galileo by Cesi on 2 June 1612, or in the *Praescriptiones* published in 1624.[112] Total dedication to the Academy is underlined in Cesi's *Discourse on the natural desire for knowledge*, but once again without any mention of a ban on members of the regular religious orders.[113] Though never explicitly laid down in the statutes, the rule was rigorously imposed. Cesi probably feared that to draw too much attention to such a regulation might have aroused suspicions regarding the attitude of the Academy towards the ecclesiastical institutions.

Other episodes indicate that often the views and priorities of Cesi did not coincide with those of the Tuscan scientist. In the letter preceding that in which he explained the reasons for rejecting Castelli's nomination, Cesi had sent Galileo a proposal regarding the official titles to be used when Academy members exchanged letters. In response to Della Porta's pressures, Federico had devised a long list of formulae to confer ceremonial solemnity to the correspondence, distinguishing the role of the *emeriti* from that of the simple *studiosi*.[114] Like Markus Welser,[115] Galileo admitted to some perplexity, which led Cesi to illustrate the reasons behind his proposal.[116] In the end he had to give up the idea, although he had those formulae examined and approved in two official meetings of the Academy specially convened for the purpose.[117]

---

111 Ibid., p. 193.

112 OG XI, p. 312 and Cesi 1624. However, both the *Ristretto* and the *Praescriptiones* refer to the ban on participation in political and theological matters. On the relations between *Lynceographum, Ristretto* and *Praescriptiones*, see Guardo-Orioli 2014, pp. 125–43.

113 Altieri Biagi 1969, pp. 71ff.

114 OG XI, pp. 507–8, letter of 11 May 1613.

115 'I do not know what to say about the titles of the Linceans [...]; I confess that I would have advised to move slowly and after careful reflection, because it seems pretty new to me [...] and will be bound to provoke discussion', letter to Faber, 25 July 1613 (CL, p. 372).

116 OG XI, p. 515, letter of Cesi to Galileo, 30 May 1613.

117 Two academic meetings were held to discuss the question of titles. At the first, held on 10 March 1613, on the proposal of Faber those present (besides Faber they were Cesi, Valerio and Stelluti) agreed that it was appropriate to draw up a list of precise formulae. At the second, held on 24 April, which was attended by the same academicians, a formula was proposed and unanimously approved. At the suggestion of Cesi, however, it was decided that, before its formal approbation, it should be sent to the non-resident members for their opinion (Contributi I, pp. 525–6).

The 'Lincean Telescope'

Between 1612 and 1613, Cesi orchestrated a subtle strategy to elaborate and disseminate an official academic reconstruction of the invention of the telescope. In his letter of 4 February 1612 he wrote to Galileo that he intended to publish a collection of letters to celebrate the telescope underlining its direct and intrinsic connection with the Academy. The volume would contain contributions by Faber and Della Porta 'in confirmation of the truth of the telescope'.[118] Faber diligently went to work under the watchful eye of the Prince. A note among the papers of Martin Fogel, the seventeenth-century German scholar who planned to write a history of the Lincean Academy, indicates that at a certain point Cesi had decided to write a text on the invention of the telescope himself. It would be based on a collection of written and oral testimonies of Galileo and other unspecified authors, who would certainly have included Della Porta.[119]

An undated letter of Della Porta, which can be assigned to spring-summer 1612, written in reply to a lost letter of Faber, provides illuminating clues as to Cesi's strategy. Reconstructing the history of the telescope was not only a controversial issue, given the many candidates to the title of inventor, but also an extremely delicate one inasmuch as it could lead to conflict between the two protagonists of the Lincean Academy: its putative father Della Porta, and the new leading light Galileo.[120] In approaching Della Porta for his testimony, Faber must have expressed his surprise at the failure of the Neapolitan to claim priority. This leads us to suppose that Faber and Cesi were convinced that the merit of having invented the telescope, at least on a theoretical level, was due to Della Porta. The Neapolitan replied that he had constructed a telescope before hearing of Galileo's instrument and had shown it to Cesi in May 1610 during the Prince's visit to Naples in the company of Stelluti.[121] As we recall, in the letter to Federico discussed in the initial pages of the first chapter of this book, Della Porta had not disguised his conviction of having

118 OG 11, p. 271.

119 'Ex collatione et auditione Galilaei et aliorum virorum Principis Caesis', Contributi I, p. 292. Cf. also Partini 1986, p. 64.

120 Cf. below, pp. 145–7.

121 'Quod specillum demonstrasse memini Principi nostro Lynceo [...]' (ibid., p. 308).

played a determinant role in the invention of the instrument.[122] In proposing to write a Lincean history of the invention of the telescope Cesi aimed at satisfying three fundamental objectives. The first was to recognise that Della Porta should be credited with the discovery of the theoretical principles of telescopic magnification. That discovery preceded by many years and facilitated the actual construction of the earliest instruments. The second was to certify Galileo's independence in the construction of the telescope and his extraordinary merits in perfecting the instrument and directing it at the sky. The third was to demonstrate the important role played by the Academy in giving the instrument invented and perfected by the two most illustrious Linceans the name of telescope, as well as to highlight the significant contributions offered by its Princeps.

It is not surprising that the ambiguous expression 'Lincean telescope' often recurs in the correspondence of Cesi and of his closest collaborators. The same expression appears on the printed sheet of 1611 which updated the list of Della Porta's works published the previous year in his *Elementorum curvilineorum libri tres*. The document informs us that Della Porta was compiling a booklet on

122 Cf. above, p. 10. In his unpublished *Taumatologia* (Cesi made every effort to publish its list of contents, but in vain), Della Porta had compared the status of discoveries obtained through mathematical demonstrations with the secrets revealed by the natural magician. To this end, at the opening of the *Taumatologia*, he had inserted the chapter 'Della prospettiva' in which he proudly laid claim to having invented the telescope: 'invented by me and described in my *Magia* [...] and attributed by many to the Flemish or others' (Della Porta 2013b, p. 4). It has to be stressed that Della Porta qualified the instrument as 'Lincean'. This text provided a curious justification for the reasons that had induced him not to concentrate on the construction and diffusion of the instrument, even though he had designed it decades before the Dutch craftsmen and Galileo. He had not held in consideration that 'piece of nonsense' (as he had defined it to Cesi) having decided to dedicate himself to an incomparably more ambitious project: the design of an eyeglass whose performance would have far outstripped that of the telescope. Like many of the other secrets described in the *Taumatologia*, he provided a description of that sensational invention that is too summary to allow us to understand what he had in mind: 'They [the lenses] are fitted into the telescope and enable a magnification of 1000x that exceeds every marvel' (ibid., p. 4). For additional information, he referred to chapter 11 of book XVII of his *Magia Naturale* (Della Porta 1611). In that chapter of the Italian version of the *Magia*, the Neapolitan described 'the mirror of Ptolemy, or rather lens, which could see ships six-hundred miles away before they became visible [to the naked eye]' (Della Porta 1577, p. 492). Della Porta's cryptic description – he justified its obscurity with the fact that the secret was not something 'for all the common herd to read' (Della Porta 1611, pp. 648–9) – might lead one to suppose that his sensational mirror or lens of Ptolemy had a parabolic profile.

the 'Lincean telescope'.[123] It is also important to stress that the manuscript of the first two chapters of the work dedicated to the history of the telescope by Nicolò Antonio Stelliola who, after his induction into the Academy, had been asked to write it by Federico with the intention of publishing it with the emblem of the lynx,[124] bears the title *De telescopio linceo*.[125]

However, the presentation to the public of this official academic version would have to await the publication of the *Thesaurus Mexicanus*. Johann Faber inserted the Prince's reconstruction as a digression in the zoological section of this work, which he compiled between 1625 and 1628. That the impulse came directly from Federico is demonstrated by the fact that Faber submitted the draft to him to make sure that he had expressed himself in accordance with the Prince's expectations:

> I wanted to ask your Excellency to take a look only at what I have written about the new inventions of Signor Galileo to see whether I have included everything or whether there is something to be removed, as you see fit.[126]

Cesi read it carefully and introduced a few corrections.[127] These accentuated the importance of the theoretical contribution of Della Porta and made more explicit Kepler's endorsement of the priority of the Neapolitan. At the same time, Faber's mention of the fact that Galileo had heard about the instrument constructed in Holland was attenuated.[128] The Prince then returned

123 CL, pp. 1269–70. The printed sheet is dated Rome 1 September 1611: 'De Linceo telescopio opusculum. Quod praeclarum hoc Perspicillum, iam pridem ante triginta annos ab ipso [Della Porta] inventum [...]' (p. 1270).

124 'On the subject of your request to me to write about the telescope, I sent five pages followed by three more via [Fabio] Colonna, which contain the completion of the first Book [...]' (ibid., p. 568, letter to Cesi, 13 May 1616). Stelliola's work was finally published (Stelliola 1627) after his death and did not bear the symbol of the lynx.

125 Ms. Archivio Linceo XXIV, Parte Prima. In his letter to Galileo on 17 August 1613 (OG XI, p. 551) to mark the anniversary of the foundation of the Academy, Stelliola declared that Galileo's fundamental contribution had not been the invention of the telescope, 'which has been shared with others', but to have 'applied it to celestial observations with admirable diligence'. The model of the Lincean version of the history of the telescope is evident in Stelliola's reconstruction.

126 CL, p. 1038, letter of 13 April 1625. Cf. also Ricci 1996.

127 Cesi's corrections are contained in Ms. Archivio Linceo 2 (cc. 50r bis and 51r) and Ms. 4 of the Library of the Accademia Nazionale dei Lincei. For a comparison of the manuscript versions with the text published in the *Thesaurus Mexicanus*, see Guardo 2011, esp. pp. 63–82 (the two versions are reproduced on pp. 73–5).

128 Faber's 'de hoc inaudisse saltem' was corrected by Cesi in 'vix audito rumore' (ibid., p. 81).

the corrected manuscript to Faber, who published it without further modifications.[129]

Faber's digression on the telescope gave pride of place to Giovan Battista Della Porta: 'He was the first to precisely describe the theory and composition [of the telescope] in his *Magia Naturalis*, and its mathematical foundations in his book *De refractionibus* more than forty years ago'.[130] After emphasising that German or Dutch craftsmen had constructed the first telescopes, Faber introduced Galileo. His version adhered to that published by the Tuscan scientist in the *Assayer*:

> When he had barely heard any rumour of it, Galileo not only constructed the telescope on his own, but perfected it to such an extent that he was the first to train this telescope on the heavens and to reveal all those astonishing phenomena to all mortals.[131]

Faber's genealogy of the telescope also reserved an important role for Cesi. The German Lincean reported that having heard of the new instrument from Belgium and before Galileo's visit to Rome, the Prince had himself constructed some exemplars, distributed them among several prominent figures in the city, and 'devised and gave it the name Telescope'.[132] Years later, Stelluti was to present the same version in his edition of the satires of Persius Flaccus.[133] Finally, in the *Apiarium*, the only work published under Cesi's name during his lifetime, the Prince presented the telescope as 'ours'.[134]

In attributing primacy in the invention of the telescope to Della Porta, Faber followed the authoritative testimony of Kepler, who had denied in his *Conversation with Galileo's Sidereal Messenger* that the telescope had been devised 'by Belgians' and had hailed Giovan Battista as its inventor. Kepler had also cited the passage from the *Magia Naturalis* describing the effects of the combination of a concave and a convex lens.[135] He had drawn a clear distinction between the 'Belgian' instrument, constructed without any knowledge of the optical laws governing its operation, and the telescope with which Galileo had discovered sensational novelties in the sky. Kepler therefore gave full credit to Galileo, who in the *Sidereus Nuncius* had claimed total independence in the construction of his instrument, declaring that he had deduced it from the

129 TM, pp. 473–4.
130 Ibid., p. 473.
131 Ibid.
132 Ibid.
133 Persius Flaccus 1630, p. 27.
134 Cesi 2005–2006, II, p. 39.
135 OG III[1], p. 108.

principles of perspective.[136] For Kepler, while the construction of the Belgian instrument was due to chance or to an unacknowledged appropriation of information contained in Della Porta's works, Galileo had transformed a theoretical principle into a performing instrument that enabled him to make epochal discoveries: 'I know the difference – he stated – between rational conjectures and ocular experience; between Ptolemy's discussion of the Antipodes and Columbus' discovery of the new world'.[137]

In the ode to Galileo that opened the *Assayer*, Faber had anticipated the Lincean genealogy of the instrument that he later illustrated in detail in the *Thesaurus Mexicanus*: 'Porta has the first reign; may you, O German, have the second; the third, Galileo, is your labour'.[138] As we will see, those verses were later used by the Jesuit Father Orazio Grassi to insinuate that Galileo had expropriated the true inventors of the instrument.[139]

Galileo must have shown signs of irritation at the Lincean history of the telescope proposed in Faber's gratulatory ode. Following the Tuscan scientist's bitter complaints about the very large number of typographical errors in the first print run of the *Assayer*, the Prince decided to reprint the work with a list of no less than 136 errata prepared by Galileo. The reprint no longer contained Faber's ode, nor that of Stelluti. If one takes into account the fact that some exemplars of the first print run also lack the poems, it becomes all the more likely that these omissions were due to Galileo's protests against the Lincean reconstruction of the invention of the telescope that he considered false and harmful to his reputation.[140]

The version elaborated by Cesi and Faber created a perfect balance between Cesi's three objectives. Moreover, it received the blessing of Della Porta who, in the letter sent to Faber in 1612, showed his willingness to accept the role of 'theoretical' inventor, while recognising with grandiloquent expressions of praise Galileo's role in perfecting the telescope and training it on the sky to make sensational discoveries.[141] In spite of the fulsomeness of his tone, however, Della Porta differed from Faber in omitting to declare that Galileo had constructed the instrument independently. That omission obviously contributed to emphasise the determinant role played by the Neapolitan, insinuating

136 Ibid., pp. 60–2.

137 Ibid., p. 109.

138 OG VI, p. 205.

139 Cf. below, pp. 148–9.

140 This interpretation was advanced by Drake and O'Malley in their annotated anthology on the cometary controversy of 1618 (Drake-O'Malley 1960, p. 360, n. 25).

141 '[...] non enim simplici sed duplicibus et doctissimis, Galileis ad tam arduum et excelsum facinus experiendum opus erat', CL, p. 309, undated letter.

that the works in which he had precociously formulated the optical principles of telescopic enlargement had been Galileo's source of inspiration.

It is worth considering also the controversial question of the real inventor of the term 'telescope'. According to the testimony of Giulio Cesare Lagalla, one of Cesi's guests on the Janiculum on that magic evening in the spring of 1611, it was on that occasion that Giovanni Demisiani, soon to be a Lincean fellow, gave the instrument that name.[142] This attribution clashes with Faber's later reconstructions in which Cesi was designated as the person responsible for naming the instrument 'telescope'. As Edward Rosen showed, after the death of Demisiani in 1614, Cesi carried out a subtle substitution of his own name for that of the academician of Greek origins.[143] Federico intervened in the draft version of the text sent him by Faber prior to its publication in the *Thesaurus Mexicanus* in order to present himself as the inventor of the name.[144] Further confirmation of the Prince's design is offered by the fact that on the same page of the *Thesaurus Mexicanus* Faber listed those in attendance on the Janiculum when the instrument was named leaving Demisiani out, even though we know from other sources that he was present.[145]

These are telling signs of the value assigned to public recognition of the Lincean genealogy of the telescope and to the determinant role played by the Princeps of the Academy. It was probably for this that Heckius glorified Galileo for having put 'such adventurous lenses' to the penetrating eyes of the lynx.[146] And it is not surprising that the Prince was indignant after a lecture at the Collegio Romano in which, not content with contesting the Galilean interpretation of the celestial novelties, the speaker had had the nerve 'to give a different name to the telescope'.[147] The letter reveals Cesi's concern at the growing hostility of

142 OG III[II], p. 927. Lagalla described Demisiani as 'a man educated in every kind of discipline with an outstanding knowledge of Greek and Latin, to whose keen mind we owe the new and highly appropriate name of Telescope for that instrument [...]'.

143 Rosen 1947, pp. 57–64.

144 'In Rome at that time [...] when he had only heard a rumour from Belgium, Federico Cesi constructed that same instrument and distributed it among several prominent figures in the city, and devised and gave it the name Telescope' (TM, p. 473).

145 Ibid.

146 CL, p. 473, letter to Galileo, 1614. This was the only letter that Heckius wrote to Galileo on returning to Italy after his long journey through Europe.

147 OG XII, p. 12, letter to Galileo, 3 January 1614. He was also irritated by a Dutch writer (whose identity was unknown to Cesi) who had failed to mention Galileo and had called the telescope a Dutch invention (*Batavica dioptra*) in a treatise on the sunspots (OG XII, p. 404, letter of Cesi to Galileo, 29 September 1612). That Dutch writer was Willebrord Snell (Snell 1612).

the Jesuits to the Lincean genealogy of the telescope. A few months later another Jesuit, Vincenzo Figliucci from Siena, would strike a similar chord.[148]

That is not all. In the *Discourse on the natural desire for knowledge* that Cesi delivered in the presence of Galileo in 1616, the Prince insisted on the role of Della Porta as inventor of the theory of the instrument, the Dutch craftsmen as builders of the first rudimentary exemplars, and Galileo as the one to have perfected the telescope and pointed it at the heavens.[149]

Galileo could hardly have been content with the derivation of *his* telescope from Della Porta, or with the picture dear to Cesi of the substantial continuity between the initial stages of the Academy, of which the Neapolitan was the most conspicuous symbol, and his own appearance on the scene. It is revealing that, in his detailed reconstruction in the *Assayer* of the process by which he had managed to construct an instrument so advanced that it could be used to explore the heavens in depth, Galileo did not mention either Cesi or Della Porta. He did admit that, when he was in Venice, he had received news of the telescope presented to Count Maurits by a Dutchman: 'Upon hearing this news, I returned to Padua, where I then resided, and set myself to thinking about the problem. The first night after my return I solved it and constructed the instrument'.[150] Galileo emphasized his total independence in the matter, also avoiding any reference to the theoretical contribution of Della Porta, or to the Lincean origin of the name of the instrument.

In the digression in the *Assayer* in which he proudly claimed paternity of the invention, he also replied with irritation to Sarsi's malign insinuation in the *Astronomical balance*. The mask of Father Grassi had called the telescope

---

148 In his booklet of *stanze*, published at the beginning of 1615 (Figliucci 1615), the Jesuit showed a willingness to minimise the role of Galileo in the invention of the telescope and in the celestial discoveries in favour of the determinant part played by the Society of Jesus. This sally surprised and annoyed both the Florentine Dominican Luigi Maraffi, who was close to Galileo, and Cesi. Maraffi wrote to Galileo on 12 December 1615: 'My opinion is that he [Figliucci] has been very sparing in praise to whom it is due [...]. Several times he insinuates that the telescope was invented in Flanders and improved in Italy, but he does not say by whom; that on the occasion of the stars of Jupiter others, such as the Jesuit mathematicians of the Collegio Romano, observed other stars; that it is uncertain who first observed the sunspots, whether the Jesuit going by the pseudonym Apelles or yourself [...] and considers that both were the first, one in Germany and the other in Italy [...]' (OG XII, p. 210). Cesi had already informed Galileo about Figliucci's work and had expressed similar recriminations: 'I have seen that he mentions you, but not as much as he should, and that he gives Apelles a part in the discovery of the spots' (ibid., p. 136).

149 Altieri Biagi 1969, pp. 88–9.

150 Drake-O'Malley 1960, p. 211 (OG VI, p. 258).

not a son (*foetus*) but simply a pupil (*alumnus*) of Galileo.[151] The Tuscan scientist probably interpreted that insinuation as a reference to Federico's thesis, backed by the authority of Kepler, of the Dutch origin of the instrument and above all of its theoretical foundations on the part of Della Porta: in a word, the disquieting spectre of the 'Lincean' telescope. His annoyance at Grassi's interpretation of the telescope as his pupil, not his son, increased when he read the Jesuit's reply in his *Ratio*.[152] Why are you irritated, Father Grassi asked, since I have done nothing but repeat the reconstruction proposed by your fellow Lincean Faber in the poem that he dedicated to you in the *Assayer*?[153]

Another symptom of Galileo's distaste for the Lincean appropriation of the telescope emerges from his coldness before Cesi's repeated requests to dedicate himself to the construction of lenses of a different curvature from the spherical one. On 21 March 1614 the Prince wrote: 'I would be particularly pleased to hear that you were trying [lenses] in other shapes that are considered better for magnification [...]'.[154] It did not escape Galileo that this was an implicit criticism of his instruments and that the Prince's request was inspired by the texts of Della Porta and Kepler. It had certainly been Della Porta who pressed Cesi to ask Galileo to grind parabolic lenses. Ever since October 1608 Della Porta had disclosed to Federico the method for constructing parabolic mirrors able to produce extraordinary burning effects.[155] After the appearance of the telescope, Della Porta was convinced that the parabolic profile was the most effective, not only for burning mirrors, but also for the lenses of the telescope. Thus he began to experiment with methods for grinding glass in that shape, assisted by Fabio Colonna, who was particularly skilful in mechanical operations. The results of those experiments cannot have been satisfactory, otherwise Cesi would not have urged Galileo to grind parabolic lenses. After Cesi, Fabio Colonna also asked Galileo to help him make parabolic lenses and mirrors,[156] but the Tuscan scientist did not reply to either of them.

151 Ibid., p. 127.

152 Ibid., pp. 257–9.

153 Ibid., p. 411: 'Why do you get upset if I called the telescope the pupil of Galileo, with almost the whole world laying claim to it? You did so before me, Faber, truly sincere, truly a brother, never more tenacious in your frankness than when you match your words to the truth. Nor do you fear the expression on the face of Galileo, skilfully depicted by a Daedalan hand on the very threshold of that book, where you sing like a white swan: 'Porta has the first reign; may you, O German, have the second; the third, Galileo, is your labour'.

154 OG XII, p. 41.

155 CL, p. 112, letter of 9 October 1608. See also the letter from Stelluti to Magini of 18 February 1611, in which he stated that he had seen a parabolic burning mirror in Della Porta's house when he visited him with Prince Cesi in Naples (ibid., pp. 155–6).

156 OG XII, pp. 88–9, letter of 29 July 1614.

That the two Neapolitan Linceans were intensely applying themselves to the grinding of parabolic lenses, also at the request of Cesi, is confirmed by a letter to Galileo of 26 September 1614, in which Della Porta explicitly mentioned the project of constructing that telescope with parabolic lenses in which Cesi had vainly called upon the Tuscan scientist to collaborate:

> With Fabio Colonna, who is very ingenious and skilful in mechanics, we are constructing a new form of telescope which will magnify 100x more than usual; while the latter permits vision up to the eighth sphere, this one will enable you to look into the empyrean. God willing, we will peer on what is going on there and produce a *Nuncio Empireo*.[157]

Galileo will not have found this quip amusing. The new telescope with parabolic lenses threatened to turn his own instrument into a plaything, opening the path to further celestial discoveries that could outstrip those announced in his *Nuncius*.

Neither Cesi nor the two Neapolitan Linceans were discouraged by Galileo's uncooperative attitude. While Della Porta was soon to disappear from the scene, Fabio Colonna continued to experiment with new types of lenses with the assistance of a Neapolitan craftsman, Francesco Fontana, who was soon to become one of the most innovative constructors of high-performance astronomical telescopes.[158]

The protagonism of Cesi and his Academy are apparent also in the naming and fortune of the other instrument introduced to the Linceans by Galileo and called a 'microscope' by Faber. In the *Apiarium* Cesi referred to the telescope and the microscope qualifying them both as 'our' and without mentioning Galileo. In the *Thesaurus Mexicanus* Faber introduced the instrument remarking that he had given it the name microscope by analogy with the telescope.[159] He recognised that Galileo had been the first in Italy to make a compound microscope and to introduce it in Rome in 1624. That Faber indicated the geographical limits within which Galileo could claim priority is a sign of his hesitation regarding his absolute primacy. Besides, he added that he had not managed to get the instrument given to him by the Tuscan scientist to work properly. To observe the wonders of nature revealed by the microscope, he had to turn to German craftsmen 'who offered precious collaboration, and showed

157 Ibid., p. 101.

158 CL, pp. 1124, 1131, 1205.

159 TM, p. 474.

us several microscopes that attracted admiration in the whole city'.[160] These craftsmen were the brothers Abraham and Aegidius Kuffler, who had trained in Cologne under Drebbel, a strong competitor of Galileo in the bid for priority of the invention of the microscope.[161] Moreover, a few pages further on, Faber claimed a fundamental role for himself and for the Academy, not only in the naming, perfecting and putting to innovative use of the instrument, but even in its invention:

> I have given the name Microscope to the system of lenses, by which the tiniest things are magnified 30,000x or more [...]. This optical instrument *was deduced by me* from arithmetical and geometrical principles [...].[162]

Faber went on to glorify the discoveries made by Cesi with the aid of that instrument, which, at this point, was bound to appear intrinsically Lincean to the reader. In this deft sleight of hand, Faber omitted the name of Galileo for the ultimate purpose of exalting the Prince and his Academy.

Cesi intervened in the text in which Faber illustrated the invention and naming of the microscope. He accentuated Galileo's role in the invention of the instrument, but without attenuating the merits of the craftsmen from Cologne who had perfected the microscope and diffused it in Rome.[163]

At this point it is not surprising that Galileo was not mentioned by Fabio Colonna in the letter in which he informed Faber that he had met one of the Kuffler brothers in Naples. Colonna claimed to have constructed a microscope of greater power of magnification than the one the German craftsman had displayed in Naples. He proposed to call it an 'Enghiscope' (an eyeglass for seeing things from nearby), evidently unaware that Faber had already given it a different name.[164]

Also among the Jesuits there were those who presented the microscope as a Lincean innovation, probably to deprive Galileo of the merit of having invented it. In his *De florum cultura*, Father Giovan Battista Ferrari, who had been in contact with Orazio Grassi for a long time, defined the instrument as 'the Lincean explorer, that is to say, the microscope, a small viewing glass enclosed in a tube'.[165]

---

160 Ibid.

161 Contributi I, p. 352.

162 TM, p. 757, emphasis added.

163 Cf. Guardo 2011, p. 81.

164 CL, p. 1047, letter of 6 June 1625.

165 '[...] lynceum exploratorem, nempe microscopium, sive conspicilium tubulatum consuluimus [...]', Ferrari 1633, p. 481.

Galileo was bound to be vexed by these reconstructions, which passed over the fact that, even though he had only presented it in 1624, he had devised the microscope many years earlier at the same time as the construction of the first telescopes.[166]

166 Galileo declared in the *Assayer* that when any object is observed through a modified telescope (i.e. a microscope) it displays the colours of the rainbow: 'Let him [Sarsi] take any material, whether stone or wood or metal, and holding it up to the Sun gaze at it most intently. There he will see all colours appearing in most minute specks; and if in looking at this, *he will make use of a telescope set for looking at very near objects*, he will see what I mean very much more distinctly and without any need of these bodies being resolved into dew or moist vapors.' (Drake-O'Malley 1960, p. 245, emphasis added: OG VI, p. 290). It should be recalled that the manuscript with the definitive version of the *Assayer* reached Rome at the end of October 1622. We also know that the text was not modified during revision and printing except for some toning down by Cesarini and Cesi. Galileo was not even given the opportunity to see the proofs. This is important because it shows that by the autumn of 1622 Galileo had already designed and used the instrument that the Linceans would later call the microscope. The invention should be antedated even further. The testimony of a student of Galileo in Padua, the Scotsman John Wedderburn, author of a detailed rejoinder to the *Brevissima peregrinatio contra Nuncium Sidereum* of Martinus Horky (Horky 1610), stated that, immediately after the construction of the telescope, Galileo had conceived 'a telescope modified to see objects from nearby'. Wedderburn also informed us that Galileo used it to observe the organs of the tiniest insects and mentioned those experiments to his colleague in Padua Cesare Cremonini: 'I heard him a few days ago telling the most excellent philosopher Cremonini various interesting things, including how he could perfectly distinguish the organs, motions and senses of the tiniest animals through a *perspicillum*; in particular those of a certain insect with both eyes covered in a thick membrane but which nevertheless permits vision as it is pierced by seven apertures like the helmet of a soldier's metal armour' (Wedderburn 1610: OG III[1], pp. 163–4; see Camerota 2004, p. 181). Wedderburn took this use of the microscope as further proof of the objectivity of the images generated by the telescope. That Galileo had constructed early prototypes of the microscope that he took to Rome in 1624 and that he made exclusive gifts of them to important persons is also demonstrated by the letter sent to him from Rome by Cardinal Francesco del Monte on 3 May 1620. The Cardinal, who had already had the opportunity to observe the marvels of the Galilean device (probably during Galileo's 1616 visit), asked the Tuscan scientist for one: 'The painter who presents to you this letter will show you an eyeglass which brings objects closer made in imitation of yours. And because the quality of this eyeglass led me to want a similar one, I beg you to give consideration to favouring me with one of your instruments [...]' (OG XIII, p. 36). Galileo immediately satisfied the request of the influential prelate, who thanked him: 'The eyeglass that you sent me is to my entire satisfaction' (ibid., p. 40). As Gilberto De Angelis pointed out, Gilberto Govi, author of an early study on the invention of the telescope, was perfectly correct in maintaining that Galileo had invented the compound microscope as early as 1610 and applied it 'to the examination of some tiny creatures' (Govi 1889, p. 15; cf. De Angelis 2012a, pp. 21–36).

CHAPTER 5

The Copernican System versus Holy Scripture

The Spectre of Giordano Bruno

The first signs of Galileo's concern at the denunciations of the incompatibility of the Copernican system with Holy Scripture emerge from the reply of Cardinal Carlo Conti on 7 July 1612 to a (lost) letter of the Tuscan scientist.[1] The Cardinal, who was on friendly terms with Cesi, was answering Galileo's question as to 'whether Holy Scripture supports the principles of Aristotle on the constitution of the universe'. Before giving his opinion, the Cardinal pointed out the need to focus on the precise significance of the Tuscan scientist's query: was he referring to 'the incorruptibility of the heavens, as his letter seemed to suggest in stating that new phenomena are discovered in the sky every day', or to the 'two motions of the Earth'?[2]

Conti's position on the first point was peremptory:

> There can be no doubt that Scripture does not support Aristotle; indeed it tends rather to the opposite, as it was the common opinion of the Fathers that the heavens were corruptible.[3]

1 Galileo presumably sent his questions to Cardinal Conti from Florence at the end of June.

2 OG XI, p. 354. On Galileo's relations with the Cardinal and on the Cardinal's letter of 7 July 1612, see Poppi 1996–1997. Poppi's article contains a careful analysis of the sources used by Galileo to write the *Copernican letters* and the development of his strategy to confront the accusations of incompatibility between the Copernican conceptions and Holy Scripture. According to Poppi's analysis, Galileo had adopted a substantially concordist strategy in the *Letters on sunspots*, but replaced it in the *Copernican Letters* with the claim of the independence of the Book of Nature from Holy Scripture. Poppi disagrees with Vincenzo Ferrone, according to whom Galileo's discussions with Cesi in Rome in 1611 and their correspondence in the immediately successive years contributed significantly to defining the way in which he dealt with Holy Scripture in his *Copernican Letters* (Ferrone 1984, pp. 247–53). Ferrone's suggestion should be borne in mind in reconstructing the Tuscan scientist's shift from the instrumentally concordist positions of the *Letters on sunspots* to claiming the full autonomy of natural philosophy from theology and Holy Scripture in the *Copernican Letters*, in general, and in the *Letter to Benedetto Castelli* in particular.

3 OG XI, p. 354.

 | DOI 10.1163/9789004342323_006

He was, however, more cautious as to whether the novelties discovered in the sky demonstrated its corruptible nature. Referring to the sunspots, Conti stressed that they could not be considered conclusive proof since it was possible to advance interpretations, such as that of Scheiner, that saved the incorruptible nature of the Sun.[4] The Cardinal urged Galileo to exercise prudence on this point and above all to consider the uncertainty of human knowledge. The sense of his message was clear: the thesis of the corruptible heavens was plausible and even authoritative, not because it was suggested by telescopic observation of celestial bodies and phenomena, but because it had the backing of Holy Scripture.

The second horn of the dilemma, the motions of the Earth, seemed to Conti less in accordance with Scripture. He underlined the opposition to it of all the Church Fathers with the exception of the commentary on the *Book of Job* by Diego da Stunica, 'although his interpretation is not generally followed'. The only way to reconcile the Copernican vision with Holy Scripture would be to maintain that the sacred texts speak 'to be understood by the common people'. He stressed, however, that this kind of interpretation should not be admitted 'unless under great necessity'.[5]

Conti was aware of the latest celestial discovery, that of the sunspots, and he had read Scheiner's *Tres epistulae*.[6] Like Cesi and Bellarmine, he was convinced that the heavens were fluid; but he hesitated, as did Federico, before the Copernican hypothesis. His position was similar to that of many other prelates and natural philosophers, not only on the Roman scene, after the shockwave

4 In his successive letter to Galileo of 18 August, Conti mentioned again that he considered plausible Scheiner's hypothesis according to which, as small planets cross the Sun they produce shadows that to us on Earth look like dark spots: 'It might be argued [...] that those spots – the Cardinal wrote – are produced by stars so minute that no gaps can be seen between them, and that their mass produces the appearance of those spots, and that they are so many in number and have such diverse movements around the Sun that, by conjoining in different shapes, they produce that diversity of spots [...]' (ibid., p. 376). It should be noted that Cesi had told Galileo that Scheiner's interpretation of the spots was finding favour in Roman circles: 'The Peripatetics all jump at it, nor are they ashamed to say that those invisible little stars are fixed in some tiny spheres or rather celestial crusts, which conjoin and disjoin in their movements, because they are just as reluctant to lose the diamantine solidity of the heavens as the privilege of incorruptibility' (CL, p. 284, letter of 3 November 1612). Federico took a dim view of Scheiner's interpretation and invited Galileo to examine and refute it. Galileo welcomed the Prince's proposal. He sarcastically demonstrated the weakness of Scheiner's hypothesis in the third of his *Letters on sunspots* (Galilei-Scheiner 2010, pp. 291ff.: OG V, pp. 231ff.).

5 Ibid., pp. 354–5.

6 Scheiner 1612.

provoked by the telescope. One of these was Monsignor Agucchi, who explicitly supported the fluidity and even the corruptibility of the heavens, but baulked at the prospect of a new system of the world. All the same, Conti did not adopt a stance decidedly contrary to Copernicus. Almost two years later he wrote to Galileo that, in spite of his perplexity, he was inclined to follow him in his heliocentric convictions for the high opinion that he held of the Tuscan scientist's competence.[7]

Galileo can hardly have been surprised by the Cardinal's reply. His contacts in Rome with Cesi and the discussions with some Fathers of the Collegio Romano, especially Grienberger, as well as with Monsignor Agucchi, had made him realise that the thesis of the fluidity of the heavens would be tolerated and their corruptible nature would be a topic for discussion, albeit a controversial one, while the motions of the Earth would meet with resolute opposition.

Given its early date, we can consider Galileo's lost letter to Conti from the summer of 1612 as the first sign of the opening of a new stage in the intellectual life of the Tuscan scientist.[8] It was to see him engaged in the defence of Copernicanism from the accusation of threatening religious faith until 5 March 1616, the date of the publication of the anti-Copernican decree by the ecclesiastical authorities.

The incompatibility of the Copernican system with Holy Scripture had already been denounced by Galileo's Florentine adversaries in the *Dianoia* of Sizi and the *Contro il moto della Terra* (*Against the motion of the Earth*) of Ludovico delle Colombe,[9] as well as in the writings of the Aristotelians who were engaged in bitter controversy with Galileo on floating bodies. It is symptomatic that Galileo's letter asking Cardinal Conti for his opinion on the question of compatibility of the fluid and corruptible heavens with Holy Scripture

7 OG XII, p. 47, letter of Conti to Galileo, 11 April 1614.

8 Galileo's position on these delicate themes in his work on sunspots seems to show the influence of Cesi's inclination to interpret Holy Scripture in such a way as to demonstrate its agreement with the celestial novelties. Galileo's letter to Dini of 1615 shows traces of that Neoplatonist philosophy – with at its centre the divine function of light – permeated also by hermetic and even Telesian influences, that Prince Cesi had placed at the heart of the Academy's agenda right from the start. However, for Galileo the concordist approach was purely instrumental and never fully convinced him. He regarded that option as a move that might help overcome the obstacles to the recognition of freedom of thought in natural philosophy. The resolute censorship of the concordist passages in the *Letters on sunspots* convinced him that such a strategy was impracticable. Thus he abandoned it and engaged in a courageous battle for the full autonomy of natural philosophy from the guardianship of theology and Holy Scripture.

9 Sizi 1611; Delle Colombe 1610–1611.

accompanied a copy of his own discourse on floating bodies published in May 1612.[10] The position adopted in Delle Colombe's text against the motion of the Earth, with which Galileo was familiar, was crystal clear:

> Perhaps the wretched supporters of the fluidity and corruptibility of the heavens and of the mobility of the Earth like Galileo will fall back on different interpretations of the literal meaning of the Scripture? No – for all the theologians are unanimously agreed that when the Scripture can be understood literally, it must never be interpreted in any other way.[11]

Delle Colombe had listed a considerable number of passages from Holy Scripture and from the Church Fathers that unequivocally contradicted the fluidity and corruptibility of the heavens and above all the motions of the Earth. In doing so, he unwittingly helped Galileo, who will certainly have taken note of those references in order to examine them carefully and to prepare his own response. Sizi, on the other hand, while concentrating mainly on the objections to the *Nuncius* based on the principles of Aristotelian natural philosophy, had stressed that the existence of the satellites of Jupiter should be rejected not only for astrological reasons, but above all because Holy Scripture declared that God had created seven planets.[12]

A few months before Galileo penned his letter to Cardinal Conti, Giulio Cesare Lagalla, in his *De phaenomenis in orbe Lunae* published in the spring of 1612, had also invoked the authority of Holy Scripture to reject the celestial novelties. Drawing on both biblical and patristic sources, Lagalla had denied that the heavens were fluid and corruptible because it was inconceivable that God had not created a perfect world. Moreover, Lagalla had denounced one of the implications of the Copernican reversal of traditional cosmology: the enormous increase in the dimensions of the celestial spaces. This was an alarming notion that seemed to revive the heretical conception of Giordano Bruno, justly condemned and punished in exemplary fashion by the Church.[13]

10 OG XII, p. 355.

11 OG III[1], p. 290. Delle Colombe's text, written between the end of 1610 and the beginning of 1611, was not published.

12 Sizi 1611; OG III[1], p. 212.

13 'Since such worlds are created and dissolved – Lagalla wrote – there will be an infinite succession of them, as Democritus stated; therefore there will be many terrestrial globes in many worlds [...]. This infernal idea was revived a few years ago by an author who claimed that every planet is a globe like our Earth [...]. The new observation of four planets around Jupiter by Galileo has been suspected, as Kepler noted in his *Conversation with Galileo's Sidereal Messenger*, of seeking to confirm such a view' (OG III[1], p. 349).

As we have seen, in the same text Lagalla provided a detailed account of the lively discussions of cosmological questions on the occasion of Galileo's visit to Rome in the spring of 1611. In all likelihood, he referred to one of the events organised by Cesi in which Galileo had maintained, certainly seconded by the Prince, the elementary and corruptible nature of the heavens. Lagalla had energetically argued against him. While displaying respect and admiration for Telesio as a protagonist of the renewal of natural philosophy, he had resolutely opposed the Telesian conception of the fiery nature of the heavens vigorously defended by Antonio Persio on that occasion.[14] Lagalla's main concern was his awareness that the Telesian thesis necessarily entailed the elementary and corruptible nature of the heavens. For this reason he resolutely reaffirmed, quoting texts from Aristotle and Augustine, that the heavens could not be subject to alteration. He based his argument on the need to attribute an incorruptible nature to the Prime Mover, for if the Prime Mover were subject to corruption, the stars would not display the perfect regularity that we observe in their movements. Nor would that regularity be possible if the substance of which they are composed underwent continuous mutations.[15]

Lagalla reported that Galileo had not excluded that the Prime Mover was inalterable and incorruptible, provided 'we admit that it coincides with the Sun'. Lagalla replied that, since the stars and the celestial spheres were composed of the same matter as the Sun, the heavens must necessarily be incorruptible.[16]

Lagalla's account confirms that during Galileo's stay in Rome in 1611 discussions were not confined to the objectivity of telescopic vision, but focused also on the fluidity and corruptibility of the heavens. It shows that in those lively debates, in which both Cesi and Persio participated, particular attention was paid to the Telesian interpretation of nature.

Between 1611 and 1612 the Roman scene had become the decisive front for Galileo's Copernican battle. Although he could not ignore the insidious attacks made on him in Florence by adversaries supported by influential representatives of the Medici court and encouraged by local ecclesiastical authorities, he was convinced that the outcome of the confrontation would be decided in the Eternal City.

---

14 Lagalla wrote: 'I recall once when I discussed this issue in Rome with Galileo, in the presence of the very learned Antonio Persio and the erudite and illustrious Federico Cesi' (ibid., p. 366).

15 Ibid., pp. 377ff.

16 Ibid., p. 366.

We find valuable evidence of those tough clashes and of the active participation of Federico Cesi, particularly with regard to the question of the fluidity and corruptibility of the heavens, in a text written in June 1611 by the eminent prelate Monsignor Giovan Battista Agucchi, to whom Galileo was introduced by Luca Valerio during his stay in Rome.[17] Penned after his reading of the *Sidereus Nuncius*, Agucchi's text bears the eloquent title *Della penetrazione e corruttibilità del cielo* (*Of the permeable and corruptible nature of the heavens*).[18] Equally significant is the dedication to Francesco Ingoli, who was to play a prominent role in the first trial of Galileo and in the years before and after the issue of the anti-Copernican decree.

Agucchi's reflections were stimulated by his discussions with Galileo who, when the prelate composed the text, must have still been in Rome or had left the city shortly before. Moreover, his conclusions are in such perfect agreement with the conceptions of nature and the composition of the celestial spaces held by Cesi that it is difficult to imagine that the Prince had not read the prelate's manuscript. As Bucciantini has demonstrated, there are too many direct assonances in Agucchi's text with arguments and passages from Galileo's successive Copernican letters for this to be no more than a coincidence.

Articulating his reasoning in twenty arguments, Agucchi maintained that 'the thesis of the permeable and corruptible nature of the heavens is not contrary to Holy Scripture'.[19] Furthermore, he was severely critical of Aristotle and even went so far as to declare that the Greek philosopher would have done better to follow the doctrines of Democritus and the atomists. He was hesistant, however, before the corruptibility of the celestial bodies and spaces and did not believe that the Copernican hypothesis could be reconciled with the sacred texts. On this point he was a resolute defender of tradition. Still, he accepted the celestial novelties without reservations and manifested a great admiration for Galileo. The influential prelate was to become one of the most diligent and passionate observers of the sunspots and to engage in an intense dialogue with Galileo on them. In his view, the celestial novelties seemed the logical culmination of a process of revision of Ptolemaic cosmology that had begun not with Copernicus but with Tycho Brahe.[20]

---

17 See below, p. 331 and n. 106.

18 Agucchi's text was discovered by Massimo Bucciantini, who published it (Bucciantini 1999, pp. 426–42).

19 Ibid., p. 427.

20 See ibid. for a detailed analysis of Agucchi's text and its striking analogies with the Copernican letter to Benedetto Castelli written by Galileo four years later.

Two years later, in 1613, the prelate added to his manuscript tract an appendix in which he presented the results of his reflections after reading Galileo's book on sunspots.[21] This appendix shows that his position had changed: he now seemed convinced also of the corruptible nature of the celestial spaces, although he was still reticent about the Copernican system of the world. In a letter sent to Galileo during the same weeks in which he was writing the appendix, Agucchi finally came out into the open on the question of the structure of the universe. Under Galileo's insistent pressure, he had given careful consideration to the Copernican hypothesis. But the Polish astronomer's system of the world failed to convince him.[22] He adduced three arguments to motivate his contrariety.

The first was 'the authority of Holy Scripture, which affirms the opposite in many places and very clearly'. Although aware that it was possible to interpret those passages (as he had himself done in the text of 1611, and as Cesi and Galileo would certainly have suggested to him) in such a way as to remove this obstacle, he had concluded after mature reflection that it was a false trail: that method 'does not satisfy most individuals and, in particular, the most Catholic or pious people who frown upon contrived ways of interpreting Holy Scripture's crystal clear meaning'.[23]

Like so many others in those years, Agucchi accepted the discoveries and praised their author, but he refused to follow Galileo as to their cosmological implications. His retreat must have been an alarm signal to Galileo. The prelate's arguments echoed the very motivations with which the censors of the *Letters on sunspots* had opposed the publication of those passages where Galileo stressed the compatibility of Scripture with the corruptible nature of the heavens.

The second argument that prevented Agucchi from subscribing to Copernicus was the authority of all 'the most esteemed mathematicians after Copernicus [...], who have not only refused to accept his system as true, but have vigorously opposed it'.[24] For the prelate, it was better to fall back on the less subversive system of Tycho (for whom he expressed the greatest admiration), which saved all the celestial novelties.

His third reservation was based on 'reason itself', which instinctively considers the Copernican hypothesis absurd because of the vast dimensions

21 Ibid., pp. 441–2.

22 OG XI, pp. 532–5, letter of Agucchi to Galileo, 13 July 1613. On Agucchi's attitude towards the Copernican conception, see also below, pp. 329–32.

23 Ibid., p. 533.

24 Ibid.

that would have to be assigned to the universe: an enormous space that, devoid of stars, would be useless and therefore difficult for the human intellect to grasp.[25] This final observation seemed to evoke the disquieting spectre of Bruno while highlighting the fetters imposed on the freedom of thought by the anthropocentric conception of the Creation. Agucchi shared this view with Cesi and with so many others who were by no means hostile to the novelties but did not accept the thesis of the total indifference of nature vis-à-vis human expectations.

In closing, Agucchi advised Galileo: 'to not say anything in public about the truth of this opinion without having solid arguments to prove it; because if it cannot be demonstrated by mathematical and necessary proofs, it will be no easy task to convince the world with only probable arguments of a thing that is difficult for the human intellect to understand'.[26]

His analysis was similar to that which Cesi had expressed and would continue to express to Galileo: the Copernican system is hard for the intellect to digest because of its implications for the position of man in the universe and his relation with nature.

Two years after the work in which Agucchi had declared that Holy Scripture confirmed the fluid nature of the heavens, clear signs indicated that the climate had changed. Like so many others in those months, Agucchi submitted to authority, abandoning the open attitude to the novelties displayed in his text of 1611, written under the spell of the *Sidereus Nuncius*.

Celestial Animals

A letter of Francesco Ingoli of the summer of 1613 eloquently shows that the discussions in Roman circles sparked by the celestial novelties lost none of their liveliness after Galileo's return to Florence. This document is also important because it informs us that they took place during meetings convened by Cesi:

> I was invited to Prince Cesi's palace on Wednesday, where there were many scholars discussing various mathematical, philosophical and theological questions. There were Peripatetics, Paracelsians and Telesians. You can imagine whether there was any agreement. One of them was

25 Ibid., p. 534.
26 Ibid., p. 535.

> fiercely attacked for maintaining that the heavens are animated. I commented that, first, this opinion had been condemned as erroneous by the Sorbonne of Paris; second, that in the first chapter of *Genesis* are listed the ends to which God made the creatures and that as to the celestial bodies it was established that they 'serve as signs to mark sacred times' and not as intelligent beings, because this is the principal end of the intellective soul that He gave only to the heavens. Although he put up a good defence by saying that what is written in the book of *Genesis* was not the principal purpose, I replied that it could not be so because the principal purposes for which the other creatures were created are clearly stated there.[27]

The letter was addressed to Cardinal Bonifacio Caetani, a passionate student of astronomy and a supporter of recognition on the part of the Church of the legitimacy of practising rational astrology.[28]

The question of whether or not the heavens were animated was once again at the centre of the debate in a subsequent meeting, as confirmed by another letter that Ingoli sent to Cardinal Caetani on 13 September. He informed the powerful prelate that on that occasion the debate on the animated heavens had been rekindled, and that the Jesuit Francesco Diotallevi had opposed the arguments put forward by the advocates of such a dangerous conception.[29] According to Ingoli, Diotallevi won the day by stating that that doctrine was 'erroneous, if not heretical'. Ingoli stressed that the notion of animated heavens echoed the unacceptable views of 'the school of Cremonini'.[30] The Paduan philosopher had proposed this conception in his *De coelo*, the arrival of which in Rome had been announced to Galileo by Cesi, who hastened to inform the Tuscan scientist that the book was 'frowned upon by the censors for its celestial animals or animated heavens'. Although the Prince also believed that the heavens were animated, he disliked Cremonini's work: 'I continue to read it – he wrote to Galileo – as you advised me, amused by the abstruse sky

27 The letter, dated 9 August 1613, was discovered and published by Massimo Bucciantini (Bucciantini 1999, pp. 411–2).

28 On Cardinal Caetani's dedication to the translation and dissemination of Ptolemy's *Tetrabiblos* and to the promotion of rational astrology, which he sharply distinguished from the evil judiciary astrology, see Bucciantini 1995, pp. 141–5.

29 On Father Diotallevi, an unconditional admirer of Bellarmine, see Sanfilippo 1991.

30 Bucciantini 1995, p. 86, n. 26.

that the Peripatetics have fabricated'.[31] That Galileo had invited his friends to read the *De coelo,* not because he valued it but to show the absurdities to which adherence to Aristotle led, is confirmed by the letter sent to him from the Dominican Maraffi on 10 January 1615. The Father stated that Galileo had once spoken to him about Cremonini's work. He had not himself read it, but a friend in whom he placed his confidence had called it a 'worthless book'.[32]

One might be tempted to identify Prince Cesi as the advocate of animated heavens at those meetings. As we shall soon see, in his letter *De caeli unitate* to Bellarmine, Federico was to maintain that the celestial sphere is at rest, while the planets are moved by driving souls.[33] However, it was not Cesi, even though it is certain that he did not side with the opponents of that thesis either. Ingoli revealed that the advocate of animated heavens was Giulio Cesare Lagalla. This identification is confirmed by the minutes of the Lincean Academy meeting of 8 July 1613: 'We had lunch in the palace of Cardinal Cesi, where Lagalla delivered a lecture *De Animabus* [sic] *caeli*'.[34] This indicates that the question of the animated heavens had become topical during those months, no doubt as a consequence of the energetic promotion of the Copernican vision by Galileo and perhaps of Kepler's demolition of the solid spheres in his *Astronomia Nova*.

A few years later, Lagalla presented a detailed account of those debates in his *De coelo animato disputatio,* in which he dwelt particularly on those with the Jesuit Father Francesco Diotallevi.[35] It is not surprising that he had difficulty in obtaining authorisation to publish his book. In the letter to the reader he explained that it was intended as a rejoinder to a booklet published in the previous year that had criticised his conception of animated heavens.[36] Lagalla does not provide us with clues to identify the author of that work, but he might have referred to a rare publication by Cornelio Ghirardelli which had

---

31 OG XII, pp. 29–31, letter of 1 March 1614. In the same letter, Cesi informed Galileo that in his theatrical performance for the wedding ceremony of Federico's cousin, Princess Peretti, Jacopo Cicognini 'had elegantly placed among the planets the Medicean ones in a choir around Jupiter. Everyone was delighted by the spectacle [...], although I did hear some prominent Peripatetics who could not refrain from snarling as dullards because they are hostile to every new thing' (p. 29).

32 Ibid., p. 128 ('un libraccio').

33 See below, pp. 228–9.

34 Contributi I, p. 527.

35 Lagalla 1622.

36 Ibid., p. 24.

denounced the thesis of animated heavens as idle superstition.[37] One of the physical arguments on which Lagalla relied to demonstrate that the heavens were animated was derived from observation of the motions of the planets. Only the assumption that angels or some other animated principle kept them in their stable orbits could explain the complexity of their movements, which astronomers had desperately tried to save by conjectures based on intricate arrangements of epicycles and eccentrics. They had taken their folly to such lengths as to devise the absurd hypothesis of the motion of the Earth.[38] These statements seem to indicate that, at this point, Lagalla had abandoned the hostility to the notion of the fluidity of the heavens displayed in the Roman debates of 1611. However the fact that the celestial spaces were fluid did not mean for him that they were homogeneous with those of the sphere of elements. The sublunary sky, he stressed, was not composed of the same matter as that of the spaces beyond the sphere of the Moon.[39]

The thesis of animated heavens had its roots in antiquity and could count on the authority of Aristotle. Telesio, Patrizi and Kepler had relaunched that conception. As the epistolary discussion with Galileo on questions *de systemate mundi* in 1612 shows, Cesi had embraced it before Lagalla had publicly adhered to it.[40] He must therefore have been particularly delighted to discover that Lagalla had abandoned the solid heavens. A few months later he enthusiastically reported to Galileo that the pressure he had been exerting on Lagalla to give up traditional cosmology seemed to be producing results. The tenacious defender of Aristotle had displayed manifest signs of capitulation in his treatise 'on celestial and nocturnal redness':

> I think he wants you to read his text in order for you to see that he is beginning to free himself from the Peripatetic philosophy, admitting that those big ethereal fires are totally ridiculous.[41]

Cesi assigned to the conversion of Lagalla great value and expediency on the Roman scene. For this reason he had invested a great deal of energy in trying to

37 Ghirardelli 1621. The author condemned all superstitions based on the conviction that the stars and constellations could be taken to predict human destiny. Among those superstitions he had included that 'of those who said that the heavens were animated' (p. 8). I am grateful to Sara Bonechi for drawing this work to my attention.

38 Lagalla 1622, pp. 24–5.

39 Ibid., p. 36.

40 Ibid., p. 39.

41 OG XII, p. 14, letter of 18 January 1614.

convince Lagalla, even tempting him with the prospect of induction into the Academy. He did not conceal his personal satisfaction when Lagalla's resolute opposition to the thesis of the solid and incorruptible spheres showed signs of weakening: 'It is delightful to hear how the other Peripatetics are calling him a heretic in philosophy'.[42] Cesi's efforts had begun during Galileo's stay in Rome in 1611. As he wrote to the Tuscan scientist on 20 August 1611, he and Johannes Schreck had been trying to break down Lagalla's obstinate loyalty to Aristotle, but without making any headway:

> All day long we have discussed his opinion in the company of Terrentius: he says he will write to you about it. But so far we have been unable to shake his [Lagalla's] adhesion to the perfection of the spheres of Aristotle.[43]

Lagalla's *Disputatio* also informs us about the concerns that prevented the Jesuit Francesco Diotallevi from accepting the thesis of the animated heavens. If angelic spirits drove the planets across the celestial spaces, the Father asked, should they be invoked in litanies and prayers? And what would happen to the driving souls of the planets on the day of the Last Judgement: would they also find a haven in the empyrean heaven, the abode of the blessed? Lagalla reassured him. The expression recurring in all the litanies 'Omnes Sancti Beatorum spirituum ordines intercedite pro nobis' included the angelic intelligences that governed the motions of planets. Nor could there be any doubt that, on the day of the Last Judgement, they would find themselves in the highest heaven.[44]

From another letter of Ingoli to Cardinal Caetani, written a few weeks later, we learn of another meeting of that composite philosophical circle, held once again in the palace of Federico Cesi.[45] The minutes of the Academy contain a concise account of it:

> On Wednesday morning Prince Cesi offered a sumptuous meal to messrs Cesarini and other gentlemen and prelates with a handful of the leading scholars of this city in his palace near St Peter's. Their discussions continued into the night.[46]

42 Ibid.

43 OG XI, pp. 174–5.

44 Lagalla 1622, p. 39.

45 Bucciantini 1999, p. 412, letter of 30 August 1613.

46 Contributi I, p. 528. The minutes are dated 'August 1613'.

These meetings assumed an intense frequency during the summer of 1613.[47] They show the strong motivations of the participants in that informal academy to tackle the delicate problems deriving from the novelties discovered in the sky by Galileo.

Questions of Character?

By August 1613 the Galileo affair had been out in the open for several months – at least since 2 November 1612, when the Dominican Niccolò Lorini had pronounced his anathema against Galileo and his followers from the Convent of San Matteo in Arcetri. Galileo was stunned by the violence of the attack. In a letter written two days later to Cesi, he told him that he had interrupted composition of the *Letters on sunspots* because of the agitated state in which he had been left by 'the many things that run through my head for other reasons too'.[48]

During the months that separate Galileo's stay in Rome in 1611 and the surge of rejection of the cosmological implications of the celestial discoveries representatives of the Church and followers of traditional cosmology moved towards positions far removed from those hoped for by the Tuscan scientist. It has been repeatedly stated that responsibility for that dramatic conflict must be largely attributed to Galileo's incapacity to size up the situation and to the obstinately aggressive nature of his quarrelsome, intolerant and stubborn character.

There are at least two reasons for rejecting this biased interpretation. Firstly, we instinctively tend to judge Galileo's persistent attempts at overcoming the opposition of such a powerful institution as the Church because we know how the affair ended. But if we put ourselves in Galileo's shoes, the reasons that led him to fight for the affirmation of the new system of the world appear perfectly comprehensible and far from being inevitably destined to that failure that only his presumption prevented him from foreseeing.

Ever since publication of the *Nuncius*, Galileo had been perfectly aware that followers of traditional natural philosophy and the religious hierarchies, in Florence as well as in Rome, would stubbornly oppose the new cosmology. At the same time, he knew that the opposition was divided. Personal experience had shown him that a significant number of influential members of

47 Meetings are recorded in that year on 8 July, 9 August, another unspecified day in August, 30 August and 13 September.

48 OG XI, p. 426.

the Roman aristocracy (the Orsini, Barberini and others) and of the religious hierarchies were working towards a compromise. The thesis of the fluidity of the heavens, one of the fundamental principles of his vision of a unitary cosmos, had met with the consensus of many who, in the light of the sensational new discoveries, especially that of the sunspots, were also prepared to admit that the heavens were corruptible. Not only personalities of the stature of Prince Cesi, Monsignor Agucchi and Cardinal Conti, but also powerful Jesuits, like Bellarmine, who made no secret of his conviction that the heavens were fluid, and Father Grienberger, whom the Princeps of the Linceans cultivated with particular attention, encouraging him to resist the Society's pressures to observe discipline and step into line. Certainly, none of them – not even Cesi – seemed prepared to publicly stand up for the new Copernican system of the universe. However, many had shown interest in the views of Tycho and had felt the fascination of Kepler's *Astronomia Nova,* a transparently Copernican work widely read and discussed in Rome at the beginning of the 1610s.

If we take this scenario into account, disregarding our knowledge of how events unfolded, we should conclude that, in the eyes of Galileo, emerging from the confrontation as victor appeared arduous but not impossible. This helps us to understand his reluctance in heeding the exhortations to prudence from Cesi and others in Rome who, while admiring him, were not prepared to follow him in his Copernican campaign. The firm determination with which Galileo pursued that ambitious objective was due not to his inability to foresee that the challenge he faced was doomed to failure, but to his lucid evaluation of the risks involved in undertaking a journey that, while fraught with difficulties, would not necessarily lead to a precipice.

In the second place we must not forget that, unlike almost all the other actors on the stage, Galileo was prepared to put himself at risk to secure acceptance of his convictions from authoritative individuals and institutions while fully conscious of their power to crush him. We detect a strong ethical motivation behind his attitude both at that time and in later years. He *had* to fight that battle in the name of the supreme value of truth and of his personal dignity.

Observed from this perspective, Galileo's attitude appears not irresponsible but simply courageous: the courage of truth and of confidence in the final triumph over the errors of pseudo-philosophy. To present his struggle and the ensuing defeat as the inevitable – and therefore to some extent deserved – consequence of his aggressive and intolerant character, of his culpable blindness towards the delicate situation in which he found himself, of his obstinate perseverance in forging ahead against all odds instead of slackening his pace to allow the Church to assimilate the celestial novelties, is to draw a caricatural portrait of the man.

For a balanced reconstruction of the battle that Galileo waged between the end of 1613 and the spring of 1616 – and, once again, several years later – it is necessary to pay careful attention to why he considered the objectives for which he was fighting viable. This does not mean depicting him as a saint of reason, nor presenting him as an exemplary model of moral conduct. He was an ambitious man, eager for recognition and success. But he pursued these objectives while advancing innovative ideas destined to change the world which he would neither trade for the rewards of a courtier, nor retract in supine submission to the powers that be. Had he been motivated solely by a desire for success, he could have satisfied it quite easily without running any risks or encountering any obstacles – all he had to do was bow his head in obedience. There was no need to make life so difficult for himself: he had already secured a dazzling reputation and generous economic benefits after the publication of the *Sidereus Nuncius* in 1610; his had become a household name; he was at the service of an influential ruler from whom he received an extremely high salary; he was admired by the most authoritative persons in the city of Rome, even by the Pope, who had granted him a personal audience on more than one occasion, as well as in the rest of Europe.

Fortunately for us he turned a deaf ear to the calls to lay down his arms.

The League of Pigeons Launches the Attack

The climate was changing, and not only in Rome. The first evidence of the manoeuvres organised by his enemies in Florence to discredit the celestial novelties and their discoverer is provided by the letter to Galileo from Ludovico Cigoli of 16 December 1611.[49] Cigoli had been informed by Luigi Maraffi, a Dominican Father of Santa Maria Novella very fond of Galileo that

> a certain bunch of persons who are malign and jealous of your virtue and merits meet at the residence of the Archbishop [Alessandro Marzimedici]. They furiously search in the hope of finding a way of denouncing you as a supporter of the motion of the Earth or something else, and one of them asked a preacher to denounce from the pulpit that you said bizarre things.[50]

49 OG XI, pp. 241–2.

50 Ibid.

A year later the situation had further deteriorated. When Galileo expressed his concern at the hardening of the attitude of some of his fellow citizens towards him, Cigoli attempted to lift his spirits. He wrote from Rome that he had thought up a witty device for the head of the 'league of pigeons', as he sarcastically called Galileo's opponents in Florence headed by Ludovico delle Colombe:

> I have thought up a device for the league and its head: a fireplace without an air vent so that, when you light a fire in it, the smoke cannot escape, but returns and fills the room in which those persons for whom night precedes evening meet.[51]

In the course of 1612 other signals indicated that a substantial and composite group of opponents was preparing the attack in Florence. Besides some of the Fathers of the Dominican convents of Santa Maria Novella, supported by the Archbishop, the growing army of adversaries included influential members of the Ducal family: not only Prince Giovanni de' Medici, but also the Grand Duchess Christine, whose religious zeal and generous goodwill towards the Society of Jesus were well known. In Florence, as in Rome, Galileo had to confront the menacing alliance of opponents who rose to the defence of the traditional natural philosophy and cosmology and authoritative representatives of the clergy who denounced the perils to the faith of the new system of the universe.

The strategy that Father Maraffi had disclosed to Cigoli was soon put into effect: the offensive was launched on 2 November 1612 with a sermon delivered by Niccolò Lorini from the Convent of San Matteo in Arcetri. The Dominican denounced the incompatibility of the Copernican system with Holy Scripture. After his outburst, Lorini hastened to apologise to Galileo, probably at the instigation of some influential protector of the Tuscan scientist at court; but his were crocodile tears. Galileo hastened to inform Cesi of the threatening episode. The Prince expressed his solidarity and urged him not to surrender: the final victory would be his.[52]

---

51 Ibid., p. 476, letter of 1 February 1613. 'League of pigeons' is a play on the surname of Delle Colombe ('of the doves').

52 Ibid., p. 428, letter of Cesi to Galileo, 30 November 1612.

CHAPTER 6

Images of Nature: Book or Theatre?

Prohibition 'Is Also Done in Case of Doubt'

Between 1614 and 1615 Cesi repeatedly urged Galileo to moderate what he regarded as his excessive determination in fighting for the affirmation of the Copernican vision. As soon as he heard of the virulent attack launched by Tommaso Caccini from the pulpit of Santa Maria Novella on 21 December 1614, he sent a long and impassioned letter from Acquasparta to Galileo. He was above all sorry not to be in Rome, where he would have been able 'to judiciously test the waters and then to act as satisfactorily as possible'. However he begged Galileo not to reply to 'those perfidious and rabid persons' who sought only to distract him from his studies, his 'heroic and most useful discoveries and works'.[1] Cesi added to the exhortation a detailed personal analysis of the situation, from which he concluded that the Tuscan scientist should absolutely refrain from reacting to the provocation. It is a lucid letter, far from the sometimes pompous and emphatic prose of the Prince; it recalls the direct and dry style of Machiavelli's analyses of the phenomenology of power and of the powerful: 'You would not obtain what is needed by replying, and might stir up the others even more if you do not act with great caution'.[2] To dispel Galileo's illusion that he could win the Copernican battle, Federico communicated to him the peremptory judgement of Bellarmine:

> Regarding the opinion of Copernicus, Bellarmine himself, who is one of the heads of the Congregations in these matters, told me that he considers it heretical and that the motion of the Earth is without any doubt contrary to Scripture: take note.[3]

Cesi had received from the Cardinal authoritative confirmation of the conviction that he had held for some time:

1 OG XII, p. 129, letter from Acquasparta, 12 January 1615.

2 Ibid.

3 Ibid.

 | DOI 10.1163/9789004342323_007

> I have always suspected that if, in due course, the Congregation of the Index were to examine Copernicus, it would prohibit him out of hand.[4]

The only defence strategy that appeared feasible to the Prince was to denounce Caccini for having accused not only Galileo but the entire class of mathematicians of maintaining doctrines threatening to the faith. Still, various considerations led him to doubt the efficacy of even that move: the Dominicans would close ranks to defend a member of their Order; the Fathers would justify the attack on the entire class of mathematicians by arguing that Caccini had merely got carried away; finally, even if his superiors were to reprove him for his excessive animosity, 'the resulting sanction would be mild and secret'.[5] The only practical way to react to the violent attack was to invite the persons present at the outburst, but not mathematicians by profession and without mentioning Galileo, to testify that Caccini had declared 'that mathematics is a diabolical art and that mathematicians, as the authors of all heresies, should be expelled from every State'. This move would be even more effective if 'the mathematicians who were members of this religion' especially those who had been direct witnesses of the denunciation, were to protest.[6]

If Galileo still insisted on denouncing Caccini in Rome, Cesi suggested that the case be put to the Congregation of 'the Cardinals which judges the bishops and regular clergy' because it did not include supporters of the Dominican. Immediately afterwards, however, he repeated the need to act with prudence and, above all, to avoid referring to Copernicus in any way: 'to prevent this from becoming an opportunity for another Congregation to decide whether his opinion is to be permitted or condemned'. In which case, he warned, the cause 'would be lost, in the light of what has been said and of the large number of Peripatetics'.[7]

For Cesi, the situation called for restraint in speech and in writing, at least 'until the opinion of Copernicus has been examined and harmonised with Holy Scripture with approved theological arguments'. He concluded with a warning:

> Remember that it is the easiest thing to prohibit or to suspend, and it is also done in case of doubt. Telesio and Patrizi are prohibited; and unless other reasons are at hand, this decision always prevails, that since there

4 Ibid.

5 Ibid., p. 130.

6 Ibid.

7 Ibid. Cesi is here warning Galileo of the risk that the case might be transferred to the Congregation of the Index to decide whether Copernicus' book should be prohibited.

> are books in excess, those that are good and safe should be read; and those that are contrary to Aristotle are the most hated of all.[8]

The Prince's words proved prophetic. Matters proceeded as he had feared. His recommendation to keep a low profile until someone had demonstrated the compatibility of Copernicanism with Holy Scripture was not, however, the result of his prophetic vision. In his mind that 'someone' had a name and surname: that of the Carmelite Friar Paolo Antonio Foscarini, whom he knew to be actively involved in that delicate operation.

Cesi had not only foreseen the final sentence; thanks to his profound knowledge of the mechanisms that governed the processes of decision-making in the Eternal City, he had already written its motivation. Moreover, if he had ever had need of it, he had received precise confirmation from Bellarmine, who, as he knew perfectly well, would play a major role if Galileo refused to abandon his struggle.

Cesi's letter induces us to resume the question of the irresponsible and aggressive attitude which, according to some interpreters, was the principle cause of Galileo's severe punishment on the part of the Church. Indeed, reading a document like Cesi's letter with our knowledge of how the affair ended might lead us to conclude that Galileo, blinded by presumption, was a victim of his own arrogance. The dramatic scenario sketched by his friend and sincere admirer must have appeared perfectly credible to his eyes, based as it was on the Prince's profound familiarity with the Roman scene. Moreover, Federico's menacing prospect was confirmed by the explicit statement of Bellarmine. Galileo rejected the Prince's advice not because of his aggressive and quarrelsome character, nor because he failed to understand the situation. He had decided to fight a battle without quarter for the affirmation of the truth of the Copernican vision inspired by a fundamental question of principle. To give up would have meant failing to live up to the task to which he assigned an essential value. Although Galileo recognised that Federico's friendly warnings were well founded, he decided to take up the challenge, conscious of the risks involved, but driven by the conviction that in the end he would succeed if not in imposing his own ideas, at least in obtaining recognition of the freedom to profess them.

Paradoxically, if we were to accept the logic of hindsight, we would inevitably conclude that Galileo's analysis of the situation was not only lucid but actually prophetic: his defeat was rapidly transformed into an enduring triumph,

8 Ibid.

one that left an indelible black mark on the institution that decided to violently oppose his reform of cosmology and natural philosophy. Cesi's prophecy of how the clash with the Church would end should be juxtaposed with the prophetic anathema that Galileo launched against the judges who condemned him in 1633:

> And who can doubt that it will lead to the worst disorders when minds created free by God are compelled to submit slavishly to an outside will? When we are asked to deny our senses and subject them to the whim of others? When people devoid of whatsoever competence are made judges over experts and are granted authority to treat them as they please? These are the novelties which bring about the ruin of Republics and the subversion of States.[9]

It is as though Galileo had looked beyond the dramatic situation in which he found himself at that moment, foreseeing that, in the not too distant future, the accusatory finger would be pointed no longer at him but at his judges and at the Church that had armed them:

> Take note, you theologians, that in wishing to make propositions concerning the motion and rest of the Sun and the Earth a matter of faith, you run the risk of perhaps with time having to condemn for heresy those who assert that the Earth is still and the Sun is in motion: *with time, I say,* when it has been proved by necessary evidence and demonstrations that the Earth moves and the Sun is still in the centre of the universe.[10]

Galileo received a fresh and most heartfelt call to moderation from Giovanni Ciampoli on 21 March 1615. Begging him to stop insisting on Copernicus, the future Lincean invited Galileo to untiringly repeat declarations of his full submission to the authority and teachings of the Church, as well as to refrain from engaging in the interpretation of Holy Scripture. Ciampoli fully shared Cesi's concern and was in agreement with the cautious strategy that the Prince had suggested. All the same, Ciampoli had a reassuring message for the Tuscan scientist, who had anxiously asked him about the reception of Caccini's denunciation in Rome: nobody in the Eternal City was talking about it; the most

---

9 OG VII, p. 540.

10 Ibid., p. 541, emphasis added.

authoritative Cardinals and even the hierarchy of the Dominican Order had no time to waste on this nonsense.[11]

Ciampoli's optimism was based on the conviction that Galileo would heed the Prince's advice not to defend the Copernican vision as the true system of the world or to mix cosmology with theology. The same optimism pervades a previous letter from Cesi in which he informed Galileo that a Carmelite Father had published a book in defence of 'the opinion of Copernicus'. This work could not have appeared at a more favourable moment, Cesi declared, whose only fear was that the initiative of the Friar might 'increase the fury of the adversaries'. He concluded nevertheless that he was confident that it would help the cause.[12] Both Ciampoli and Cesi optimistically assumed that Galileo would accept the advice to avoid categorically proclaiming the truth of the Copernican vision and would adopt the concordist strategy of Father Foscarini. However, aware of the impossibility of talking about Copernicus without running great risks, the Prince reproached Foscarini for having improvidently declared that the Linceans adhered compactly to the Copernican vision:

> He [Foscarini] considers Copernicans all of our company, while that is not the case, commonly professing as we do only the freedom of thought in natural philosophy.[13]

This letter reinforced Galileo's conviction that the Prince would not follow him in the battle that he intended to wage. Cesi was not prepared to commit either his own person or his Academy to the defence of the truth of the Copernican system. He would vigorously uphold Galileo's right to profess it, but on a precise condition: that he did not attempt to proclaim the superiority of mathematical demonstrations and of the evidence provided by observations and experiments to the words of Holy Scripture.

Cesi's priority was to interpret Scripture using the most refined philological instruments to show that it was in perfect agreement with the celestial novelties. Galileo couched the question of the relation between nature and Scripture in very different terms: there was no need to devote oneself to demonstrating that Scripture was in perfect agreement with the truths of nature; the sacred texts should simply disappear from the horizon of discussions on natural philosophy.

---

11 OG XII, p. 160, letter of 21 March 1615.

12 Ibid., p. 150. Cesi was referring to the *Lettera sopra l'opinione de' Pittagorici* (*Letter on the opinion of the followers of Pythagoras*) of Paolo Antonio Foscarini (Foscarini 1615).

13 Ibid.

This distinction helps us understand the meaning that Galileo attributed to the *Copernican letters* and the distance between his strategy and that envisaged by Cesi. Their ways had parted, at least on this question. And they had done so because Galileo had decided not to accept the reassuring compromise proposed by the Prince. They were both fighting for the affirmation of freedom of thought in natural philosophy, but they understood that freedom in different ways.

The optimism of Ciampoli and Cesi was soon to be replaced by the keenest apprehension. When they sent the above-mentioned letters to Galileo, neither of them had yet seen the Copernican letter that Galileo had addressed to Benedetto Castelli.[14] No sooner had Cesi, Ciampoli, Monsignor Dini and the other exponents of the Lincean circle read it, than they became aware of the misunderstanding: Galileo had not the slightest intention of following the course carefully outlined by the Prince and the Lincean circle.

On the same day, 7 March, on which Cesi wrote to Galileo sending him the letter of Father Foscarini, another letter was sent to Florence from Rome, this time from Piero Dini. Unlike Cesi, he had already read the letter to Castelli and had had several copies made and distributed – no doubt on the instructions of Galileo – to exponents of the Roman clergy, including Father Grienberger and Cardinal Bellarmine. Dini had also exchanged views with Bellarmine on the contents of Galileo's letter to Castelli. As he reported to the Tuscan scientist, the Cardinal had predicted how this delicate *affaire* would end if Foscarini and Galileo did not retreat from their positions:

> As for Copernicus, His Excellency [Bellarmine] declares that he cannot believe that his book will be prohibited, but the worst that can happen, in his opinion, is the need to add some marginal notes stating that his doctrine was introduced to save the phenomena […].[15]

Dini went on to inform Galileo that, according to the Cardinal, it was impossible to harmonize the numerous affirmations of the centrality and immobility of the Earth in Scripture with the Copernican vision. To Dini's objection that those passages might be interpreted in such a way as to reconcile them with the heliocentric hypothesis, the Cardinal replied drily 'that it was

14 In his letter of 7 March 1616, Cesi had asked Galileo to send him a copy of the Copernican letter to Castelli: 'It will be very useful and of particular satisfaction to me to have the letter that you say you have written on the subject […]' (ibid.).

15 Ibid., p. 151.

not something that could be permitted'. He invited Dini to advise Galileo to submit to the supreme authority of the Church: if 'he remembers to submit to the determinations of the Holy Church [...], it will be found to help him considerably'.[16]

The next day Dini went to see Father Grienberger in the hope of receiving more encouraging signs, but in vain. Even the most receptive of the Jesuits to the Galilean discoveries demonstrated a prudent alignment. He used an argument that Bellarmine would repeat soon afterwards: it would only be admissible to interpret Scripture after an incontrovertible physical demonstration of the truth of the Copernican system had been provided.[17]

A week later, Dini passed on the opinions of other influential persons to Galileo, all of them in agreement with the course indicated by Bellarmine. The voice of Cardinal Maffeo Barberini joined the chorus not long afterwards: Galileo 'should speak with caution and only as a professor of mathematics'.[18]

It seems evident that Cesi and Dini considered Bellarmine as the privileged interlocutor to understand whether there was room for the legitimization of the Copernican vision based on the interpretation of Holy Scripture. Although with an amiable tone, the Jesuit expressed his clear opposition to the idea. He manifested his conviction that the heliocentric vision was groundless or even absurd not only on theological but also on physical grounds. It followed that he would be bound to oppose any attempt at interpreting Holy Scripture in such a way as to make it agree with the Copernican conception. It was above all this signal from the Cardinal that alarmed Cesi and Dini, prompting them to redouble their efforts in braking the impetus with which Galileo maintained the central position of the Sun and the mobility of the Earth not as a hypothesis but as the real structure of the universe.

In the same weeks Bellarmine, after reading Foscarini's *Letter*, had not hesitated to convey his perplexity directly to the Carmelite. Knowing that Foscarini was in contact with the Tuscan scientist through Cesi, the Cardinal included a message for Galileo in the letter that he sent to the Friar:

> It seems prudent to me for you and Signor Galileo to be content with speaking *ex suppositione* and not absolutely, as I have always believed that Copernicus spoke.[19]

16 Ibid.

17 Ibid.

18 Ibid., p. 555, letter of 14 March 1615.

19 Ibid., p. 171, letter of 12 April 1615.

The mathematician should know his place; it was not for him to establish the real meaning of Holy Scripture: proposing interpretations of the Bible at odds with those established by the Church Fathers would not be permitted. The Cardinal gave Foscarini – though once again the real target was Galileo – a menacing warning:

> With your prudence you should consider whether the Church can allow giving an interpretation to the Scriptures in a sense contrary to the Holy Fathers and all the Greek and Latin exegetes.[20]

It was Cesi, Dini and persons within the Lincean circle who managed the contacts between Foscarini, Galileo and Bellarmine during those weeks in the hope of finding a compromise that would avoid a disastrous clash with the Church.

By now the cards were on the table in both Rome and Florence. Galileo had become fully aware of the direction which Cesi and the Roman friends, having heard the opinions of leading clergy, wanted him to take. On the other side, after reading his Copernican letter to Castelli, Galileo's Roman supporters had realised his true intentions.

Other important events, which took place between March and April 1615, contributed to increase the concerns of Federico and the Linceans. On 20 March Tommaso Caccini gave depositions against Galileo before the Tribunal of the Inquisition.[21] A few days later, on 28 March, Ciampoli informed Galileo that he had read the letter that the Tuscan scientist had recently sent to Piero Dini in Rome.[22] The attention of Ciampoli and Dini had been particularly drawn to the pages in which Galileo had proposed a bold interpretation of the passages from Psalm 18 to dispel Bellarmine's conviction that they were absolutely incompatible with the Copernican doctrine. While admiring the ingenuity of Galileo's interpretation Ciampoli, speaking also on behalf of Cesi, considered it essential not to insist on that course of action:

> It is perhaps best not to dwell too much on it [the interpretation of Scripture]; this was the opinion of Prince Cesi to avoid being accused of wanting to sound out the defence where no one is on the offensive.[23]

---

20 Ibid., p. 172.

21 OG XIX, pp. 307–11.

22 OG V, pp. 297–305; see Galilei 2009, pp. 17–32.

23 OG XII, p. 163.

A few days earlier Ciampoli had recommended that Galileo not become involved in the interpretation of Holy Scripture. He was seconded by Cardinal Del Monte, who had formed this opinion after a long discussion with Bellarmine. Ciampoli warned Galileo not to follow the example of Foscarini, whose book 'because it strove to accord [...] Scripture with the new cosmology runs the great risk of being suspended at the first Congregation of the Holy Office, which will be held in a month's time'.[24]

The Copernican letter to Dini had been conceived by Galileo as a response to Bellarmine's reservations regarding the Copernican system, above all because of its incompatibility with Holy Scripture. The Tuscan scientist had asked the Roman friends to convey the letter to the Cardinal, but they hesitated. It was not until 2 May that Dini summoned the courage to inform Galileo that the letter had not been delivered. Dini assigned responsibility for the decision to Cesi, with whom he had had a long exchange of views:

> The Prince thought that I should not present that letter to him [Bellarmine], for, as he and many other authoritative clerics were Peripatetics, he was afraid of irritating them on a point that had already been won, that is, that it is admissible to write as a mathematician and by way of hypothesis, as they claim that Copernicus has done. Even if this is not conceded by his followers, it is enough for the others, and the result is the same, that is, to grant freedom of expression as long as one does not, as has been said on previous occasions, enter the sacristy.[25]

A few weeks later, Prince Cesi repeated the reasons that suggested proceeding with the greatest caution. After several efforts 'and having studied the lie of the land', he had arrived at the conclusion 'that neither the first author [Copernicus], nor the letter of the Father [Foscarini], nor the [Copernican] opinion itself will run any risk as long as due caution is observed'.[26] Presumably Cesi was here referring to reassurances received from Bellarmine, who had explicitly invited Foscarini and Galileo to speak *ex hypothesi*. Cesi added that, after the *Letter on the opinion of the followers of Pythagoras*, Foscarini was engaged in writing another text to show with new arguments the agreement between Holy Scripture and the celestial novelties. Until publication of the Friar's book, the Prince added,

24 Ibid., p. 160, letter of 21 March 1615.

25 Ibid., p. 175.

26 Ibid., pp. 189–90, letter of 20 June 1615.

> it will be a necessary precaution [...] to be silent here [i.e. in the *Copernican letters*], not dwelling on this opinion any further, and skirting around it elsewhere [...], and in case others speak about it, to say that it is not a truthful doctrine, but, stepping aside and leaving it to the judgement of the superiors, that it is only used *ex hypothesi* [...] and to avoid declaring it to be true.[27]

Acting in this way, everything will fall into place without any trauma: 'the opinion will remain admissible and approved so fully that whoever will wish to maintain it will be able to do so freely, *as is the case with matters of a purely physical and mathematical nature*'.[28] The sting was in the tail: like Dini, Cesi sugared the bitter pill, making it appear a great success to have obtained from Bellarmine and from the keen-eyed guardians of orthodoxy the promise that it would be admissible to speak about Copernicus and also to teach his doctrine as long as it was *ex hypothesi*.

When he sent this letter to Galileo, Cesi had not yet read the latter's reply to Dini's letter of 2 May. Had he done so, he would have changed his tone. Galileo's (undated) reply to Dini was probably sent to Rome towards the end of May. It is an important document that marks his clear distancing from Cesi and the Roman circles close to the Prince and his Academy.

For Dini, Cesi and for Galileo's most sincere friends in the Eternal City to be able to speak freely about Copernicus, even if only *ex hypothesi* as a mathematician, was, as Dini put it, 'a point that had already been won'.[29] Galileo's perception of the situation was completely different from that of the presumed victory that Federico and his companions were triumphantly claiming. This time he did not mince words in his reply to the umpteenth recommendation to remain within the hypothetical dimension of mathematical reasoning. He wanted Cesi and Dini to realise once and for all that he would not follow their friendly advice.[30]

The letter sent to Dini in May 1615 was that of a man tormented by physical ailments and bowed down by the dark clouds that he saw gathering on the horizon towards which he had decided to proceed with determination. He replied to the reservations of his friends regarding his bold sallies into the minefield of biblical exegesis that he had not been 'the first mover'. He would never have

27 Ibid., p. 190.

28 Ibid., emphasis added.

29 Ibid., p. 175.

30 Ibid., p. 185.

spent his time on scriptural exegesis if Bellarmine, other representatives of the clergy and the Aristotelians had not driven him to do so. He had been forced to engage on that terrain because those obstinate adversaries had sought in every way to prevent him from demonstrating the greater strength of the Copernican vision by invoking its incompatibility with Scripture and menacing him with the threat of prohibition and persecution. Copernicus' book is 'admitted by the Holy Church', he proudly declared. How could they maintain, so many years after its publication, that it contained 'propositions contrary to the faith'? On what grounds do they assume the right to prohibit the interpretation of Scripture, shutting me up because 'I want, as much as I can, to demonstrate that it is perhaps they who are mistaken'?[31]

For Galileo, the only way to confront and settle the question of the compatibility of the Copernican vision with Holy Scripture was

> to show with a thousand proofs that it [the Copernican system] is true and that the contrary cannot be in any way maintained; therefore, since two truths cannot be contrary to one another, it is necessary that it is at one with Holy Scripture.[32]

---

31 Ibid., p. 184.

32 Ibid. Paolo Rossi stressed the inadequacy of the interpretation of the passages of the *Copernican letters* on the two Books of Nature as evidence of the convinced adoption of a concordist strategy on the part of Galileo (Rossi 1978). At the conclusion of his analysis of the texts in which Galileo ventured into the field of biblical exegesis to demonstrate that Holy Scripture was not incompatible with the heliocentric structure of the universe, Rossi stated that the Tuscan scientist's concordist endeavour 'was actually quite short and fragmentary' (p. 66). Galileo's main purpose in those writings was to state that Holy Scripture has no authority when it comes to understanding the structure of the universe. As Rossi recalled (pp. 68–9), as more recent studies of the *Copernican letters* confirm, and as Galileo's explicit declarations make clear ('willingly I leave the task of interpreting Holy Scripture to those who are infinitely more capable than I am', OG V, p. 300), he agreed to adopt a concordist approach on very few occasions, and exclusively to evade the threatening accusations levelled against him. He was driven in that direction also by Cesi and other Linceans, who had made serious efforts to demonstrate the agreement between the celestial novelties and Holy Scripture through a rigorous exegesis of those scriptural passages that, when taken literally, seemed to contradict the new conception of the universe. Galileo's rare concordist statements are at odds with the primary aim that he had assigned to the *Copernican letters*: to claim the autonomy of natural philosophy from theology. He would stubbornly continue to pursue that aim even after the anti-Copernican decree of 1616. In Galileo's eyes, the relation between the Book of Nature and Holy Scripture should be based on the principle of their distinct fields of competence. This implied that natural philosophy was not subordinate to theology. On the question of Galileo's alleged concordist strategy, see the introduction in Galilei 2009.

This is why he challenged his opponents: let them have the courage to listen to my arguments; if they are as absurd as they claim, they will have no difficulty in demolishing them. These bold declarations brought it home to Cesi, Dini and Ciampoli that the strategy that Galileo intended to follow diverged from the one that they had been recommending to him.

Galileo repeated his resolute position in the *Letter to Christine of Lorraine*,[33] which soon reached his friends in Rome. Their reading of this document increased their concern. Although among the *Copernican letters* that addressed to the Grand Duchess of Tuscany is the most accommodating to the prudent Lincean paradigm, Galileo emphasised once again his conviction of the physical truth of the Copernican system.[34] The very way in which he had striven to follow the concordist model suggested to him by his friends showed the radical difference of his conception of the relations between truths of nature and Holy Scripture. Even the most explicit step in the concordist direction taken in the letter to the Grand Duchess – the interpretation of the divine miracle that stopped the Sun in its course at Gibeon to allow the Israelites to triumph in battle over their enemies[35] – offers eloquent evidence of the distance of his hermeneutic method from that recommended and practised by the Prince. Galileo based his analysis on mathematical principles, by means of which he demonstrated that it was necessary to interpret the literal meaning of that passage if it

33 One may wonder why Galileo decided to address to the Grand Duchess Christine of Lorraine the most elaborated and cautious of the *Copernican letters*, a document that could be defined as an open letter conceived for wide diffusion. Since the discussions on the implications of the Copernican system had arisen during the dinner offered by the Grand Duchess in Pisa to a group of professors of the local university – as Castelli had informed Galileo (OG XI, pp. 605–6, letter of 14 December 1612) – it is probable that the Tuscan scientist considered it opportune to reply to her perplexities in the first person. He will also have thought that the involvement of Christine of Lorraine could help to shield him in his clashes with the vigilant Roman authorities and protect him from the attacks of his resolute Florentine enemies, for many of whom the Grand Duchess was an important point of reference. Unfortunately we do not know whether the dedication was the result of negotiations with the court, nor how Christine received it. There are many grounds to consider the Grand Duchess a key figure in the delicate balance between supporters and opponents of Galileo in the Medici court. It would therefore be important to reconstruct her role in the cultural life of Medici Tuscany and her attitude towards the events of the battle in which Galileo was engaged.

34 The letter to Christine was awaited eagerly in Rome. In his letter to Galileo of 16 May 1615, Dini urged Galileo, also on Cesi's behalf, to hasten the completion of the letter to the Grand Duchess. Galileo's Roman friends were confident that he would adhere in that text to the cautious strategy that they had been suggesting to him (OG XII, p. 181).

35 *Joshua* 10: 12–13.

were not to become nonsensical.[36] His message was anything but accommodating: the miraculous nature of the divine intervention could only be understood by those who knew perfectly the rigorous way in which nature operates.[37]

Galileo invited his Roman friends not to insist in their attempts to dissuade him from following his chosen path. While thanking them for their advice and for the warnings of the risks entailed, he informed them that he could not accept to talk and write about the Copernican system as a simple hypothesis. He considered the threat of the prohibition of *De revolutionibus* to be catastrophic also for the reputation of the Church itself. It seemed equally inconceivable to him that the Church could pronounce a formal sentence to establish that Copernicus was speaking in hypothetical terms:

> To declare that Copernicus did not consider the mobility of the Earth as a truth of nature, but only used it, as an astronomer, as a useful but false hypothesis to account for the celestial phenomena, and that it is therefore admissible to use it as such and to prohibit belief in its verity [...] would mean confessing to not having read his book.[38]

While declaring that he understood the reasons that had induced Cesi not to convey to Bellarmine his Copernican letter to Dini, Galileo expressed his concern that the Cardinal could interpret the fact that he had not replied to his anti-Copernican arguments as a retreat from the arena. That was something that he absolutely could not countenance:

> I would not like some important Roman authorities to think that I do not applaud the positions of Copernicus except as an astronomical but untrue hypothesis, nor that they might think that, being I myself considered one of the closest adherents to the doctrine of this author, all his other followers would think the same, thus concluding that it is erroneous regarding the truth of nature. This conclusion, if I am not wrong, would perhaps be a mistake, because the truth wins over all other demonstrations.[39]

Galileo had not the slightest intention of abandoning his struggle for the affirmation of the true constitution of the universe. He begged Dini to inform

36 OG V, pp. 345–7.
37 Cf. Redondi 2009, p. 126.
38 OG XII, p. 184.
39 Ibid., p. 185.

Bellarmine of his decision not to accept the escape route that had been offered him. He threw down the gauntlet, inviting his prelate friend to show the Cardinal the resolute letter that he had written to him: 'I wish that you show it to him and read it together'. He also urged Dini to insist that Bellarmine give his arguments a hearing before passing sentence on Copernicus. In doing so the Cardinal would help the Church avoid the risk of condemning a doctrine that 'sooner or later could be demonstrated to be true' with irreparable consequences for those who had formally declared it false.[40]

His desire to make a desperate attempt to halt the feared drift towards the prohibition of Copernicus led Galileo to compile a new text in defence of the heliocentric system: the *Considerazioni sopra l'opinione copernicana* (*Considerations on the Copernican opinion*).[41] This text, written between the spring and the summer of 1615, was conceived as a systematic and carefully thought-out reply to the objections against the Copernican system raised by Bellarmine in the letter to Foscarini. In the *Considerations* the arguments against the physical truth of the Copernican system were openly confronted and the scientific and theological statements of the Cardinal contested in respectful but firm terms. Bellarmine was wrong to suppose that Copernicus had conceived his system only as a mathematical hypothesis. Galileo invited the Cardinal to read carefully the dedicatory letter of *De revolutionibus* to Pope Paul III: he would be made perfectly aware that the Polish astronomer was writing and thinking as a natural philosopher.[42] Like Ptolemy, Copernicus considered astronomy a discipline that investigates the true structure of the universe. How could the Cardinal justify the 'juridical' condemnation of that doctrine solely because it contradicted the literal sense of a few passages of Holy Scripture, without demonstrating its falsity by means of rigorous physical demonstrations?[43]

---

40 Ibid.

41 OG V, pp. 349–70; see too the annotated new edition in Galilei 2009, pp. 85–110.

42 Galileo did not know the identity of the author of the anonymous preface to *De revolutionibus*, but he was absolutely convinced that it had not been written by Copernicus since it completely altered his vision. In the *Considerations on the Copernican opinion* he supposed that that anonymous text had been added by the publisher: 'On the reverse of the title page of the book of Copernicus there is a certain note to the reader, which is not by the author, since it refers to him in the third person and does not name him [...]. I think that this text was added by the bookseller to facilitate the sale of the book, fearing that it would have been considered a fantastic chimera by everybody [...]' (OG V, pp. 360–1; see also Galilei 2009, pp. 99–100).

43 OG V, p. 364.

Galileo took up the challenge launched by Bellarmine: the Cardinal had declared that it would be admissible to interpret Holy Scripture after incontestable physical demonstration of the truth of the Copernican structure of the universe. Only at that point (which the Jesuit was convinced would never be reached) would the theologians have to accept that new system of the world, attributing the cause of the previous misunderstanding to 'the weakness of our intellect [...] which had not penetrated the true meaning of the Scriptures'.[44] Galileo turned the tables on Bellarmine by challenging him to provide a scientific demonstration of the falsity of the Copernican vision:

> Now that rigorous demonstrations have raised doubts about the true meaning of some passages of Scripture, we can no longer take it to be evident and secure in order to refute the same proposition [the Copernican system], but it is necessary to undermine the demonstrations and to discover their fallacy with other arguments, experiments and more precise observations [...]. [Only then, and not before] could we be assured of the true meaning of the Scriptures and make use of them with certainty. The secure way, therefore, is to begin with the demonstrations, confirming the true and refuting the fallacious ones.[45]

This was the dogged way in which Galileo examined and refuted all the objections raised against Copernicus by Bellarmine. In spite of Cesi's recommendations to be prudent, Galileo had not budged an inch: if anything, in the *Considerations* he was even more resolute in affirming the truth of the Copernican vision than he had been in the letters to Castelli, Dini and Christine of Lorraine.

The Tuscan scientist ignored the advice of Cesi and the Roman friends to proceed with caution in yet another text: the letter-treatise on the causes of tides that he had addressed to Cardinal Orsini immediately after his arrival in Rome at the end of 1615. Galileo had personally delivered to the Cardinal the letter of the Grand Duke requesting Orsini to offer collaboration and support to his Philosopher and Mathematician engaged in a particularly delicate matter. On that occasion Galileo explained to Orsini his own conviction that the tides were the products of the combined effect on the seas of the diurnal and annual movements of the Earth. At the Cardinal's request, Galileo wrote the *Discourse on the ebb and flow of the sea* and sent it to Orsini on 8 January.

---

44 OG XII, p. 172, letter of Bellarmine to Foscarini, 12 April 1615.

45 OG V, pp. 365–6.

It was the most openly Copernican writing that Galileo had composed to date. It is difficult to imagine that Cesi had not read it beforehand, or had not at least been informed of the arguments it contained; nor will he have hesitated to renew the pressure on Galileo to tone down the categorical affirmations of the physical truth of the diurnal and annual movements of the Earth. Galileo heeded those recommendations in his own way: at the end of the *Discourse* he put a declaration of submission to 'the most eminent knowledge' of the supreme authority of the Church in the event that the arguments in this text, founded on 'philosophical and astronomical reasonings and observations,' should be declared 'fallacious or erroneous' on the basis of theological considerations.[46] It will have escaped no one that the declaration of submission to the supreme authority of theology was simulated.

The note that accompanied the dispatch of a copy of the *Discourse* to Archduke Leopold of Austria two years later shows that, even after the anti-Copernican decree, while hiding behind the fragile mask of the declaration of submission to the supreme authority of the Church, Galileo continued to underline the cosmological implications of the celestial discoveries. He wrote to the Archduke that, in accordance with 'the determinations of the superiors', he should consider 'this present text [...] as a poem or a dream [...], a chimera that I have been the first to conceive'.[47] It will be recalled that Descartes was to use the semantically analogous metaphor of the 'fable' to dissimulate the explicitly Copernican positions adopted in his *Monde*.

Although we have no direct evidence of Cesi's reactions to the *Considerations* and the *Discourse* addressed to Cardinal Orsini, it is not difficult to imagine that, after leafing through the pages of these texts, he must have once again shivered with apprehension. Nor do we know whether the two texts ever reached the hands of Cardinal Bellarmine, for whom they had been conceived. It is probable, as in the case of the letter to Dini and, years later, of Galileo's reply to Francesco Ingoli, that the Prince exercised a friendly censorship. Federico was perfectly aware of the reaction that a reading of those texts would have produced in the mind of the Cardinal. He knew the man and his determined defence of orthodoxy too well to harbour the slightest hope that those writings could achieve the objective envisaged by Galileo. The *Considerations* and the letter-treatise on the tides were a further confirmation to Dini, Cesi and the Lincean circle that Galileo refused to play the game according to the rules that they considered the most reasonable and safe.

46 Ibid., p. 395.

47 OG XII, p. 391, letter of 23 May 1618.

On the other hand, it was evident to them that to insist in their recommendations would irrevocably compromise relations with the most illustrious member of the Academy. So it is not surprising that, after these lively discussions, the correspondence between Galileo and Cesi slowed down, while that with Dini came to an end.[48]

However, during the six months spent in Rome between the end of 1615 and the late spring of the following year Galileo had frequent opportunities to meet Cesi. During their meetings Federico would have continued to manifest his own scepticism regarding the outcome of the battle engaged by Galileo. Reasons of prudence prompted him to take a step back after the stern decision of the Church to suspend Copernicus and to prohibit the *Letter* of Father Foscarini.

The Carmelite paid a higher price than Galileo, who emerged from the anti-Copernican decree of March 1616 formally unscathed. Notwithstanding the fact that Foscarini had promptly heeded the recommendations to prudence of Bellarmine and Cesi, his letter had been printed, which made it appear particularly dangerous and called for its being sanctioned. Remus Quietanus, an authoritative witness to the events in Rome during those delicate months, also placed responsibility for the process that led to the condemnation of Copernicus on the shoulders of Foscarini.[49]

Having taken stock of the violent reactions to his *Letter on the opinion of the followers of Pythagoras*, and obviously paying due attention to the peremptory request from Bellarmine to refrain from maintaining the physical truth of the Copernican doctrine even before the emission of the anti-Copernican decree, Foscarini had begun to write a new work in which he considerably moderated his tone, although still emphasizing the need to interpret Holy Scripture in order to harmonize it with natural truths. The Friar had informed Galileo of this new enterprise, specifying that he intended to proceed 'with the circumspection that is required'.[50] He would adopt a prudent formula: that of a dialogue between a supporter of Ptolemy and one of Copernicus, without explicitly

48 Although he lived until 1625, Dini disappeared from Galileo's horizon, while Cesi's letters to Galileo after the latter's return to Florence can be counted on the fingers of one hand.

49 Ibid., p. 481, letter of 13 August 1619: '[Foscarini] diffused this opinion among the people in Italian in a printed text [considered aggravating factors] leading to perilous consequences and opinions'. Remus believed that Galileo too had contributed to the precipitation of the situation by the determination with which he had defended his own convictions. Cf. Bucciantini 1995, p. 125.

50 OG XII, p. 215, letter of Foscarini to Galileo (undated, but end 1615-beginning 1616).

assuming a position in favour of either. It was a model that Cesi had recommended several times to Galileo, who would adopt it many years later in the *Dialogue concerning the two chief world systems*. Foscarini was confident that this option would enable him to write about Copernicus without running the risks of which Bellarmine had warned him.

Although in his letter to Galileo Foscarini manifested his sympathy for the Copernican vision, he did not adopt a clear position in favour of the physical truth of the heliocentric system. Galileo must have felt somewhat embarrassed when he discovered that his ally intended to demonstrate the superiority of the Copernican doctrine not only with philosophical reflections, observations and mathematical reasonings, but also with arguments:

> [...] drawn from the deeply hidden allegories of the most ancient fables of the first and oldest poets from which every philosophy began [the same method that Cesi was to systematically apply], from the oracles of the pagan gods and the Sibyls and others, from many hieroglyphs of the Egyptians, from many mysterious images [...].[51]

Galileo must have been equally dismayed to learn that Foscarini was convinced that of the four Aristotelian elements, only water followed the Earth in the diurnal motion attributed to it by Copernicus, while air and fire remained unaffected. Finally, the question with which Foscarini concluded his letter must have made him laugh:

> Would a man rising to the spheres of the Sun or Moon or Venus or Mercury turn his feet towards that sphere and his head towards the Earth, and, vice versa, in descending from there would he do the opposite?[52]

It should also be noted that, in order to preserve the Sun's perfection, Foscarini had declared that the sunspots discovered by Galileo were not situated on its surface.[53]

It is not surprising that, after reading the *Letter on the opinion of the followers of Pythagoras*, Benedetto Castelli had been scandalised by the Carmelite's ignorance on matters of cosmology.[54]

51 Ibid., p. 216.

52 OG XIII, pp. 219–20.

53 Foscarini 1615, p. 45; see Caroti 1987, pp. 106–7.

54 Ibid., p. 165.

The new work that Foscarini was planning – a sort of encyclopaedia of nature – reveals the same programmatic integration of erudition, interpretation of ancient knowledge and the search for occult patterns linking natural phenomena that characterise the greater part of Cesi's research programme. Foscarini had published in 1613 the *Institutionum omnis generis doctrinarum Synthaxis*, a work that will have certainly come to Federico's attention. It dealt with many of the topics at the heart of the *Mirror of Reason* and the *Theatre of Nature* that the Prince had delineated in the *Indicatio*.[55] In the face of this evidence Galileo was bound to be assailed by doubts as to whether the members of the party who supported him had really understood the spirit and objectives of the battle that he had decided to fight and, above all, whether they were prepared to accept its implications.

The Natural Desire for Knowledge

At the beginning of December 1615 Galileo moved to Rome in the attempt to reverse the outcome of the battle for the affirmation of the Copernican truth which, judging by the news from his friends in the Eternal City, he feared could prove futile.

As in 1611, Galileo was on an official mission, authorized and funded by the Grand Duke, with the usual dispatch of letters of recommendation to Cardinals and exponents of the patriciate close to the Medici court inviting them to lend full support to the scientist's cause.

The picture that Galileo was forming of the situation in the course of his long stay in Rome can be reconstructed from his intense correspondence with Curzio Picchena, the Secretary of State of the Grand Duchy, to whom he sent frequent and detailed updates. His letters to Picchena display a view of the unfolding of his delicate mission that may appear, but only considering it in hindsight, excessively optimistic. Galileo had every confidence in the strength of his arguments. Moreover, he had realized that he could easily gain the upper hand in debates with even the most hostile of his adversaries.

During the same weeks other actors of the Roman scene were sending to Florence radically different accounts of the situation and of its evolution. A comparison of the testimonies of attentive observers with the accounts Galileo sent to Picchena seems to reveal an almost schizophrenic character. On the one hand, we recognise the familiar direct and polemical style with which he

55 See above, pp. 120ff.

confronted his adversaries. Eloquent testimony of this is provided by the letter of Antonio Querengo to Alessandro d'Este describing the bitter verbal duels between Galileo and his opponents. In spite of the fact that these conflicts concluded with the mortification of those who contested the Tuscan scientist's arguments, he failed to breach their obstinate resistance:

> Your Illustrious Excellency would have been delighted to hear Galileo debating, as he often does, amid fifteen or twenty interlocutors who make cruel attacks on him, now in one house and now in another. But he is fortified enough to make fools of them all; and although the novelty of his opinion fails to convince, he demonstrates the vanity of the majority of the arguments with which his opponents seek to bring him down [...]. What pleased me most was that, before replying to the contrary arguments, he amplified them and reinforced them with new and grandiose reasons thus making his adversaries appear all the more ridiculous when he demolished them.[56]

On the other hand, in his letter to Picchena only three days after the one sent by Querengo to Alessandro d'Este, Galileo informed the Secretary of State that the resolute opposition he was encountering and the risk of exposing himself unduly had led him to change strategy, assuming a more circumspect attitude and, above all, avoiding direct verbal confrontation with his adversaries. He hoped that, in proposing his ideas not orally but on paper, they would creep almost imperceptibly into their minds. As he confided to Picchena, this subtle strategy was based on dissimulation, circumvention, recourse to intermediaries to convey the information that he wanted to communicate without revealing himself as its source, and the circulation of writings that ended up 'by chance' in the hands of those whom he was most keen to win over to his cause. Behind his decision to assume a more cautious and indirect approach we recognize the effects of the advice received from trusted friends like Cesi and the Linceans. At this stage Galileo seems to have taken into account their understanding of how extremely delicate matters of this kind had to be conducted in the context of the Eternal City.

He explained to Picchena that he had engaged himself in 'operations [...] accidentally more difficult and prolonged than what their nature demanded', not only to facilitate the success of his mission, but also to protect his friends, who 'cannot offer any support to me without incurring the heaviest censure'.

56 OG XII, pp. 226–7, letter of 20 January 1616.

The meaning of this message is clear: he felt responsible for protecting the Princeps of the Academy and his fellow Linceans from suspicions of complicity. To avoid compromising his associates, he could not even use them as ambassadors.

> So I have to proceed with great labour and care, seeking third persons who, without even knowing why, serve as my intermediaries with the principal personages, in order that, incidentally and almost at their request, I may have the occasion to speak and explain the details of my interests. Moreover, I have to explain some points on paper and ensure that my writings secretly reach those whom I desire should read them, finding often an attitude more open to dead writings than to the spoken voice. Those writings are more easily received and contradicted without embarrassment, and finally their conclusions accepted when we have no other witnesses to our mental discourses than ourselves. We do not behave in such a way when we defend our opinions in public.[57]

The letter reveals his penetrating analysis of the psychology of his sceptical interlocutors, who would be more likely to change their minds after reading a text by themselves than in the presence of numerous witnesses under the insistent pressure of Galileo's dialectical arguments.

However, the Tuscan scientist soon wearied of the patient method suggested by his Roman friends and resumed the explicit and direct attitude that was more congenial to him. We find confirmation of this in the letter of 4 March from the Tuscan ambassador Piero Guicciardini to Cosimo II. This document reveals both the ambassador's disappointment as well as that of Cesi and the Linceans at fresh evidence of Galileo's recklessness and his confidence in the strength of his own arguments: 'Galileo has put his own opinion – Guicciardini wrote – above that of his friends'.[58]

In the documents relating to Galileo's stay in Rome in 1615–6 the name of Prince Cesi is hardly ever mentioned. We therefore lack clues as to the manner in which Cesi and the Linceans closest to him supported Galileo in those delicate weeks. It should be noted, however, that many of his meetings with the Roman prelates, exponents of the patriciate and learned personalities took place in the palace of the young Duke Virginio Cesarini, a cousin of the Prince and a future Lincean. It was probably Cesi himself who had asked Cesarini to

57 Ibid., pp. 227–8.

58 Ibid., p. 241.

lend his support to Galileo. The young Duke was closely linked with the Jesuits and enjoyed the unconditional admiration of Cardinal Bellarmine, who regarded him as a new Pico della Mirandola for his extraordinary erudition. Furthermore, he had strong ties of friendship and esteem with Giovanni Ciampoli, another sincere friend and admirer of Galileo and a future Lincean, who acted as Cesarini's personal secretary.

We do not even know how the Prince reacted to the publication of the decree of 5 March. The entire Lincean circle must have been bitterly distressed by the decision of the Church, above all from the awareness that that dramatic conclusion could have been avoided if Galileo, following the advice of Cesi and the Roman friends, had accepted the compromise of maintaining the Copernican system merely as a hypothesis. This must have been the tenor of the recriminations that Johannes Faber expressed to the ex-Lincean Johann Schreck immediately after receiving news of the emission of the decree. Faber's letter is lost but we can deduce its contents from Schreck's reply:

> I am surprised that Galileo is urging the motion of the Earth so strongly, whereas it would be enough to say that it is a hypothesis to facilitate astronomical calculation, whatever the truth itself may be.[59]

The only documents that provide information about Galileo's meetings with Cesi during the Roman mission of 1616 are the minutes of the two sessions of the Academy in which the Tuscan scientist participated. The first of these, held in Cesi's palace on 26 January, was attended by Francesco Stelluti, Luca Valerio and Johannes Faber. The Prince invited his fellow Linceans to propose new candidates for induction into the Academy. He himself began by nominating the Palermitan physician, mathematician and painter Domenico Valguarnera, the Neapolitan physician and philosopher Mario Schipani, the naturalist Francesco Imperato, also from Naples, the Roman Marquis Carlo Muti, a devotee of mathematics and poetry, and the young Duke Virginio Cesarini.

While failing to mention the fact that Cesarini was his cousin, Cesi presented a portrait of the Duke that hardly seems appropriate to support his candidature for membership of the Academy: 'extremely learned in scholastic philosophy and a follower of the Peripatetic sect'.[60] To justify his proposal, Cesi

59 CL, p. 564, letter of 6 April 1616.

60 Contributi I, p. 534.

immediately added that Virginio 'also paid attention to other [philosophers]'.[61] He knew that his cousin, an admirer of Lucretius, was attracted by the corpuscular philosophy that was at odds with the scholastic tradition.

We learn from the minutes of this academic meeting that Galileo intervened to champion Cesarini's candidature. It is likely that he had found the Prince's presentation lukewarm and, above all, contradictory. Moreover, he probably wanted to emphasise that he had gained the impression that the very learned young man, whom he had frequented assiduously during those Roman weeks, was displaying a growing interest in the celestial novelties and their cosmological implications. He was therefore absolutely in favour of his enrolment also to encourage him along that path. As the minutes record,

> Galileo praised the intellect of this almost adolescent very highly and joined his vote to that of the Prince so that Virginio be encouraged to pursue the freedom of thought and the search for truth.[62]

The most important event of that meeting, indeed the reason why it had been called, in addition to taking advantage of the exceptional presence of Galileo, was the presentation of the 'lucid and learned' lecture – the *Discourse on the natural desire for knowledge* – delivered by the Prince on the 'organization and purpose' of the Lincean Academy.[63] The *Discourse* was the fruit of long and careful meditation. Together with the later *Apiarium,* it is the only writing of the Prince to have the appearance of a completed work.[64] He expounded in it the founding principles and ends which inspired the intricate web of norms,

61 Ibid. The difference of opinion between the two cousins also emerges from Federico's notes for the Lincean biography of Cesarini. He recalled the early youth of Virginio as characterised by unconditional adhesion to Scholastic philosophy (he called him 'an extremely keen Scholastic philosopher'). He went on to say that he, Ciampoli and Galileo had made every attempt to convert him to the new philosophy. Finally, after their repeated insistence, Cesarini recognised 'that the direction proposed to him [by Cesi] was the true philosophical path to arrive at cognition of nature and of mathematical certainties' (ibid., pp. 783–4; see also Bellini 1997, pp. 281–3, and 2009, pp. 108–57). Cesi's draft was used by the Lincean Josse de Rycke, author of the official Lincean biography of Virginio (Rycke 1629). On the Flemish Lincean, see Contributi II, pp. 1133–220.

62 Ibid., p. 534.

63 Ibid., p. 533.

64 However, the *Discourse* was never published. The only writings of the Prince to be printed during his lifetime are the *Apiarium* and (although in an incomplete form) the *Tabulae Phytosophicae.*

regulations, procedures of enrolment and organisation that he had been compiling and refining for many years in the *Lynceographum*. Only a decade later were a limited number of those norms published in a booklet entitled *Praescriptiones Lynceae Academiae*. This severe mutilation of the ambitious project outlined in the *Lynceographum* is proof of Federico's growing realization of the impracticality of the utopian motivations that had led him to found the institution and to draw up its statutes.[65]

The *Discourse on the natural desire for knowledge* provides eloquent testimony of the non-linear course followed in the thirteen years since the foundation of the Academy.[66] Abandoning the prescriptive style of the *Lynceographum*, the text reveals Cesi's recognition of the changes that had taken place with regard to the ideals that had marked its beginnings. The Linceans were no longer a secret circle. They did not refrain from campaigning in public for the reform of knowledge and the values on which the search for truth should be based.[67] They no longer used the symbolic academic names of the early days and had abandoned the rule of communicating with one another in code. The image of the lynx had also undergone a transformation: from a symbol of the search for hidden truths to a naturalistic representation of the keen-eyed feline.

Within the sheltered walls of the Academy, Cesi proposed a lucid analysis of the dramatic situation of knowledge in post-Tridentine society. In pages characterised by a lofty style and an elevated moral tone, he denounced the degeneration of the culture of the time, inclined to the pursuit of material goods, to the dissimulation of thought, to the supine acceptance of traditional knowledge. University studies, which ought to reward excellence, ended with a degree which 'crowns all those who complete the course without any attention to whether they limp or proceed straight ahead [...]. As a consequence, the doctorate blocks the road to knowledge for many'.[68] Besides, the most flourishing studies were those of law and medicine to acquire professional skills with

65 The publication of the *Praescriptiones* marks the definitive abandonment of Cesi's original project to rigorously discipline the public and private conduct of the academicians (cf. Contributi I, p. 234). The *Praescriptiones* have been recently re-published, with introduction and commentary by Marco Guardo (Guardo-Orioli 2014).

66 I have used the text of the *Discourse* edited in Altieri Biagi 1969.

67 On the ambiguous attitude of the Prince and the other Linceans regarding the public or secret nature of their activities, see Olmi 1981.

68 Altieri Biagi 1969, p. 64.

a market value,[69] while the disciplines 'more able to satisfy the natural desire for knowledge, those that confer more cognition, perfection and ornament – sublime philosophy, mathematics and philological and poetic erudition' – were increasingly abandoned.[70]

Once students had completed their university education, every process of acquiring knowledge came to a halt: after graduation, students were left to their own devices. Cesi insisted at length on the degeneration of teaching institutions: the university lecture theatre was like a court, where a stale and hollow culture was represented with a worn-out ceremonial. Moreover, philosophers were more inclined to frequent the courts, where they played the role of 'parasite, clown or flatterer', than to teach in the universities.

Unlike the practical arts, which had made, and continued to make, impressive progress, theoretical disciplines and philosophy were anchored to a passive veneration of knowledge acquired in antiquity.

Cesi's speech was a passionate appeal to restore sense and dignity to the mission of schools and school teachers, urging them to impart knowledge that was neither crystallised nor fragmented; to appreciate the value of learning as an ongoing process (acting as 'philosophers and not philodoxes')[71] and as a

---

69 Galileo shared Cesi's criticisms of university education. When writing to Paolo Gualdo in Padua to inform him of the imminent publication of the *Letters on sunspots*, the Tuscan scientist explained his reasons for writing the work in the vernacular: 'because I must have everyone able to read it [...]. I am driven to do this by the sight of young people sent through the universities indiscriminately to become physicians, philosophers, etc. Many who apply themselves to these professions are quite incapable, while others, who would be suitable, remain occupied with family affairs or other activities that have nothing to do with the study of human letters. Although furnished, as Ruzzante [author of plays in Paduan dialect] says, with innate talent (*bon snaturale*) they do not know Latin and are persuaded that, in those books that are inaccessible to their understanding, very important things of logic and metaphysics are hidden. I want them to see that nature, just as it has given them eyes to see its productions, has given them just as good a brain as that of the logicians and metaphysicians to be able to grasp and understand them.' (OG XI, p. 327).

70 Altieri Biagi 1969, p. 66.

71 Ibid., p. 65. Describing in the *Discourse* the facilities that he intended to make available to the Linceans and to deserving scholars in the central and peripheral branches of the Academy, Federico insisted on the fundamental function of encyclopaedias and dictionaries: 'those best forms of repertories, methods and synopses' (ibid., p. 73). In the previous pages (pp. 64–5) the Prince had underlined the importance of good dictionaries and encyclopaedias in helping to form a panoramic vision of the links and hierarchies of the various branches of knowledge, attributing the degeneration of studies also to a declining use of these resources.

value in itself rather than a means to secure practical benefits; to firmly defend the freedom of thought in the study of nature.

He laid great stress on the rigorous imperatives that applied also to moral conduct based on these principles. In the *Discourse* one clearly perceives not only echoes of the radical criticism of traditional knowledge by the Rosicrucian movement but also of the fierce denunciation of the degeneration of the 'false age' in the *Advertisements from Parnassus* of Traiano Boccalini. It is telling that the anonymous author of the manifesto of the Rosicrucians, the *Fama fraternatis* – not by chance a book to be found in Federico's library – had translated entire passages of Boccalini's work into German.[72] It is also worth mentioning that Boccalini was an admirer of Galileo. While in exile in Venice, he sent to the Tuscan scientist, through their common friend Girolamo Magagnati, greetings and messages of appreciation for his work.[73] Cesi's *Discourse* also bears traces of the violent polemics of Giordano Bruno against the moral and religious degeneration of his day.[74]

At the heart of the *Discourse* is a passionate reflection on values. The claim to the freedom of thought in natural philosophy and the invitation to promote the natural desire for knowledge were for Cesi the foundations not only of a new way of approaching the contemplation and study of nature, but also of achieving the ends assigned to mankind by his Creator.

These orientations and priorities, which were the cornerstone of the original project of the Academy, took on even stronger cogency in the dramatic situation of those weeks. Federico's awareness that Galileo's battle would end in defeat induced him to place even greater emphasis on the need for a radical reform of knowledge, its ends and the method to acquire it. His insistence on the deliberate decision to keep the Academy aloof from political and religious questions was based on a severe moral judgement on the practices and goals of

72 Boccalini 1948, pp. 258–86 and Ricci 1994, pp. 46–7 and 89–91. An English translation of Boccalini's book by Henry Carey (Boccalini 1656) was published as *Advertisements from Parnassus* in 1656.

73 OG XI, p. 104 and esp. p. 527 (letter of Magagnati to Galileo, 22 June 1613).

74 Ricci (1994) found evidence of the influence of Bruno in the *Lynceographum* and the *Discourse*. The Prince's sympathies with Bruno are also confirmed by the enthusiasm with which he welcomed Nicolò Antonio Stelliola into the Academy. Stelliola, who was a follower of Bruno and shared his Copernican convictions, intended to publish a vast encyclopaedic compilation in which the Pythagorean conception of the universe was openly defended. Like Bruno, he soon came under the watchful eye of the Inquisition on charges not unlike those levelled against the luckless Friar, including that of interpreting Holy Scripture in a free and personal way.

those activities. The definition of political matters in the *Discourse* as 'disagreeable' helps to understand that the declarations of submission to the superior political authority contained in all the normative documents of the Academy written by Cesi were dictated by considerations of opportunity and prudence. His denunciation, on the other hand, of the degeneration of religious practices was expressed, although implicitly, by the stress on the genuinely religious zeal of the foundational ideals of the Linceans, who set out to dedicate all their energies to the search for truth, renouncing all the temptations of worldly pleasures.[75]

Having analysed its contents, let us turn our attention to the way in which the *Discourse* was received by Galileo, guest of honour at that academic meeting. He will surely have noted the continuous evocation of themes and questions of particular concern to him, especially during those weeks that were to prove so fateful. In the words pronounced by the Prince he will have found further confirmation that the strategy that the Academy intended to follow was far removed from the position that he had decided to adopt in his confrontations with the resolute opponents of the new image of nature.

As numerous passages show, when giving the *Discourse* its definitive form, Cesi must have had Galileo's presence at the meeting in mind. To begin with, he solemnly declared the need to 'submit matter to mathematical rigour', an absolutely central theme for Galileo. Moreover, in contrasting the indisputable nature of mathematical conclusions with the flexible nature of the pronouncements of jurists, adaptable as they were to circumstances and opportunities, he seemed to echo the sharp distinction between the objective truth of mathematical demonstrations and the disputable nature of legal procedures and commercial transactions outlined by Galileo in the Copernican letter to Christine of Lorraine; a text that Cesi had read over and again with great attention:

> There is a great difference between giving orders to a mathematician or philosopher, on the one hand, or to a merchant or jurist, on the other.

75 Gilberto De Angelis detected the influence of Franciscan spirituality in the strong ethical motivations of Prince Cesi (De Angelis 2007). He argued that it was his mother Olimpia Orsini, a fervent follower of the teachings of St Francis of Assisi, who had guided him in this direction. De Angelis demonstrated that the eleven-year-old Federico and his two brothers became tertiary members of the Order of St. Francis (ibid., p. 48). This early spiritual orientation may have prompted the Prince's decision to put the pursuit of knowledge through divine love at the centre of the agenda of the Academy and of the *Lynceographum*.

> It is not so easy to change the conclusions rigorously demonstrated on the phenomena of nature and of the heavens as it is with respect to opinions on what is or is not allowed in a contract, a financial statement or an exchange of goods.[76]

With his passionate eulogy of the 'noble Academy of Florence under Lorenzo the Magnificent', which had protected and promoted the work of such great authors as Pico, Poliziano and Ficino, the Prince no doubt wanted to please a guest who had come to Rome boasting the title of Philosopher and Mathematician of the Grand Duke of Tuscany. The Prince admired Marsilio Ficino whose *De vita* had been a fundamental source of inspiration also for the compilation of the *Discourse*.[77]

Federico's emphasis on the material benefits that derived from the disinterested pursuit of knowledge likewise reflected Galileo's insistence on the useful practical applications produced by an in-depth understanding of the mathematical structure of nature.[78] Cesi knew that Galileo had invented a sophisticated instrument, the proportional compass, that could be put to military use and was capable of performing a vast series of practical operations. He knew too that during those same days in which the Tuscan scientist was devoting all his energy to dissuading the Church from condemning Copernicus, he continued to pursue the plan to devise a viable method to solve the dramatic problem of the precise determination of longitude at sea: an eloquent example of the practical advantages generated by the disinterested study of nature.

Still motivated by his desire to please his Tuscan guest, Cesi stressed that a continuous growth of knowledge was produced not only by 'remarkable instruments, the most rare and effective medicaments, fire-arms, weapons, defences, machines, devices to raise water' but also by extraordinary inventions like that of the telescope 'that brings the remotest things within our view in an instant'.[79]

---

76 OG V, p. 326.

77 Altieri Biagi 1969, pp. 82–3. Cesi's eulogy of the Florentine Academy and its patron was also intended to suggest the substantial continuity between the noble and prestigious Florentine enterprise, which had renewed an ancient but exhausted tradition, and the Lincean Academy. Angelo De Filiis had similarly referred to Florentine humanism as a fundamental source of inspiration in the dedicatory letter of the *Letters on sunspots* (Galilei-Scheiner 2010, p. 375: OG V, p. 77). On the significance to Cesi and his associates of the image of a direct filiation of Lincean Rome from Medici Florence, see Garin 1986, esp. pp. 18–20.

78 Ibid., p. 59.

79 Ibid., pp. 89–90.

Also the Prince's strong affirmation of freedom of research as a fundamental moral principle reflected a concept underlined by Galileo in the *Copernican letters*: that of keeping the study of natural phenomena from the watchful guardianship of the theologians and ecclesiastical censors.

However, in Federico's insistence on the need to guarantee freedom of thought in the study of nature, Galileo must have felt some discomfort. The freedom to which Cesi referred must have appeared limited to the Tuscan scientist in comparison with the value and significance that he himself attributed to that principle. The experience that Galileo was going through had convinced him that in the Prince's eyes, the demand for freedom of thought came to a halt before the barrier erected by political or religious authorities. In the case of conflict with those authorities, Federico was prepared to retreat, confining the free confrontation of different visions of nature to within the safe walls of his Roman palace in the Via della Maschera d'oro. Galileo had realised this even before listening to the *Discourse*: Cesi's continuous calls to be on his guard in the face of the ecclesiastical warnings had brought home to him that their respective interpretations of the freedom of thought in natural philosophy did not coincide.

Federico's repeated references in the *Discourse* to the Book of Nature, a metaphor dear to Galileo, mark another of the themes aimed at pleasing the Florentine Lincean. Using expressions extraordinarily close to those employed by Galileo, Cesi vigorously emphasised the need to base on observation and experiments the study of 'this great, truthful and universal Book of the world'.[80] In a later passage he drew a sharp distinction, as had the Tuscan scientist, between the truthful 'Book of the universe' and the writings of Aristotle. However, elsewhere in the *Discourse* Federico resorted to the metaphor of the Theatre of Nature: in explaining the reasons for choosing the lynx as the symbol of the Academy, he defined it as

> a spur and reminder to acquire that keenness and penetration of the mind's eye that is necessary to know and to look minutely and diligently, at and inside [...], all the objects and creatures of this great Theatre of Nature'.[81]

Galileo will have noted the ambiguity of the shift from the image of nature as a Book to that of nature as a Theatre. The expression 'Theatre of Nature', and

80 Ibid., p. 59.

81 Ibid., p. 71.

others that the Prince regarded as equivalents,[82] often recur in his writings, almost always with different semantic values from those attributed to the 'Book of Nature' by Galileo. The metaphor of the Theatre was widely used in frontispieces of sixteenth-century works on natural history and of texts on the art of memory, both literary genres cultivated by the Prince. The *Theatre of Nature* alluded to the multiplicity of creatures and natural substances, evoking their occult meanings, their 'signatures', highlighting the influence that they exerted on man. As we have seen in our analysis of Federico's *Indicatio*, the *Mirror of Reason* (explicitly cited in the *Discourse* of 1616)[83] and the *Theatre of Nature* were conceived as synoptic representations of the universal logic governing the human mind and the hierarchical organisation of the natural world. Cesi's attempt to define in those writings the position of living creatures and of inorganic substances along the continuous Ladder of Nature was inextricably entangled with a mass of symbolic interpretations and the unveiling of the truths concealed in the myths and fables of ancient authors. On the contrary, Galileo's reading of the Book of Nature focused on the rational, universal and necessary mathematical principles that govern its operation.

The Prince's words confirmed the conviction that Galileo had formed through the intense discussions with Cesi and Dini in previous weeks. Federico supported his battle for the affirmation of freedom of thought in natural philosophy with energy, conviction and passion, but on condition that it did not trespass into the mine-field of questioning the supreme authority of Holy Scripture. Galileo was bound to be disappointed also by the Prince's declaration of the submission of natural philosophy to the absolute sovereignty of theology. This principle seemed to have been deliberately formulated by the Prince to distance himself from the conception of the clear division of the fields of competence between theology and natural science that the Tuscan scientist had energetically defended in the *Copernican letters* to Castelli and Christine of Lorraine. To make matters worse, Federico's declaration on this point in the *Discourse* no longer took the form of a piece of friendly advice, but was proclaimed as a founding principle of the Academy: the Linceans 'by their particular constitution have banished from their studies any controversy outside the study of nature and mathematics'.[84]

---

82 Like *Frontispiece* (which obviously referred to the Book of Nature), *cabinet* ('armadio'), *tribune*, as well as *synopsis*, *syntaxis*, etc. The Prince also frequently used the terms *factory* ('opificio') and *workshop* ('officina') to emphasise the incessant activities of transformation of nature (cf. Guerrini 2006, pp. 14–5).

83 Altieri Biagi 1969, p. 64.

84 Ibid., p. 79.

Prince Cesi's noble lecture opened Galileo's eyes completely. Cesi was a generous supporter who sincerely appreciated his work and who had conceived an ambitious project to reform traditional knowledge. However, his social status as a member of the Roman patriciate, programmatically allied with the Church and its hierarchies, prevented him, and would prevent him in the future, from fully adhering to Galileo's project.

When the Tuscan scientist attended a second session of the Academy on 24 March, the situation had changed significantly. On 5 March a decree of the Congregation of the Index had prohibited Foscarini's *Letter on the opinion of the followers of Pythagoras* and suspended Copernicus' *De revolutionibus* until it had been corrected.[85] A few days before the emission of the decree, Galileo received in private and confidentially, out of respect for his position as the servant of a Christian prince, the warning not to maintain or to teach the Copernican doctrine.[86] The decree caused consternation among the ranks of the Linceans too. One event in particular seemed to pose a serious threat to the institution: on learning of the suspension of Copernicus and the prohibition of Foscarini's *Letter*, Luca Valerio, one of the most competent and highly respected members of the Linceans, censor and editor of the publications of the Academy, tendered his resignation.[87] Valerio declared that he could no longer be associated with an institution that compactly adhered to the Copernican vision after it had been formally declared heretical.[88] Fearing the consequences

---

85 OG XIX, pp. 322–3.

86 Ibid., p. 321.

87 For Valerio's biography and his relations with the Jesuits see Baldini-Napolitani 1990 (esp. pp. 9–18). Valerio joined the Society of Jesus in 1570, but obtained permission to leave it ten years later; see also Contributi I, pp. 835–64.

88 Baldini and Napolitani (Baldini-Napolitani 1990, pp. 55ff.) challenged what they believed to be Gabrieli's conviction that the poetess Margherita Sarrocchi, Valerio's lover, was responsible for his break with the Lincean Academy after the publication of the anti-Copernican decree in 1616. The two authors pointed out that Gabrieli's reconstruction was not supported by convincing documentary evidence. They stressed that Margherita's hostile attitude towards Galileo dates from the years 1612–3 for reasons that had nothing to do with the Tuscan scientist's Copernican convictions nor with the condemnation of heliocentrism. However, Baldini and Napolitani attributed erroneously to Gabrieli the thesis of the negative role played by Margherita Sarrocchi in Valerio's decision to leave the Academy. It was in fact Cesi, not Gabrieli, who suspected the poetess of malign intervention: in a letter to Faber of 6 November 1617, in which he commented on the recent death of Sarrocchi, the Prince expressed the hope that Valerio, free of her negative influence, would be able to enter the ranks of the Academy again. Although he referred to Cesi's opinion on the role of Margherita in a footnote, Gabrieli made it clear that he considered the Prince's suspicions to be unfounded (CL, p. 618).

of that decision, Cesi immediately set out to remedy the situation by calling a meeting of the Academy[89] which was attended by Galileo, Francesco Stelluti, Angelo De Filiis and Johannes Faber. On that occasion it was unanimously agreed to suspend Luca Valerio.[90]

Cesi motivated this decision on three grounds. Firstly, Valerio had violated the essential principle of the Lincei, that of respecting the freedom of thought in natural philosophy, which obliged members to accept the diverse opinions of their fellows. Secondly, he had declared that the entire Academy adhered to the Copernican hypothesis, while it was driven solely by the desire to promote the free and disinterested study of nature. Thirdly, he had 'falsely accused Galileo of a serious error and crime', that is of maintaining the Copernican doctrine to be true, while 'Galileo held it simply as a hypothesis'. The Prince knew perfectly well that this was false, but he said it for the sake of the Tuscan scientist and, above all, to protect the Academy.

Galileo certainly did not appreciate Cesi's final motivation, even though he knew it to be a white lie. But the wound inflicted by the anti-Copernican decree and by Bellarmine's private admonition led him to accept without protest a portrait in which he did not recognise himself.

Simulated versus True Religion

In spite of the heavy blow, Galileo had no intention of giving up the fight for the affirmation of the new system of the world. The Tuscan ambassador in Rome, Piero Guicciardini, applied every possible pressure to get Galileo to admit defeat and to return as soon as possible to Florence. In the letter of 13 May 1616 in which he informed the Secretary of State of the Grand Duke of Tuscany that Galileo had led 'a crazy life' in Rome, incurring exorbitant expense for the public treasury, the ambassador stressed that, in spite of the condemnation of the Copernican doctrine, Galileo persisted in his 'dogged determination to wear down the friars and to fight with those against whom he is bound to lose; but sooner or later your Excellencies will hear that he has fallen down some towering precipice'.[91]

89 See Contributi I, pp. 535–6.

90 Valerio was suspended, not expelled. Having sworn lifelong allegiance to the institution, Linceans could not resign from the Academy.

91 OG XII, p. 259. Appended to the autograph letter preserved in the archive of the Guicciardini family in Florence is a note of the expenses incurred by Galileo during his stay in Rome in 1616 (see Galilei 2015, pp. 166–9).

The insistent invitations of the Tuscan Resident – who had already predicted in 1611 how the clash would end – were not enough to convince Galileo to return to Florence. An authoritative intervention on the part of the Medici Court was called for. The Secretary of State of the Grand Duchy, Curzio Picchena, ordered Galileo to return at the earliest opportunity. He motivated this resolute decision and its urgency with a disquieting message: 'rumours are going round that displease us and the friars are omnipotent'.[92] Guicciardini adopted an even more direct tone: Galileo must leave Rome for his own personal safety and, above all, to protect the interests of the Grand Duchy, whose image in the Holy Seat was in danger of being seriously compromised.[93]

Once again it should be stressed that Galileo's obstinate refusal to give up the fight and his underestimation of the risks he ran were not due to his terrible character, his presumption of superiority, or his inability to understand the situation in which he found himself. His resolute attitude was inspired by the desire to gain consensus for the radical reform of traditional knowledge on which he had embarked. He saw that task as a moral mission, motivated by an almost religious sentiment. That reform was aimed not only at establishing a new conception of nature, but above all at restoring priority to the search for truth. He could not abandon that enterprise without losing his sense of dignity. This was the state of mind from which Galileo had drawn determination and courage to face the clash with his adversaries, fully conscious of their obstinacy and superior power. The same motivations inspired his confidence that in the end his enterprise would be crowned with success.

Precise evidence shows that these were the feelings that led him not to abandon the hope of a favourable outcome, in spite of the many signs to the contrary. In the numerous reports that he sent from Rome to the Secretary of State in Florence in the days preceding the adoption of the anti-Copernican decree, Galileo repeatedly expressed his confidence that it was still possible to obtain the legitimation of Copernicanism. He put his trust in the influential protection of the Grand Duke of Tuscany and in his own ability to easily overcome the objections of his opponents while demonstrating, at the same time, the sincerity of his own religious feelings and of his respect for the authority of the Church.

Moreover, he was sure that, however the affair designed to ruin his reputation might end, there would not be consequences for his person. He provided the Secretary of State with precise assurances on this point in the letter of

92 Ibid., letter of 23 May 1616.

93 Ibid., pp. 241–3.

13 February 1616: 'as for my person, I had been assured that all the superiors had remained absolutely convinced of my sincere religious zeal'.[94] This comforting certainty led him to ignore the recommendation of the Tuscan Resident to refrain from provoking the friars and to cultivate the illusion that he could change the mechanism of power in Rome.

The fact that he had saved his skin was not enough for Galileo. Nor did it induce him to withdraw from the battle for the affirmation of the Copernican truth. His determination was nourished by reflections that went beyond the question of the true structure of the universe. What was at stake was the crucial problem of recognition of the freedom of research, a principle to which Galileo, like Cesi, attributed a profound moral significance. He was convinced that the affirmation of the new science implied at the same time a reform not only of values, but even of the way of conceiving religious sentiment. My adversaries, he wrote in those weeks, disguise themselves 'with simulated religion to make me appear bereft of the true one'.[95]

In 1616, and later in 1623, the *annus mirabilis* of the election of Pope Urban VIII and of the publication of the *Assayer*, and again in 1633, when the most severe and humiliating condemnation was issued against him, Galileo was driven to continue his battle by his moral duty to fight for the affirmation of a new conception of nature and the cosmos. It was a fight that could not be subordinated to the risk of having to pay the consequences, however serious they might be, with his own person.

On 6 February 1616, a few days after the meeting of the Academy at which Cesi had read his *Discourse* and a month before the promulgation of the anti-Copernican decree, Galileo disclosed to the Secretary of State of the Grand Duchy the motivations that led him to continue the struggle, even though 'transactions are completely finished in that part concerning my personal affairs'.[96] He could not abandon the field because

> to my cause is connected a question [the truth of the Copernican doctrine] that no longer concerns me in person but all of those who in the past eighty years [...] have supported or still support a certain doctrine and opinion known to your Excellency, on the determination of which discussions are going on at the moment to take [on the part of the Church] the most just and appropriate deliberations.[97]

94 Ibid., p. 232.

95 Ibid., p. 251.

96 Ibid., p. 230.

97 Ibid.

He felt compelled to make every effort to avoid the prohibition of a conception that had universal value in his eyes. It did not just explain the structure of the universe; it showed the perfection of the divine work and helped to understand the role and ends of man in Creation. He was driven not to surrender not only by the desire that truths of which he was absolutely convinced should be affirmed, but also by the ethical and even religious implications of the Copernican vision:

> As one who happens to be able to be of some service [in the positive conclusion of the Copernican question] for that part which depends on the knowledge of the truth provided by the sciences that I profess, *neither can nor must I deny that assistance which is imposed by my conscience as a zealous and Catholic Christian.*[98]

A little further on, in the same letter, he explained that he could not retreat in a question 'directed towards a just and religious end'.

The same feeling of a task imposed by the superior demands of the search for truth and the duty to protect it from 'three very powerful smiths, ignorance, envy and impiety' (emphasising once again the sharp distinction between true and false religion), was stressed in another letter written a few days later, once again to Curzio Picchena. Galileo upheld the conviction that his personal exoneration did not permit him to shrink from the obligation to combat his enemies with every weapon at his disposal:

> [although] they see themselves totally defeated and exterminated in their power to offend me, nevertheless they do not cease to use every kind of machination and iniquitous stratagem to take it out at least on the works of others [...], seeking not only to obscure their fame, but to annihilate their works and studies which are so noble and useful to the world'.[99]

Although presented in an extremely negative light, the same moral and religious motivations are revealed in the account of Galileo's behaviour sent to Florence by Piero Guicciardini only two days before the formal emission of the anti-Copernican decree which the Tuscan Resident already knew would be adopted. This time, as previous alarm signals sent to the Secretary of State, whom Guicciardini regarded as too weak and accommodating with Galileo,

98 Ibid., emphasis added.

99 Ibid., p. 234.

had failed to produce any effect, he bypassed him by writing directly to the Grand Duke. This breach of protocol was imposed by a situation that he considered an emergency: if the Grand Duke did not recall Galileo at the earliest opportunity, the consequences would not only be very serious for the scientist himself, but would prejudice the political relations of the Grand Duchy with the Church. Guicciardini portrayed the Tuscan scientist as animated by a sacred fire that made him dangerously imprudent:

> He is inflamed in his opinions, has an extreme passion, and little strength and prudence to keep it in check [...]. He is impassioned from within and does not perceive how he should behave [...]. He is so vehement, obstinate and passionate that it is impossible for anyone near him to escape from his hands [...].[100]

A similar opinion, which certainly reflected that of Cesi and the other Linceans, was formulated by Remus Quietanus a few years later in the letter with which he informed Kepler of the insertion of the *Epitome of Copernican astronomy* in the Index of prohibited books as a consequence of the anti-Copernican decree of 1616. While he attributed the main responsibility for that infamous decision to Foscarini, Remus declared that Galileo was also to blame because of the excessive determination with which he had defended his own opinions: 'Galileo argued his cause too forcefully at that time in Rome'.[101]

Picchena and the Grand Duke were alarmed not only by the news from Rome. In Florence the 'league of pigeons', as Cigoli had mockingly called the resolute Florentine opponents of Galileo, were engaged in equally menacing initiatives. They had barely finished exulting at the publication of the anti-Copernican decree in March 1616 than they took measures to open fire again on their hated fellow citizen. In June 1616, immediately after his return to Florence, Galileo informed Cesi that his enemies in the city were sharpening their knives in readiness for a new crusade. Their objective this time was to have the reply of Galileo-Castelli to the objections of the followers of Aristotle to the Tuscan scientist's discourse on hydrostatics[102] put on the Index of prohibited books.

100 Ibid., pp. 242–3, letter of 4 March 1616.

101 Ibid., p. 481, letter of 13 August 1619, cf. above, p. 185 and n. 49.

102 Galileo-Castelli 1615. Although signed by Castelli, the work was to a large extent the fruit of Galileo's pen.

It was not the first time that his fellow citizens had tried to have that text silenced. Two years earlier, while Castelli's text was being printed, Galileo had caught wind of their plans to prevent its publication. We learn of his suspicions from a letter from Castelli in Pisa of 19 March 1614 written in reply to a (lost) letter of Galileo.[103] That his fears were well founded is confirmed by the message that Galileo sent a few weeks later to Michelangelo Buonarroti. He informed his trusted friend that the printer (Cosimo Giunti) had returned the manuscript sent to him for typesetting, adding that he would personally pass by to explain the reason to him: 'I cannot imagine – Galileo wrote to Buonarroti – what this story is about'.[104] To get to the bottom of the ambiguous situation, Galileo asked his friend to carry out a delicate mission with the utmost discretion:

> I beg you to drop in at his [Cosimo Giunti's] workshop as soon as you can, concealing your knowledge of this fact, but to enter and ask if any part of it [the text of Galilei-Castelli] has already been printed. Use your skill to find out what he is thinking, the reason for these delays and obstacles and where they come from.[105]

It is easy to understand the reasons that, after the emission of the anti-Copernican decree, induced the 'league of pigeons' to reopen hostilities against the work published under the name of Castelli in the previous year. In that text, in replying to the insinuations regarding the Copernican inclinations of Galileo advanced by Ludovico delle Colombe in his *Apologetic discourse*, Galileo-Castelli explicitly admitted that the Tuscan scientist adhered to the Copernican system:[106]

> I asked him [Galileo] why he considered the Copernican system much closer to the truth than the Ptolemaic or Aristotelian ones, and why he had not replied to your [Delle Colombe's] objections.[107]

103 Ibid., p. 40. Castelli's comments reveal that Galileo had informed him of the attempts to prevent publication: 'We shall have to do our best to get it through [i.e. to obtain ecclesiastical permission to publish the *Discourse*] because there will be no lack of intrigues of every kind to prevent it from being printed'.

104 Ibid., p. 62, letter of 15 May 1614.

105 Ibid.

106 OG IV, pp. 356 and 516.

107 Ibid., p. 587.

In a later passage, after declaring that the Copernican system 'is considered true by him [Galileo]', Galileo-Castelli concluded that if Delle Colombe had had a 'knowledge of mathematics', he would have realised for himself 'how impossible it is to understand him [Copernicus] and not to agree with his opinion'.[108]

The letter in which Galileo informed Cesi of the renewed hostility of the Florentine opponents is lost, but its content can be reconstructed from the Prince's reply on 25 June 1616:

> It is not surprising that those who are malign and envious of you continue their usual practice; since as it is impossible for them to change their condition, or rather nature, it is extremely difficult for them to acquiesce. Let us leave them to bark in vain and to incur the due penalty and mortification for the great frenzy they show. I have discussed the question [that worries you] with Signor Butio and he will try to understand what is going on [in Rome].[109]

The signor Butio mentioned by Cesi was the physician Antonio Bucci from Faenza, a consultant of the Congregation of the Index who was familiar with the Master of the Sacred Palace. On behalf of the Master, he had acted as official reviser of the *Letters on sunspots* in 1613.[110] He had earlier expressed a favourable opinion on another publication supported by the Princeps of the Linceans, the *Elementorum curvilineorum libri tres* of Della Porta.[111] Bucci, who was on friendly terms with the Linceans, particularly with Johannes Faber,[112] immediately took action and sent a reassuring message to Cesi, who hastened to inform Galileo: 'Signor Butio has already verified that there was no order here concerning Father Castelli's book of 1615'.[113]

While Cesi's *Discourse* of 1616 represents the culmination of his reflections on the disinterested pursuit of knowledge and the mission of the learned, Galileo was to go further, indicating that the new system of knowledge that he was

108 Ibid., pp. 58–9.

109 Ibid., pp. 573–4.

110 Galilei-Scheiner 1610, p. 371 (OG V, p. 74). Bucci's revision was full of unconditional praise for the 'new and marvellous observations hitherto unknown and unheard of, explained in a courteous and elegant style'.

111 Della Porta 1610c.

112 On Bucci's cordial relations with the Linceans and especially with Faber, see Brevaglieri 2009, pp. 123–7.

113 OG XII, pp. 271–2.

elaborating made it necessary to redefine the relations between man and nature. The Copernican universe, with the Earth shifted from its central position and transported in space by a twofold movement governed by strict mathematical laws, as well as the replacement of the Aristotelian elements and their qualities by a homogeneous substance that, rarefying and condensing in infinitely diverse ways, generated the multiplicity of bodies, called into question fundamental aspects not only of the traditional conception of nature, but also of the Christian vision of a universe created to the measure of and in function of man. These were serious threats to doctrines on which the Christian faith was founded. As Castelli wrote to Galileo, it was precisely for those implications that the Archbishop of Pisa was horrified at the mention of Copernicus, and even more so of Galileo:

> Every creature having been made in the service of man, it was clear as a necessary consequence that the Earth could not move like the stars; and if on these questions I [Castelli] had had the intelligence of understanding this concept properly, perhaps I would have changed my mind.[114]

Galileo paid no heed to the Archbishop's advice. In the *Assayer* he stressed that the anthropocentric prejudice that characterised the traditional conceptions of the universe should be banished from our minds. Written in the form of a letter to Virginio Cesarini, by now vice-Princeps of the Linceans, the *Assayer* was a work of a clearly revolutionary character. Anti-miraculism, rejection of the edifying illusion of a nature conceived for human ends and needs, the rigorously rational structure of the world, Copernicanism (transparent behind the dissimulation), the relativistic theory of motion, the atomistic conception of matter and the affirmation of the subjective nature of sense perceptions formed an explosive cocktail that threatened the authority of the traditional conception of the relation between man and nature. That new body of knowledge was seen by many to be dramatically destabilising and it triggered resolute defence mechanisms. The sentence against Galileo in 1633 represented not only a magisterial pronouncement by the Church on a technical problem of celestial mechanics. What was condemned was Galileo's boldness in calling into question the Christian vision of the centrality of man in Creation and the reduction of nature to a subservient role in which God intervened with continuous miracles to modify its course. From this point of view, Galileo's teachings appeared more threatening than those of Copernicus and Kepler, as they were

114 Ibid., p. 154, letter of 12 March 1615.

both convinced, albeit in different ways, that the world had been created by God for man and to his perfect measure.

It is no coincidence that, from the end of the nineteenth century, to exorcise Galileo's demolition of the anthropocentric vision, a number of authoritative philosophers have been conducting what might be called a new trial of Galileo – a secular trial this time, which continues to find supporters who pass severe judgement on the negative consequences of Galileo's intellectual legacy.

These authors hold that Galileo shoulders most of the responsibility for the dramatic process of degeneration triggered by his wicked interpretation of nature as a purely mathematical dimension and by his reduction of human sense perceptions to subjective phenomena. These absurd conceptions are alleged to have produced the intellectual catastrophe that has thrown the European conscience into profound crisis.

The charges levelled against Galileo are extremely severe: he separated the world of nature from that of life; he rendered the facts of nature indifferent to man, to his passions, his beliefs, his fears and his hopes. Hannah Arendt, Arthur Koestler, Jürgen Habermas, Max Horkheimer and Theodor Adorno, Paul Valéry, Hans Jonas, leading personalities of psychoanalysis, Karl Löwith, scholars of the calibre of Ernst Bloch as well as important Marxist authors, have denounced Galileo as responsible for the 'disenchantment of the world' and the destruction of the concept of living nature. In Koestler's words, Galileo bears the blame for having opened up a process that has led to a 'progressive spiritual desiccation'.[115]

Heidegger's condemnation was final: 'science is born, thought disappears'.[116] No less dramatic was the conclusion of Husserl, author of the most systematic and articulated condemnation of the disaster caused by Galileo, by Descartes and by the protagonists of the mathematisation of nature: 'Galileo is at once a discovering and a concealing genius'.[117] With this striking expression Husserl delineated a caricatural image of Galileo as responsible for the irreducible opposition between a mathematised nature and the world perceived by human senses, with the result that the latter was completely obscured while science and philosophy suffered a dramatic parting of the ways. Science chose as its exclusive field measurable reality, retreating from the world of life, of values and of the spirit. Galileo's purely geometric and quantitative Book of Nature concealed what cannot be suppressed for man: the sense of things, depriving

115 Koestler 1968, p. 540, but see the whole of the *Epilogue*.

116 Heidegger 1939, p. 39.

117 Husserl 1970, p. 52.

them of any transcendental teleology. His conception of nature is responsible for the drama of the modern world: it has emptied existence of meaning; has caused the disappearance of the intuitive world; has annihilated the individual and his aspirations.

Did Cesi and his companions realise the implications for the traditional conception of the relation between man and nature of the new Galilean cosmology and physics? A nature that follows transparent and rigorous rules leaves no room for those miracles, secrets and *arcana* that had held such a fascination for Cesi and his companions from the very start of their adventure. We cannot forget the presence in the foundational texts of the Lincean Academy of visions of nature very distant from that of its indifference to human modes of thought and expectations on which Galileo was to insist. For Cesi it was necessary to free the mind of the contamination produced by a corrupt conception of knowledge and its ends. But at the conclusion of that process of purification, man would regain the primary role assigned to him by God in the plan of the Creation. As we have already stressed, Galileo reminded Cesi that:

> we should not wish nature to accommodate itself to what seems better ordered and disposed to us; but we should rather accommodate our own intellect to what nature has made, in the certainty that this is the best and only way.[118]

There are good grounds for doubting whether Cesi would have followed Galileo on this terrain, recognizing that the Copernican vision inevitably implied a new way of understanding the position and purpose of man in nature. Federico was not prepared to accept the consequences of this radical change of perspective. Nor is it surprising that the Prince had on several occasions expressed the greatest admiration for the cosmological conception of Kepler, for whom man remained the ultimate end of the Creation.

Among the Linceans, only Fabio Colonna seems to have shared Galileo's revolutionary vision of the relations between man and nature:

> We have held – the Neapolitan Lincean wrote – that we should believe more in the observation of natural things than in the things imagined and supposed on the basis of a single principle, without taking into account the means and end of the thing itself [...]. The thing perfectly observed is what provides the method; it is not the method which will make the

118 OG XI, p. 344, letter of 30 June 1612.

> thing conform to its presupposed method. The nature of things cannot change at the caprice of man nor make itself cognisable according to his way of thinking. It is man who should adapt his caprice to the nature of things, exactly observed, and derive his method from them if he can.[119]

Miraculous Rains

Let us now shift our attention to the progress of the research projects delineated by the Prince in the *Indicatio*. It should be noted that the outline of the *Discourse on the natural desire for knowledge*[120] immediately follows the *Taumatombria* (*On miraculous rains*) to emphasise the close connection that Cesi established between the study of natural phenomena and the reform of moral conduct. The *Discourse* should therefore not be read as an autonomous text, but as a central chapter of Federico's ambitious encyclopedic treatise.

We owe to Luigi Guerrini the first detailed analysis of the holograph manuscript in the Orto botanico of the University of Padua containing the most advanced version of the *Taumatombria*.[121] According to Guerrini, Cesi wrote this text between 1612 and 1613,[122] that is in the same months in which he was working on the *Coelispicium*.

His immersion in the themes of the *Taumatombria* thus coincided with a phase of intense study of celestial questions as well as with his commitment to supporting the Galilean programme, to protecting from severe censorship the *Letters on sunspots* and to trying to halt the tide of events that would lead to the publication of the anti-Copernican decree of 1616.

The rather forbidding Latin title of the treatise on miraculous rains evoked Giovan Battista della Porta's *Taumatologia*[123] which Cesi had long tried, unsuccessfully, to publish under the Lincean crest. Even more apparent is the affinity of Federico's treatise with the Neapolitan's *De distillatione* and above all with his *De aeris transmutationibus*,[124] a work that had been published at Federico's expense.

119 Colonna 1618, p. 16.

120 Contributi I, pp. 49–50.

121 Guerrini 2006, esp. pp. 28–68.

122 Ibid., p. 27.

123 The manuscript containing the index of the *Taumatologia* is in the Biblioteca dell'Accademia dei Lincei (Ms. Arch. Linceo 9). For the index of secrets of the *Taumatologia*, see Della Porta 2013b.

124 Guerrini 2006, p. 27.

A glance at the *De aeris transmutationibus* is enough to recognize the profound impression that it made on the young founder of the Lincean Academy. In this work Della Porta proposed to illustrate the true nature of meteorological phenomena, clearing the ground of the misunderstandings that had derived from centuries of passive imitation of Aristotle's treatise on meteors. Anyone who reads *De aeris transmutationibus* before immersing himself in what remains of Cesi's *Taumatombria*, or in his *Apiarium* and *Tabulae Phytosophicae*, will be struck by the influence of Della Porta's text on the orientations and research methods of the Princeps of the Linceans, to whom *De aeris transmutationibus* was dedicated.

For present purposes it is sufficient to underline those aspects that show most clearly the attention with which the young Federico read that text, which long occupied a prominent place on his desk. The opening pages of the Neapolitan's treatise contained a *Synopsis* (fig. 15) offering a hierarchical and relational classification – from left (general principles) to right (filiations and derivations) – of the variety of transmutations of the element air. Della Porta's synopsis displayed the same page layout as that adopted by Cesi in the *Indicatio*. As we shall see, that graphic model recurs in Federico's *Tabulae Phytosophicae*. It has also to be stressed that Della Porta's term *synopsis* would be repeatedly used by Cesi. Moreover, the initial pages of *De aeris transmutationibus* presented a collection of statements on meteorological phenomena of ancient authors to which the Neapolitan appended his own interpretation (*ratio nostra*);[125] Federico was to adopt a similar method in his *Apiarium*. Della Porta's text paid special attention to the analysis of dew, honey, manna and resinous and glutinous oils – the same substances on which Cesi focuses in the Paduan manuscript. The Neapolitan stated that nature generates these subtle substances through processes of distillation – exactly as Federico was to do in his writings on miraculous rains. During the dog-days, Della Porta declared, the Sun's heat and the influences of the stars extracted 'moist vapours, which are the dew, oil and resins,' from grasses and plants.[126] In rarefying these moistures, the Sun 'draws them to the ether diffusing them in the air'. Condensed by the cold of the night, they fall to the ground at dawn covering grasses and plants. The densest moistures assume the form and composition of honey, rubber and glutinous latex, which find their way into leaves, trunks and even stones. These substances are variously absorbed by the different vegetal species: 'thus each moisture retains the innate properties of its plant'.[127]

125 Ibid., pp. 93ff.
126 Ibid., p. 112.
127 Ibid.

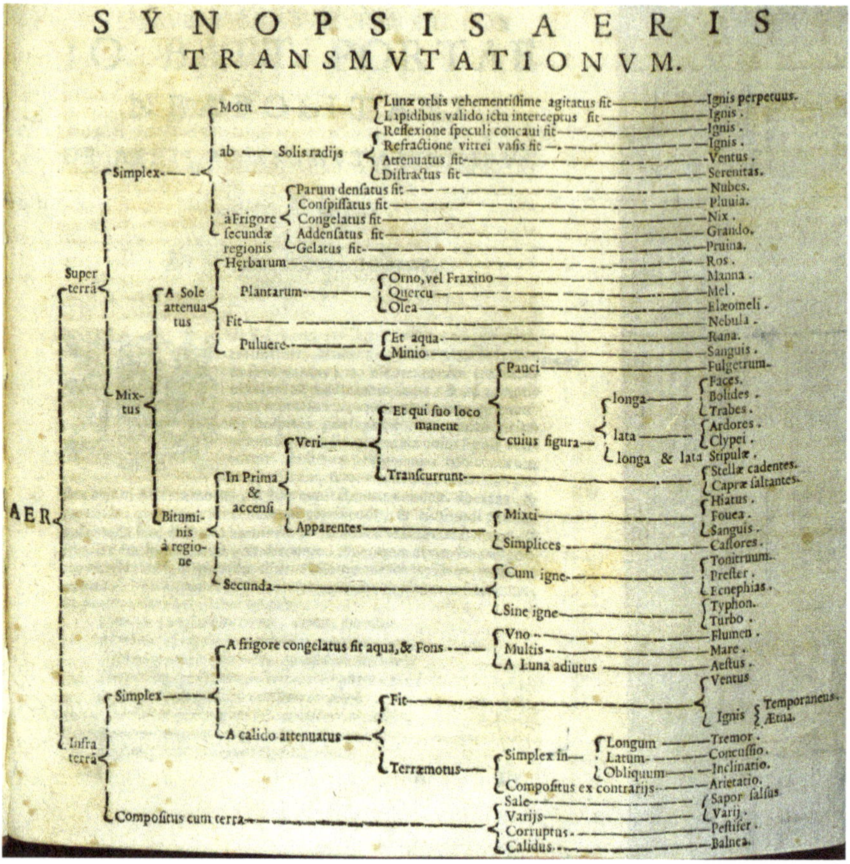

FIGURE 15 *Della Porta's* Synopsis *of the transmutations of air*

For Della Porta, as well as for Cesi, the processes of natural production of honey-like fluids were cyclical, governed by the alternation of the heat of the day and the cold of the night. In their analyses of the nature of honey both stressed the fundamental role of bees that extract the dew exuded by plants and carry out a further distillation in their intestines from which honey is produced.[128] For Della Porta, the variety of dewy fluids that the bees suck from the plants produces the different composition and various flavours of honey.[129]

128 '[...] e frondibus, floribus et truncis sedula opera suos sudores exugunt, suis intestinis fovent et percoquunt, mel conficiunt', ibid., p. 119.

129 Ibid.

In *De aeris transmutationibus* Cesi could also find reflections of great interest for his researches on miraculous rains, including rains of frogs and blood,[130] on subterranean fires, on the nature of bituminous substances and on many other questions that he would ponder throughout his life.[131]

In composing the *Taumatombria* Cesi intended to relaunch the Neapolitan's programme of a radical reform of Aristotle's treatise on meteors new in its principles, in the determination of the true causes of atmospheric phenomena, and in the interpretation of the nature of the various substances between the Earth and the sky.

Federico's principal innovation with respect to the model of Della Porta lay in attributing a central function not only to the Sun but to the entire rarefied celestial element in the production of the precious substances that fall from the sky and nourish plants. This function was obviously based on the conception, dear to the Prince, of the unitary and homogeneous nature of the celestial spaces.

Cesi was convinced that research on meteorological phenomena should no longer be confined to the dimension of the sublunary sky. If the sky is fluid and homogeneous, terrestrial exhalations do not encounter any obstacles in their ascent. Consequently, they come into contact with the more subtle matter of the heavens, becoming more rarefied and purified. Condensed by the cold of the night, this subtle matter falls to the Earth in the form of very fine rain: 'miraculous rain', as Cesi called it, for the extraordinary natural productions that it contributed to generate.

The Prince regarded the cyclical interaction between terrestrial exhalations and celestial matter as an operation of chemical transformation. It was essential to understand it in order to grasp the phenomena of life on Earth.

130 Ibid., pp. 119–22. Among the miraculous rains which aroused Cesi's curiosity there were even those of sulphur. In his letter of 2 July 1620 the Prince asked Faber to gather any news about a rain of sulphur in Vienna: 'Federico [Cesi's relative Federico Bianchi] informs me that last month there was a rain of sulphur in Vienna. You know how keen I am to have minute details of the truth of things like this, and how relevant it is to my treatise on natural prodigies; you will do me a great favour to make every effort to procure detailed and true information' (CL, p. 719).

131 Ibid., pp. 241–3. A passage of the *Indicatio* referring to the 'Ignificium Puteolanum', that is, the solfatara of Pozzuoli (Contributi I, p. 45), among the workshops of nature ('naturae officinae') is evidence that Cesi had personally visited the solfatara, probably accompanied by Della Porta, in 1604 or 1610.

In the Paduan manuscript, the sequence of the three closely interconnected Cesian texts[132] corresponds exactly to the index of the section *De succis oposcopia* (autoptic analysis of natural fluids) delineated in the Frontispiece of the *Indicatio* containing the *Taumatombria*. That index showed Cesi's intention of conducting a close analysis of honey, manna, sugar, wax, resin and *laserpitium*.[133] As Luigi Guerrini has demonstrated, the work on *laserpitium* (a plant credited with extraordinary therapeutic properties, supposed to have first appeared in the Garden of the Hesperides after a black rain of pitch) was planned as a commentary on and revision of the descriptions of the mythical plant by Pliny and Theophrastus.[134] The Prince's keen interest in that miraculous plant stemmed from his conviction that it was a substance characterised by a twofold nature: it grew on the Earth, but was the result of a celestial operation. For Cesi it was tangible proof of the reciprocal permeability of the sky and the Earth. Only by accepting the principle of the fluid and penetrable nature of the celestial spaces was it possible to declare that rains that fell from the sky, even from its furthermost reaches, could confer extraordinary properties on vegetal essences that grow on Earth.

The homogeneous nature of the cosmos is therefore the essential premise on which the analyses developed in Cesi's Paduan manuscript were based. Thanks to its extraordinary heat, the Sun raises the purest substances from the Earth, distilling them. Condensed by the cool temperatures of the night, these substances fall back to the Earth to beneficially moisten plants and minerals. In the course of their passage through the lower and denser levels of the atmosphere the powers and properties of these subtle substances are weakened by contamination. At this point Cesi introduced a new category of skilled distillers: bees, that extract from the plants the dew that has fallen from the sky, refine it in the chemical laboratory of their digestive system and at the end of the process extrude the purest honey.[135] The celestial rains, the heat of the Sun, the cold of the night, the bees and the fluids of the plants (their vital internal principle) are all involved in the process of production of the most

132 *De laserpitio et laserpitii pluvia; De melle, manna et saccharo mellisque pluviis; De cera, cerisque et pluvia cerea.*

133 Contributi I, p. 40.

134 For a detailed analysis of Federico's text on the *laserpitium*, see Guerrini 2006, pp. 28–45. See also Pliny, *Historia Naturalis* XIX, 42 and Theophrastus, *Historia plantarum* VI.

135 Ibid., pp. 76ff. A publication of the *Oposcopia* and of the other texts by Cesi in the Paduan manuscript is being prepared by Luigi Guerrini on the initiative of the Accademia Nazionale dei Lincei.

perfect substances: honey, sugar, manna, wax and *laserpitium*. For Cesi it is in the power of man to achieve the same extraordinary results by imitating those natural chemical processes with refined techniques of artificial distillation.

The twofold process of celestial and terrestrial distillation produces the purest substances, the ultimate principles of nature, the primordial forces that permeate the organic and inorganic world. Cesi defined these substances as intermediate (*mezzane*) because they are produced by both celestial and terrestrial elements and processes. We can detect an almost religious note in the Prince's declaration of the presence in nature of the divine imprint in those vital principles that can only be revealed by means of subtle operations of distillation.

These pages of Cesi's *Oposcopia* are evidence once more of the influence of Della Porta and of his conception of a nature that can only be investigated by isolating its ultimate principles through processes of chemical analysis. Cesi's fascination with Della Porta's *De distillatione* began early.[136] In that work – especially in the first chapters of Book I – the Neapolitan had advanced a sort of general theory of vegetal matter, based on the results of processes of distillation of grasses and plants. These chemical experiments had led Della Porta to the conviction that he had managed to isolate the essential humours present in vegetals, which he had reduced to three fundamental types: nutritional, substantial and vital.[137]

Although Cesi's borrowings from the Neapolitan are substantial and extremely transparent, Della Porta is not the only author from whom Cesi drew inspiration in composing the *Taumatombria* and *Oposcopia*. In these texts one can clearly perceive the influence of his reading of Telesio and his familiarity

136 Cesi had borne the expense of publishing the *De distillatione*. Della Porta's dedication of the treatise to the Prince was a fulsome eulogy of his young patron and of the ancient and noble *gens Caesia* from which he descended (Della Porta 1608).

137 For Della Porta the nutritional humour is produced by the distillation of the roots that transmit nourishment to the plant. The substantial humour, which plays an essential role in the vitality of plants, is the result of the almost complete dissolution of the vegetal matter. The vital humour is the seat of spirit and heat, essential factors for the life of the vegetal world. This humour is common to all plants. Unlike the first two, it is not a vapour but a glutinous oil, which guarantees enduring cohesion between the different parts of the plant. 'The whole substance, quality and property' of the plant – Della Porta wrote – resides in the vital oil, where 'scent, flavour and quality are found in the highest degree of excellence' (ibid., pp. 34–5). In that oil 'the whole essence is preserved, and if it is removed, all that is left is a cadaver, a body entirely devoid of life'. On the three fundamental humours produced by distillation see Guerrini 2006, pp. 46–8.

with Antonio Persio. As we have seen, Cesi particularly appreciated Telesio's vision of a unitary and substantially homogeneous cosmos, a fundamental principle for the radical reform of Aristotle's treatise on meteors that Federico had undertaken. In his *De fulmine*, Telesio had tried to establish the nature of an aerolite that had fallen in Castrovillari, near his native Cosenza, in January 1583.[138] Rejecting the Aristotelian explanation of meteors (stones swept up by the winds and dropped at great distance) he maintained that they were terrestrial exhalations (*fuligines*) that had ascended into the higher zones of the sky where they were condensed before falling back to earth.[139] This interpretation was based on the principle of a continuous exchange between the cold terrestrial matter and the hot substance of the heavens.

During their meetings in Rome in the spring of 1611, Cesi had certainly informed Galileo of his studies on meteorological questions. On the occasion of one of their conversations, Galileo told the Prince of the recent fall of a meteorite in Tuscany and of the investigations of its nature that the Grand Duke had asked the Professors of the University of Pisa to carry out. This news impressed Federico who, more than two years later, requested from the Tuscan scientist detailed information about that singular phenomenon:

> I beg you to do me the favour of informing me in detail and with all the circumstances of time, place, appearance, weight etc., about the stone that fell from the sky in Florence, which you mentioned to me when you were here, saying that the Grand Duke had sent it to Pisa for those philosophers to examine; and I would be extremely grateful if it were possible to obtain a drawing of it to include among the objects of my many speculations and treatises on natural prodigies.[140]

It is only natural to wonder whether Persio was a party to the Roman conversation on atmospheric phenomena, and whether Cesi, perhaps encouraged by the Abbot, had read not only Telesio's *De fulmine* but also the more detailed analysis of meteorological phenomena in another text that Persio had inserted into the collection of Telesian works published in Venice: the *De iis*

138 *De fulmine quod Lucana in terra decidit* was inserted by Persio into the collection of works by Telesio published in Venice (Telesio 1590). It was reissued by Francesco Fiorentino (Fiorentino 1894, pp. 323–74) and Luigi De Franco (Telesio 1981, pp. 162–87). See De Franco 1995, pp. 346ff.

139 Ibid., p. 349. Telesio was not surprised that the aerolite was scorching, coming as it did from the hot higher region of the atmosphere.

140 OG XI, p. 562, letter of 7 September 1613.

quae in aere fiunt et de Terremotibus.[141] In that short treatise, in which Telesio once again parted company with Aristotle, he had attributed to the Sun the extraction from the soil of exhalations and vapours which, condensed by the cold at night, fell to Earth in the form of rain.[142] In both of those treatises by Telesio, as well as in his *De rerum natura*, Cesi would have found analyses and principles entirely convergent with his own vision of the heavens permeated by substances exhaled by the Earth and of the Earth continuously nourished and enriched by celestial rains.

In this connection it is worth considering another short treatise by Persio[143] in which Telesio's critique of Aristotle's theory of the four elements was taken up again. Joining with Telesio in emphasising that there are only two principles in nature, heat and cold, Persio claimed that the only *active* principle is heat. To this end he attempted to demonstrate, against Aristotle, that water is hot by nature:

> As an agent acts upon matter to give it form, no one can doubt that it is the Sun that acts upon the Earth through heat to exhale something from it; and, as we perceive, what is detracted is progressively converted to its own nature, starting with what is closest to its original nature, and then moving on step by step in transforming it into its own appearance and nature. Whence we may conclude that the vapours that rise from the Earth are rendered closer to the Sun and the heavens in their nature, and are converted into waters, which cannot be of a different nature from when they were vapours, and are therefore hot like the Sun. There is no point in saying that they must be cold because that is the nature of the Earth. It appears that they are so changed by the nature of the Sun that nothing at all of their similarity with the Earth remains.[144]

Note Persio's insistence on the operation of the heat of the Sun which extracts vapours from the Earth; condensed by the cold of the night, these vapours fall back in the form of a watery liquid, which still retains the active principle of heat. Cesi might have drawn inspiration also from texts like these to define the cyclical process, based on the alternation of heat and cold, sublimation and condensation, which governs meteorological phenomena generating the miraculous rains that produce honey, sugar and manna.

141 Cf. Telesio 1981, pp. 14–5 and 45–79.

142 Telesio 1990, pp. 34–6.

143 Artese 1987.

144 Ibid., pp. 455–6.

Further evidence of the Prince's careful reading of Telesio's works can be detected in his impassioned battle for the recognition of freedom of thought, a principle vigorously claimed by Telesio in the preface to his *Liber positionum*.[145] Also the conception of the soul of the world, the *spiritus* (which Telesio considered a very refined material substance) that permeates the entire universe, enlivening every corner of it, as well as Persio's attribution of a sensitive soul to plants and even to minerals and stones in his *Trattato* of 1590 certainly attracted the attention of the founder of the Lincei.[146] Cesi shared the idea that a vital principle is present in minerals and that life is in all things. Moreover, in Persio's works Cesi would have found the Platonic vision of a continuous Chain of Being marked by imperceptible gradations showing the ubiquitous presence of the vital principle of spirit/heat: a concept to which Federico was to assign fundamental importance in his interpretation of the operations of nature.

Other resonances can be perceived in the reflections on the nature of light, heat and fire of Telesio and Persio. They both assigned to light the task of spreading motion through the entire universe and proclaimed it the supreme physical principle, expression and image of divinity itself. The exaltation of light had assumed almost religious tones in the passionate prayer to the Sun that concluded Persio's *Trattato dell'ingegno dell'huomo* (*Treatise on human ingenuity*).[147]

As we shall soon see, other central features of the natural philosophy of Telesio and Persio fascinated Cesi. Above all, the rejection of the Aristotelian theory of qualities, based on the distinction of the two active principles of heat and cold from the simply accessory qualities of wet and dry, as well as the conception of heat as the agent of spontaneous generation through putrefaction. Moreover, Telesio's radical criticism of the interpretations of Aristotle and Galen on the nature of flavours[148] will be explicitly echoed in Cesi's classification of flavours and in his quantitative treatment of their therapeutical virtues in the seventh of his *Tabulae Phytosophicae*.[149] Also Telesio's idea that the false natural philosophy of Aristotle and the dogmatic character that it had assumed in the Scholastic tradition were an obstacle to understanding the real meaning of the moral message conveyed by God in Holy Scripture was shared

145 Telesio 1576.

146 Persio 1576b.

147 Sirri-Torrini 1992, pp. 65ff.; Bondì 1997, p. 93.

148 *De saporibus* was reissued by De Franco (Telesio 1981, pp. 337–67).

149 See below, pp. 443–59.

by Federico.[150] Finally, it was probably from Telesio's attempt to interpret the nature of sleep and waking, that Cesi drew inspiration for his project to subject those phenomena to a specific analysis, as shown by the *Indicatio*.[151]

We shall not follow Cesi in the arduous pages of his treatise on the *laserpitium* and of the *Oposcopia*. Nor shall we insist on the Prince's efforts to find in the myths and writings of the most authoritative ancient authors and even in numismatic iconography the veiled anticipation of the conclusions at which he was arriving through his direct investigation of nature. The same model of integration and convergence between observational data (this time with the assistance of the microscope) and the testimonies of mythology and the ancient literary tradition (not only texts of natural philosophy, but also of poetry, history and sacred texts) will recur in the *Apiarium*.

It is worth stressing that in the same years in which Galileo was elaborating his own mechanistic corpuscular conception of matter, the Prince too was engaged in an effort to isolate the ultimate components of nature. However, unlike Galileo, he did not conceive of them as corpuscles of homogeneous matter endowed with diverse shapes and velocities, but rather as an oily and glutinous substance, charged with vital energy, capable of producing an infinity of diverse effects and creatures through processes of chemical combination or solution, sympathy or antipathy, affinity or contrariety. Cesi can be ranked among the atomists, as he has often been, only if his atomism is clearly distinguished from Galileo's corpuscularism.[152] In Cesi we observe the combination of the Aristotelian theory of the elements and qualities with a conception of the structure of matter characterised by a chemical corpuscularism. Cesi's gluten is the result of chemical operations, while Galileo, who also resorts to gluten to account for the cohesion of bodies, attributes exclusively mechanical actions to it.[153] The same distinction holds for the atoms of fire: for Cesi they

150 See Redondi 2004, pp. 132–3 and 135.

151 Cf. Telesio 1590 (*sub indice*) and 1981, pp. 379–414. For Cesi's inquiry on sleeping and waking see Contributi I, p. 35.

152 One of the earliest and most explicit declarations of Cesi's inclination toward a corpuscular conception of nature can be read in the section of the *Zibaldone* entitled 'De minimorum magnitudine': 'Our dictum: pay attention to the smallest if you want the greatest. For everything seems to consist of minimal elements. They should be studied with care, as the greatest will never result in any way except from the smallest parts [...]. Telling examples: the greatest fires from a tiny spark; rivers from water dripping drop by drop; a vast mound from crumbs and grains, etc.' (ibid., I, p. 47).

153 Cf. Guerrini 2006, pp. 46–7, 52, 70 and 73; Galluzzi 2011, esp. pp. 44ff.

operate through chemical operations, while for Galileo they act via mechanical collisions and penetrations at high speed.

One may reasonably suppose that, if the Tuscan scientist had been able to read Cesi's *Taumatombria*, he would have detected in those pages the fumes of Della Porta's natural magic and alchemy. Cesi must have been aware of this fact. For this reason he refrained from sending Galileo either the *Indicatio* with its outline of his personal study programme, or the *Laserpitium* and the *Oposcopia*. Nor is it surprising that, in spite of the Prince's continuous promises, Galileo was never given the opportunity to read the *Coelispicium*, so intrinsically linked to the analyses developed by Federico in the Paduan manuscript.

Cesi continued to work on the *Coelispicium* after the decree of 1616, always conceiving it as a fundamental object of study for his new science of meteors, focusing in particular on the exchanges between the Earth and its creatures, on one hand, and the celestial bodies and spaces, on the other. There are numerous traces of this continued application in the correspondence of Cesi, especially in the intense exchange of letters that he maintained with Johannes Faber between 1616 and 1624. During most of that period the Prince was in Acquasparta, engaged not only in honing his treatise on celestial matters but also in his beloved botanical studies and in the interpretation of the nature of the singular findings in the fossil deposits close to his residence in Umbria.

CHAPTER 7

Confronting the New Scenario

Cesi-Bellarmine: Attempts at Dialogue

As we have seen, after the decree of 1616 Cesi's correspondence with Galileo slowed down considerably. In his rare letters the Prince deliberately avoided dealing with cosmological questions, confining himself to offering Galileo his affectionate solidarity and informing him of the embarrassment of some authoritative Jesuits, especially Father Grienberger, who had been 'dismayed' by the 'outcome of the past negotiations'.[1] Likewise, after returning to Florence, Galileo's correspondence with the Prince dropped to such a level that, not receiving any reply to his letters, Federico feared that their epistolary exchanges were being intercepted on the orders of the Inquisition.[2]

Cesi had retired to the ancestral palace in Acquasparta, where he dedicated most of his time to his programme of studies and to naturalistic explorations of the surrounding territory. In his letters to Faber of 1617 and 1618, the Prince often mentioned his intense commitment to the compilation of the *Coelispicium*. I am 'immersed again in my contemplations', he wrote to Faber on 8 August 1617, expressing the conviction that he was now very close to 'completing the treatise'.[3] He also informed the German Lincean about his work method: he recorded his reflections and his collation of passages from the sources that he consulted on cards that he would use in composing the text.[4]

A few days later he wrote to Galileo that, in spite of the amount of time taken up by practical matters, he still managed to enjoy himself 'now and then with my contemplations of nature'.[5] Note that, while the Prince gave Faber precise information about the nature and progress of his research, he systematically avoided doing the same with Galileo.

1 OG XII, p. 285, letter of 8 October 1616. Cesi wrote to Galileo that, besides Grienberger, his fellow Jesuit Father Guldin had also shown regret at the anti-Copernican decree.

2 Ibid., p. 340, letter to Galileo of 22 August 1617: 'I have not received a letter from you for four months and wonder about interceptions [...]'.

3 CL, p. 610.

4 'Cogitationes interim schedulis raptim interserere non piget ad opus postmodum digerendas', ibid.

5 OG XII, p. 341, letter of 22 August 1617.

 | DOI 10.1163/9789004342316_008

He referred again to his work on the *Coelispicium* in a letter to Faber of 6 November: 'As for my studies, I am totally absorbed by celestial questions, and sometimes particularly ponder the arguments of the Stoic philosopher Justus Lipsius'.[6] The mention of Lipsius, protagonist of the revival of the Stoic vision of man and of nature, reveals once more the close connection between study of nature and philosophical and moral reflections that had, from the outset, inspired the Prince's vision.

The compilation of the *Coelispicium* advanced slowly until the summer of 1618, when it underwent a sudden acceleration. In June Cesi asked Faber to get Ciampoli to procure him a work by Kepler (probably the Imperial astronomer's reply to Ingoli),[7] a clear sign of his application to the treatise on celestial matters. A month later the *Coelispicium* had made encouraging progress. This is suggested by an excited letter of the Prince from Acquasparta in which he informed Faber that he had completed a text (which he enclosed) that he intended to send to no less a person than Cardinal Bellarmine.[8] The choice of this eminent addressee may appear surprising in view of the fact that the Jesuit had made it abundantly clear to the Prince that he opposed the implications of the celestial novelties, anticipating the fate that would befall Galileo if he insisted on maintaining the truth of the Copernican system. Cesi was also aware of the Cardinal's leading role in the severe condemnation of the new cosmology. However, boasting a long and cordial familiarity with Bellarmine, he knew that, while the Jesuit had taken up a hostile position regarding the physical truth of heliocentrism, he had nevertheless shown an open attitude to the proposed reform of traditional astronomy based on the demolition of the solid spheres of Aristotle and Ptolemy and on the recognition of the fluid nature of the heavens.[9]

The Prince had had the opportunity to touch upon these themes with Bellarmine on many occasions. In all probability, the last face-to-face discussion between them had taken place in the late spring of 1618 in Rome, where Cesi had been residing from the beginning of the year, having taken a long period of leave from Acquasparta. In the letter addressed to the Cardinal, Cesi referred explicitly to Bellarmine's invitation to write down and to send him in

6 'Quoad studia totus in coelestibus recreor, et quandoque Lipsianis praesertim Stoicis', CL, pp. 617–8.

7 Ibid., p. 641, letter of 13 June 1618. Cf. below, pp. 240–1.

8 'En opusculum seu epistola a me ad Cardinalem Bellarminum perscripta', ibid., p. 646.

9 Cf. Baldini U. 1984 and 1992 (pp. 285–345).

confidence the arguments on which he based his conviction of the fluidity and unitary nature of the celestial spaces.

The text sent to Bellarmine should not be confused with the *Coelispicium*. It contained a preliminary exposition of only some of the themes that the Prince intended to develop in that treatise on which he was keen to obtain the opinion of an authority on whose discretion he knew he could rely. The Cardinal's reactions would furnish him with valuable indications as to whether conditions were favourable for him to complete the *Coelispicium* and to eventually publish it.

In sending Faber the draft of his letter to Bellarmine, the Prince asked him to edit the text and, above all, to elevate its literary style: 'return it to me in a polished, corrected and more elegant form at the earliest possible date'.[10] Conscious of his limits in the epistolary genre, he called on the assistance of his friend. He prided himself on having some competence only in speculations *de rerum natura*.[11]

The Prince added a second request. When the right opportunity arose, he wanted Faber to read, but 'almost furtively', to Cesarini and Ciampoli the letter addressed to the Cardinal. The German Lincean should make sure to tell them that he had taken that initiative without the Prince's knowledge.[12] This request for confidentiality was probably motivated by the desire to protect Bellarmine. After the decree of 1616, the Cardinal neither could, nor wanted, to expose himself to any suspicion. For this reason he had probably asked the Prince to avoid mentioning their exchange of opinions to third parties.

Cesi provided Faber with another precious piece of information. After editing the draft that he had sent him, he had in mind the idea of composing yet another letter 'of the same substance but on other themes'.[13] As Gabrieli pointed out,[14] there is no record of a second letter to Bellarmine on celestial questions. The concluding remark of the postscript, 'enough now of fluidity' (*sufficiat nunc de fluiditate*), suggests that the Prince would not have insisted

10 '[...] tersam, correctam concinioribus quandoque verbis ornatam elegantemque ad me remittat statim statim', CL, p. 646 (letter from Acquasparta, 30 July 1618).

11 '[...] contemplationum me esse fateor et conscriptionum physicarum quantuluscumque sum', ibid.

12 'Legat [the letter] D.V.ª ex se, quasi me inscio, et quid dixerint rescribat', ibid., p. 647.

13 Ibid.

14 Ibid., n. 2.

in it on the fluidity of the heavens. Other passages in the correspondence with Faber of the successive weeks indicate that Cesi did effectively devote himself to this new project, but offer no clear indication of its contents.[15]

It is unlikely that in the second letter Cesi intended tackling questions bearing on the structure of the universe. The desire to keep clear of that minefield had been explicitly declared from the time of the outline of the *Coelispicium* in the *Indicatio*, that is, from before the publication of the anti-Copernican decree.[16] The decree could only reinforce that determination. It is more probable that in the second letter Federico would have proposed his reflections on the question of the corruptible nature of the heavens, a problem barely touched upon in the extant letter to Bellarmine. This question was essential to show that direct connection between celestial and terrestrial phenomena on which he had fixed his attention for some time and that was to play a key role in the writings to which he would devote himself in the following years. As we emphasised in analysing Agucchi's manuscript treatise of 1611,[17] the thesis of the corruptibility of the heavens appeared to the supporters of traditional cosmology more difficult to digest than that of their fluidity.

Faber took great pains to satisfy the Prince's requests: only two weeks separated the dispatch of Federico's message to the German Lincean and the day on which Cesi signed his letter addressed to Bellarmine.[18] The Prince's epistolary treatise is known by the title *De caeli unitate et tenuitate fusaque et pervia stellarum motibus natura ex Sacris Litteris* (On the unity and tenuousness of the heavens and on their homogeneous and penetrable nature as well as on the motions of the stars according to Sacred Scripture), which is absent from the copy with autograph corrections that has come down to us.[19] The title was introduced in 1630 on the occasion of the first publication of the document,

---

15 It is probable that the attachments (which have not survived) mentioned in the letters from Cesi to Faber of the successive weeks referred to the project of a second intervention on celestial questions; cf. CL, p. 660 (letter to Faber of 21 August) and p. 665 (letter to Faber, July-August). In the latter, Cesi informed his fellow academician that he had 'somewhat enriched' (*locupletata alquanto*) the letter to Cardinal Bellarmine.

16 Cf. above, p. 125.

17 Cf. above, pp. 158–60.

18 CL, pp. 648–56. The letter to the Cardinal is dated 14 August 1618.

19 See Gabrieli, ibid., p. 648.

with the approval and direct collaboration of the Prince, in the *Rosa Ursina* of the German Jesuit Christoph Scheiner.[20]

As Cesi declared in the opening passage of his letter-treatise, during the course of their discussions[21] he and the Cardinal had been in full agreement on the fluid nature of the heavens:

> When you wished to hear my opinion, you declared that you agreed with my proposal of the tenuousness and permeability of the heavens and the need to remove the many and complex masses of large and small spheres from the purity of nature, and you affirmed that it was in full agreement with Holy Scripture.[22]

This account of the unconventional cosmological convictions of the Jesuit helps to understand why Cesi had urged Faber to respect the utmost confidentiality. The exchange of opinions between the Prince and the Cardinal illuminated by the *De caeli unitate* took place during the critical months in which, pending the correction of the *De revolutionibus*, many authoritative persons were asking themselves how much freedom of expression would be left to them after the adoption of the anti-Copernican decree. If for some of

20 The Latin letter *De caeli unitate*, sent by Cesi to Bellarmine on 14 August 1618, and the Cardinal's reply were first published in the *Rosa Ursina* of Scheiner (Scheiner 1626–30, pp. 777–82). Gabrieli republished the two letters (CL, pp. 648–58). These documents were reprinted in the original Latin with an Italian translation by M.L. Altieri Biagi and Bruno Basile (Altieri Biagi-Basile 1980, pp. 1–38). For an Italian translation very close to that of the two above-mentioned authors see Cesi 2003, pp. 55–74.

21 That it was the intense conversations with Bellarmine that led Federico to compose the letter on the fluid and unitary nature of the heavens is confirmed by Cesi in his reply to the request of Scheiner (passed on to him by Faber) to recount the occasion that had led to the writing of the *De caeli unitate*. The Prince mentioned the Cardinal's orientation in favour of the fluidity of the heavens but he was still hesitant – Cesi wrote – because of 'the opposition of the schools'. Cesi stressed the insistence with which Bellarmine had asked him to put his arguments against the solid spheres down in writing. He also stated that he had exchanged other letters with the Cardinal on these themes besides the *De caeli unitate* (CL, p. 1167, letter of Cesi to Faber of 1 June 1628).

22 CL, pp. 648–9. According to Lattis, while a Stoic influence can be detected in Bellarmine's ideas on the structure of the universe (Lattis 1994, pp. 94–102), he drew the conclusion of the fluid nature of the heavens from Holy Scripture. He would thus have subscribed to Patrizi's idea that the celestial bodies move like fish in water or birds in the air. A strong interest in the Neo-Stoic positions of Justus Lipsius was shared by Cesi, Faber, and the entire Lincean circle. On Bellarmine's cosmological conceptions see Baldini U. 1984.

them it was simply a question of knowing what position to assume, others, like Cesi and Bellarmine, found playing the part of passive onlookers to be out of keeping with their role. They therefore reflected on the strategy to prevent the decree from completely stifling the possibility of a cautious and gradual reform of traditional cosmology after the shockwave produced by the celestial discoveries. Both feared that the conception of the unitary and fluid nature of the heavens might also be prohibited. The situation of uncertainty as to how to interpret the limits imposed by the decree before the publication of the emendations to the *De revolutionibus* helps to understand the goals that Cesi pursued in his epistolary treatise and the significance that he assigned to the confidential involvement of Bellarmine in the discussion on the fluid heavens.

The *De caeli unitate* was conceived and sent to Bellarmine in the same weeks in which Clavius' successor in the Collegio Romano, Father Christoph Grienberger, was occupied with revising the work of a fellow Jesuit, the *Cosmographia* of Giuseppe Biancani.[23] The Cesi-Bellarmine discussion and Grienberger's revision of Biancani's manuscript are the expression of the same uncertainty that prevailed within the Society of Jesus and the Lincean Academy regarding the attitude to adopt on cosmological matters after the 1616 decree. It is therefore necessary to read those documents in parallel.

The *De caeli unitate* offered the Cardinal a foretaste of the *Coelispicium*, in which the Prince would illustrate in detail the principal opinions of classical and modern writers on the nature and substance of the heavens. In this first essay, he focused attention on the interpretation of the passages in which these problems were mentioned in Holy Scripture, to which he assigned, without hesitation, supreme authority:

> I am torn from discussion of these lesser matters to that supreme and first truth of Holy Scripture, whose fullest authority it is sinful to question, and from which several confirmations are derived that it would be impious to oppose.[24]

This explicit declaration reveals once again the distance between Federico's approach and the position assumed by Galileo in the *Copernican letters*. For the Prince, the mathematical and physical reasons that called for a profound

23 Biancani's work was published in 1620. On the publication of the *Cosmographia*, see Baldini U. 1992, p. 245. Baldini published documents that throw light on this highly revealing episode of the Society's systematic censorship of the Father's manuscripts before publication (ibid., pp. 217–50).

24 CL, p. 649.

reform of traditional cosmology could be accepted only if confirmed by the interpretation of the real meaning of the sacred texts. For the Tuscan scientist, the conclusions arrived at after rigorous processes of observation and mathematical demonstration had the status of absolute truths, independently of what Holy Scripture affirmed: it would be the task of the theologians to interpret the sacred texts and to reconcile them with the demonstrative conclusions. Bellarmine, Cesi and many others proceeded in the opposite direction: the interpretation of the real meaning of Holy Scripture was the highest court, to whose judgement the conclusions *de rerum natura* must necessarily bow.

It was not only as a consequence of the 1616 decree that Cesi had assumed that position. As we have seen, well before the issue of the decree he had become convinced that this was the path to follow and invited Galileo to pursue it with him.

Convinced that the battle for the affirmation of the Copernican system was lost, the Prince feared that the decree might drag conceptions like the fluidity and corruptibility of the heavens in its wake. Those conceptions formed the basis not only of his unitary vision of the cosmos, but also of his investigations of the organisation of nature and the phenomena of life. This was why he made every effort to persuade Cardinal Bellarmine to support that cause.

While confident that Galileo too would benefit from his initiative, Cesi was perfectly aware that the Tuscan scientist would not approve of it. For this reason he avoided the slightest mention to him of his exchange of opinions with Bellarmine. Galileo had to be kept completely in the dark about this operation: his disagreement would jeopardise its chance of success, with even more serious consequences for himself.

The *De caeli unitate* is packed with citations from the Church Fathers and learned glosses on the genuine meaning of Holy Scripture based on Cesi's in-depth philological analysis of the terms and contexts of the original Hebrew text. In developing his extremely refined analyses, the Prince exploited his knowledge of ancient Hebrew as well as of Greek and Latin exegetical traditions. If one applies the instruments of the most rigorous exegesis to the sacred texts, Federico affirmed, all the obstacles to acceptance of the fluidity and unitary nature of the heavens are overcome.

Cesi endeavoured to demonstrate that the firmament is immobile and fluid. He argued that, when correctly understood, Holy Scripture affirmed that not the solid spheres but the stars and planets are in motion. The heavens are not contained in a sphere, given that Holy Scripture and the Church Fathers define the universe with terms like expansion, dilation and diffusion. Moreover, the number of stars in the cosmos was infinitely superior to the 1022 recorded in Ptolemy's catalogue. According to the Prince, that number 'did not fall short, but was barely the beginning of their enumeration'.[25] The obstinate defence of the Ptolemaic heavens by the astronomers of his time was based on their disregard for observation and experiment. They had even gone so far as to despise the telescope with which Galileo had revealed 'so many things previously hidden in the heavens: new planets, new fixed stars and new configurations of the celestial bodies'.[26] The term *raquia* to qualify the heavens in the Hebrew text, he continued, unequivocally defined its extremely rarefied nature, similar to our air, although of a much finer consistency. The traditional conception of the solid spheres and the clear break in continuity between the sphere of the elements and the rest of the universe had no foundation in the biblical texts. On the other hand, was not the vision of the fluid heavens more logical, simple and natural than that of the solid spheres? Was it not a concept that our minds were able to grasp with intuitive clarity? Contrary to any logic or necessity and, above all, profoundly altering the real meaning of Holy Scripture, supporters of traditional cosmology had confined us

> inhabitants of the Earth to a multi-layered bark-like shell, surrounded on all sides by very hard orbs [...]. Will the mind not driven by reason, nor inspired by any sense of beauty, cling so closely to them [the traditional ideas] that it diligently tries to accommodate the things themselves, the reasons and the sacred texts in order not to depart or be dislodged from them in any way?[27]

Having removed the spheres and established the fluidity and permeability of the heavens, Cesi had to explain what sustained the celestial bodies in the heavenly spaces and caused them to cyclically describe their orbits with extraordinary regularity. He advanced an interpretation that he claimed to have discussed with the Cardinal on other occasions. According to Federico, as many passages of the Bible show when correctly interpreted, the celestial bodies are the seats of the angels. The latter did not cause the celestial spheres to rotate like mill-stones, as stated by Aristotle, but drove the stars and planets along routes preordained by the supreme mind of the Creator.[28]

Before analysing Bellarmine's reply, it should be noted that in the *De caeli unitate* one perceives clear echoes of Patrizi's criticism of the solid spheres

25 '[...] non mancus mutilusque, sed vix ullius numerationis initium', ibid., p. 651.

26 '[...] tam multa in caelo priscis abscondita, novos nobis Planetas, nova fixa, novas astrorum facies detexit', ibid.

27 Ibid., p. 654.

28 Ibid., pp. 652–6.

in which, according to the traditional cosmology, the planets were fixed like knots in a plank. In his *Pancosmia,* Patrizi had denounced those astronomers who, instead of describing the true aspect of the heavens and the real motions of the planets, had conceived the most extravagant and absurd hypotheses. In Patrizi's eyes, that attitude had transformed the heavens into a 'new chaos' (*novum aliud chaos*).[29] Having cleared the field of the solid spheres, Patrizi, taking up theses that had been maintained not only by Plato but also by Aristotle and the hermetic tradition, had stated that the celestial bodies are living beings, guided in their motions by a soul-mind in perfect resonance with the soul of the world. Using a suggestive metaphor, he had declared that the celestial bodies proceeded in an orderly manner through the fluid spaces of the heavens like flocks of cranes.[30]

It is hard to imagine that Cesi was not aware that among the censures of the inquisitors against Patrizi's *Nova de universis philosophia* was that of having supported the thesis of animated heavens, based not only on the authority of Plato and Aristotle, but also on that of St Thomas. The censors had been equally alarmed by Patrizi's thesis of the infinity of the universe.[31] Moreover, as we have seen, the Prince also knew of the tenacious defence of the animated heavens by Lagalla, who had rested his case on the authority of Aristotle and Alexander of Aphrodisias.[32]

It should also be borne in mind that the question of the true nature and form of the heavens had been at the centre of the discussions between Patrizi and Telesio. As an adolescent the Prince must have heard them evoked several times in his conversations with Persio. Patrizi had vigorously refuted the Telesian thesis of the planets borne along by the spheres in which they were embedded: 'Are we to suppose – he wrote – that they [the planets] are fixed in the heavens like knots in a plank and are entrained by its movement, rather than as living beings endowed with will and inclination?'[33] And a few lines further on: 'What objection is there to their [the planets'] moving, not like knots in a plank, but as entities existing by themselves, endowed with a principle of self-preservation?'[34] Was it not more logical, he insisted, to suppose that the

29 Patrizi 1591, p. 91r.

30 Ibid., p. 90v. On Patrizi's cosmological conceptions see Rossi 1977 (pp. 109–47) and Donahue 1981 (pp. 47–50).

31 Patrizi 1970, pp. 217–8.

32 Cf. above, pp. 160–2.

33 Fiorentino 1872–4, II, pp. 376–7.

34 Ibid. Cf. Rossi 1977.

planets and stars move freely through the fluid celestial spaces like birds in the air and fish in water?[35]

Telesio replied to Patrizi that he was not opposed to assuming that the Sun rotated around its own axis (*a seipsis circumvolvi*)[36] independently of the motion of the sphere in which it was fixed. But he refused to give up the spheres, nor was he tempted by Patrizi's heliocentric orientations.

The welding of the thesis of the Sun spinning around its own axis, proposed by a resolute supporter of the central position of an immobile Earth like Telesio, to the heliocentric vision would be effected by Campanella in his *Philosophia sensibus demonstrata*.[37] On 13 June 1611, Campanella wrote to Galileo that, while admiring the work of Copernicus, he had formulated some criticisms of the Polish astronomer in the treatise on metaphysics that he was confident would soon be published. The most serious was that of not having explicitly assigned to the Sun the motion of rotation around its own axis:

> The Sun was to be posited as highly mobile, set in very rapid motion by heat, which embraces and thins everything [...]; and since it is not borne around any centre [as Copernicus affirms], it must spin around itself.[38]

As tangible evidence of its rotation, Campanella cited the observation of the image of the Sun reflected on a smooth surface of water, which made that movement evident. Moreover, he added, since the Sun is the perfect body *par excellence*, it was absurd to conceive it as immobile.

Campanella's remarks in his letter to Galileo were actually quotations from Telesio. The argument of the Sun reflected on the surface of the water had been formulated by the latter in the second and third edition of his *De rerum natura*,[39] and was to return to the forefront in his discussions with Patrizi.[40] Likewise the concept of heat as the principle of life and movement was

35 It should be recalled that, probably influenced by the lecture notes of the natural philosophy courses given by the Jesuit Fathers in the Collegio Romano, Galileo too, in his so-called *Juvenilia*, had compared the motion of the planets in Ptolemy's system to that of the knots in a plank (OG I, p. 103). Cf. Besomi-Camerota 2000, p. 34. The same metaphor can be found in Scheiner (Scheiner 1626–30, p. 733).

36 Fiorentino 1872–4, II, p. 393.

37 Campanella 1591.

38 OG XI, p. 22; Campanella 2010, p. 318.

39 Telesio 1965–74, I, p. 318.

40 For the discussion between Telesio and Patrizi on the spinning of the Sun on its own axis, see Lerner 1975 and 1980 and Dollo 1995.

a distinctive feature of Telesio's conception of nature. It should also be noted that Campanella informed Galileo of his conviction of the rotation of the Sun around its own axis before the Tuscan scientist announced the discovery of the sunspots, from which he deduced the definitive proof of that phenomenon.

There is no doubt that Cesi and the other founders of the Academy were familiar with the lively debates, permeated by keen expectations that the demolition of traditional cosmology would herald a process of moral and religious reform, between Telesians, who adhered to geocentrism, and supporters of Platonic philosophy, like Patrizi, who displayed a clear inclination towards heliocentrism.

Let us return to the dialogue between Cesi and Bellarmine. In the concluding remarks of his letter Federico declared that he submitted his reflections to the judgement of the Cardinal. If he secured his approval, 'I will consider that I have accomplished a task that has not been easy for me and will apply myself to completing my work on celestial matters with greater fervour and enthusiasm'.[41] Bellarmine was not slow to reply. On 25 August he sent the Prince a short letter that cannot have pleased him. Having expressed his appreciation of the thesis of the fluidity of the heavens, of which he had been convinced for some time, he advanced his reservations. Above all, he could not accept that the heavenly spaces were without limits and, as a consequence, deprived of a definite form. He manifested the profound conviction that the cosmos was contained in a sphere because of the credit that had to be given to the explicit affirmations in this sense in *Ecclesiastes*, and because 'there is no doubt that the sphere is the most perfect shape and we know that the works of God are perfect (*Dei perfecta sunt opera*)'.[42] The inquisitor from the trial of Giordano Bruno had been disturbed by Cesi's insistence on the absence of limits to the cosmos. Nor will it have escaped his notice that the Prince's arguments echoed the thesis of Patrizi, who in his *De nova universis philosophia* – a work that had been suspended pending emendation – had denied that the heavenly spaces had a spherical shape, thus revealing his inclination towards an infinite universe.

Furthermore, the Cardinal complained that Cesi had not expressed his opinion on a point which he considered to be of crucial importance:

> I wanted to learn from you how the movements can be saved of the Sun and the stars, particularly the fixed stars, which proceed always together

41 CL, p. 665.

42 Ibid., p. 662.

> and make their circles larger or smaller according to whether they are further or closer to the pole; as this is the reason why the philosophers and astrologers assign seven heavens to the seven planets, and one heaven to all the fixed stars; and, above all, I wanted to know how the diverse motions of the same star can be saved if there is nothing but an immobile heaven.[43]

While polite and conciliatory, as befitted a friend who had been invited to reveal his own ideas in confidence, Bellarmine did not conceal his embarrassment at reading the *De caeli unitate*. He felt that Cesi's reply to his questions had been evasive. He had hoped to learn from him what guided the stars and planets in their regular movements through the fluid spaces of the heavens. Cesi had shrugged off the problem by attributing that delicate task to the angelic intelligences, while Bellarmine had looked forward to an explanation based on physical arguments. He confessed that as a young man he had given much thought to these problems, obviously starting out from a geocentric conception, but without arriving at conclusive results:

> I wanted to learn these and similar things from you, to know whether you had perhaps made particular considerations on how to save the motions of the stars in such a way as to save the opinion of the Holy Fathers at the same time, who claim that the heavens are immobile and the stars mobile.[44]

Bellarmine was in search of a solution that would save both the motions of the stars and the passages of Holy Scripture on the structure of the universe. Having asked Cesi for his opinion *de systemate mundi* in confidence, he did not disguise his disappointment at the diverse direction that the Prince's argument had taken.

The Cardinal concluded by inviting Federico to quieten his mind. His letter had confirmed that the solution to those problems lay beyond the scope of the human intellect. The only course open to them was to seek refuge in the safe harbour of the faith: 'Let us endeavour to live in sacred fear of God in order to reach heaven, where everything will be revealed in an instant'.[45]

43 Ibid.

44 Ibid., p. 663.

45 Ibid.

It was an eloquent way of closing the discussion. Cesi was let down in his hope of receiving encouragement from the authoritative prelate to continue with his battle for the affirmation of the unity and fluidity of the heavens.

Both men were perfectly aware of having touched on delicate questions which would have irritated more than one pious conscience. It was certainly the fear that it might fall into the hands of some bitter enemy of the freedom of thought in cosmology and natural philosophy that prompted Bellarmine to return the manuscript of the *De caeli unitate* to its sender, motivating this unusual decision with a clearly instrumental excuse: 'I return your manuscript in case you may not have kept a copy and by chance it might get lost among my innumerable notebooks'.[46]

It was an elegant leave-taking and a clear invitation to exercise caution. Cesi understood the message perfectly. He abandoned the idea of writing a second letter to the Cardinal. There no longer seemed to be room for the affirmation of that freedom of thought in natural philosophy for which Galileo had faced a resolute battle without seeking short-cuts or accepting compromises.

'The Time Has Come to Grant Greater Freedom of Thought'

In spite of his disappointment, Cesi trusted that his letter would encourage Bellarmine to speak up for the authorities of the Church and the hierarchy of the Society of Jesus to recognise the thesis of the fluidity of the heavens at least as admissible.

The Prince was aware of manifestations of concern among the ranks of the Jesuits because of the ever more severe internal censorship after the anti-Copernican decree. The Father General of the Society, Muzio Vitelleschi, was violently opposed not only to the celestial novelties but also to the slightest deviation from Aristotelian physics. Cesi knew that this attitude embarrassed those Fathers directly involved in scientific debates with the most gifted natural philosophers outside the Society. Thanks to his frequent contacts with Father Grienberger, he had also learnt that the Jesuits were not united behind the decision of the Society to impose blind obedience to traditional knowledge. However, the 1616 decree had weakened the voice of those Fathers who were pressing for a process of renewal, however gradual.

During the same months in which Cesi and Bellarmine were holding their discussions, Father Grienberger ran up against these difficulties in carrying

46 Ibid.

out the task assigned to him by his superiors of revising before publication the work of a fellow Jesuit: the *Cosmographia* of Giuseppe Biancani.[47] Grienberger's evaluation of that work helps us to understand the delicate scenario in which the dialogue between the Prince and the Cardinal took place.

The German Father had subjected Biancani's text to a close reading. This had led him to manifest a number of reservations, which were the direct consequence of the inescapable limits imposed on freedom of expression by the Society of Jesus. The censorship of the *Cosmographia* thus became an explicit defence of the freedom of thought *in naturalibus*, to use the expression dear to Cesi. For Grienberger, Biancani had not made any real mistakes, but the quality of his work was compromised by the reticent attitude that he had adopted with regard to the most controversial debates in astronomical research of recent decades. In reply to Biancani's claim that Copernicus conceived his own system simply as a hypothesis, Grienberger curtly stated: 'Copernicus does not speak *ex hypothesi*, but makes every effort to prove that his system of the world is true'.[48] It was for this very reason that the Polish astronomer's work had been suspended. Grienberger felt that Biancani showed excessive restraint also with regard to the real existence of epicycles. Moreover the author of the *Cosmographia* had not taken up a clear stance on the question of the corruptibility of the heavens. Since Biancani had declared their fluidity possible, Grienberger maintained that their corruptibility should be considered equally admissible: 'I do not see why anyone who admits the fluidity of the heavenly spaces will not readily accept also their corruptibility, an opinion which finds favour with a good many philosophers and theologians'.[49] Grienberger invoked Bellarmine himself to corroborate the positive inclination that many theologians had shown towards the fluidity and corruptibility of the heavens:

> I know that it [the corruptibility] generally finds favour with them all, in particular with the illustrious Cardinal Bellarmine, who does not deny that the planets move without being fixed to solid spheres, admits of corruptibility in the heavens and believes that this opinion is more in conformity with Holy Scripture and its exegetes.[50]

47 Biancani 1620. Grienberger's censorship (preserved in the Archivio Storico della Compagnia di Gesù) was published in Baldini U. 1992 (pp. 216–50).

48 Ibid., p. 232.

49 Ibid., p. 233.

50 Ibid.

Grienberger went on to criticise Biancani for failing to give due credit to the celestial novelties that had demonstrated the untenability of Ptolemy's cosmology: the satellites of Jupiter, the orbits of Venus and Mercury around the Sun, and even the sunspots, 'which perhaps move on the surface of the globe of the Sun'.[51] He also stressed that the extreme probability that the heavens were fluid and corruptible, as well as that there were bodies that moved along orbits that did not have the Earth as their centre, was confirmed not only by the new celestial observations, but also by the essential principle of the attribution to nature and to its Creator of modes of operation characterised by the greatest simplicity and economy: 'Nature could be justifiably accused of superfluity if it had created such immense and multiple orbs, some of them concentric and some eccentric, for one astral body or another'.[52] It was the same argument proposed by Cesi in his correspondence with Galileo of 1612 and repeated by the Prince in the *De caeli unitate*.

For these reasons Grienberger found the solution of animated planets – again the same thesis that Cesi had formulated in his letter to the Cardinal – more simple and logical: 'There seems to be no reason why the stars cannot traverse those same routes through the heavens, either by themselves or driven by some intelligent being, which they would describe if they were carried by the motion of several spheres'.[53]

So if we want to banish from the heavens not only the solid spheres but also the intricate congeries of eccentrics and epicycles, Grienberger added, it would be sufficient to declare 'that a star is moved by a single irregular movement that is composed of, or rather results from, a variety of different motions, like the spiral motion of the Sun'.[54] These were the directions that the most recent astronomical research had been taking, leading to a series of new discoveries that made it impossible to blindly maintain the traditional system of the world. The ban on discussing those new findings imposed by the hierarchies of the Society had put the Fathers in the greatest difficulty, with the risk of losing the prestige and credibility that they had taken such pains to achieve:

> In this respect I sympathise with my fellow Fathers: for it is humiliating to declare one's ignorance, or to defend what others have defended; for this reason they [the Jesuit Fathers] often prefer to reject mathematical

51 Ibid. This was a clear way of distancing himself from Scheiner's interpretation of the phenomenon.

52 Ibid.

53 Ibid., pp. 233–4.

54 Ibid., p. 234.

> observations, even those that have been approved by the greater number of the most renowned astronomers. And it is even more humiliating and grave when such observations are denied by Fathers with expertise in mathematics.[55]

Grienberger's direct expressions denounced the marginalisation of the Fathers in the most advanced scientific debate, due to the severe prohibitions on taking any new discovery into consideration. The Jesuit confessed that even his own teaching activities at the Collegio Romano suffered from this situation. He found himself constantly embarrassed by the questions raised by his fellow Jesuits and students on the celestial novelties. These difficulties would not end, he resolutely declared, 'until permission is granted to reply more freely'.[56] The restrictions imposed by the Jesuit hierarchies obliged him to exercise constant self-censorship. He had to conceal his own interpretation of the sunspots, dissimulate his opinion on the movements of the comets beyond the sphere of the Earth and avoid pointing out how Tycho's observations of the circumsolar circuit of Mars imposed abandoning the conception of the solid spheres of the planets.

After his vigorous denunciation of this untenable situation, Grienberger addressed an impassioned plea to the hierarchies of the Society:

> It seems to me that the time has come to grant greater freedom of thought to the mathematicians and philosophers in this matter; since the fluidity and corruptibility of the heavens are absolutely not contrary to theology or philosophy, and even less to mathematics.[57]

The Jesuit Father was boldly overstepping the limits of Cesi's horizon. It was not a question of urging theologians to make every effort to interpret the passages of Holy Scripture describing the structure of the universe: the face of the heavens had changed as a result of the celestial discoveries. While Grienberger agreed that theologians should tread carefully before demolishing the entire edifice of traditional knowledge, he condemned the error of resorting to authority to oppose convictions on the real structure of the cosmos based on incontrovertible evidence.

55 Ibid.

56 Ibid.

57 Ibid., p. 235.

Grienberger expressed his reservations also on Biancani's citation of Galileo as a supporter of the physical truth of the Copernican hypothesis. After the 1616 decree, Grienberger observed, 'Galileo [...] does not defend Copernicus any more, nor can one find in his writings any explicit pronunciation on this matter: thus he should not be mentioned here'.[58] Biancani accepted his colleague's criticism and modified the text in the printed version.[59]

In summing up his verdict, Grienberger returned to what he saw as the most serious limitation of the *Cosmographia*: the excessive caution displayed by the author with regard to the new celestial discoveries and their implications, especially concerning the fluidity and corruptibility of the heavens. Still, he excused his colleague's reticence: 'For he had his hands tied just like us'.[60] The *Cosmographia* would have been different – Grienberger insisted – 'if he [Biancani] had been accorded more liberty [...]. That is, if he could assert the fluidity and corruptibility absolutely, or at least conditionally'. That Biancani considered the fluidity of the heavens as extremely probable was demonstrated by what he had written on the subject in his preceding treatise on Aristotle's mathematics.[61] Due to the limitations imposed by the hierarchies of the Society, the *Cosmographia* was inevitably condemned to ambiguity. Grienberger hoped that, once those restrictions were removed, or at least relaxed, Biancani would have to rethink the work entirely:

> It will be necessary for the author to go through the whole work again and to discuss this matter in a manner to satisfy not only beginners but also experts in mathematics and philosophy.[62]

It is not surprising that Biancani was irritated by Grienberger's criticisms. Why do you take it out on me, he replied, when you know perfectly well that

58 Ibid.

59 Ibid., p. 236: 'After warnings from others too I have deleted Galileo. I approve your argument against the revolution of the Earth and I shall add it here under your name.' The concluding part of Biancani's reply referred to Grienberger's invitation to moderate the tone when mentioning the followers of Copernicus, bearing in mind that they are more numerous than his adversaries: 'Is it not true – Grienberger had remarked – that all the others [astronomers] reject it [the Copernican hypothesis] as an absurdity, for there are more mathematicians today who are in favour of Copernicus than against him, though there are perhaps only a few who think that his system corresponds to the real constitution of the world' (ibid., p. 235).

60 Ibid.

61 Biancani 1615.

62 Baldini U. 1992, p. 236.

my lips are sealed? And why do you not stick your neck out, you who have more authority than I do, to convince the superiors to recognise the admissibility of the fluid and corruptible nature of the heavens? 'You – he added – in whom the superiors have greater trust should propose those arguments [...] to them'.[63] Grienberger's reply confirmed his reservations regarding the excessively obsequious prudence of his colleague. Biancani had let slip an important opportunity; that of composing a text that, while not taking up an explicit position on the question of the system of the world, would present students with an objective picture of the state of the art in cosmological research:

> In my opinion in this treatise more space should have been reserved for the illustration of the hypothesis that has been recently introduced as well as of the novelties discovered in our age, although without approving them.[64]

I am sure that Cesi knew of the noble and courageous claim to the freedom of thought *in naturalibus* addressed by Grienberger to his superiors. Nor will the German Jesuit have been unaware of the discussions between Cesi and Bellarmine and of their disappointing conclusion.

The many points of similarity between Federico's *De caeli unitate* and Grienberger's revision of Biancani's *Cosmographia* suggest that they had consulted one another, united in the strategic objective of convincing Bellarmine to become the authoritative paladin of the need to guarantee a space for free discussion of the celestial novelties and their cosmological implications. It is significant that they had both attached particular importance to the fact that the Cardinal believed in the fluid and corruptible nature of the heavens.

However, the most striking aspect of these discussions is the total extraneity of Galileo. All the evidence suggests that the Tuscan scientist had no inkling of Cesi and Grienberger's subtle operation. To account for his surprising exclusion it should be borne in mind that the battle to which Cesi dedicated himself in 1618 was not the one for which Galileo had fought and intended to continue fighting. Aware as he was that Galileo would have considered his strategy and its objectives limited, Federico feared that he would not have hesitated to manifest his own radically different vision, thereby compromising the outcome of the operation. The secrecy with which Cesi moved was not

63 Ibid.

64 Ibid., p. 238.

only motivated by the desire to protect his own person and the delicate position of Cardinal Bellarmine. It was also intended to keep Galileo in the dark, with the connivance of Faber, Cesarini and Ciampoli, who strictly observed the injunction to discretion and did not betray the slightest hint of what was taking place.

This episode throws further light on the nature of the relation between Cesi and Galileo, especially after the anti-Copernican decree, as well as on the distance between their objectives and the methods to attain them. Cesi's circumspect battle for recognition, even if under tutelage, of a space for freedom of thought *in naturalibus* did not correspond to the goal doggedly pursued by Galileo: to release the study of nature from the vigilant guardianship of the theologians, affirming the right to make pronouncements on the laws of nature and the structure of the universe without risking persecution.

Kepler Enters the Scene

There is a third episode in those months of 1618 that seems to be intertwined with Cesi's letter to Bellarmine and Grienberger's censorship of Biancani's *Cosmographia*. This is Kepler's reply to Francesco Ingoli's *De situ et quiete Terrae contra Copernici systema disputatio*,[65] a work composed in the turbulent weeks preceding the adoption of the anti-Copernican decree. As stated by its author, the *Disputatio* had been prompted in 1616 by an animated discussion with Galileo on the system of the world in the Roman home of Lorenzo Magalotti, legate to the ruling family of Tuscany and future Cardinal.[66] Magalotti was on familiar terms with Cesi and it is extremely likely that the Prince was present on that occasion. At any rate, he certainly knew about the discussion, which had obviously concluded without the participants reaching unanimous agreement. At the end of the debate, Ingoli – who would soon be appointed consultor of the Congregation of the Index[67] – had promised to commit to paper the reasons that made it impossible to accept the Copernican hypothesis, so that

65 OG V, pp. 399–412. Ingoli's *Disputatio* was first published by A. Favaro (1891, pp. 149ff.). It was reissued with Italian translation and commentary by Bucciantini and Camerota (Galilei 2009, pp. 163–88). For an English translation of the *Disputatio*, see Graney 2015.

66 Cf. Galilei 2009, p. 163.

67 Ingoli was to be charged with the task of emending the *De revolutionibus* of Copernicus as well as with the formulation of the motivations for placing Kepler's *Epitome* on the *Index*. Cf. Bucciantini 1995, pp. 207–12. On the prohibition of Kepler's *Epitome* (Kepler 1612), see ibid., pp. 127–47.

Galileo could consider them and respond.[68] Ingoli kept his promise, while Galileo was discouraged from respecting the obligation to reply by the publication of the anti-Copernican decree.

That task was undertaken two years later by Kepler. The Imperial astronomer had no difficulty in demolishing Ingoli's weak arguments[69] and also vigorously opposed his strenuous defence of the subordination of astronomy to theology.[70] It had been Tommaso Mingoni, one of the main channels of communication between Kepler and the Italian peninsula, who had sent him Ingoli's text in June 1617 with a request to let him know what he thought of it.[71] Kepler complied with that request a year later, dispatching to Mingoni his own observations on the *Disputatio*.[72] In his letter of May 1618, Mingoni told Kepler that he had transmitted his severe review to Ingoli.[73] A few weeks later, Mingoni asked Kepler for a copy of the *Epitome astronomiae copernicanae* fresh from the press, 'so that I can post it to a Roman friend who is very keen to have it'.[74] It is logical to assume that Mingoni had previously sent Kepler's *Responsio* to the *Disputatio* of Ingoli to that same friend. That Roman friend was in all likelihood Johannes Faber, the point of reference of the German community in the Eternal City. This clue suggests that the text of Kepler owned by Ciampoli[75]

68 For a reconstruction of the activities of Francesco Ingoli on the Roman scene between the 1610s and 1620s, of his function as consultor of the Congregation of the Index and an upholder of the traditional cosmology and philosophy, see ibid., pp. 79–147. Bucciantini also stressed the attention which Cesi and the Roman Linceans paid to Kepler's severe rejoinder to Ingoli's *Disputatio* (OG V, pp. 403–12). He also discovered and published Ingoli's reply to Kepler's text: *Replicationes Francisci Ingoli Ravennatis* [...] *contra Copernicum ad Joannis Kepleri impugnationes* (Bucciantini 2005, pp. 177–205).

69 Ingoli had used the argument (taken from Tycho) of the absence of parallax of the comet as proof of the falsity of the Copernican hypothesis and of the truth of the geo-heliocentric system of the Danish astronomer. Kepler replied that both the trajectory of the comet towards the north and the absence of parallax derived from the combination of the rectilinear motion of the comet and the circumsolar motion of the Earth (cf. Kepler 1937-, XX^1, pp. 168–80; see also Bucciantini 2003, pp. 261–2).

70 For a detailed exposition of the arguments with which Kepler refuted Ingoli's affirmations in the *Disputatio*, see Bucciantini 2003, pp. 106–10.

71 Kepler 1937-, XX^1, p. 168; see Bucciantini 1995, pp. 103–5.

72 Mingoni received Kepler's rejoinder to Ingoli in May 1618 (Kepler 1937, XVII, p. 268; letter of Mingoni to Kepler of 31 May 1618).

73 Kepler 1937-, XVII, p. 268.

74 Ibid., p. 269.

75 CL, p. 641, letter of 13 June 1618. A manuscript copy of Kepler's *Responsio* is preserved in the Biblioteca dell'Accademia dei Lincei (Ms. Archivio Linceo 1).

that Cesi asked Faber to obtain for him on 13 June 1618 was the Imperial astronomer's reply to Ingoli.[76] We learn from Federico's letter that a copy of that work was in the hands of Giovanni Ciampoli. The tone of the Prince's request indicates the importance that he attached to that text and, at the same time, the circumspection with which he expressed his desire to read Kepler's severe criticisms of Ingoli's *Disputatio*. He urged Faber to ask Ciampoli for the loan of Kepler's text before measuring his finger for the emerald ring due to him as a newly inducted Lincean. Federico added that he left it up to Faber's discretion whether to act in the presence of the inseparable Cesarini or to do so when Ciampoli was alone.[77] Once again we note a climate of wary prudence in Cesi's relations with the two Roman associates. As we have already stressed, his recommendation to Faber to be guarded in dealing with Cesarini and Ciampoli was to reappear a few weeks later in the request to sound them out with extreme caution to ascertain their reactions on reading the *De caeli unitate*.[78] These clues seem to indicate the emergence of tensions, or at least of a climate of suspicion, within the Lincean core, in particular between Cesi and Cesarini.

The request for Kepler's text, the haste that he imposed on Faber, and the call to act with discretion all show that the Prince felt it necessary to take into account the Imperial astronomer's arguments against the *Disputatio* in the event of using them in the composition of the letter to Bellarmine on the fluid and unitary nature of the heavens.

Cesi was well informed about the increasingly strategic position that Francesco Ingoli was assuming on the Roman scene. He also knew that Ingoli would reply to Kepler's objections – as indeed he did – taking advantage of the occasion to convey to Galileo and his Roman friends the message that he had not been intimidated by the authority of the Imperial astronomer: there was no possibility, in spite of Kepler's explicit request in the *Epitome*, that the anti-Copernican decree of 1616 would be revoked.

There is a final piece of information that helps to reconstruct the complexity of the scenario that was taking shape in the course of 1618 with the active protagonism of the Lincean nucleus. This is the appearance, in rapid succession, of three comets in the skies of Europe. We shall soon tackle the opening of this new front in the conflict between Galileo and the Linceans, on one side, and the Fathers of the Collegio Romano, led by the Jesuit Orazio Grassi,

76 Cf. above, n. 7. Bucciantini supposed that Kepler's text was the reply to Ingoli (Bucciantini 1995, p. 114).

77 Ibid.

78 Cf. above, p. 222.

on the other. The Jesuits were not the only ones to maintain that the comets offered conclusive proof against the Copernican system. When the Linceans informed Galileo that his adversaries in Rome were triumphantly declaring that the comets passed the death sentence on the hypothesis of the motions of the Earth, they were in all probability referring not only to the *Astronomical Disputation* of Father Grassi, but also to a text by Francesco Ingoli, in which he adhered to the interpretation of the motion of comets by Tycho Brahe.[79] In that text Ingoli declared that he had taken part in the animated debates on the comets in the palace of the Cardinal of Santa Susanna, in the course of which the discussion had centred on 'whether the Copernican motions of the Earth are refuted by the motion of this comet'.[80] Ingoli stated that those debates were attended by very learned men in the house of the influential Cardinal, whom Galileo would frequently meet during his stay in Rome in 1624. Even though Ingoli does not name those '*dottissimi uomini*', it is difficult to imagine that Cesi and/or exponents of the Lincean circle were not present, or at least that the echo of those discussions did not reach their ears. After all, Cesi was on very familiar terms with the Cardinal of Santa Susanna, Scipione Cobelluzzi, as shown by his request, addressed to Faber on 13 April 1620, to pay his respects to the Cardinal on his behalf, reminding him 'of the true and much obliged servant that I always remain'.[81]

As far as is known, Galileo was not informed of the lively debates taking place in Rome in the second half of 1618. He certainly knew that Kepler had replied to Ingoli and would soon be able to read the Imperial astronomer's text. He also managed to obtain, though only with difficulty, an exemplar of Kepler's *Epitome*. He had been unable to find the work in Florence because it had been promptly put on the Index on 10 May 1619 after Francesco Ingoli's severe revision, in his capacity as consultant of the Congregation of the Index.[82]

79 The manuscript of the *De cometa anni 1618 tractatus* (Biblioteca Apostolica Vaticana, Ms. Ottob. Lat. 2484 II, cc. 208r-220) was rediscovered and published by Bucciantini (Bucciantini 1995, pp. 177–205).

80 Ibid., p. 152.

81 CL, p. 714.

82 On behalf of Archduke Leopold, it was Remus Quietanus who had asked Kepler for a copy of the *Epitome* to be sent to Galileo, adding that, although the work was banned, the Tuscan scientist was confident that he would be able to obtain permission to read it (OG XII, p. 469, letter of Remus to Kepler, 29 July 1619).

CHAPTER 8

Relaunching Copernicanism

Ariosto versus Tasso

Back in Florence after a long and unsuccessful stay in Rome, Galileo immediately immersed himself in a campaign of observations to complete the periodic tables of the satellites of Jupiter on which he had been working since the second half of 1610. He was confident that those tables would enable him to devise a method of establishing with precision longitude at sea: a long cherished ambition from which he also hoped to derive considerable material benefits.[1]

The pace of Galileo's correspondence with his Roman friends slackened from the middle of 1616. We have no letters from him to Cesi between August 1616 and 19 October 1622. Although they were not six years of total silence[2] (the greater part of Galileo's letters to the Prince have not passed down to us), Cesi's continuous complaints about the dearth of replies to his letters on the part of the Tuscan scientist confirm that the latter was not a diligent correspondent.

Until at least February 1619 the Prince's letters to Galileo were brief and sporadic too. They show Federico's willingness to maintain relations of esteem and friendship with the Tuscan scientist while avoiding celestial matters or advice on how to deal with the hostility of the Peripatetics and ecclesiastical authorities to the new doctrines. The Prince sent his best wishes for a speedy recovery from the ill health that continuously plagued Galileo, news about his own health and that of the other Linceans, requests for his Florentine friend's opinion on proposed candidacies, and little else. Federico also asked Galileo to help him in obtaining from the Grand Duke of Tuscany the privilege for the *Thesaurus Mexicanus*, whose publication he overoptimistically considered to be imminent. Galileo acted promptly to satisfy his request.[3]

At the end of May 1618, Galileo decided to go on a pilgrimage to the Sanctuary of the Madonna in Loreto, perhaps to fulfil the vow made in the hope

---

1 Cf. Galluzzi 2008, pp. 120–31.

2 That Galileo did write to Cesi during these years is confirmed by the Prince's letter of 15 February 1619: 'Having been without news of you since July, I recently received two letters in succession, almost simultaneously, and, a little earlier, one to Stelluti, which is here with me' (OG XII, pp. 438–9).

3 Ibid., pp. 380–1. Galileo wrote on 19 April 1618 to the Secretary of State Curzio Picchena to support the Prince's request.

 DOI 10.1163/9789004342323_009

of receiving some relief from the infirmity that had kept him bedridden for many months. The suggestion to go on that pilgrimage may have come from Cesi himself, who visited the sanctuary three times.[4] Before setting out, Galileo promised to pay a visit to the Prince in Acquasparta, not far from Loreto. Federico responded with enthusiastic and affectionate signs of encouragement.[5] In the end, however, to the bitter disappointment of the Prince and Stelluti,[6] Galileo chose a route that took him far from Acquasparta. It is difficult to know whether that decision and the sharp decline in their correspondence are signs of a cooling of their relations after the dramatic epilogue of 1616.

In the summer of 1618, Cesi co-opted two new members to the Academy who were particularly dear to Galileo: Giovanni Ciampoli and Virginio Cesarini.[7] On 1 October the latter sent a letter to the Tuscan scientist filled with expressions of gratitude for having opened his eyes to the falsity of traditional natural philosophy. Cesarini expressed his admiration for the remarkable integration of science and eloquence of which Galileo had offered unparalleled examples:

> I have always admired in you your knowledge and the way in which you explain it, not only casting light with your intellect on the most obscure contemplations, but even illuminating clouded minds, professing an eminence of intellectual clarity that is granted only to few.[8]

In considerate and affable terms, the Roman nobleman pointed out however, that they did not share the same literary inclinations. Unlike Galileo, who had a predilection for Tuscan poetry, he and Ciampoli admired the Greek and Latin poets, who perfectly combined stylistic elegance with solid content.[9]

---

4 CL, pp. 1268–9 (see esp. n. 3), letter of Cesi to Francesco Stelluti n.d. (1608?). On Cesi's pilgrimages to Loreto cf. Contributi I, p. 196.

5 OG XII, p. 388, letter of Cesi to Galileo, 11 May 1618.

6 Ibid., pp. 396–7, letter of Cesi to Galileo, 10 July 1618; pp. 406–7, letter of Stelluti to Galileo, 10 August 1618.

7 Contributi I, pp. 537–8. The association of Ciampoli and Cesarini was deliberated on 4 June. The Lincean rings were conferred on 1 July.

8 OG XII, pp. 413–5, letter of 1 October 1618.

9 Cesarini had had the opportunity to admire Galileo's eloquence during the meetings with the nobility, prelates and *literati* that he had hosted in his ancestral palace in 1616. He was struck above all by the Tuscan scientist's ability to combine eloquence with science: 'by means of one of which you know the most mysterious phenomena of nature, and by means of the other you render the same accessible to another's understanding' (ibid.).

Not surprisingly, with the collaboration of Ciampoli, the young Duke was working intensely during those months on a never to be published edition of the *De rerum natura* of Lucretius, an eloquent example of a conception of the poetic genre as a medium to convey philosophical and moral doctrines.[10]

A few years later, Cesarini underlined once again to Galileo the value of poetry in transmitting elevated concepts:

> I am spending my time with them [the Muses], making every effort to ensure that my poetic compositions are not devoid of philosophical doctrine, and as far as possible I endeavour to leave in them a living testimony of the respect and reverence that I pay to the eminent virtues.[11]

Cesarini's opinion was fully shared by his cousin, Prince Cesi, who had included the practice of poetry, provided it had a moralising content, among the aims of the Lincean Academy. It will be recalled that in the letter sent to Stelluti the day after the foundation of the Academy, Cesi had encouraged his friend to devote himself to the poetic genre, inviting him to enrich it with content consistent with the programme and objectives of the Linceans:

> I greatly applaud the study of poetry. Being of a wandering nature, it will be necessary for your Saturn to weigh it down, not with a verse here or there (as other poets are wont to do), but with some composition made of Lincean themes or interwoven with the Lincean achievements.[12]

The Prince was to stress the value of poetic productions dense with moral and philosophical content also in the *Lynceographum*.[13]

Cesarini and Cesi's conception of the nature and ends of poetry reveals a new scenario in which the Tuscan scientist's inclinations and goals did not coincide with those of the Princeps of the Academy and his closest associates.

---

10 Cf. Ardissino 2007, p. 165 and Bellini 1997 and 2009.

11 OG XIII, p. 88, letter of Cesarini to Galileo, 7 May 1622. On the conception of the nature and ends of poetry held by Cesarini and Ciampoli, see Contributi I, pp. 774–5. For the poetic production of Cesarini, see Raimondi 1966, pp. 44–72, Bonaccorsi 2001 and Bellini 1997 and 2009.

12 CL, pp. 39–40, letter of 17 July 1604. Saturn was the zodiacal sign of Stelluti, to which he had referred in choosing the Academic name of *Tardigradus* ('slow-paced').

13 Cesi recommended Greek (preferably) and Latin poetry based on non-trivial content above Tuscan or any other vernacular (Lynceographum, p. 70).

Galileo considered poetry as a genre based above all on a free and direct expression of the imagination that could not be reduced to didactic functions. Unlike Cesi and Cesarini, he believed that poetry pursued goals incompatible with the objectives of natural and moral philosophy. He considered the *Gerusalemme liberata* a clumsy poem precisely because Tasso had pursued an impossible hybridisation of poetic discourse and rational analysis of the human and natural world. According to his first biographer, Niccolò Gherardini, Galileo 'used to compare the reading of *Gerusalemme liberata* with the acidic flavour of cucumbers, all the more distasteful if compared with the sweet flavour of melons which the sensations produced by Ariosto's *Orlando furioso* resembled'.[14]

The Copernican System Overthrown?

A more intense flow of letters from Cesi and the other Linceans to Galileo was gradually resumed after the end of 1619, prompted by new celestial phenomena. The last of the three comets that had appeared in rapid succession became visible around the middle of November 1618. It was a difficult period for Galileo, whose poor health prevented him from observing the heavens. However, he was kept informed about the course of the comet by his pupils and assistants with whom he discussed the nature and causes of the phenomenon.

For the first time a comet could be observed through the lenses of a telescope, an exercise in which astronomers all over Europe were engaged. The results, however, were disappointing: when seen through the telescope, the body of the comet did not appear appreciably magnified.

Particularly lively debates broke out in Rome, involving some Linceans and, of course, the Collegio Romano. In the absence of Cesi, who was in Acquasparta, and of Stelluti, who had withdrawn to Fabriano, the newly elected Lincean Virginio Cesarini trained the telescope on the comet and noted down his observations. Aware of his modest competence in celestial matters, he consulted the mathematicians of the Collegio Romano before sending an account of his observations to Galileo on 1 December 1618:

14 OG XIX, p. 645. Cf. Battistini 2011, p. 15. Battistini provided a detailed bibliography on Galileo's conception of poetry.

> Not trusting my lack of experience, today I discussed my observations, enclosed in this letter, with some Jesuit mathematicians who paid me a visit at home. They confirmed their accuracy.[15]

On 25 December, Stelluti, who had rejoined Cesi in Acquasparta, also sent Galileo a report of their observations. Both he and Federico had been astonished 'to see with the telescope [the comet] grow not at all, or only a very little more than it appears to the naked eye'.[16] They had deduced that the comet was at a very great distance, 'beyond all the planets'. The Linceans were eager for Galileo to explain why the telescope did not magnify the comet 'as it does with Jupiter and Saturn'.[17]

At Cesi's request, on 12 January 1619, the mathematician of Archduke Leopold of Austria, Johannes Remus (Johann Ruderauf), sent Galileo a detailed description of his own observations, including his idea about the material composition of the comet:

> Its matter consists of sunspots, gradually clustered, dispersed and coagulated again in a certain region of the deep sky.[18]

Remus appended to his letter a series of considerations on what, according to common opinion, the comet presaged: 'a pestilence of the whole universe' and a 'sign of past, present or imminent evils'.[19]

In mid-February, Cesi sent to Galileo from Acquasparta a passage of a new letter on cometary questions sent to Faber by Remus. From the Prince's letter we learn that Galileo had previously informed the Roman friends of his intention to intervene in the debate on the nature of the comets.[20] The document also indicates that Cesi, 'absorbed more than ever in celestial contemplations', had developed his own interpretation of those celestial phenomena. Galileo had informed the Prince that, prevented from conducting observations due to his ill health, he had asked Mario Guiducci to follow and to record the course

15 OG XII, pp. 422–3. The enclosed document is not extant.

16 Ibid., p. 431.

17 Ibid.

18 Ibid., p. 434.

19 Ibid.

20 Ibid., p. 439, letter of 15 February 1619: 'I will be very pleased to see the discourse on the comets, for truly the spectacle of these past months has been splendid, and I would have greatly liked to be in your company and to share my thoughts with you [...]'.

of the comet.[21] On 2 March Giovan Battista Rinuccini wrote from Rome asking Galileo to send him 'the discourse that they say you have promised on the comet'. 'They' must be a reference to the Lincean circle, particularly to Cesarini and Ciampoli. Rinuccini also passed on information that must have sounded alarming to Galileo:

> The Jesuits have composed a Problem, which is in press, and firmly maintain that it [the comet] is beyond the sky of the Moon; and some apart from the Jesuits are spreading word that this overthrows the system of Copernicus and that he has no better contrary argument than this.[22]

The booklet whose imminent publication was announced by Rinuccini was the *Astronomical disputation on the three comets of the year 1618* (*De tribus cometis disputatio astronomica anni 1618*),[23] a text penned by the Jesuit Father Orazio Grassi, although it did not bear the name of the author.

Without mentioning Galileo, the Father emphasised in the opening page that the telescope, which had revealed innumerable celestial novelties, proved entirely useless when it came to investigating the nature of the comet, since, when seen through that instrument, it did not appear magnified at all: 'only comets have remained aloof from these lynx eyes' he concluded, in an unveiled allusion to Galileo and the Lincean appropriation of the telescope.[24] Grassi further insinuated that the scrupulous observations of the comet demonstrated that it moved around the Sun beyond the sphere of the Moon. Moreover he implied that the observation of the celestial course of the comet suggested that the true constitution of the universe was that delineated by Tycho Brahe, who had refuted the absurd Copernican conception.

A copy of Grassi's booklet soon reached Galileo, no doubt thanks to the diligence of his friends in Rome, who were closely following the development of

21 This can be deduced from Francesco Stelluti's reply of 22 February to the (lost) letter that Galileo had sent him a week earlier: 'If your friend [Guiducci] has completed his observations on the comet, I would like to have a copy' (ibid., p. 442).

22 Ibid., p. 448.

23 Grassi 1619a. The *Astronomical disputation* was published at the end of March. It was republished with facing Italian translation in Galilei-Guiducci 2002, pp. 254–87. For the English translation of the text, see Drake-O'Malley 1960, pp. 3–19 (OG VI, pp. 21–35).

24 Drake-O'Malley 1960, p. 6 (OG VI, p. 25). In Grassi's words we find new evidence of the fortune enjoyed by the image of the 'Lincean telescope'.

the situation. His reading of Grassi's text and the news from the Eternal City on the favourable reception of the anti-Copernican interpretation of the comet convinced the Tuscan scientist to enter the fray without delay to oppose a prospect that he considered particularly menacing.

In May Mario Guiducci delivered a lecture in the Florentine Academy expounding the theses that would be published a few weeks later in the *Discourse on the comets*,[25] which opened with a dedication to Leopold of Austria dated June 8. The *Discourse* was to a large extent the work of Galileo, who, acting for once on Cesi's advice, had decided that it was better for one of his disciples to formally throw down the gauntlet.[26]

Father Grassi had based the argument that the comet moved at a great distance from the Earth, far beyond the sky of the Moon, on the absence of parallax. He had also maintained that its progressive movement towards the north confirmed the system of Tycho Brahe and his interpretation of the comets: the comet was a real body that moved around the Sun and was transported by the great luminary, like all the planets, around the immobile Earth at the centre of the universe. It was composed of dark celestial matter, illuminated by brilliant light for a limited period of time.

In the *Discourse* Galileo-Guiducci seized on Grassi's adhesion to Tycho's theory of the comets. They vigorously contested above all the interpretation of the absence of parallax proposed in the Jesuit's *Astronomical disputation*. Opposing the Danish astronomer's cometary conception, they affirmed the illusory nature of these celestial phenomena: comets were formed of exhalations and vapours emitted from the Earth which, risen to great heights and illuminated by the Sun, produced the deceptive appearance of a real body. Since the comet was not a physical body, they concluded, it was a completely useless exercise to strive to measure its parallax:

> I shall not believe that parallax has really any place in comets, until it is first proven that comets are not reflections of light, but are unique, fixed, real and permanent objects.[27]

25 For the reconstruction of the process that led to the publication of the *Discourse*, see the introduction by Besomi and Helbing in Galilei-Guiducci 2002, pp. 15–42. See also Favaro's *Avvertimento* in OG VI, pp. 5–19.

26 Ibid., pp. 8ff.; see also Galilei-Guiducci 2002, pp. 25–9.

27 Drake-O'Malley 1960, p. 39 (OG VI, p. 71).

Galileo had also been irritated by Grassi's malign insinuation regarding his lack of competence in optics.[28] However, in the *Discourse* the most violent attack was launched against the adhesion of the anonymous author of the *Astronomical disputation* to Brahe's interpretation of the nature and motion of the comet. Galileo was alarmed above all by the Jesuit's explicit declaration that the Copernican system had to be rejected not only because it was condemned by the Church, but also because it was refuted by the most diligent observations of the motion of the comet.

After the anti-Copernican decree, increasingly clear signs had been emerging of the inclination of the Jesuits to adopt Tycho's geo-heliocentric system, even though they had opposed it at length as its author belonged to the damned Protestant sect.[29] This explains why Galileo and Guiducci regarded the confutation of the Tychonian system as a fundamental step in their strategy to relaunch the Copernican conception of the universe and why they spared no effort in demonstrating the absurdity of the Danish astronomer's system of the world.

---

28 In the *Astronomical disputation* Grassi proposed his own interpretation of the functioning of the telescope. The magnifying power of the instrument decreased, he claimed, as the distance from the bodies under observation increased, until being nullified, as in the case of the comet: 'it has been discovered by long experience and proven by optical reasons that all things observed with this instrument seem larger than they appear to the naked eye; yet according to the law that the enlargements appear less and less the further away the objects are removed from the eye, it results that fixed stars, the most remote of all from us, receive no perceptible magnification from the telescope' (Drake-O'Malley 1960, p. 17: OG VI, p. 33). Grassi's words 'I know that this argument is of little significance to some' were certainly directed against Galileo, as was the explicit condemnation of the incompetence in optics of those who refused to accept that principle: 'But perhaps they have given little consideration to the principles of optics, which, it must be understood, play a very important part in what we are considering' (ibid.). Galilei-Guiducci replied vehemently to this insinuation in the *Discourse* (Drake-O'Malley 1960, pp. 40–6: OG VI, pp. 73–9).

29 The Jesuits officially subscribed to Tycho's hypothesis only after the anti-Copernican decree of 1616. However, before the decree, a number of Fathers – and not the least competent or authoritative among them – had shown their sympathy for Tycho's system, although encountering the opposition of the hierarchies of the Order. The case of Cristoforo Borri is emblematic in this connection. Father Borri was one of the first Jesuits to follow Tycho. For this reason in 1612 he was compelled to leave the Chair of Mathematics of the Collegio Braidense in Milan after the formal injunction of the General of the Order, Claudio Acquaviva. See Besomi-Camerota 2000, in particular pp. 119ff. On Father Borri as promoter of Tychonian cosmology, see also Baldini U. 1992, p. 169; Lerner 1995, pp. 157–63; Lattis 1994, pp. 208–10 and Dollo 1997, pp. 162–5.

The *Discourse* paid particular attention to Father Grassi's anti-Copernican argument based on the deviation of the comet northwards in the final stage of its visible trajectory. Galilei-Guiducci replicated that Grassi's explanation of that phenomenon, based on Tycho's system, was neither the only one possible nor the most logical and economical. To account for that deviation, it was enough to suppose that the Earth moved around the Sun. Having stuck his neck out so far, Galileo pulled back: though he found it completely coherent and plausible, he did not dare to subscribe to this hypothesis (he did not mention Copernicus, but it was obvious to whom he was referring), based as it was on principles prohibited by the supreme authority of the Church:

> I shall not pretend here not to know that, if the material in which the comet takes form has only a movement straight and perpendicular to the Earth's surface (that is from the centre toward the sky), the comet should appear to be directed exactly toward the zenith; yet it did not appear so, but declined toward the north. This forces us either to change what has been said or else to retain that but to add some other cause for this apparent deviation. I cannot do the one, *nor should I like to do the other.*[30]

The *Discourse* was the first in a series of publications issued by Pietro Cecconcelli, who adopted for his printshop the device 'Alle stelle medicee', an obvious tribute to Galileo and the reigning Tuscan dynasty. Encircled by the six balls of the Medici crest, the device (fig. 16) shows Jupiter (in the form of a star framing the astronomical sign of the planet) surrounded by its four moons. Beneath the crest was the inscription 'Solemque suum sua sidera norunt' ('the Sun is nourished by its own stars'). As Besomi and Helbing have pointed out, the device and the inscription (taken from Virgil's *Aeneid*) were intended to exalt Galileo's discovery of the 'Medicean stars', and, at the same time, to propose a flattering comparison of the Medici with Jupiter at the centre of the revolutions of its four planets.[31] If, as it seems logical to suppose, Galileo and Guiducci played a prominent role in the elaboration of the device of the new printshop,[32] it is extremely likely that they intended the combination of the

30 Drake-O'Malley 1960, p. 57, emphasis added.

31 Galilei-Guiducci 2002, pp. 373–4; see also Tognoni 2001, pp. 20–1. The Virgilian passage (*Aeneid* VI. 641) describes Elysium as having its own Sun and lit by its own stars.

32 As suggested by Longo (Longo 1997–1998).

DISCORSO
DELLE COMETE
DI MARIO GVIDVCCI
FATTO DA LVI
NELL'ACCADEMIA FIORENTINA
NEL SVO MEDESIMO CONSOLATO

SOLEMQVE
SVVM SVA
SIDERA NO
RVNT

IN FIRENZE
Nella Stamperia di Pietro Cecconcelli, Alle Stelle Medicee. 1619.

CON LICENZIA DE' SVPERIORI.

FIGURE 16 *The device of the printshop 'Alle stelle medicee'* (GUIDUCCI 1619)

visual representation of Jupiter and its moons with the message of the central function of the Sun transmitted by the Virgilian verse not only to convey a courtly message but also to suggest the Sun as the driving force governing all celestial movements.

The *Discourse* reached Grassi and the Linceans who were in Rome (Cesi had retired to Acquasparta) at the beginning of July. Giovanni Ciampoli was the first to deliver his impressions and those of Virginio Cesarini to Galileo. They had both read and reread the *Discourse* with the greatest attention and found it extraordinarily penetrating. However, Ciampoli cautiously confessed that he had not found all the theses expounded in that book fully convincing. After insisting at length on the novelty of the work and the originality of its arguments, he begged the liberty ('since you ask me freely') to formulate a reservation on his own behalf and on that of Cesarini:

> I will tell you of something that has not gone down well here, and that is your taking issue with the Collegio Romano, which has shown so much respect for you in public. The Jesuits have taken great offence and are preparing their reply; and although I know and am aware of the solidity of your conclusions, I am displeased that their goodwill and esteem of you have decreased so much.[33]

Ciampoli's letter presents a scenario which is familiar to us by now. Galileo rushed headlong into the fray, attacking the Jesuit who had written a text with an interpretation of the nature and motion of the comet that seemed devised to demolish the Copernican system and to credit that of Tycho. His friends were concerned that their patient work of mediation to avoid Galileo's relations with the Collegio Romano from being definitively jeopardised, not only to his own detriment but also to that of the Academy, would founder. Had Galileo forgotten Cesarini's close ties with the Society of Jesus and his endeavours to prevent the 1616 decree from compromising relations with those Fathers, who, like Grienberger, had shown their inclination towards a gradual reform of traditional natural philosophy and cosmology?

While feigning ignorance of the identity of the author of the *Astronomical disputation on the three comets*, Ciampoli and Cesarini were convinced that the Jesuits would close ranks and answer back with determination involving Galileo in a polemic from which nothing good could come. The same concern was expressed in the letter sent to the Tuscan scientist from Rome on 3 August by Virginio Cesarini: 'I have nothing to add to what you will have heard from Giovanni [Ciampoli] regarding the reactions provoked by that text [the *Discourse*]'.[34] Although worried by his imprudence, Cesarini assured Galileo of his support: 'among the persons of greater authority in this city, I am and shall remain as much a defender of you and of the doctrine that you have published as any affectionate pupil should'.[35]

33 OG XII, p. 466, letter of 12 July 1619.

34 Ibid., p. 473.

35 Ibid.

In the same weeks the Linceans distributed the *Discourse* to influential members of the Roman clergy and nobility and to the most qualified astronomers and natural philosophers. As previously mentioned, the *Discourse* was dedicated to Leopold of Austria, who had met Galileo and admired his personality. The Archduke asked his personal mathematician, Johannes Remus, to give him an account of the work. The German mathematician had spent many years in Rome, where he was a frequent visitor to Cesi's residence and regularly met with the Lincean circle. It was through Cesi and Faber that Galileo had obtained the treatise on the comet that Remus had published in 1618.[36] The review that Remus, after consulting Kepler, sent to Galileo was extremely severe.[37] Hardly any of the theses contained in the *Discourse* convinced him: 'truly there is no lack of reasons to doubt'. He seized particularly upon Galileo's justification for the absence of parallax on the ground that the comet consisted only of subtle exhalations illuminated by the Sun's rays.[38] Remus was not the only one to express reservations on the *Discourse*: Sagredo and Baliani also conveyed their perplexities to the Tuscan scientist.[39]

The Linceans were aware of the objections to the *Discourse* raised by authoritative mathematicians and natural philosophers, as well as of the resentment that it had produced among the Jesuits, who would promptly prepare their counter-offensive. Remus too was informed of the Fathers' determination to respond. He wrote to Galileo that he had heard from an authoritative source – Father Scheiner – that the author of the *Astronomical disputation* ridiculed by Galileo 'will pay you back in the same coin'.[40] He had also confided to Galileo that the author of that text was the Jesuit Orazio Grassi.[41] Galileo received further important information about the Jesuit strategy from the letter sent to him on 24 September by another freshly inducted Lincean, Carlo Muti:

36 CL, p. 682, letter of 25 February 1619, in which Cesi informed Faber that he had sent Galileo the text on the comet of 1618 (Quietanus 1619) that Remus had sent to him to be forwarded to the Tuscan scientist (OG XII, pp. 433–5, letter of Remus to Galileo, 12 January 1619).

37 Ibid., p. 465, letter of Remus to Kepler, 23 July 1619.

38 Ibid., pp. 484–8, letter of Remus to Galileo, 24 August 1619.

39 'The opinion of your friend the Academician [Guiducci] on the comet appears rather hard to me […]. I would be almost tempted to refute it' (ibid., p. 465, letter of Francesco Sagredo to Galileo, 6 July 1619). 'Regarding your subtle consideration on the matter, place and movement of the comet […] I would not be so ready to concede your claims' (ibid., p. 476, letter of Baliani to Galileo, 8 August 1619).

40 Ibid., p. 487. Scheiner had received the *Discourse* from Archduke Leopold of Austria.

41 Ibid.: 'I have heard that it [the *Astronomical disputation*] is the work of Father Grassi from Genoa'.

> We shall be awaiting the text of the Jesuit Fathers: I understand that he [the use of the singular indicates that Muti knew that Grassi was the author of the *Astronomical Disputation*] has gone to Perugia to have it printed there. I have great expectations because they are very learned men.[42]

Like Cesarini, Muti expressed his admiration for the Fathers of the Collegio Romano. He seems to have sensed that a violent storm was about to break. Some of those 'learned men' had revealed to him the arguments that would be used in the work gone to press:

> Some of them told me that he considered the movement [of the comet] important evidence and said something in favour of the parallax, and that with an experiment he cast doubt on the example of the candle alight inside the bowl [...] and complained that you had no reason to attack him so violently [...].[43]

Why did the Roman Linceans keep the identity of the author of the *Astronomical disputation* from Galileo for so long? If even Remus knew, they must have known too. Perhaps they hoped for an accommodation, avoiding a personal conflict that would have definitively compromised the situation. It was one thing to engage in polemic with a masked interlocutor, as Galilei-Guiducci had done in the *Discourse*, but it was a very different matter to attack a Jesuit in flesh and blood, and moreover such a prominent figure as Father Orazio Grassi, the Mathematician of the Collegio Romano.

If, after the information received from Remus, Galileo still entertained doubts as to the identity of the author of the *Disputatio*, the letter he received from Ciampoli on 18 October dispelled them completely: 'The Jesuit Father Grassi, who recently returned from Perugia, brought us [sc. Ciampoli and Cesarini] his Discourse on the comet this evening'.[44] This was the *Astronomical balance* (*Libra astronomica et philosophica*),[45] which Ciampoli claimed not to have read to comply with Galileo's request to forward the volume to him as

42 Ibid., p. 492.

43 Ibid. Note that Muti again avoided naming the author, probably to keep a promise made to the Fathers of the Collegio. In the *Libra* Grassi refuted the experiment of the flame of a candle in a spinning bowl illustrated by Galilei-Guiducci in the *Discourse* (cf. Drake-O'Malley 1960, pp. 109ff: OG VI, pp. 156–9). Cf. below, pp. 259–60 and note 63.

44 OG XII, p. 494.

45 Grassi 1619b.

soon as he received it. Father Grassi had delivered it personally to Cesarini and had adopted extremely accommodating tones during his visit to the Duke: 'The Father tells me he has put forward his arguments as best he could, but that he has always mentioned your name with due respect'.[46] Less reassuring news was contained in the letter sent to Faber on 2 November from Acquasparta by Francesco Stelluti on behalf of Cesi. They had both received and read the *Astronomical balance* and knew the identity of its author:

> It is the work, I believe, of the Mathematician of the Collegio, Father Orazio Grassi, although it goes under another name, that is, his name with the letters changed [...]. He is furious with Galileo for that treatise on the comets by Mario Guiducci, professing to refute all his doctrine.[47]

It was a danger signal, also because in the *Astronomical balance* Grassi cited Cesarini, Vice Princeps of the Academy, as a direct witness of the repetition of an experiment proposed by Galilei-Guiducci. The Jesuit argued that, when conducted with due care, that experiment produced results completely different from those illustrated in the *Discourse*.[48] To make matters worse, Grassi had referred to the Academy in terms that appeared offensive to Cesi and his associates: it is 'as if Galileo, as a Lincean – the Father had written – has not been able to see properly'.[49]

After reading the *Astronomical balance*, Cesi's first concern was not that of weighing up the scientific solidity of Grassi's arguments against the *Discourse*. He was preoccupied above all by the involvement of the Academy in a dispute between one of its members and an authoritative Father of the Society of Jesus. As Stelluti wrote to Faber, Federico was determined to stop Galileo from reacting impulsively:

46 OG XII, p. 494.

47 Ibid. The work appeared under the name of 'Lothario Sarsi Sigensanus', the imperfect anagram of the real author's name (Horatius Grassius Salonensis).

48 Among the witnesses present at the repetition of the experiment with the candle, Grassi singled out Virginio Cesarini. He declared that the Duke had been convinced by the Jesuit's arguments (Drake-O'Malley 1960 p. 111: OG VI, p. 157).

49 CL, pp. 698–9, letter of Stelluti to Faber, 2 November 1619. One gets the impression that Stelluti and Cesi were over-sensitive. The mention of the 'lynx-eyed' Galileo in the *Astronomical balance* was heavily ironic but not offensive: 'Even with his telescope, the lynx-eyed astrologer – the Jesuit wrote – cannot look into the inner thoughts of the mind' (Drake-O'Malley 1960, p. 71: OG VI, p. 116). Galileo glossed that text: 'I do not claim to have ways of penetrating intimate secrets, neither with my eyes nor with my ears' (ibid.).

> The Prince wants to know how Galileo has taken it, and whether he will get Guiducci to reply on his behalf: in this case I do not think that the reply will cause him any trouble. Be sure to deliver the enclosed letter and find out what is required.[50]

The absence of any letters from Galileo during this period makes it difficult to reconstruct the strategy that he was defining to reply to the assaults of Father Grassi in the *Astronomical balance*.

In the opening pages of the Jesuit's book Galileo was reproached for having violently attacked Lothario Sarsi, a pupil of a *magister* of the Collegio Romano, Orazio Grassi, even though the Father had expressed his appreciation of the Tuscan scientist 'in private conversation and in public disputations'.[51] Had Galileo forgotten the honours paid him by the entire Collegio in 1611, when he visited Rome after the celestial discoveries made with the telescope? Did he no longer recall the generous praise that had caused him to blush contained in the *Sidereus Nuncius Collegii Romani* read by Father Van Maelcote at the public meeting in the Collegio in April 1611? How had he been able, in the *Discourse* signed by Guiducci but penned by him, to hurt 'the dignity of the Collegio Romano'?[52]

Galileo's marginal annotation on the page of the *Astronomical balance* containing these recriminations shows that his spirit remained undaunted:

> Let me ask how I have become so degraded that, while they [the Jesuits] had once borne my things with great pomp, they can now write entire books to tax them [...]. Thus the first was mockery and adulation; so according to your philosophy it is possible to praise and condemn the same doctrines, depending on whether you are swayed by peace or anger.[53]

It was not only the declaration of resentment on the part of the entire Collegio Romano that caused concern. In the *Astronomical balance* Grassi insinuated that Galileo's arguments made one suspect him of adhering to the Copernican system, which had been justly condemned by the Church. The Father insisted several times, with malicious allusions, on this delicate issue, vigorously rejecting Galileo's criticisms of Tycho. Whom should Sarsi have followed?

---

50 Ibid., p. 699.

51 Drake-O'Malley 1960, p. 71 (OG VI, p. 115).

52 Ibid.

53 Ibid.

Ptolemy, whose system had been demolished by Tycho, thanks to his scrupulous observations of the orbit of Mars? Or should he have followed in the footsteps of Copernicus?

> Every pious man who is dutiful will rather call everyone away from him, and will equally reject and spurn his recently condemned hypothesis.[54]

The imputations of Galileo's adherence to heterodox doctrines continued: he had used the dangerous conceptions of Telesio and Cardano, authors of 'a sterile and unhappy philosophy'.[55] In a later passage, he drew the reader's attention to the fact that Kepler too, well known to be a supporter of the Copernican doctrine, had attributed – like Galileo – a rectilinear movement to the comet, explaining its final deviation northwards as a consequence of the motion of the Earth. Sarsi declared that he had not taken into consideration this hypothesis because the pious were prohibited from professing it, while it appeared that Galileo had followed the Imperial astronomer in this fallacious opinion.[56] Moreover, the Jesuit stressed that Galileo's cosmological vision was based on the conception of the fluidity of the heavens, implying that that doctrine was further confirmation of Copernican sympathies.[57] Finally, he criticised the author of the *Discourse* for having written that the apparent contradiction between the hypothesis of the vertical motion of the comet and the observation of the deviation of its trajectory northwards could be overcome by supposing the annual motion of the Earth. The Jesuit saw in this statement further confirmation of the author's adhesion to the hypothesis of the revolution of the Earth around the Sun. Grassi's denunciation of Galileo's Copernican convictions was interspersed with sarcastic remarks on the cowardly attitude that had led the authors of the *Discourse* to disguise them:

> Certainly when Galileo says that the motion that should be added would only cause the apparent deviation of the comet, he agrees very clearly that this motion has to be placed elsewhere than in the comet-vapour,

54 Ibid. In one of his marginal notes on the *Ratio* of Grassi, Galileo would return the insinuation to sender: 'The author partly follows the opinion of Tycho, who was a Calvinist, but this is not to say that he follows the Calvinists' (OG VI, p. 684).

55 Drake-O'Malley 1960, p. 74 (OG VI, p. 118). Cf. above, pp. 73–4.

56 Ibid., p. 120.

57 Drake-O'Malley 1960, p. 75 (OG VI, p. 121): 'What if it [the motion of the comet] be not even elliptical but entirely irregular, since in Galileo's system it would be able to move freely without any hindrance?'

> given that it desires to move only northward. If this is so, I do not see of what body this will be the motion.[58]

The suspicion that the movement to which Galileo alluded was that of the Earth, Grassi continued, was bound to appear founded:

> I hear something or other softly and timidly whispered in my ear about the motion of the Earth. Away with the word dissonant to the truth and harsh to pious ears! It had better be whispered with lowered voice. But if the matter were so, Galileo's opinion would have proclaimed that which is considered as a false foundation by others than he. For if the Earth is not moved, this straight motion does not agree with the observations of the comet.[59]

As a consequence of a holy decree that all Catholics were obliged to obey, the Father concluded, it must be asserted without hesitation that the Earth is at rest. It followed from this that the comet does not proceed in a straight line. And for the final blow: 'I do not think that that ever came into Galileo's mind, whom I have always known as a pious and religious person'.[60] In a later passage, Grassi repeated the same point: it is not permitted to the pious to maintain that our planet moves.[61]

The *Astronomical balance* was not composed only of nasty insinuations. Grassi also countered the opinions expounded in the *Discourse* with arguments based on experiments and mathematical demonstrations. As mentioned above, Galilei-Guiducci had maintained that the earthly exhalations rose vertically. The rise of vapours and exhalations – they had stated – would not be affected or deflected by the subtle celestial matter in perpetual motion around the Earth at rest at the centre of the universe. To confirm this affirmation they had presented the results of an experiment: the flame of a small candle placed inside a hemispherical container remained erect without showing any alteration, when the container was rotated.[62] Grassi reported in the *Astronomical balance* that, on repeating the experiment, he had obtained completely different results from those described in the *Discourse*: when the container was

58 Drake-O'Malley 1960, pp. 97–8, translation amended (OG VI, p. 145).

59 Ibid.

60 Drake-O'Malley 1960, p. 98 (OG VI, p. 146).

61 Ibid., p. 148.

62 Ibid., pp. 53–4.

spinning, the flame of the candle leaned slightly in the same direction. He had repeated the experiment several times in the presence of authoritative witnesses who could confirm the results. He was eager to cite one in particular: the Lincean Virginio Cesarini, who, he stressed, had been astonished at the outcome of the test.[63] It had been precisely the mention of Cesarini that had alarmed Stelluti and Cesi. Galileo must have been equally surprised at reading that page of the *Astronomical balance*, as well as the letter in which Ciampoli informed him that Grassi had invited Cesarini to the Collegio Romano in order to show him his counter-experiment with the candle.[64] Had the Roman Duke indeed approved the results of the tests with which Father Grassi claimed to have demolished the experiment illustrated in the *Discourse* to demonstrate the vertical rise of the exhalations?

The final pages of the *Astronomical balance* analysed and rejected the arguments with which Galilei-Guiducci had criticised the Aristotelian interpretation of the comets as terrestrial exhalations kindled by the whirling motion of the sky of the Moon with which they came into contact at the end of their ascent through the sphere of the elements. While affirming that he did not share the Stagirite's interpretation of the comets, the Jesuit defended the Aristotelian conception of heat as produced by motion (*motus est causa caloris*). The pages in which Galilei-Guiducci demolished this corner-stone of Peripatetic natural philosophy appear the most effective and convincing to a modern reader. On the contrary, among the arguments put forward in the *Astronomical balance*, those with which Grassi defended the Aristotelian theory of the generation

63 Ibid., p. 157. Grassi had gone to Cesarini's house, where Ciampoli was also present, to inform the Duke that he had repeated the experiment of the candle lit in a rotating bowl and had come up with different results from those reported by Galilei-Guiducci in the *Discourse on the comets*. On 24 August 1619 Ciampoli wrote to Galileo: 'Father Grassi came here recently, to report that he had found a way of conducting a more accurate experiment regarding the motion of the air contained in the bowl. Being curious to see it, I accompanied him to the Collegio' (ibid., p. 425). Ciampoli followed this up with a detailed account of the experiment, or rather experiments, because the Jesuit had shown not only that with the candle, but also an equivalent, replacing the candle with two little paper wings suspended in the centre of the bowl. Ciampoli informed Galileo that both the paper wings and the flame of the candle leaned when the bowl was rotated. Ciampoli did not mention the presence of Cesarini at the experimental performance in the Collegio. Nor did he betray his own personal opinion, although his embarrassment at the discrepancy between the results of Grassi's tests and the account of the experiments with the candle in the *Discourse* appears clear. See the commentary of Besomi and Helbing in Galileo 2005, pp. 590–3, and Bellini 1997, p. 60.

64 See previous note.

of heat are the most fragile and embarrassing. The Jesuit did not support his thesis with mathematical demonstrations and experiments, but based his argument on a blind adherence to the most incredible statements of Aristotle and his commentators, and on a prolix listing of the most improbable testimonies of classical literary and poetic sources on which he conferred indisputable authority.[65]

A Delicate Balance

Though they did not participate directly, the members of the Lincean colony in Rome played a conspicuous role in the discussions on comets. As we have seen, the protagonists of the cometary debates (even those taking place outside Italy, as shown by Kepler and Remus' interventions) regarded the Roman Academy as an important point of reference. Cesi, Cesarini, Ciampoli and Faber endeavoured to support Galileo in his dispute with Father Grassi, doing their best to avoid jeopardising the Academy's relations with the Collegio Romano.

Matters did not proceed as Cesi expected. Once again, as in 1616, the affair slipped from his grasp.

The Prince and his associates were well aware that Galileo's entrance on the scene had caused the situation to precipitate. Federico was reluctant to engage in open battle. He sincerely appreciated the fascination and moral value of Galileo's endeavour, but as a member of the Roman patriciate and relative of a Cardinal, he could not openly align the Academy with the Tuscan scientist. He did his best to moderate Galileo's determination and above all his impetuousness. He trusted that this strategy would not only protect a man whom he held in the highest esteem, but would also gradually foster the conditions to make the programme delineated in the institutional documents of the Academy public without ambiguity or dissimulation.

The *Astronomical balance* with its heavy insinuations led Cesi and his companions to reflect carefully on the strategy to adopt in replying to the Jesuit. This time they could not ignore the challenge: the attack had been violent and involved the Academy directly. A response was called for, albeit expressed in such a way as not to definitively compromise the relations between the Linceans and the Fathers of the Collegio Romano.

The Prince was concerned above all that Galileo might react to the provocation instinctively and caustically. His resentment would have to be tempered.

---

65 Drake-O'Malley 1960, pp. 122–32 (OG VI, pp. 160–79).

For this reason Cesi, Cesarini and Ciampoli pressured him to relinquish the idea of entering the conflict in person. As we have seen, Ciampoli, acting also on behalf of Cesarini, expressed in no uncertain terms his disapproval of the attack that Galilei-Guiducci had launched against the Jesuits. Cesi was soon to voice his own opinion from Acquasparta, in perfect agreement with the uneasiness of his Roman friends. Stelluti advised Galileo, on behalf of Cesi, not to reply in person to Sarsi and to absolutely refrain from venting his anger on the Collegio Romano:

> It would be a never ending battle to take it out on the Fathers who, as they are so many, would draw in a whole world and then, even though they are wrong, would want to be right; and that could not help but would cause damage, since they are enemies of new opinions as they are all Peripatetics there.[66]

They were difficult months, in which the effects of the anti-Copernican decree of 1616 made themselves felt. Suspecting that the correspondence with Galileo was intercepted, Cesi used Faber in Rome and Stelluti in Fabriano as sorting centres for the exchange of letters to and from Florence. Galileo sent Stelluti his lost reply to the Prince's recommendations to be prudent. As Cesi wrote to Faber, Galileo did not agree with the proposed strategy: 'Galileo finds some difficulties in putting into effect my advice on the reply'.[67] Cesi begged Faber to insist with the Tuscan scientist. His advice had now become a peremptory order, dictated by sheer necessity: 'I insist that in any event the reply be made by one of his disciples'.[68]

The indirect dialogue continued in the successive weeks. We learn from the letter Cesi sent him on 4 March that Galileo had asked Guiducci to reply to Sarsi. But the Florentine advocate objected that, if he were to do so, he would be unable to refrain from adopting a caustic and resentful tone. Given Guiducci's reluctance, Galileo was inclined to reply himself, adopting a conciliatory tone to avoid further tensions.[69] Cesi hastened to approve the decision to keep clear of ' satirical and bitter tones, which demonstrate and accompany an excess of emotions and passions rather than the soundness of a doctrine'. Nonetheless, he was convinced that a direct engagement was not the best solution.

66 Letter from Fabriano (27 January 1620), OG XIII, p. 21.

67 Ibid., p. 24, letter of Cesi to Faber, 23 February 1620.

68 Ibid.

69 Ibid., p. 25.

If Galileo decided to write the reply to Grassi in person, he would have to shun a face to face confrontation with his interlocutor:

> If it is impossible at present to reply through the intermediary of some disciple, perhaps it would not be bad to do what is required in a simple letter from you to Guiducci himself, or to some other friend from there who has discussed this matter with you, offering you the occasion to comment on it again.[70]

'Some other friend from there': that is from Florence – not a Roman friend, and certainly not a Lincean!

On 4 April, at the request of Cesi, Stelluti – who was still in Fabriano – insisted once more. They were firmly convinced that a reply was necessary, but without a direct confrontation with Father Grassi or his mask:

> You could address the reply to some friend of yours and not to that Lothario, much less to Father Grassi, to avoid getting into a dispute with a student, nor with that Father [...]. In writing to a friend, you will be able to manifest your opinions and reasonings freely [...]; and on that occasion you will also be able to express your philosophical thoughts, as in that reply of Father Castelli against Ludovico delle Colombe [...]. Write to a third person, pretending to have been asked by him to do so [...].[71]

The message was clear: return to work on the foundations of natural philosophy and keep your obsession with attacking all those who provoke you at bay. Cesi was joined a few days later by another Lincean, Carlo Muti, inspired by the same muse.[72]

The resolute efforts to get Galileo to abandon his intention of settling accounts directly with Father Grassi reached a climax in the middle of May. A favourable circumstance – Cesarini's stopover in Acquasparta on his way to Bologna to consult a famous physician from whom he hoped to receive an effective remedy for his serious chest complaint – made it possible for Federico, Virginio and his companion and alter ego, Giovanni Ciampoli, to meet. There are no minutes of that meeting, which can be considered an official academic session, a rare event in those years, due also to the Prince's long absences from

70 Ibid.

71 Ibid., p. 31.

72 Ibid., pp. 32–3, letter of 18 April 1620.

Rome. However, it is not difficult to imagine the questions at the centre of the agenda. The topics discussed and the conclusions reached were conveyed by Federico to Galileo in his letter of 18 May. The tune had not changed:

> We have together given full consideration to the reply to the *Astronomical balance* and it has seemed to us as necessary for it to be issued soon as it is for you to avoid engaging in the duel directly, but either through Guiducci, provided he is persuaded to abstain from caustic and bitter tones, or by writing the letter yourself, even though it be very long, at the request of a friend [...].[73]

To ward off any misunderstanding, on the same day and from the same palace in Acquasparta, Giovanni Ciampoli repeated the same recommendation. Galileo should stay out of the skirmish, not only for reasons of prudence: crossing swords with a 'masked person' would compromise his dignity.[74]

Contrary to what the correspondence between Galileo and the Roman friends of those weeks seems to suggest, Guiducci abandoned his reservations and decided to take up his pen against the author of the *Astronomical balance*. On 20 June 1620, exactly a month after the meeting of the three Linceans in Acquasparta, Guiducci sent a letter to the Jesuit Tarquinio Galluzzi, under whom he had studied in the Collegio Romano.[75] He expressed his resentment at the insinuation of the author of the *Astronomical balance* that, in the *Discourse*, he had been nothing but the mouthpiece of Galileo. Firmly repudiating the accusation of having launched a violent attack on the Collegio Romano, Guiducci explained that his target had not been a Jesuit, but Lothario Sarsi, who was simply a student of Father Grassi. Moreover, he claimed the right to adhere to one opinion or another in natural questions, an evident allusion to the theme so dear to the Linceans of the freedom of thought *in naturalibus*. He concluded with the announcement that Sarsi would soon receive the reply of a person of much greater authority, his own Master, 'from whose writings, which are soon to be published, Sarsi will discover to his cost how different Sig. Galileo's style is from mine'.[76]

---

73 Ibid., pp. 37–8.

74 Ibid., p. 39.

75 For the letter to Father Galluzzi see OG VI, pp. 180–96. An English translation of the letter is found in Drake-O'Malley 1960, pp. 133–50.

76 Drake-O'Malley 1960, p. 148 (OG VI, p. 195).

Guiducci's letter to Father Galluzzi carefully avoided mentioning the insinuations of Father Grassi on the Copernican convictions of Galileo, affirming simply that it was for Sarsi to interpret the real meaning of Holy Scripture, whose authority had been invoked by the mask of Father Grassi to maintain the transparency of fiery bodies.[77] Though expressing his reservations on the interpretation of the biblical passage mentioned by the author of the *Astronomical balance*, Guiducci left the final judgement 'to the declarations and explanations of the sacred doctors and masters in divinity', whose prerogative it was to say 'whether or not it is commendable to cite Holy Scripture in this way'.[78]

In informing Cesi, on 19 June, of the letter that he would send to Father Galluzzi on the following day, Guiducci had avoided drawing the Prince's attention to the fact that by now Galileo had taken the decision to reply to Grassi in person. He went so far as to intimate that the Prince's counsel to avoid the direct involvement of the Tuscan scientist had been followed to the letter. Cesi was therefore dumbfounded when he read in the letter of the Florentine advocate to Father Galluzzi the announcement of Galileo's imminent entry into battle. Once again, the Tuscan scientist had made his own decision, ignoring the appeals of Federico and the Roman Linceans.[79]

A few days later, Galileo communicated his decision directly to the Prince, who hastened to pass the news on to Cesarini and Ciampoli.[80] From Ciampoli's letter to Galileo of 17 July, written, as usual, also in the name of Cesarini, we learn that the project had already assumed a definitive shape, to the point that the Tuscan scientist had proposed publication of the work under the emblem of the lynx. He had therefore turned to his two fellow academicians in Rome for their suggestions on the most suitable person to whom the work should be dedicated.[81]

Having consulted his patron, Ciampoli proposed to Galileo the name of Virginio Cesarini. He motivated this nomination with two considerations. The first, a purely instrumental one, was suggested by the fact that in the *Astronomical balance* Cesarini was cited as a witness to the repetition of an experiment which, according to Sarsi, demolished the conclusions drawn from it by

77 Cf. below, pp. 376–7 and n. 106.

78 Drake-O'Malley 1960, pp. 145–8 (OG VI, pp. 193–4).

79 OG XIII, p. 41.

80 Galileo's letter is lost, but its contents can be reconstructed from Ciampoli's reply to him of 17 July 1620 (ibid., pp. 43–4).

81 Ibid.

Galilei-Guiducci. The more relevant motivation, however, was that the choice of Virginio would embarrass the Jesuits for the prestige that Cesarini enjoyed among the most authoritative exponents of the Order (Bellarmine, his great admirer, was still alive) and for his cordial relations with the Fathers of the Collegio Romano:

> In spite of the fact that Virginio is greatly loved by the Fathers, he will speak ardently in defence of those doctrines that appear admirable to him and that he does not believe can be conceived by other than your own eminent intellect.[82]

Ciampoli would soon consult Cesi, who was still in the dark regarding the proposal, but he was convinced that the solution envisaged would please him.[83]

Ciampoli had barely completed the letter to Galileo when he received a missive from Acquasparta addressed to the Tuscan scientist. The Prince asked Ciampoli and Cesarini to read the letter and consider the proposal it contained concerning the dedication of the *Assayer* (evidently Galileo had also asked Federico for his opinion) before transmitting it to Florence:

> Prince Cesi has sent me the enclosed letter open: it contains a note in which he explains the reasons why he thinks that the work should be dedicated to Father Bamberger [i.e. Grienberger].[84]

It is not difficult to imagine what had led the Prince to propose a candidate who had waged a courageous battle to persuade the Society of Jesus to loosen the reins of censorship. Cesi trusted that his involvement would make it clear that, in spite of the insinuations of Sarsi in the *Astronomical balance*, the Tuscan scientist (and consequently the Linceans) had no intention of picking a fight with the Fathers of the Collegio Romano. It was simply a lively discussion between Galileo and a student of Father Grassi on a purely scientific question, which did not imply any structural contraposition between the Academy and Galileo, on the one hand, and the Collegio Romano, on the other.

The nomination of Grienberger indicates that Cesi was by now out of touch with the political dynamics of power in Rome. It was on the basis of first-hand information deriving from their familiarity with the context ('we being here

82 Ibid., p. 44.

83 Ibid.

84 Ibid.

on the spot')[85] that Ciampoli and Cesarini expressed their negative opinion regarding the Prince's proposal. The Roman Linceans were perfectly aware that Grienberger's battle to obtain an opening for liberty of thought in the Society of Jesus had ended in defeat. The Jesuit Father had learnt his lesson and rejoined the ranks. Involving him in this delicate matter would not serve the cause; it would only exacerbate the hostility of the Collegio Romano towards Galileo and the Academy. Moreover, it would render the good Father a terrible service: 'We consider it absolutely not advisable to do that to avoid putting that poor Father in serious trouble, as we certainly know *ex exemplo* would follow'.[86]

'Ex exemplo' referred to the negative consequences that the Father had suffered on preceding occasions for adopting positions not in alignment with the rules imposed by the hierarchies of the Order.

The determination with which Grassi had moved against Galileo and the Linceans in the *Astronomical balance* and a careful evaluation by Ciampoli and Cesarini of the most suitable dedicatee for the *Assayer* show that the climate in Rome had become stifling.

Awareness of the particularly hostile context also emerges from the tone of the letter of 2 August in which Ciampoli replied, as usual on behalf of his patron, to the letter (not extant) with which Galileo had formally invited Cesarini to accept the dedication of the *Assayer*. The Roman nobleman declared himself honoured and seized the opportunity to convey to Galileo his own reflections on the strategy to be followed in composing the work in order not to offend the susceptibility of the Fathers of the Collegio Romano.[87] It was Galileo himself who had asked for Virginio's opinion on how best to confront his interlocutor. Should he address Father Grassi directly, or reply to his mask? And what strategy did Ciampoli and Cesarini suggest to prevent his response to the *Astronomical balance* from being perceived as an affront to the entire Collegio Romano?

Before expressing his personal opinion, as well as that of Cesarini, Ciampoli made sure to remind Galileo that he had asked advice not only from friends and admirers, but also from persons who were perfectly informed about the delicate balance governing political and cultural life in the Eternal City.[88] Their suggestions, he added, were the fruit of long and meditated reflection. Like Galileo, they considered that:

---

85 Ibid.

86 Ibid.

87 Ibid., pp. 46–7.

88 Ibid., p. 46: 'We are on the spot and we know well the present situation'.

> [...] the affection that you wish to show towards the Fathers would appear too feigned if you pretended not to know and not to want to believe that one of them [the Jesuits] was the author of the *Astronomical balance*.

The Fathers of the Collegio had not concealed the identity of the author of the *Astronomical balance*. On the contrary, 'they have publicly glorified him and sung his victories'. Galileo should therefore declare that, after having learnt 'that a book was being written against him by the Jesuit Fathers', he had been torn between 'two very great and contrary difficulties': his obligation (from which he could not shrink) to defend his own reputation; and his desire to demonstrate 'his constant respect for the Fathers, that he had chosen to profess throughout his life'. For this reason, he had 'raised his hands to heaven and thanked God when he saw the *Astronomical balance* appear under the name of Lothario', who declared himself a simple student of the Collegio, 'as it then appeared to him that he could defend himself without offending the Fathers, who as persons not only intelligent but also just could not possibly be displeased if he fought for his defence and for truth'.

To make quite sure, Ciampoli recommended an act of extreme compliance on the part of Galileo: he should declare that his reverent submission to the authority of the Fathers of the Collegio was such that he 'would have bowed his head' without hesitation if the book had been published under their aegis. Moreover, he should take care to declare the greatest admiration and the highest respect for 'the Jesuit mathematician', with whom he personally would never have dared to cross swords.

Proud of this extraordinary demonstration of cynical simulation, Ciampoli concluded triumphantly: if we adopt these tactics, 'I believe that the Fathers can take the least possible offence'.[89] This was an effective way of communicating to Galileo his conviction that the response to the *Astronomical balance* would in any event be regarded as an act of hostility towards the Collegio Romano. It was an easy prophecy, also because Ciampoli feared that Galileo would not follow to the letter the subtle strategy conceived by the three Linceans.

The letter in which Galileo asked Cesarini for advice was sent from Rome to Acquasparta along with Ciampoli's reply to the Tuscan scientist. Cesi declared himself to be in perfect agreement regarding the dedication of the *Assayer* and the attitude to adopt vis-à-vis the Collegio Romano. Moreover, he confirmed the intention to publish the work under the emblem of the Academy, as established together with Ciampoli and Cesarini in Acquasparta in March. On

89 Ibid., p. 47.

that occasion they had concurred that the formal decision to publish the text that Galileo was preparing under the emblem of the lynx should be taken in an official meeting of the Academy. Unable to move from Acquasparta, Cesi asked Faber to invite the Vice Princeps Cesarini to call the meeting.[90] However, the plan could be put into effect only months later, in the spring of 1621, after Cesi had returned to Rome. The meeting – the first in years – was held in the palace of Cesarini, who took part in it from his sickbed. Although not explicitly mentioned in the minutes,[91] there was unanimous accord on the need to exhort Galileo to compose the reply to Sarsi with the utmost urgency.

The Ebb and Flow of Fortune

Galileo's serious bouts of illness during those months disrupted the continuity of his work. His condition was aggravated by the 'mental anguish' provoked by the relentless attacks of his fellow citizens, of which he often complained to Cesi.[92]

In May 1620 – four years after the emission of the anti-Copernican decree – the Congregation of the Index had finally published the corrections to Copernicus' *De revolutionibus.*[93] The decision was taken on the grounds of a proposal of Francesco Ingoli who, in a document sent to the Fathers of the Congregation, had indicated the criteria to be adopted for the emendation of that text.[94] Given that the entire *De revolutionibus* was based on the presupposition of the movements of the Earth, it was impossible to emend that book with the introduction of a few corrections.[95] For this reason he suggested emending only

90 Ibid., p. 46, letter of Cesi to Faber, 11 August 1620.

91 The minutes of that meeting held in May 1621 and attended by no less than seven Linceans – the highest number ever – indicate that the proposals of numerous candidates for induction in the Academy (including Peiresc and Remus) were examined. See Contributi I, pp. 539–41.

92 Cf. below, pp. 286ff.

93 For the English translation of this document, see Gingerich 1981, pp. 51–2. Bucciantini convincingly attributed it to Francesco Ingoli (Bucciantini 1995, p. 145).

94 On Ingoli's censorship of *De revolutionibus*, see Gingerich 1981 and Bucciantini 1995, pp. 207–9.

95 As Ingoli emphasized, 'the emendation of Copernicus would not be a correction, but rather the total destruction of his system' (Gingerich 1981, p. 52 and Bucciantini 1995, p. 207). Paradoxically, some of Ingoli's expressions seemed to echo, though drawing the opposite conclusion, Galileo's affirmation of the impossibility of purging from the *De revolutionibus*

those passages of the *De revolutionibus* in which Copernicus spoke as a natural philosopher. Thus the reader would confer exclusively hypothetical value on the cosmological conclusions of the Polish astronomer, giving full credit to the statements on the conjectural statute of the heliocentric system contained in the anonymous letter that opened the book. In Ingoli's opinion this solution would help to re-establish perfect harmony between Holy Scripture and natural truth:

> Because the subject which Copernicus is dealing with is astronomy, whose most distinctive method is to use false and imaginary principles for saving appearances and celestial phenomena, as is established by the epicycles, eccentrics, equants, apogees and perigees of the ancients. If certain of Copernicus' passages on the motion of the Earth are not hypothetical, make them hypothetical: then they will not be against either the truth or the holy writ.[96]

Reading the emendations to the *De revolutionibus*, which he had been expecting for some time, Galileo must have felt even more depressed. The text of the admonition declared that Copernicus' work contained many affirmations 'repugnant to the true and Catholic interpretation of Holy Scripture, that is hardly to be tolerated in a Christian'.[97] These expressions seemed conceived to exorcise the very theses that the Tuscan scientist had advanced in the *Copernican letters*.

The arguments proposed in Ingoli's rejoinder to the *Responsio* of Kepler contributed towards deepening his pessimism. It is likely that when Galileo expressed his concerns to Cesi, he was fresh from the reading of that text, in which his claim to the superior authority in natural questions of mathematical demonstrations and sense experience over the affirmations of Holy Scripture was denounced in threatening terms:

> With regard to the alleged freedom to interpret Scripture against the shared opinion of the Fathers when it is not a question of faith or morals,

the explicit declarations in favour of heliocentrism: 'As far as Copernicus is concerned, in my opinion he cannot be toned down, as the key point and foundation of his whole doctrine is the mobility of the Earth and the stability of the Sun: his doctrine has to be condemned as a whole or left in its entirety […]' (letter of Galileo to Piero Dini, 23 March 1615: OG V, p. 299).

96 Cf. Gingerich 1981, p. 52 and Bucciantini 1995, p. 208.

97 Cf. Finocchiaro M.A. 1989, p. 200. See also Gingerich 1981, pp. 60–1 and Bucciantini 1995, p. 108. The emendations to the *De revolutionibus* were published on 15 May 1620.

> I consider this to be a highly dangerous proposition [...]. For as the interpretation of Scripture is the preserve of the Church and of the Supreme Roman Pontiff [...], I do not understand how others claim to assume this task without great and pernicious temerity, especially against the unanimous opinion of the Holy Fathers.[98]

The accusation was serious and direct: it was from these seeds of rebellion against the tradition and magisterial mission of the Church and the Pope that those heretical doctrines, which had so dramatically disrupted the unity of the Christian world and provoked bloody conflicts in Europe, had sprung.[99]

Reading those passages, Galileo must have felt the blood freeze in his veins. Nor can it have escaped him that the peremptory proclamation of theology as queen and mistress of all human sciences was not only directed against Kepler's claim of the autonomy from Holy Scripture of natural philosophy and cosmology, but denounced all those who, like Galileo, had dared to call that principle into question:

> Theology imposes its rule on philosophers and those who study nature, to avoid opposing or obscuring truth with their reasonings, as it is a superior science and one based on principles revealed by the first and highest source of truth [...].[100]

Ingoli's affirmations led Galileo to fear that an influential party – it is worth recalling that Ingoli was the theologian of Cardinals Caetani and Ludovisi as well as closely associated with Santa Susanna[101] – was machinating in Rome for his personal position to be re-examined in relation to the 1616 decree, from which he had emerged formally unscathed.

From his Lincean friends in Rome, however, he received reassuring news. Paul V had passed away and the conclave had very rapidly elected Cardinal Ludovisi, who assumed the title of Pope Gregory XV.[102] As Giovan Battista

98 Ibid., p. 204.

99 Ibid. Cf. below, pp. 323–7.

100 Ibid., p. 205.

101 On Ingoli see ibid., pp. 84ff.

102 Alessandro Ludovisi was elected Pope Gregory XV on 9 February 1619. Cesi attended the extremely short conclave (it lasted only one day!) as assistant to his uncle the Cardinal. The Prince wrote an extraordinary account of that event, showing the lucidity with which he observed the behaviour of the elderly Cardinals and the cynical negotiations undertaken by various parties to favour the success of their preferred candidate (Cesi 2003, pp. 97–125).

Rinuccini informed Galileo, it was hoped that the new Pope would tolerate a freer and more lively cultural activity.[103] It seemed therefore that Galileo could resume his work, ill health permitting. However, an alarming sign will not have escaped his notice: Ingoli was among the first to receive an important promotion from the new Pope.[104]

In the meantime, the Roman Linceans did not ease their insistent pressure on Galileo to honour his obligation to respond to Sarsi. Having hesitated for so long, it now seemed that Cesi and his associates were in the grip of a feverish impatience.

On 23 June, Cesarini wrote to Galileo on behalf of Cesi and the entire Academy to express their hope that the work would soon see the light of day: 'During the last meeting [...] it was resolved that I should encourage you to publish the reply to Sarsi'.[105] The reputation of the Florentine Lincean had to be defended against the 'ignorant calumnies of ill-wishers', who took advantage of his silence to proclaim their own triumph. It was not only the reputation of Galileo that was of concern to Cesi and his associates. They considered it fundamental for the honour and dignity of the Academy offended by the Jesuit that the Tuscan scientist should unmask his false arguments:

> Your followers and friends consider bridling the tongues of the common herd a real victory [...] I assure you that nothing could please our Academy more.[106]

In the successive weeks, Galileo's silence led all the Linceans to make their voices heard. Cesi, either with personal letters or through Faber, spurred Galileo to action on behalf of all the members of the Academy.[107] A few days earlier, Carlo Muti had begged him 'not to delay publication of his text'.[108] Ciampoli – who

103 OG XIII, pp. 59–60, letter to Galileo, 27 March 1619. Rinuccini spoke of a 'most auspicious' beginning of this pontificate, 'in which [...] a public profession had been made to promote persons of merit'. He was also confident that the new Pope would adopt an attitude towards 'geometry and the more profound sciences' quite different from that of the 'previous pontificate', during which these studies had been reduced 'to something approaching clowning' (p. 59).

104 In 1622 he was appointed Secretary of the Congregation *De Propaganda Fide*, established by Gregory XV (ibid., p. 85).

105 OG XIII, p. 68.

106 Ibid., p. 69.

107 Ibid., p. 76, letter of Cesi to Faber, 28 August 1621.

108 Ibid., p. 74, letter of Carlo Muti to Galileo, 15 August 1621.

in the meantime had been appointed *Segretario dei Brevi* by the new Pope[109] – was particularly insistent: he and Cesarini were eagerly awaiting 'the copy of the Sarseid'.[110] This epistolary avalanche shows the importance attached by the Linceans to refuting Father Grassi's arguments. Cesarini repeatedly appealed to Galileo's wounded pride. With equal insistence he recalled him to the task that fell to him as a member of the Academy to defend it from the defamatory insinuations of the Jesuit:

> Although basking in glory you may look down on this unequal contest, it is nevertheless owed to the public name of the Linceans offended by Sarsi and by other malevolent persons [...].[111]

These expressions indicate that Cesi and the academicians had noted the venomous barb of the *Astronomical balance*. That barb had brought 'other malevolent persons' boldly out into the open. It was therefore necessary to react without hesitation. The Jesuits had declared war not only on Galileo but on the entire Academy.

More than a year was to pass before Galileo finally satisfied these reiterated requests. The manuscript of the *Assayer* reached the hands of Virginio Cesarini on 28 October 1622. The Duke responded immediately, declaring that he was 'resolved to print it as soon as possible to avoid postponing a work that was useful to the world'.[112] However, the rules of the Academy required that the text first be read and approved by all the members – a long and tiresome process at a time when there were no photocopiers, faxes or scanners. Cesi, who was in Acquasparta, had to wait until the end of the year to receive the manuscript from Cesarini, who drew his attention to certain passages that he considered too aggressive:

> I beg you, after pointing out the things that will appear to you to be too caustic, or any concept that you do not approve, to send the manuscript back to me so that we can have it printed without delay [...]. Ciampoli and I have noted some things that will be accommodated or corrected [...].[113]

---

109 He was responsible for composing apostolic letters to worldly rulers and for drafting the Pope's Latin speeches.

110 Ibid., p. 84, letter to Galileo, 15 January 1622.

111 Ibid., p. 89, letter from Cesarini to Galileo, 7 May 1622.

112 Ibid., p. 99.

113 Ibid., pp. 102–3, letter of Cesarini to Cesi, 22 December 1622.

There was no time to waste. The adversaries were manoeuvring to block publication of the *Assayer*: ' the Jesuits have already found out about it'.[114] Nothing was secret in Rome: news reached the Collegio Romano – as we would say today – in real time.

The desire to proceed as rapidly as possible was thwarted by the laborious procedure to secure the approval of the Linceans. By 7 January 1622, Ciampoli had still not read the 'Sarseid'.[115] Five days later, Cesarini wrote to Galileo that he had not yet been able 'to obtain the opinion of all the Lincean fellows' in order to communicate to him 'what they unanimously wanted moderated, changed or suppressed'.[116] He informed him that Cesi had already sent 'some minor observations, to which I will add some trifling circumspections made by me and Giovanni Ciampoli'. Note the extreme tact with which Cesarini advised Galileo that corrections and in some cases excisions would be suggested to him. By now they were familiar with Galileo's determination: 'minor observations' and 'trifling circumspections' were carefully studied expressions designed to prepare Galileo to view their emendations favourably. However it would be up to Galileo – Cesarini assured him – to decide whether or not to accept their amicable advice.

Once these final obstacles had been removed, the work would be published in Rome under the device of the Academy, which had no intention of retreating before the powerful enemies of truth:

> We certainly want to publish the work, and we want to do so in Rome, in spite of the power of the adversaries, against whom we shall arm ourselves with the shield of truth and the protection of our patrons.[117]

The Linceans were ready to fight a pitched battle, albeit not without some hesitation before what they knew would be a bitter struggle: 'There is no doubt – Cesarini concluded – that we will be resolutely opposed; but I am certain that we shall be victorious'.[118]

Fears that the publication might be obstructed were not unfounded. Father Grassi and the Jesuits of the Collegio Romano knew that the manuscript of the *Assayer* was in Rome in the hands of Cesarini. The Duke told Galileo that the

114 Ibid.
115 Ibid., p. 104, letter of Ciampoli to Galileo, 7 January 1623.
116 Ibid., p. 105, letter of Cesarini to Galileo, 12 January 1623.
117 Ibid., p. 106.
118 Ibid.

Jesuit Fathers had received that information from their fellows in Florence. Virginio painted an accurate picture of the eagerness of the Jesuits of the Collegio Romano to get their hands on the manuscript:

> They are thirsty and anxious, and have had the boldness to ask me for it [the manuscript]; but I have refused because it would have helped them to block its publication.

In spite of these troubling signs, Cesarini's enthusiasm was not abated: the work would close 'the mouth of every kind of person and perhaps even that of Sarsi himself'. He dreamed of the European distribution of the *Assayer* in a Latin translation that he intended to take upon himself.

However, in the final paragraphs of his letter, Cesarini sketched a less promising scenario. The appearance of threatening clouds on the horizon had been provoked this time by the publication of Campanella's *Apologia pro Galilaeo*:[119]

> Some envious persons have taken advantage of this opportunity to revive calumnies against you that had been rejected and quashed some time ago; in any case there is no shortage of protectors and friends ready to defend your name and reputation.[120]

As had been the case with Cesi, Cesarini too now feared that his correspondence with Galileo was being intercepted. He therefore expressed himself with calculated ambiguity on the most delicate cosmological matters. This explains why he wrote that, in his *Apologia*, Campanella had defended the thesis of the motions of the Earth that Galileo had 'shared at that time', that is, before the anti-Copernican decree. He added that the Tuscan scientist had nothing to fear, having 'always respected the decree of the Holy Congregation'. In spite of this show of confidence, Virginio's peace of mind was disturbed by an anxious doubt: 'I trust that we will be able to obtain the permit to print the letter sent to me against Sarsi' notwithstanding the resolute opposition of influential persons. Cesarini was naturally alluding above all to the Jesuits, who in the opening lessons of their courses at the Collegio Romano had violently denounced 'the discoverers of new things in science, and with long speeches have tried to persuade students that there is no truth outside Aristotle'.[121] The Duke went so

119 Campanella 1622.
120 OG XIII, p. 107.
121 Ibid.

far as to define the Collegio's stance as an 'excommunication'.[122] He declared, however, that he was confident that the 'noble speculations' of the Tuscan scientist 'will have free circulation and approval in Rome'.

We may wonder what Galileo felt when he read this letter full of light and shade, caresses and cold showers. He must certainly have been pleased to have finally managed to get the Academy to come into the open. Cesarini had even requested that he send him the expanded version of his treatise on the tides, a work which the Duke knew to be overtly Copernican.[123] Moreover, the letter left no doubt as to Cesarini's personal commitment to opposing the manoeuvres of the Jesuits with every means at his disposal. However, in spite of Virginio's optimism, Galileo saw a return of the spectre of the polemics and denunciations that had caused him indescribable pain only a few years earlier.

Cesarini and Ciampoli continued to devote themselves to the positive outcome of the operation. A first success was the designation of the Dominican Father Niccolò Riccardi ('Father Monster') as revisor of the *Assayer*. He had met Galileo in Florence and had been fascinated by his personality. The enthusiastic and outspoken comments with which he formulated his favourable opinion on the work clearly show that his was not an impartial revision:

> I believe our age is to be glorified by future ages not only as the heir of works of past philosophers but as the discoverer of many secrets of nature which they were unable to reveal, thanks to the deep and sound reflections of this author in whose time I count myself fortunate to be born – when the gold of truth is no longer weighed in bulk and with the steelyard, but is assayed with so delicate a balance.[124]

At the beginning of March, after repeating that the modifications to the *Assayer* were irrelevant ('the change is not anything substantial, only the accommodation of certain words'), Cesarini informed Galileo that the printing of the work was finally under way.[125]

---

122 Cesarini was referring to the violent expressions used in the inaugural lectures in the Collegio Romano: 'at the start of this year's courses, those Fathers in the Collegio Romano, in their public discourses, have detested the discoverers of new things in the sciences [...]' (ibid.).

123 Ibid.

124 Drake-O'Malley 1960, p. 152 (OG VI, p. 200). Father Riccardi's comparison between the *Astronomical balance* (steelyard, or ordinary balance) of Sarsi and Galileo's *Assayer* (precision balance to assay gold) implied a severe mortification of Father Grassi.

125 OG XIII, p. 111, letter of 20 March 1623.

In the meantime, Galileo had informed Ciampoli that his Florentine adversaries had received assurances from Rome that the *imprimatur* would be denied. To show that their malicious insinuations were unfounded, on May 27 Ciampoli sent Galileo the first printed sheets of the work 'to enable you to disappoint those who, through obstinate malignity or jealousy, do not want to believe that the licence will be obtained'.[126]

Halfway through printing of the *Assayer*, an unexpected turn of events brightened a scenario that had appeared rather bleak, both in Rome and in Florence. On 8 July, the Ludovisi Pope Gregory XV, during whose brief pontificate the influence of Ingoli had grown appreciably, passed away. Cesi immediately returned to Rome from Acquasparta to follow the delicate negotiations of the Conclave,[127] which concluded with the election of the Florentine Cardinal Maffeo Barberini, who assumed the name of Urban VIII. The news reached Galileo in Florence with extraordinary rapidity. On 10 August Sister Maria Celeste replied to her father, who had informed her of the election, delighted that the papal throne was to be occupied by someone of whom she knew very well 'the affection that he bears you and the esteem in which he holds your virtues'.[128] So that his beloved daughter could perfectly understand the great hopes that he nourished after this election, Galileo had sent her the letters, full of attestations of admiration and praise, that he had received the previous year from Cardinal Maffeo Barberini. In the same days he also expressed his great expectations in a letter to Cesarini.[129]

Galileo's enthusiasm was fuelled by the excitement that pervaded his fellow academicians in Rome. On 12 August, Francesco Stelluti told him that the new Pope 'is very fond of our Princeps (that is Cesi)'. Urban VIII had confirmed Ciampoli as *Segretario dei Brevi* assigning to him also the function of *Cameriere Segreto*. Moreover, Cesarini had been appointed *Maestro di Camera*, and another Lincean, Cassiano dal Pozzo, 'will serve the nephew of the Pope, who will be made a Cardinal'.[130] The precarious balance of the previous months seemed to have shifted decidedly in favour of the Linceans: 'We now have three academicians in the Vatican, as well as many other friends. Let us pray the

126 Ibid., p. 117.

127 Ibid., p. 121, letter of Francesco Stelluti to Galileo, 12 August 1623: 'The news of the vacant seat has brought the Princeps and myself to Rome [...]'.

128 Ibid., p. 120.

129 'Your letter has made me become a Harpocrates, so much has your joy increased my satisfaction [...]. I am delighted that you are delighted [...]', ibid., p. 124, Cesarini to Galileo, 18 August 1623.

130 Ibid., p. 121. The nephew of Urban VIII was Francesco Barberini.

Lord God to preserve this Pope for a long time [...].'[131] In the following weeks, the correspondence between the Linceans in Rome and Galileo intensified. Cesarini, Ciampoli, Faber and Stelluti rivalled one another in exalting the new Pope, stressing the esteem in which he held Galileo, who had already written to him in person:

> His Holiness hears with the greatest pleasure and with great affection the singing of your praises, and both I and Ciampoli often mention your name to him. I have read him the courteous letter you sent him, which he received with great satisfaction.[132]

The Linceans in Rome realised that a new chapter was opening. The calculated caution of Cesi and his companions after 1616 was now replaced by the hope of finally bringing out into the open that project of a radical reform of knowledge that had been, from its very beginnings, the fundamental mission of the Academy: 'This election, to the satisfaction and delight of all, will be all the more reason for us to cheer with joy [...].'[133]

Throwing all hesitation to the winds, the printing of the *Assayer* had now to proceed without further delay. On 19 August Faber informed Galileo that his book was 'almost completely printed'. He added that Cesi was working to strengthen the influence of the Academy, forming alliances with the most influential personalities of the Barberini party. Federico hoped to pull off a major coup by enrolling Francesco Barberini, the powerful and erudite Cardinal Nephew, in the Academy.[134]

The auspicious ascent of Maffeo Barberini to the pontificate demanded a change of strategy. Galileo had conceived the graphic design of the frontispiece of the *Assayer* before the election of Urban VIII. He had even had a drawing made and sent it to Cesi, though we do not know what it represented: perhaps a goldsmith assaying precious stones or metals. Acknowledging receipt of that drawing on 8 September, Stelluti hastened to inform Galileo that the new circumstances suggested opting for a different solution. Since the *Assayer* would be presented to the Pope in deferential homage from the Academy, Cesarini had accepted without hesitation to appear in it not as the dedicatee, but

131 Ibid.

132 Ibid., p. 124, letter of Cesarini to Galileo, 18 August 1623.

133 Ibid., p. 125, letter of Ciampoli to Galileo, 18 August 1623.

134 Ibid., pp. 125–6.

simply as the addressee of Galileo's treatise. The work would be dedicated to the Pope on behalf of all the Linceans:

> We shall not make use of the design you sent because we have decided to make the frontispiece entirely in copper and to dedicate the book in the name of the Academy to the Pope. It will bear his device and the device of the Academy, with two statues, one representing natural philosophy and the other mathematics.[135]

The design of the new frontispiece had already been executed and the engraver was transferring it to copper. Stelluti informed Galileo that six hundred copies of the text, which included twenty images, would be printed. His letter shows that excitement was at fever pitch: he and Cesi were convinced that Cesarini would soon be elected Cardinal; moreover, the ring with the emerald was being prepared for the finger of Francesco Barberini, who had enthusiastically accepted induction into the Academy.

A month later, further good news was dispatched to Florence by Stelluti, who kept Galileo constantly up to date on the unfolding of events in Rome: Francesco Barberini would shortly be elected Cardinal, 'so we shall enjoy the protection of a powerful Cardinal'.[136] On hearing the news, Galileo hastened to write to the nephew of the Pope, enthusiastically expressing his satisfaction at belonging to an Academy glorified by such an eminent member.[137] On 9 October, the day that he wrote to the Cardinal Nephew, the printing of the *Assayer* had not yet been completed. On 10 October Galileo received the frontispiece and approved it. On the same day he took up his pen to reveal to Cesi a dream that he was harbouring, induced by the rosy horizon that stretched out before him. Excited by the encouraging news that he was receiving from Rome, he declared that he was convinced that the 'marvellous conjuncture' of the election of Urban VIII provided the most favourable occasion to relaunch his long cherished project. 'As far as my expectations are concerned', he confided to Federico, it will certainly not be possible 'to ever find another opportunity like it'.[138] It is not difficult to imagine what Galileo was referring to when he informed Cesi that his thoughts were constantly focused on 'things of great moment for the republic of letters'. He was not talking about comets: that chapter

135 Ibid., p. 129, letter of Stelluti to Galileo, 8 September 1623.

136 Ibid., pp. 132–3, letter of 30 September 1623.

137 Ibid., pp. 134–5, letter to Francesco Barberini, 9 October 1623.

138 Ibid., p. 135.

had been closed. On his desk there were now writings in which he analysed the cause of the tides. For months Galileo had been assiduously working on that project, conceived many years earlier, convinced that his interpretation of the flux and reflux of the sea would settle accounts with the resolute enemies of the Copernican conception. He had resigned himself to leaving that text unpublished for future generations, but now he entertained hopes of publishing it during his lifetime.

He had to seek the counsel of Cesi to realize this grand project. For this reason he asked the Prince how long he would be staying in Rome, where the Tuscan scientist planned to return at the earliest opportunity, ill-health permitting, as for many months he had been forced to spend more time in bed than at his desk.[139]

On 20 October Tommaso Rinuccini, who was in Rome as special ambassador of the Grand Duke of Tuscany to pay homage to the new Pope, informed Galileo that Urban VIII would be pleased to meet him should he return to the Eternal City. Cesi was still in town awaiting completion of the printing of the *Assayer* in order to be able to present it personally to the Pope.[140] All that was left was to print the dedication to Urban VIII, which Cesarini would release the next day.[141]

When Cesi replied to Galileo's letter on 21 October, he was on the point of leaving for Acquasparta. He confirmed that it was indeed a 'good conjuncture': the correction of 'marvellous' to 'good' was not without significance. He declared moreover that he would welcome the opportunity to meet Galileo again. Cesi was perfectly aware of what his fellow Academician had in mind. But his tone was designed to calm the Tuscan scientist's excitement: the situation was still complex and called for caution without forcing events. Federico suggested that Galileo should stop in Acquasparta on his way to Rome so that the Prince could illustrate the current balance of power in Rome and advise him on the most suitable tactics:

> I will vividly present you with the state of things at present and how much it will be necessary and possible to consider in that respect [...] and you will arrive here [sc. Rome] not new to the scene, but very well informed about what is called for.[142]

---

139 'I am resolved, as soon as my health allows, to come to pay you my respects [...] and to have the opportunity of a lengthy exchange of ideas with you' (ibid.).

140 Ibid., p. 139.

141 Ibid., p. 141, letter to Galileo, 21 October 1623. Cesi wrote that the printing of the volume 'is already finished [...], including the frontispiece [...]'.

142 Ibid., p. 140.

While confirming that the climate had changed, Cesi intimated that the obstacles that had provoked the harsh defeat of 1616 had not been removed.

On 27 October a copy of the *Assayer* was presented by Cesi to Urban VIII and the Cardinal Nephew. The next day other exemplars were in the hands of influential Cardinals. Cesi's agenda was filled with appointments to personally present the work to the most authoritative personages.[143] Reports from Rome increased Galileo's excitement, as witnessed by the letters of Sister Maria Celeste, to whom he conveyed his enthusiasm.[144]

Now that the *Assayer* was in the public domain, the Linceans were alert to any sign of a reaction on the part of Grassi and the Jesuits:

> Many of us go spying – Rinuccini wrote to Galileo – to find out how it has been received and read, in particular by those for whom it was written, or rather, who gave occasion to write it.[145]

In the meantime they had learnt from a Roman bookseller that, after acquiring a copy of the *Assayer*, Grassi had exploded with anger, 'saying that you had taken three years to write it, but that he would pay you back in three months'.[146] The testimony of Tommaso Rinuccini and Stelluti, according to whom Grassi changed colour after his first glance at the volume, suggests that his outburst was provoked by the initial pages of the work.

In the opening paragraphs of the *Assayer*, addressing Cesarini, Galileo lamented the unrelenting and violent opposition that his works had encountered from the start of his professional activity as well as the repeated attempts to steal his inventions and celestial discoveries. As we have stressed above, in these opening pages Galileo recognised that Cesi and others in the Lincean circle had had good grounds ten years earlier, on the occasion of the publication of the *Letters on sunspots*, for urging him to press his claims to priority against malign usurpers.[147] He meticulously listed the numerous polemics into which he had been drawn after the publication of his celestial discoveries, of

143 Ibid., p. 142, letter of Stelluti to Galileo, 28 October 1623.

144 Ibid., pp. 143–4, letter of Sister Maria Celeste to Galileo, 29 October 1623.

145 Ibid., p. 145, letter of Tommaso Rinuccini to Galileo, 3 November 1623.

146 Ibid. The same scene was described in Stelluti's letter to Galileo of 4 November 1623: 'The first exemplar to be seen in public was that received by the Dominican Niccolò Ridolfi, Master of the Sacred Palace, who gave it to the bookseller of the piazza del Sole. Sarsi immediately ran there, but the real one, because the fictitious one is just a name: he asked for the book, and upon reading the frontispiece he changed colour [...]' (p. 147).

147 See above, pp. 108–10.

his writings on hydrostatics, and of the volume in which he had announced the sensational novelty of the sunspots.[148] Nor did he confine himself to recalling the unpleasant experiences suffered after training his telescope on the heavens. He seized the opportunity to denounce the plagiarism by Baldassare Capra of his invention of the geometric and military compass while he was still teaching in Padua.[149]

The list of the attacks which had been directed at him served the purpose of presenting the polemic with Father Grassi on the comets as further evidence of his prejudicial persecution. As the target of the malign accusations of Sarsi, who considered him, without any foundation, as the real author of the *Discourse on the comets*, Galileo had to defend his own ideas and reputation, as well as to take down a peg those envious enemies who had not hesitated to launch violent accusations against reflections that he had presented not in printed works, but only in private writings:

> I say nothing of some of my private discourses, proofs and propositions (many of them unpublished) which have been seriously impugned or deprecated as worthless.[150]

This was a clear reference to the doggedness with which authoritative representatives of the Church and the Society of Jesus had censored his unpublished *Copernican letters* and the serious personal consequences that had ensued.

Galileo's preliminary remarks were followed by a series of statements which indicate that he had paid little heed to the suggestions of Cesi, Ciampoli and Cesarini on the prudent strategy to adopt in crossing swords with Sarsi-Grassi. His lack of attention to the recommendations of his Roman friends cannot be attributed to his hope of an imminent vindication in the new climate determined by the 'marvellous conjuncture', given that the final version of the opening pages of the *Assayer* was sent to Rome before Galileo had heard of the election of Urban VIII.

If the tone of his recriminations was resolute, at least in the form Galileo showed that he had taken into account the advice of the Roman Linceans. He pretended to believe that the *Astronomical balance* was the work of Sarsi. However, his dissimulation was evident, since he explicitly attributed to Grassi

148 OG VI, pp. 214–7. Galileo denounced with vehemence especially the attempts of Simon Mayr to usurp his priority in the discovery of the satellites of Jupiter.

149 Ibid., pp. 214–5.

150 Drake-O'Malley 1960, p. 164 (OG VI, p. 214).

the *Astronomical disputation* that had triggered the polemic.[151] Moreover he stressed that his critical remarks were not aimed at Father Grassi, but at those who attacked him while affirming that their position was supported and confirmed by the mathematician of the Collegio Romano. Having to defend himself from those who, 'strengthened by such assistance, rose up against me still more imperiously', it was impossible 'to contradict them without including Father Grassi too'.[152]

The point at which Galileo departed most from the course indicated by Ciampoli and Cesarini came immediately afterwards. Sarsi had accused him of ingratitude towards the Fathers of the Collegio Romano, who had received the celestial discoveries and his writings on hydrostatics most graciously. The Tuscan scientist's reply was terse and biting. On questions regarding the interpretation of natural phenomena, it is neither possible nor admissible to take obligations of courtesy into account, nor to tender recognition for past manifestations of goodwill. Etiquette, gallantry and even gratitude could not be invoked when dealing with questions bearing on the truth or falsity of important issues in natural philosophy. Had the Fathers of the Collegio Romano, he asked Sarsi, shown their appreciation of his discoveries out of politeness and their innate goodwill, or because they were convinced that he had demonstrated new truths?

> I ask you whether those reverend Fathers consider my ideas to be true or take them to be false? If they recognize my ideas as true and praise them as such, then it is at exorbitant interest that you demand payment of what was lent, when you expect me to extol with equal praise things I know to be false. But if they regard my ideas as foolish and still praise them, then I may indeed thank them for their urbanity, but I should have been much more grateful if they had lifted me out of my error and had shown me the truth, for I regard much more highly the utility of accurate correction than I do mere pomp and ceremony.[153]

The quest for truth does not admit of negotiations, nor can it be subordinated to considerations of opportunity, to formal rules of conduct, nor even less as

151 'I say in the first place that I have never lamented over having been maltreated in the *Astronomical disputation* of Father Grassi in which I am sure that his Reverence never thought to offend me' (Drake-O'Malley 1960, p. 177: OG VI, p. 226).

152 Drake-O'Malley 1960, p. 177 (OG VI, p. 226).

153 Drake-O'Malley 1960, p. 178 (OG VI, p. 227).

recompense for favours received. Galileo's pride, injured by the condemnation of 1616, re-emerges in these passages, as does his determination not to be intimidated by the strength of his adversaries.

Sarsi was mistaken 'in saying that I am guilty of contempt for the dignity of the Collegio Romano'. Nor could he believe that Lothario's assertions about the resentment of the Fathers of the Collegio Romano at Guiducci's severe critique of the *Astronomical disputation* corresponded to the truth. If that were the case, he would have to conclude that such a prestigious institution harboured practices of duplicity:

> If Sarsi's voice does emanate from the Collegio, I have reason to suspect that my doctrine and my reputation have been in bad odour there not merely at present but all along, inasmuch as in this *Balance* none of my thoughts are approved.[154]

He could not comprehend how praise could have been replaced by resentful hostility. The only explanation that he could find was that Sarsi had spoken in a personal capacity. He was profoundly convinced that none of his arguments were approved by the Fathers of the Collegio. The entire responsibility for the conflict thus rested on the shoulders of Lothario, whose philosophy is such

> as to give him the right equally to praise or to blame, confirm or refute the same doctrines accordingly as benevolence or wrath may move him.[155]

We are now in a better position to understand why the blood rushed to Grassi's head when he opened the first pages of the *Assayer*. The intensity of the taunting was maintained in the successive pages. Once more we encounter the strong intellectual and ethical motivation that induced Galileo to defend his deepest convictions, without being intimidated by the risks he might incur. While recognising the well-meaning intentions of his Roman friends' counsel, the principles that had turned his life into one of constant conflict with those who opposed the freedom of thought in natural philosophy did not allow him to heed their recommendations.

Father Grassi's impulsive reaction was certainly also triggered by his awareness of the reasons that had led Galileo to launch such a violent attack on Tycho Brahe's cometary hypothesis: behind his firm opposition to the Danish

154 Drake-O'Malley 1960, p. 179 (OG VI, p. 228).

155 Ibid.

astronomer lay Galileo's will to defend the Copernican conception of the universe.[156]

A month after his outburst before the Roman bookseller, Father Grassi appeared to have overcome his anger. Tommaso Rinuccini, who had come across the Jesuit by chance, informed Galileo that, while maintaining his positions, the Father recognised that there was 'some good' in the *Assayer*. He would reply in due course and 'without mordacity'. Moreover the Jesuit expressed his intention 'to befriend him should the Tuscan scientist come to Rome.[157] This conciliatory frame of mind, however, did not last long. Grassi had flown into a rage once more on hearing that supporters of Galileo were exulting in Florence because the *Assayer* had 'closed the mouths of all the Jesuits who would not know what to reply'.[158]

Rinuccini passed on to Galileo another important piece of information: 'This morning I heard from a Jesuit that they are strictly prohibited from discussing this text [...]'. The election of Maffeo Barberini to the pontificate had compelled the hierarchies of the Order to change their attitude. The party of the Linceans and Galileo, their protégé, had acquired extremely influential positions which dissuaded from attacking them. The predictions of Ciampoli and Stelluti seemed to be coming true: the Academy would finally be able to raise its head. It is telling that, in spite of his promise to reply to Galileo within three months, Grassi had to wait two years before his rejoinder, the *Ratio ponderum Librae et Simbellae* (*Ratio of weighings made with a balance and a small scale*), would be published.[159]

---

156 Galileo may also have been spurred on to attack Tycho Brahe in the *Assayer* by his reading of an anonymous work, extremely rare today, published in Milan in 1619 (Besomi-Camerota 2000). The title runs *Assemblea celeste. Radunata nuovamente in Parnasso sopra la nova cometa. Dall'Accademico Danico il Riposato descritta in una lettera all'Accademico Italico lo Spassionato* (*Assemblea celeste*, 1619). The pseudonym 'Danish Academician' conceals the identity of an author whom it has not yet been possible to identify (Michele Camerota supposes a connection with the circles of the Collegio Braidense of Milan). The title also suggests that the work was written to defend Tycho's cometary conception. Galileo and Guiducci had the *Assemblea celeste* in their hands in 1619 and passed severe judgement on its theses (Besomi-Camerota 2000, pp. 137, n. 6 and 135–58; see also Pagnini 1933–1934).

157 OG XIII, p. 154, letter of Tommaso Rinuccini to Galileo, 2 December 1623.

158 According to the unknown friend of Rinuccini, Father Grassi replied harshly: 'If the Jesuits knew how to promptly reply to a hundred heretics, they would also know how to do it to a Catholic' (ibid.).

159 The *Ratio* (Grassi 1626) was published at the end of 1626.

Tommaso Caccini Back on Stage?

As we have pointed out, Rinuccini had provided Galileo with an important clue: Grassi knew about the resonance on the Florentine scene of Galileo's vigorous attacks on Sarsi in the *Assayer*. This is further evidence of the direct link between the opponents of the Tuscan scientist in both cities. Galileo's correspondence with the Roman friends contains numerous confirmations that, after the 1616 decree, his enemies in Florence had risen up again and were constantly harassing him. The attempt to put the *Risposta* of Galileo-Castelli on the Index has already been mentioned,[160] as has the care that the Florentine Jesuits took to keep their Roman colleagues well informed about the progress of the writing of the *Assayer* and, later on, its imminent publication.

Galileo for his part repeatedly expressed to Cesi his concern about the constant initiatives in his own city to denigrate and denounce him.[161] New documents published by Massimo Bucciantini help us to put into focus the composition of the anti-Galilean front in Florence confirming the fundamental role played by Don Giovanni de' Medici, the natural son of Cosimo I. Three weeks after the issuing of the anti-Copernican decree, Ludovico delle Colombe triumphantly informed Don Giovanni that he had seen the decree in the hands of the Dominican Niccolò Lorini. He emphasised how that official document exposed the distorted version of the success of his Roman mission that Galileo had presented to the Grand Duke: 'The edict is clear, printed and visible, while the rest is idle chatter, based on God knows what'.[162] Four days later Ludovico delle Colombe took pains to make sure that Don Giovanni had been informed about the publication of the decree.[163]

To the substantial dossier of the intimidations of the Aristotelians, Dominicans and Jesuits in Florence and of their patrons at court should be added the dissemination in both Florence and Pisa, after the issuing of the anti-Copernican decree, of the news, once again tailor-made, of Galileo's condemnation and abjuration. From Pisa Castelli informed the Tuscan scientist about the fabrications that Archbishop Francesco Bonciani was spreading: 'Here

160 See above, p. 285.

161 The letters in which he registered these complaints are lost, but they can be reconstructed on the basis of Cesi's replies.

162 Bucciantini 2005, p. 304 and Galilei 2015, p. 163: letter to Galileo, 26 March 1616.

163 Bucciantini 2005, p. 305 and Galilei 2015, p. 163: letter to Galileo, 30 March 1616. Bucciantini outlined the personality and interests of Don Giovanni and the role he played as a point of reference of the anti-Galilean party at court. On Don Giovanni de' Medici see also Guerrini 2009, pp. 24ff.

B. [Bonciani] has written that you recanted secretly in the hands of Bellarmine'.[164] The same malevolent information had been circulated at court. After receiving assurances from Galileo that the decree had not carried formal penalties for his person, the Secretary of State, Curzio Picchena, confessed that he and the Grand Duke, having been presented with a completely different scenario, had breathed a sigh of relief:

> Your last letter of the 12th was read by me to His Highnesses, who were very pleased by the news [...] *given that very different rumours had been spread here* [...].[165]

Luigi Guerrini has shown that the hostility of the Florentine Aristotelians and the Dominican Fathers towards Galileo and Copernican cosmology had been manifested before the publication of the *Sidereus Nuncius*. Guerrini has also drawn attention to a document that shows that the arrival in Tuscany of the *Sidereus Nuncius* provoked violent reactions both in Florence and Pisa. This document is the letter written in the second half of 1610 by the diplomat from Pistoia Bonifacio Vannozzi, a follower of Telesio,[166] to the Supreme Magistrate of Justice of the same city, Girolamo Baldinotti, who had asked him for information about the celestial novelties:

> Regarding Galileo I am of one mind with you, and any good theologian will laugh at those who maintain that the Earth really moves since it can never be moved [*non inclinabitur in seculum*].[167] Such things have been said on other occasions as hypotheses, not as truth. To say that the Moon is Earth-like, with valleys and hills, is as if to say that there are flocks that graze there and cowherds who cultivate it. We must stand by the Church, which is the enemy of anything new, according to the teachings of St Paul. These are thoughts from brilliant minds, but they are dangerous, and I prefer to be a theosophist rather than a philosopher, as you too seem to be, and I kiss your hands.[168]

164 OG XII, p. 234, letter of Castelli to Galileo, 20 April 1616.

165 Ibid., p. 250, letter, from Leghorn, of Picchena to Galileo, 20 March 1616, emphasis added. For the letter of the 12th, see ibid., pp. 248–9, letter of Galileo to Picchena.

166 On Vannozzi see Bucciantini-Camerota-Giudice 2015, pp. 65 and 183–5.

167 *Psalm* 104:5.

168 I have used the translation by C. Bolton in Bucciantini-Camerota-Giudice 2015, p. 183. See also Guerrini 2009, p. 31 and Galilei 2015, p. 66. The letter of Bonifacio Vannozzi was unknown to Antonio Favaro, editor of the National Edition of Galileo's *Works*.

Through which channels did the Florentine opponents of Galileo maintain contact with the authoritative Roman circles that resolutely opposed the celestial novelties, and above all their menacing cosmological implications? Ludovico delle Colombe was certainly on friendly terms with the Fathers of the Collegio Romano, with whom he exchanged views on questions raised by the Tuscan scientist. This is shown by the fact that he was so soon able to read the confidential judgement that Father Clavius and his colleagues had formulated, at the request of Cardinal Bellarmine, on Galileo's celestial discoveries.[169] It is probable that the hostility of Ludovico delle Colombe towards Galileo dated from the lively debates on the *stella nova* of 1604.[170]

The early formation of a hardened anti-Galilean party in Florence is confirmed by Cigoli's letter to Galileo on December 1611, describing the manoeuvres of 'a certain bunch of persons who are malign and jealous of your virtue and merits,' who 'meet and rejoice in the house of the Archbishop'.[171]

At the end of 1620, as a consequence of a recrudescence of their attacks in Florence, Galileo let off steam once again in his lost letter to Cesi about these interminable persecutions. The Prince tried to calm him, promising that he would ask Cesarini to make investigations and, if necessary, to intervene. The fact that Federico intended to involve Cesarini suggests that this time the attack came from the ranks of the Florentine Fathers of the Society of Jesus.[172]

Years later, once they had discovered that Galileo was working on the *Dialogue concerning the two chief World Systems* and trusted in the support of his Roman friends for its publication, the exponents of the anti-Galilean party in Florence hastened to spread the news that Ciampoli had fallen into disgrace with the Pope. Alarmed, Galileo asked Castelli for confirmation. The Benedictine clarified that it was purely a fabrication, wishful thinking on the part of his adversaries, for whose errors he invoked divine pardon:

> Monsignor [Ciampoli] enjoys the same favour as ever, to the infinite satisfaction of His Holiness [Urban VIII], who meets him two or three times a day. The Pope's favour has never diminished, not even in thought,

---

169 OG XI, p. 118, letter of Ludovico delle Colombe to Father Clavius: 'I have seen your reply to Cardinal Bellarmine, and I am pleased that you in particular do not accept that the Moon has an irregular and mountainous surface [...]'. Delle Colombe's letter is dated 27 May 1611. Little more than a month had passed since the reply of the Fathers of the Collegio, dated 24 April, to the request of Cardinal Bellarmine.

170 Guerrini 2009, pp. 19ff.

171 OG XI, pp. 241–2, letter of 16 December 1611.

172 Ibid., XIII, p. 54, letter of Cesi to Galileo of 4 January 1621.

> in spite of what some (whom I believe to be few), unworthy to share in the glory of Monsignor, have rumoured there because they hoped for it, though very far not only from the truth but from all likelihood. May God pardon them.[173]

Further evidence of the incessant attacks directed at Galileo in Florence after the 1616 decree[174] is offered by a document which has been until now neglected: the letter that Castelli sent from Pisa to Galileo on 6 December 1623. In a lost letter, the Tuscan scientist had informed his disciple of a new persecution by his Florentine adversaries, revealing the identity of the most active among them. It is no surprise to learn that it was Tommaso Caccini:

> I am sorry that Father Caccini – Castelli replied – jeopardises Princes and the Holy Office itself, if it is true that he goes around saying that, if it were not for the protection of various Princes, you would have been brought before the Inquisition, as though the Princes impede the Holy Office and protect ill-doers, and likewise the Holy Office pays respect to Princes in taking action against impiety [...].[175]

If this document is read closely, Castelli's insistence on the Holy Office cannot but appear striking. Had Galileo complained that Caccini had had the effrontery to knock on the door of the Holy Tribunal once again?

We touch here on an extremely delicate question which has provoked lively discussions in the last decades. I refer to the document – the so-called G3 document – discovered and published by Pietro Redondi, who constructed a fascinating book around it.[176] That document contains a denunciation of the *Assayer* to the Tribunal of the Inquisition, based on the concern that the conception of the subjective nature of secondary qualities proposed in that work by Galileo called into question the dogma of the Eucharist. Redondi

173 Ibid., XIV, p. 90, letter of 6 April 1630.

174 For the initiatives against Galileo of the Florentine Dominicans, supported by circles at court, at the time of the First Trial, see Bucciantini 1995 and 2005.

175 OG XIII, p. 156.

176 Redondi 1983. The anonymous document, preserved in the Archivio Segreto Vaticano (*Indice, Protocolli, EE, II.a.27*, cc. 302r-303v), bears neither date nor addressee. It is traditionally referred to as G3 because of the manuscript annotation on its first page (ibid., pp. 427–30). It was republished by Pagano with a detailed codicological description and an analysis of the interpretations of the numerous scholars who had studied this document (Pagano 2009, pp. 180–4).

argued that this is the document to which Guiducci referred in the letter of 2 April 1625, in which he informed Galileo that Cesi had decided not to pass his transparently Copernican reply on to Francesco Ingoli, because he knew that a denunciation of the *Assayer* as an explicitly Copernican work had been transmitted to the Holy Office:

> Some months ago, a pious person proposed to the Congregation of the Holy Office that the *Assayer* should be prohibited or emended, arguing that it praised the doctrine of Copernicus concerning the motion of the Earth [...].[177]

According to Redondi, Guiducci had completely misunderstood the information received from Cesi, who in reality told him that the anonymous denouncer had expressed his concern not for the adhesion of Galileo to Copernicus, but for the heretical implications for the dogma of the Eucharist of the corpuscular interpretation of sense perceptions expounded in the *Assayer*.

Numerous authoritative scholars have participated in the debates triggered by Redondi's book. Though based on different arguments they have, practically without exception, rejected his thesis. Taking his interpretation of G3 to the extreme, Redondi argued that the trial and condemnation of 1633 were the result of a 'substitution of theory': wishing to avoid using too heavy a hand to overcome the stubborn resistance of a person who counted on influential protectors by inducing him to confess his guilt, the inquisitors negotiated with Galileo the transformation of the main charge (that of sacramental heresy formulated in the anonymous denunciation) to the less serious imputation of adhesion to Copernican doctrine. This would enable them to pass a less dramatically severe sentence.

It is indeed difficult to imagine that Guiducci made the colossal blunder attributed to him by Redondi. On the other hand, there is no reason to be surprised by the fact that the *Assayer* was denounced for dissimulated Copernicanism. It is not true – as Redondi claimed – that it does not betray the Copernican inclinations of the author. On the contrary, like the *Discourse on the comets* of 1619, the *Assayer* is a transparently Copernican book, even if Galileo had to dissimulate his own adhesion to the system conceived by the Polish astronomer. It could not be otherwise, not only because he had not given up the battle for the affirmation of what he considered to be the true structure of the cosmos, but also because Grassi in the *Astronomical balance* had inserted

177 OG XIII, p. 265. For further analysis of this very important letter see below, pp. 328–32.

many insinuations regarding the Copernican convictions of the true author of the *Discourse on the comets*. It is enough to read the *Astronomical balance* to be convinced that, in the eyes of Father Grassi, the text of Galilei-Guiducci was infected by the Copernican virus.

To realise that, in spite of the decree of 1616, Galileo had not laid down arms, it is sufficient to open the *Assayer* at the pages in which he proposed the well-known metaphor of the Book of Nature written in mathematical characters. In directly confronting the question of the structure of the universe, he declared that he could not understand why Sarsi claimed that it was necessary to follow the system of Tycho instead of that of Ptolemy or Copernicus. He seized the opportunity to remind his interlocutor that it was thanks to his personal contribution that Ptolemy's system had been dealt a mortal blow. Had he not been the first to observe through the telescope the variations in the apparent size of Mars and Venus and the phases of Venus? This was rock-hard evidence that obliged everyone to recognise that the two planets orbit the Sun 'according to the Copernican system'.

In tones that could leave no doubt about what he really thought, feebly concealed beneath a transparent mask, he added:

> As to the Copernican hypothesis, had not the most sovereign wisdom removed us Catholics from error for our own good and illuminated our blindness, I do not believe that this grace and benefit could have been derived from the reasons and experiences embraced by Tycho.
>
> Thus the two systems being surely false and Tycho's being null, Sarsi should not reprehend me for desiring with Seneca the true constitution of the universe.[178]

Numerous other passages in the *Discourse on the comets* and later in the *Assayer* make it abundantly clear that the Tuscan scientist's denial of following Copernicus was purely instrumental. It will be recalled that Galilei-Guiducci had declared that, although the comet moved along a rectilinear path, the fact that, when observed from the Earth, it deviated northwards could be explained by supposing that the Earth revolves around the Sun. In the *Astronomical balance*, Grassi had underlined how that argument revealed Galileo's irresponsible adhesion to Copernicus, in spite of the ban imposed by the decree of 1616.[179]

---

178 Drake-O'Malley 1960, pp. 184–5 (OG VI, p. 233).

179 Cf. above, pp. 257–9.

In commenting on that malign insinuation, Galileo asked with what right did Sarsi take the liberty of affirming, on the basis of a simply hypothetical explanation of an astronomical phenomenon, that the author of the *Assayer* adhered to the Copernican conception of the world? Had not Sarsi declared a little earlier that both I and Guiducci hold the Copernican conception to be false?[180]

Galileo might have stopped here. But he could not miss the opportunity to stress that the hostility of Sarsi towards Copernicus had its roots in ignorance of the true movements of the celestial bodies:

> I much question whether Sarsi has even formed for himself any correct idea of the motions attributed to the Earth, or of the many and varied appearances which would have to be perceived in other world bodies as the result of it [...]. And in conclusion I may add that if the movement attributed to the Earth (which I, as a pious and Catholic person, consider to be most false and vain) lends itself to yielding explanations for so many widely diverse appearances observed in the heavenly bodies, then I should not be sure that so false a thing might not deceptively correspond with the appearances of the comet, unless Sarsi will descend to more specific considerations than those which he has thus far produced.[181]

After reading these passages, how can one possibly deny that in the *Assayer* Galileo tackled questions *de systemate mundi* from a Copernican perspective?[182] Once this equivocation is removed, the fact that the *Assayer* was denounced for its author's adhesion to the Copernican vision appears perfectly logical.

What is surprising is that, in spite of the general rebuttal of Redondi's interpretation of G3, no one has wondered about the identity of the author who had denounced Galileo to the Holy Office on suspicion of adhering to the Copernican doctrine, nor about the context in which that move took place. The

180 OG VI, p. 310.

181 Drake-O'Malley 1960, p. 269 (OG VI, pp. 310–1).

182 Galileo's Copernican convictions, although expressed in a deliberately ambiguous form, appear evident from numerous other passages of the *Assayer*. For example, Galileo invited Sarsi to demonstrate the falsity of the heliocentric hypothesis not based on the 1616 decree, but with scientific arguments: 'Nor does the excuse he puts forth seem to me at all sufficient – that since this opinion was deduced by Kepler as a consequence of the Earth's motion, a proposition that cannot be piously and religiously held, Sarsi has therefore deemed it worthless. Why, this must be so much the more motive for destroying it and demonstrating its impossibility' (Drake-O'Malley 1960, p. 192: OG VI, p. 240).

only fact on which there can be no doubt is that the anonymous denunciation produced no consequences. Guiducci reassured Galileo: thanks to the intervention of Cesi, the case was assigned to Father Guevara, a sincere admirer of the Tuscan scientist and linked to the Cardinal Nephew.[183] Guevara expressed the opinion that 'the doctrine of the motion of the Earth', even if Galileo held it to be true, was not dangerous to the faith; there was therefore no point in condemning it.[184] How could Guevara have issued such an accommodating statement if the denunciation had turned on the suspicion of Eucharistic heresy?

The information that Guiducci had received from Cesi contains an element that helps to reconstruct the context of the aborted affair. Guiducci declared that the denunciation had been formulated 'some months ago': therefore, with extreme probability, after the election of Maffeo Barberini as Pope and immediately after publication of the *Assayer*. The fact that Cesi was informed of it so promptly that he could ensure that the investigation was entrusted to an obliging censor and knew its result, leads one to suppose that the insidious operation was neutralised thanks to the authoritative intervention of Francesco Barberini. It is not fortuitous that in his letter Guiducci had begged Galileo to act with the utmost caution, especially taking into account the absence of the Cardinal Nephew from Rome.[185]

This brings us to the most delicate question. In a context so favourable to Galileo and the Linceans, who would have dared to lodge an official denunciation against the Tuscan scientist? A clue is offered by the already mentioned letter that Castelli sent to Galileo in December 1623, a few weeks after the publication of the *Assayer*. Guiducci informed Galileo of the denunciation fourteen months later – an interval compatible with the reference to an episode that, according to the Florentine advocate, had taken place 'some months ago'. As we recall, in his lost reply to the Tuscan scientist Castelli linked the name of Caccini with the Holy Office no less than three times. The Benedictine even declared that, in view of his disrespectful behaviour towards Galileo and towards the ruling Tuscan dynasty whose subject he was, it was Caccini who should have been brought before the Inquisition. Castelli also denounced the harm that Caccini had done to the Grand Duke and his family in accusing them of protecting 'ill-doers'.[186]

183 OG XIII, p. 265.

184 Ibid.

185 In those weeks Francesco Barberini was engaged in a delicate legation in France.

186 Ibid., p. 156: 'I think it is Father Caccini who deserves to be brought before the Inquisition because of his lack of due respect for the Princes'.

These clues suggest that, after the publication of the *Assayer*, it was Tommaso Caccini who once again tried to eliminate Galileo, using the same strategy as in 1615. The publication of the *Assayer* had kindled afresh the animosity of the Florentine anti-Galileans, among whom Caccini had assumed an even higher profile after the triumphal outcome of his first move. Moreover, the Dominican friar could count on powerful supporters at court. What we know for certain is that Caccini and his associates in the Grand Duchy had established a programmatic alliance with Grassi and the Collegio Romano in an attempt to stop the *Assayer* from being printed.[187]

Another element points to Caccini as the author of the denunciation. After a long absence, he had returned to Florence in the spring of 1622,[188] immediately resuming his manoeuvres against the Tuscan scientist.[189] The animosity of Caccini and of the Florentine anti-Galilean party was rekindled by the insinuations of Grassi in the *Astronomical balance* on the obstinacy with which Galileo-Guiducci refused to submit to the anti-Copernican decree in the *Discourse*. Having fought strenuously for the adoption of that pronunciation by the Church, the Dominican feared that the decree might be attenuated or even revoked by Urban VIII.

187 See above, pp. 274–5.

188 Cf. Ricci Riccardi 1902, p. 210.

189 Tommaso Caccini spent the years after the emission of the decree in Rome, except for a brief journey to Florence at the beginning of 1619. From the testimony of a Florentine Friar to the brother of Tommaso, Matteo Caccini, we learn that the Dominican had returned to Florence 'to resume his intrigues against Galileo' (ibid., p. 190). A few months later, on 8 March 1619, Matteo provided precious new information on the moves of his brother: Tommaso had expressed his desire to return to Florence on a permanent basis. Matteo 'did not see anything good in this resolution'. He explained his brother's decision with the fact that it would enable him 'to keep a closer eye on a certain person' (ibid., p. 191). It is difficult not to recognise Galileo (as Ricci Riccardi also did) in that 'person'. There is another interesting detail to consider. Tommaso Caccini returned to Florence, this time for good, in the late spring of 1622, a few months before Galileo was to express to Benedetto Castelli his own worries about the renewal of the persecutions by the Dominican. Tommaso was soon to receive an important position in Florence: Master in Theology, and then Prior of the Convent of St Mark. This was a prestigious position that could not be obtained without authoritative support (ibid., pp. 215–6). Castelli's reply to Galileo's lost letter is eloquent confirmation of the renewed efforts on the part of Caccini, immediately after his return to Florence, to destroy the reputation of Galileo once and for all. He drew support, as before, from the powerful local anti-Galilean party, while keeping open the channel with the centres of power – especially the Jesuits of the Collegio Romano – who pursued the same objective in the Eternal City.

It thus seems more than plausible that the denunciation to the Holy Office for a relapse into Copernicanism, of which Cesi, via Guiducci, informed Galileo, was conceived and transmitted to the Holy Office, not from Rome but from Florence, on the initiative of Tommaso Caccini, who certainly informed Grassi and the other Fathers of the Collegio Romano of his move.

Another episode confirms that the repeated attempts to reduce Galileo to silence by appealing to the authority of the Holy Tribunal were the product of a concerted strategy between Florence and Rome, in which Caccini, together with some of his fellow Dominicans and the Jesuit Fathers in Florence, in direct and constant communication with the Collegio Romano, played the role of principal instigators. The opening of the Vatican Secret Archive to scholars in 1996 – thirteen years after the publication of Redondi's *Galileo heretic* – made it possible to identify a new (undated and unsigned) document, from which it transpires that, at a date after that for a relapse into Copernicanism, a new denunciation against the *Assayer* was addressed to the ecclesiastical authorities.

In actual fact, the document in question does not contain a formal denunciation, but the opinion of an unnamed consultor on a (lost) anonymous complaint to the Holy Office on the danger to the dogma of the Eucharist of Galileo's interpretation of the secondary qualities in the *Assayer*. The fact that in the pagination of the volume of the Vatican Secret Archive the anonymous opinion follows G3 without solution of continuity suggests a direct conceptual and chronological connection between the two documents.[190] This is confirmed by the fact that the author of the opinion analyses the arguments of G3 one by one. Scholars who have examined this document have concluded that it is a *censura* written between 1628 and 1632 by Michael Inchofer.[191] The Jesuit Father, who was to play an important role in the trial of 1633, entirely confirmed the suspicions of the anonymous informer. Moreover, he declared himself in favour of a formal opening of an inquisitorial procedure against the author of that dangerous theory.[192] The Jesuit did not explicitly mention the *Assayer* or Galileo, who was, however, unequivocally alluded to three times

190 The document (*Index, Protocolli, EE II. a. 27*, cc. 301r-v) was published with Italian translation and commentary by Rafael Martinez (Martinez 2001, pp. 234–42). Pagano republished it, placing it in direct sequence with G3 (Pagano 2009, pp. 184–6, with detailed bibliography on p. 184).

191 See Baldini U.-Spruit 2001, Martinez 2001, Cerbu 2001, Pagano 2009, esp. pp. LXXXV-XC and 180–6. Cf. also Camerota 2004, pp. 393–8 and Baldini U.-Spruit 2009.

192 The author concluded: 'Let this suffice for the moment in view of a further inquisition to be held before the Holy Office' (Pagano 2009, p. 186).

as *Lyncaeus*: an evident sign of Inchofer's desire to cast aspersions also on the Lincean party and, consequently, on the circle of the Barberini who were protecting Galileo. On this occasion too it was most probably the Lincean-Barberini party that thwarted the new hostile initiative against Galileo.

The direct relation between G3 and Inchofer's opinion demonstrates that G3 was not written in 1624–5 as supposed by Redondi. It could therefore not be the document to which Guiducci referred in his letter to Galileo of April 1625. Guiducci had not got things wrong. The denunciation passed on confidentially to him by Cesi did indeed refer to the suspicion of Galileo's relapse into Copernicanism.

It appears evident that G3 must date from after the publication in Paris at the end of 1626 of Grassi's *Ratio*, the text in which the Jesuit explicitly insinuated for the first time that the doctrine of secondary qualities expounded in the *Assayer* was threatening for the Eucharistic dogma.

The re-contextualization of G3 opens up the possibility of identifying its author. Useful clues in this direction are provided by another document which, curiously, has not attracted the attention it deserves from the scholars who have studied these delicate episodes. Our source is again Benedetto Castelli, now settled in Rome. At the beginning of 1628 Galileo had manifested his serious preoccupations to his disciple, inviting him to attentively consider those passages in the *Ratio* in which Grassi denounced the perilous implications of the Tuscan scientist's conceptions for the dogma of the Eucharist. Galileo's letter has not survived, but its content can be reconstructed from Castelli's reply of 22 January 1628.[193] Galileo had asked him to assess the danger of Grassi's malign insinuations and, above all, to ascertain whether steps had been taken to alert the inquisitorial authorities in Rome. In particular, Castelli should sound out the Dominican Niccolò Riccardi, nicknamed 'Father Monster'.[194] A few weeks later, after failing to receive an answer, Galileo wrote again to his pupil, who finally informed him on 26 February of the encouraging outcome of the discussion that he had had with the Dominican Father:[195]

---

193 OG XIII, p. 389.

194 After assuring Galileo that he would sound out 'Father Monster', Castelli expressed his own disgust at the insinuations contained in Grassi's *Ratio* ('his stupid falsehoods make my stomach turn'). After only a quick look at the *Ratio* in Ciampoli's house, he was struck above all by the Jesuit's denunciation of the danger of Galileo's interpretation of secondary qualities for the dogma of the Eucharist. Castelli was certain that Riccardi would reassure him: the Father 'on other occasions has told me that those things did not bother him, and that in any case he had enough spirit to always take your part' (OG XIII, p. 369).

195 OG XIII, p. 391: Castelli to Galileo, 8 February 1628.

> I spoke to Father Monster, asking him to give his opinion on Sarsi's arguments. He said that your opinions were not contrary to the faith as they were purely philosophical, and that he would serve you in all that you asked him, but that he did not want to appear now, in order to be able to serve in case you were given trouble by the Tribunal of the Holy Office, where he acts as auditor [...].[196]

Galileo knew that some malevolent person, using Grassi's insinuations in the *Ratio*, was plotting against him while Castelli and the Roman Linceans were completely in the dark about this new affair. This suggests Florence once again as the place where the initiative had originated. Galileo must have heard the worrying rumour from some friends who were in confidence with his adversaries. The name of Tommaso Caccini as the author of this fresh attempt at destroying Galileo's reputation comes instinctively to mind. The Friar was by now Master of Theology in the Dominican convent of St Mark in Florence. Suspicion regarding his active role is reinforced by a clue that emerged from the palaeographical examinations of G3 by Rafael Martinez, who demonstrated that the paper on which that document was written bears the watermark of Cardinal Tiberio Muti, Bishop of Viterbo from 1611 to 1636.[197] The denunciation cannot have come from the bishop himself, uncle of the Lincean Carlo Muti, a fervent admirer of Galileo and close to the Lincean circle.[198] Between 1619 and 1622 Tommaso Caccini resided in the Dominican convent of Viterbo, after being removed from Rome as a consequence of his poor relations with Cardinal Borghese.[199] He will therefore have had the opportunity to frequent Cardinal Muti and the episcopal entourage.[200] Possibly, sheets with the watermark of Cardinal Tiberio Muti, which he then used to denounce the *Assayer* to the Inquisition, were in the luggage that accompanied him to Florence after the end of his stay in Viterbo. Alternatively, he may have convinced someone in the Viterbo episcopal circles of the grave threat to the faith of Galileo's doctrines, recurring to his collaboration later on to denounce the hated Tuscan scientist. The suspicion that we are dealing with a new initiative of Caccini is strengthened by a detail imparted confidentially by Father Monster

196 Ibid., pp. 393–4. Castelli had met Father Monster in the company of Ascanio Piccolomini.

197 Martinez 2001, esp. pp. 233–4.

198 Ibid., p. 233, n. 70. Cf. also Pagano 2009, p. LXXXVII.

199 Cf. Ricci Riccardi 1903, pp. 186ff.

200 On Caccini's years in Viterbo, see the unfortunately scanty remarks in Ricci Riccardi (ibid., pp. 196–210).

to Castelli: in taking up the defence of Galileo, he had had to face the hostility of some fellow Dominicans.[201]

To recapitulate, Galileo was denounced in late 1623, immediately after the publication of the *Assayer*, for having relapsed into Copernicanism. Neither the denunciation itself nor Father Guevara's opinion have passed to us. Those documents were probably removed from circulation on the orders of Cardinal Barberini at the request of Cesi and other exponents of the Lincean circle, which was extremely influential in the Curia at that time. The letter of Castelli to Galileo of 6 December 1624 suggests that Tommaso Caccini was the author of that denunciation.

A few years later, inspired by the insinuations contained in Grassi's *Ratio*, an anonymous informer took pains to warn the guardians of orthodoxy of the threat to the dogma of the Eucharist posed by the interpretation of secondary qualities contained in the *Assayer*. He did so in the document known as G3. To find out whether that denunciation had reached Rome and, if so, how it had been received, Galileo moved his most trusted pawns, as is shown by the letter of Castelli to the Tuscan scientist of 22 January 1628, from which it appears that Galileo was convinced that the attack had been launched from Florence. It is more than reasonable to suppose that this time too the author of the move was Tommaso Caccini, who wanted to settle accounts with Galileo once and for all. As in the case of the denunciation for relapse into Copernicanism of 1623, the new operation was hatched in Florence in concertation with influential interlocutors in Rome. Suspicion of having been accomplices to Caccini in the Eternal City falls on Father Grassi, the Collegio Romano, and their most authoritative points of reference in the institutions in charge of safeguarding orthodoxy.

It is not fortuitous that the assessment of the malign denunciation of the heretical implication of Galileo's corpuscular theories was entrusted this time to a Jesuit, whose opinion indicates the preconceived opinion that he held not only of the relevant passages in the *Assayer*, but also of Galileo's entire production, that he qualified as an extremely serious threat to be countered with the greatest determination.[202] In 1624 the assessment of the foundation

---

201 Letter from Rome of Benedetto Castelli to Galileo in Florence, 26 February 1629 (OG XIII, p. 394): 'He said that he had been strongly rebuked by his fellow friars for your sake [...]'.

202 The opening statements of the opinion reveal the preconceived hostility of Inchofer to the author of the *Assayer*: 'I have seen the discourse of the Lincean and recognised that this man never adheres to true philosophy, whether through error, ignorance or temerity' (Pagano 2009, p. 185).

of the accusations levelled against Galileo had been entrusted to a passionate admirer of the Tuscan scientist. A few years later, in the face of the renewed animosity of his adversaries, Galileo had to deal with a resolute censor of the new natural philosophy and cosmology. The dynamics of this second denunciation clearly show the change that had taken place between the prospects opened up by the 'marvellous conjuncture' of 1623 and the climate that prevailed towards the end of the decade. Other episodes will soon confirm this new scenario.

CHAPTER 9

Metamorphosis of a Conjuncture: from 'Marvellous' to 'Unfavourable'

Boating on the Lake

We left the Roman scene at the end of 1624 with Father Grassi enraged after his first inspection of the *Assayer*. He apparently calmed down in the following weeks, as suggested by the reports of his frequent and cordial meetings with the Jesuit that Guiducci sent to Galileo. The relations between the two men became intimate to the point of Grassi paying a visit to Guiducci at his home when he was confined to bed by illness. Even though at times he may have doubted the sincerity of the Jesuit Father, Guiducci seemed convinced that the reading of the *Assayer* had led Grassi to overcome his hostility and to establish cordial relations with Galileo. When Grassi learnt that the Tuscan scientist would soon be coming to Rome, he declared that he was impatient for that moment to manifest his personal esteem and admiration.[1]

However, Galileo did not meet Grassi during his stay in Rome. It was a decisive journey for the fate of the Tuscan scientist's project. It seemed that everyone was awaiting his arrival in the Eternal City, eager to pay him homage. But Galileo hoped that the visit would yield more tangible results. The priority on his agenda was the revocation of the anti-Copernican decree of 1616. Following Cesi's suggestion, he would stop in Acquasparta on his way to Rome to discuss the best means to that end.

After a series of postponements due to bad weather, health problems and practical considerations, Galileo finally left Florence at the beginning of April 1624. As in 1611 and 1616, letters to the Resident of the Grand Duchy of Christine of Lorraine and Ferdinand II Medici had preceded his arrival in Rome,

1 OG XIII, p. 161, letter from Guiducci to Galileo, 18 December 1623.

 | DOI 10.1163/9789004342316_010

even though this time it was not a state visit.[2] Galileo reached Acquasparta on 8 May and was to remain there until 21 May. Federico and Galileo had never had occasion to spend so much time together for discussing and exchanging opinions, nor would they ever do so again.

We know very little about the weeks spent in the Prince's ancestral palace. However, from the letter sent to Faber on 17 April by Johann Baptist Winther, we learn that discussions between the Tuscan scientist and Federico went on until late: 'At two o'clock at night I was summoned to the Prince, who was sitting beside the fire in the company of Galileo'.[3] Intense conversation alternated with excursions into the beautiful surroundings. Francesco Stelluti, who was also present, accompanied Galileo on a visit to the Marmore falls and on a boat ride on the lake of Piediluco.

In his detailed reconstruction of Galileo's visit to Acquasparta, although admitting that we have no precise evidence of the topics that they dwelt upon,[4] Gabrieli reasonably supposed that they concentrated above all on how to secure the revocation, or at least mitigation, of the anti-Copernican decree.[5]

Moreover, Cesi and Stelluti were certainly struck by the microscope that Galileo showed them. Given the direction of his own researches, the instrument must have appeared to Federico even more miraculous than the telescope. However, he would have to wait a few months before having one at his

2 Unlike Galileo's previous visits to Rome, which had been presented as state visits, exponents of the ruling family of Tuscany adopted a much lower profile on this occasion. The letter to Cardinal Carlo de' Medici with which Christine of Lorraine announced the arrival of the Philosopher and Mathematician of the Grand Duke in the Eternal City was rather cool. She wrote only to inform the Cardinal that Galileo had 'resolved to go to Rome for a while' and had been granted permission to leave Florence by his patrons. It was the certification of the licence rather than a recommendation to assist an official envoy (ibid., p. 163). The letter sent by Ferdinand II to the Resident of the Grand Duchy in Rome, Francesco Niccolini, was different in tone. Nevertheless, the Grand Duke too took pains to specify that Galileo was visiting Rome 'for his private affairs' (ibid., p. 167). Evidently the negative outcome of the Tuscan scientist's official mission in 1616 prompted the Tuscan dynasty to act with prudence, bearing in mind also the strong opposition to Galileo on the Florentine scene.

3 CL, p. 866, letter of Winther to Faber, 17 April 1624. Winther had arrived in Acquasparta as personal physician and scientific secretary of Cesi.

4 Besides Gabrieli's essay of 1942 (Contributi I, pp. 195–221), Galileo's stay in Acquasparta was analyzed by Rati 2005 and Biagetti A. 1974, who did not offer, however, any original contributions.

5 Contributi I, pp. 205ff.

disposal, as the microscopes that Galileo had brought with him were destined for authoritative persons on the Roman scene whom he hoped to win over to his cause.[6] Only after returning to Florence would Galileo satisfy the Prince's curiosity, accompanying the gift of a microscope with a letter describing the marvellous observations performed with the instrument.[7]

New light is thrown on the discussions held in Acquasparta by a recently published document: the letter sent by Francesco Stelluti to an unspecified addressee on 8 January 1633.[8] Although Gabrieli knew about this letter, he did not pay it the attention it deserves.[9] Stelluti's letter indicates that the conversations were not entirely concentrated on how Galileo should tackle the Roman scene. Cesi and Stelluti must have asked Galileo to explain the reasons on which he founded his conviction that the Copernican conception corresponded to the real constitution of the universe. Galileo will certainly have told the Prince and his faithful collaborator that his interpretation of the causes of the tides provided incontrovertible physical proof that the Earth is not motionless. He was, however, perfectly aware that the conclusions that he drew from the ebb and flow of the sea depended on the revolutionary principles of his new science of motion, on which he had been hard at work ever since his first teaching experience at the university of Pisa.

It may seem strange, but hardly any references to his research on motion can be found in the correspondence between Galileo and the Linceans. On the other hand, neither Cesi nor his closest associates could have assimilated the innovative concepts of that new doctrine from Galileo's published works, which touched only briefly on the subject. Now that a confrontation that he considered decisive was imminent, Galileo felt that the affectionate solidarity and generous support of the Prince and the Linceans should be backed up by

6 That Galileo had brought some microscopes to Rome as gifts for the most authoritative persons is confirmed by Faber's letter to Cesi of 11 May 1624: 'I spent yesterday evening with Galileo […]; he gave a beautiful microscope to Cardinal Zollern for the Duke of Bavaria. I saw a fly through it […]' (CL, p. 875).

7 'I send you a microscope to see the smallest things from nearby […]' (OG XIII, p. 208, letter of 23 September 1624).

8 Cf. Galilei 2015, pp. 275–9. Lino Conti (Conti 1990a) supposed that the addressee was Fabio Colonna. It has been shown with convincing arguments, however, that the addressee should be identified with the Theatine Father Giovanni Guevara (Galilei 2015, p. 279). Stelluti's letter was published for the first time by Anna Nicolò (Nicolò 1986, pp. 188–92).

9 See Contributi I, p. 28. The letter was not included in the *Carteggio Linceo* (CL) edited by Gabrieli.

a more mature understanding of the solidity of the arguments on which his Copernican battle was founded. He therefore seized the opportunity of his sojourn in Acquasparta to explain to Cesi and Stelluti the fundamental concepts of his new science of motion and its cosmological implications.

They both had difficulty in following Galileo's complex mathematical demonstrations on the nature and properties of natural and violent motion. To dispel their hesitations, the Tuscan scientist had to recur to some experiments, whose results showed unequivocally the strength of the principles on which his new science was based.

In his letter Stelluti provided a concise summary of Galileo's recently published *Dialogue*, emphasising that he had only leafed through it hurriedly at the home of a friend.[10] He focused his attention on the Second Day and, in particular, on the revolutionary principles with which Galileo demolished the objections of Tycho to the motion of the Earth. Stelluti suspected that his correspondent would be incredulous at the example presented in the *Dialogue* of a stone which, if dropped from the top of a tower, would touch the ground at its foot even in the case of the Earth's daily motion. For this reason he hastened to assure him that he had personally witnessed an experiment that demonstrated incontrovertibly the truth of that principle:

> As I was passing over the lake of Pediluco with Galileo in a six-oared rowing boat, which proceeded at a fairly fast pace, with him sitting on one side of the boat and I on the other, he asked me whether I had something heavy. I said that I had the key of my room, which he took; and while the boat was travelling rapidly, he threw the key high into the air. I thought it would be lost in the water, but although the boat had advanced sixteen or twenty feet forwards, the key nevertheless fell between him and me; because, besides its upward trajectory, it had acquired from the movement of the boat the inclination to follow it, as in fact it did.[11]

This episode shows that, while boating on the lake, Galileo and Stelluti continued to discuss questions that they had been debating with the Prince in

10 That friend was probably Benedetto Castelli, as the letter from Stelluti to Galileo of 19 June 1632 implies: 'and because he [Castelli] told me that he had it [the *Dialogue*] at home, I went there with great expectations, but I could only give it a glance, because I was unable to stay very long' (OG XIV, p. 361). It must have been again from Castelli that Stelluti learnt of the imminent prohibition of the *Dialogue*: 'if it is not prohibited, I will send it to you' (Galileo 2015, p. 275).

11 Ibid., pp. 230–1.

his residence in Acquasparta. The information provided by this document is extremely important. Stelluti and Cesi had been on familiar terms with Galileo since 1611 and had maintained a regular correspondence with him during the intervening thirteen years. Moreover, they had frequent opportunities to meet and converse with him in Rome between the end of 1615 and late spring 1616. And yet, when Galileo arrived in Acquasparta in 1624, neither Cesi nor Stelluti had a clear idea of the new science of motion that Galileo had elaborated many years before. They both also had difficulty in understanding the general principles on which it was founded, as a detail in Stelluti's letter – 'I thought the key would be lost in the water' – clearly shows.

So the intense discussions in Acquasparta ranged from the definition of the strategy to follow in Rome to the examination of Galileo's new conception of motion, its implications for the understanding of the tides, and, above all, the confirmation that it offered of the physical truth of the Copernican system. Galileo staged for his trusted friends a general rehearsal of the demonstrations that he intended to present to the interlocutors whom he would soon be facing in the Eternal City.

Cesi and Stelluti were now in a better position to grasp the reasons that had led Galileo to uphold the Copernican system as a principle that could not be reduced to a simple hypothesis. It was nature herself, when properly investigated, who showed the way things really were.

Not even on this occasion did the Prince and the Tuscan scientist show themselves to be kindred spirits ('spiriti omogenei e omeotropi'), as Gabrieli defined them. The great scholar emphasised the complementary character of their qualities while toning down the differences between their personalities and their projects: 'magnanimity, philosophical breadth and reflective Roman prudence' in the case of Cesi; 'intellectual brilliance, mathematical precision and bitter and caustic Tuscan impulsiveness' in that of Galileo.[12]

12 Contributi I, p. 204. At the end of his tentative reconstruction of the conversations between Cesi and Galileo in Acquasparta during those two weeks in May 1624, Gabrieli was so carried away with enthusiasm that he dreamed of writing a novel in dramatic form or even a filmscript. Gabrieli imagined a group portrait of gentlemen conversing (like that by Paul Pieter Rubens, now in the Wallraf-Richartz Museum in Cologne, in which he had recognised the portrait of Faber [Contributi II, pp. 1233–9]), or even a dramatised version in instalments. After Acquasparta, he would set another scene in Florence: 'a successive scene could be set in Palazzo Pitti in Florence, with Vincenzo Viviani, a disciple of Galileo, Borelli and the other members of the Accademia del Cimento' (ibid., pp. 220–1).

An Ambiguous Funeral

The conversations of the three Linceans and their excursions into the beautiful countryside of Umbria were saddened by the news of the death of Virginio Cesarini, which reached Cesi from Faber via Winther while Galileo was still in Acquasparta.[13] It was a terrible shock, not only on a personal level, but also for the life of the Academy, over which Cesarini had *de facto* presided for some time as a consequence of the founder's increasingly lengthy absences from Rome. That tragic event unhinged the strategy that Cesi and Galileo had devised, which counted on Cesarini's authority and diplomatic abilities to steer it successfully through the complex Roman scenario. Galileo also felt the blow: the reply of Sister Maria Celeste to a lost letter in which he had informed her of the loss of a patron of such great authority bears witness to the Tuscan scientist's feelings of disorientation and preoccupation. While delighted at the friendly hospitality Galileo had received from Federico Cesi, the affectionate daughter expressed her own sorrow at the loss of Cesarini, in whom she knew her father had placed great confidence:

> I know that your joy is counterbalanced by considerable sorrow, on account of the sudden death of Signor Don Virginio Cesarini, so esteemed and so loved by you. I, too, have been saddened by his passing, thinking of the grief that you must endure for the loss of such a dear friend, just when you stood on the verge of soon seeing him again.[14]

Galileo did not arrive in Rome in time to attend Cesarini's funeral. He would undoubtedly have been not only surprised but also seriously distressed by the sight of the body of his friend and fellow Lincean lying on the bier in Jesuit robes.[15] When he heard of it, he could not have regarded it as a good omen for the success of his mission. Galileo did not know that the decision had not been taken by Cesarini on his deathbed in answer to furtive pressure exerted on him by the Fathers of the Society of Jesus. Four years earlier the Duke himself had asked the Father General for permission to enter the Order of St Ignatius and had obtained his ready consent. Virginio's autograph will, penned on 1 June 1620, bears this out:

---

13 Letter of Faber to Cesi, 13 April 1624: '*Venit summa dies et ineluctabile tempus*. We have lost here on Earth our Virginio, whom we hope is in Heaven [...]' (CL, p. 863).

14 OG XIII, p. 174, letter of 26 April 1624.

15 On the fate of the mortal remains of Cesarini, see Contributi I, pp. 480–96 and Redondi 1983, pp. 117–20.

> I wish to be taken to burial wearing the religious habit of the Society of Jesus, the Reverend Father General having bestowed upon me the favour of accepting me into this religion, in accordance with my wishes.[16]

Anyone who has examined the epistolary documents of those years and is aware of the tensions between the Linceans and the Jesuits that preceded and followed the publication of the *Assayer*, dedicated to Cesarini himself, will feel disorientated. This episode offers a further lesson on the complexity of the relations between the principal centres of power of Roman society, including the Lincean Academy, its promoters and patrons.

Cesi knew perfectly well of his cousin's decision: the will was sealed in spring 1620 in Acquasparta in the Prince's residence, where Cesarini, accompanied by Ciampoli, had stopped on his way to Bologna to consult a well-known physician. Moreover, Federico was named in that document as executor of Virginio's will.[17] Not by chance, in drafting the biography of his cousin and fellow Lincean, he did not conceal the fact that Cesarini had hesitated for a long time before setting forth on the new path indicated by the Linceans and Galileo.[18] Nor could Ciampoli possibly have been ignorant of the contents of a will in which his noble

---

16 Contributi I, p. 805.

17 The will was published by Gabrieli (ibid., pp. 804–5). It reveals Virginio's dual loyalty: to the Academy, to which he bequeathed his library and manuscripts, and to the Society of Jesus, to which he left his mortal remains to be buried in the Jesuit habit in the Church of St Ignatius. The will shows that Cesarini feared that the life of the Academy might not be a long one; otherwise he would not have specified that the legacy to the Linceans was only valid if, at the moment of his death, the Academy 'will be established in some place of residence' (p. 805). If that condition were not met, the executor of the will – Federico Cesi himself, who must have crossed his fingers when he read this clause – would have to arrange for the books to be transferred 'to some college of the Jesuit Fathers to be designated by Giovanni [Ciampoli]' (ibid.). Besides the clear intention of maintaining a balance between the Academy and the Collegio Romano even at the moment of death, the will betrays the Roman Duke's awareness of the fragility of the institutional structure of the Academy. Confirmation that Cesarini occupied a sort of free zone between the two worlds can be found in his own programme of study: he planned to compile a comprehensive commentary on the 'impious' poem of Lucretius and, at the same time, to compose a work on the immortality of the soul at the request of Cardinal Bellarmine (ibid., p. 779).

18 'There was continuous divergence between them (Cesarini and Cesi) in their common studies [...]. The Prince was trying to persuade him and to draw him to cognitions founded on sense evidence [...]. This lasted a long time without any result [...]. Later, when they were in Rome, the Linceans Galileo Galilei and Giovanni Ciampoli joined the Prince in putting before the eyes of Cesarini that the most beautiful book of all was that of the whole universe and of nature written by blessed God [...]' (ibid., p. 783).

patron left him all his books and manuscripts, as well as the generous sum of 3,500 scudi.[19]

It is therefore not surprising that the official funerary speech for Cesarini was delivered in the Campidoglio by the Jesuit Alessandro Gottifredi. The Father avoided any mention of Cesarini's affiliation with the Lincean Academy or of his familiarity and kinship with Cesi, nor did he evoke the enthusiasm with which Virginio had supported Galileo.[20] He also passed over the fact that the Tuscan scientist had dedicated the *Assayer* to the deceased Duke, stressing instead the great esteem in which Cesarini was held by Cardinal Bellarmine.[21]

A second funerary oration was delivered to the Accademia degli Umoristi by the ex-Jesuit Agostino Mascardi, 'chamberlain of honour' of Urban VIII, on 5 May,[22] when Galileo was already in Rome. Although Mascardi belonged to the Barberini circle and was by no means hostile to Galileo and the Linceans, he followed the example of Gottifredi in making no reference either to the Tuscan scientist or to the *Assayer*. He also refrained from mentioning Virginio's relations with Cesi and the Linceans, while – as Gottifredi had done – he praised the perspicacity of Cardinal Bellarmine, who had recognised a new Pico della Mirandola in the young Roman nobleman.[23] Faber, who was unable

19 Ibid, p. 805.

20 Gottifredi 1624. The only hint at Virginio's membership of the Lincean Academy is contained in the reference to the lynx-like eyes of his mind ('Lynceis mentis oculis', p. 6). The frontispiece of Gottifredi's oration included the devices of the Linceans and the Umoristi, as well as the trigon of Barberini bees and a portrait of Pico della Mirandola (see Bellini 1997, pp. 275ff.).

21 Ibid., pp. 24–9. After praising Virginio's moral and intellectual qualities, Father Gottifredi stressed how, in view of the great respect that Bellarmine had for the young nobleman, the Cardinal had asked him to compose a work on the immortality of the soul. In writing that text against unbelievers, Virginio would have imitated Pico's work against predictive astrology (ibid., pp. 24–6). Gottifredi also stated that Cesarini had abandoned Aristotle in favour of the rigorous moral principles of the Stoics ('factum diceres ex Peripatetico stoicum', p. 31). The parallel with Pico was repeated by Gottifredi with regard to the Duke's decision to be buried in the vestments of the Order of St Ignatius, thus emulating the decision of Pico, who 'had left dispositions to be buried in the religious vestments of the Dominican order'.

22 Mascardi 1626.

23 'Robert Bellarmine declared that Don Virginio Cesarini was in no way inferior to Pico della Mirandola [...]'. A little later, he added: 'In everything he was held by Cardinal Robert Bellarmine in the same esteem as Pico [...]; what a glory, deserving of the envy of later generations, to have had Cardinal Bellarmine sing one's praises!' (ibid., p. 357). On the differences in tone and content between the two funerary speeches see Bellini 1997 (pp. 245–309) and 2009 (pp. 135ff.).

to attend the ceremony in person, wrote to Cesi that he had heard that Mascardi's speech was 'exquisite'.[24]

If Galileo felt dismayed by the *post mortem* appropriation of the intellectual identity of a person in whose support he had placed great trust, Cesi was upset for different reasons. More than by the manoeuvres of the Jesuits, he was irritated by the attempts of the members of the Academy of the Umoristi to proclaim themselves as legitimate heirs to the intellectual legacy of the Duke. Cesi considered the Academy of the Umoristi completely devoid of intellectual and moral motivations as well as of rigorous regulations and goals. He made every effort to ensure that the emblem of the Umoristi would not appear beside that of the Linceans on the marble plaque to be placed in the Campidoglio in memory of his cousin. Federico asked his collaborator, Bernardino Lucani, to do everything he could to make certain that only the symbol of the Linceans appeared on the marble.[25] He also asked Lucani to remind his cousin's relatives that Virginio had publicly declared himself a Lincean on several occasions (including in the *Assayer*, a few months before his death), and that, in the hierarchy of knowledge, natural philosophy, to which the Linceans applied their energies, occupied a higher position than poetry and literature, the disciplines practised by the Umoristi. Besides, the Lincean Academy was an institution structurally different from all other academies: 'ours is not a device (*impresa*) like that of the other academies, but a simple expression like a coat of arms (*arme*)'.[26] Gabrieli confessed that he was unable to understand the distinction that the Prince was drawing here.[27] In my opinion, Federico meant that the 'coat of arms' of the lynx did not indicate a particular field of activity, as was usually the case with an academic device, but constituted the direct image of the driving force behind the initiative: the desire to know, a principle shared by all the Linceans, irrespective of the different disciplines that they practised,

24 CL, p. 875, letter of 11 May 1624.

25 As a consequence of Cesi's pressures, neither of the Academies was mentioned on the marble plaque.

26 CL, p. 944. The words 'impresa' and 'arme' are underlined in Federico's autograph letter to stress their importance. On the insistence with which Cesi claimed that the commemorative inscription in the Campidoglio should only record the Lincean affiliation of Cesarini and not that to the Accademia degli Umoristi, see Alessandrini 1986, pp. 86–9, and Bellini 1997, pp. 275ff.

27 CL, p. 944, n. 3.

calling upon them to engage in the disinterested study of man and nature.[28] As we have stressed, Cesi obsessively underlined the significance of the lynx as a stimulus to knowledge, not only in the institutional documents and the *Lynceographum*, but also in the diploma that new Linceans received together with the emerald signature ring:

> We also decided – we read in the diploma – that those who deserve to enter this philosophical army should be decorated with the Lincean emblem, name and inscription, not to be borne as a sign of arrogance, as we see many others doing, but as a stimulus for our fellows to examine, scrutinise and investigate keenly those things which are completely neglected by others.[29]

Federico would emphasize again the symbolic value of the lynx as a 'continuous spur' in the *Discourse on the natural desire for knowledge*.[30]

Cesi's eagerness to mark a clear distinction between the Linceans and other academies is also revealed by the letter to Stelluti of April 1613, in which he reminded his friend that, before proceeding to their formal induction, the nature

28 Cf. De Angelis 2012b. According to Gabrieli, Cesi decided to publish the *Praescriptiones Lynceae Academiae* (Cesi 1624) as a consequence of the tensions caused by the funeral of Cesarini. That decision would have been taken to underline the difference between the Lincean Academy and that of the Umoristi (Contributi I, p. 493).

29 The diploma was published by Gabrieli (ibid., pp. 502–3).

30 As we have seen, Cesi expressed a mordant judgement on the academic institutions in his *Discourse on the natural desire for knowledge*, insisting in particular on the literary ones (like the Umoristi), 'which would yield valuable results if they concentrated on selected forms of erudition and on the virtue and usefulness of philology and poetry rather than on sonnets, madrigals, witticisms and comedies, and more on useful lectures than vain and pompous speechifying' (Altieri Biagi 1969, p. 70). Saverio Ricci justly emphasised the echo in Cesi's *Discourse* of 1616 of the severe judgement on the decadence of the academies formulated by Traiano Boccalini in his *Advertisements from Parnassus* (*Ragguagli di Parnaso*). Also the call to Princes by Cesi not to consider the Academy of the lynx-eyed useless or dangerous reflected central motifs of the *Advertisements*. Boccalini had justified the Princes who had closed down academies in which a useless and empty ceremonial had replaced the principle, dear to Cesi, of the natural desire for knowledge. In those degenerate academies, Boccalini declared, 'that burning desire to know languished and those virtuous exercises cooled so much that while at first the academies were frequented by private individuals and by Princes held in the highest repute, as time went on they were so abandoned and scorned that it often happened that they were even prohibited for being harmful rather than useful.' See Boccalini 1948, pp. 165–7 and Ricci 1999, pp. 230–1.

and ends of the Academy should be carefully explained to newly elected members: 'it is important for them to realise that it [the Lincean Academy] *is not like an ordinary academy*'.[31]

A Copernican Carriage

Galileo left Acquasparta on 21 April and arrived in Rome two days later at a late hour. The next morning, accompanied by Carlo Barberini, he had an hour's meeting that 'greatly pleased' him with the Pontiff.[32] The following day he derived 'equal satisfaction' from his lengthy conversation with the Cardinal Nephew.[33] The detailed reports to the Tuscan Secretary of State on his intensive series of visits reveal that Galileo soon manifested signs of discomfort. As he wrote to Picchena, the numerous visits he was continuously paying

> make me feel that I am old, and that to be a courtesan is a young man's job. Their physical robustness and the enticement of expectations give them the strength to endure such tasks.[34]

It had taken very little time for him to realise that it would be extremely difficult to overcome the ceremonial indifference of the Roman world in the face of his expectations. He had a feeling of *déjà vu* that reminded him of his interminable peregrinations through the palaces of Rome in 1611 and in 1616. His discouragement made him think of home: 'I wish – he wrote to Picchena – to return to my peace and quiet, which I shall do as soon as I can'.[35] The rapid

31 CL, p. 345, emphasis added. The desire to stress the originality of his academic project, distinguishing the Linceans from the plethora of academies active in Rome and in the rest of Italy and Europe, led Cesi to use a wide variety of formulae that alluded to different forms of organisation – military, religious and literary – to define its profile and goals. In the opening of the *Lynceographum*, for example, he defined the Academy 'philosophorum Ordo, seu Consessus, vel Academia, studiosorum classis seu collegium' (Cesi 2001, p. 3). Elsewhere he had recourse to the terms *phalanx*, *militia* and *compagnia*. It should be stressed that Cesi considered the Academy as the expression of all these designations together. It was the model of all models of organisation and for this reason distinct from and superior to all other academies.

32 OG XIII, p. 175: letter from Rome to Curzio Picchena, Secretary of State of the Grand Duchy, 27 April 1624.

33 Ibid.

34 Ibid.

35 Ibid.

dwindling of the expectations that had led him to undertake such a tiring journey is also evident from the reply of Federico Cesi on 30 April to a lost letter in which Galileo had expressed the same feelings and intentions.[36]

It is likely that the enthusiasm with which he had undertaken the journey had already cooled in Acquasparta due to the recommendations of Cesi and Stelluti to proceed with caution and not to take for granted that the 'marvellous conjuncture' of the election of Urban VIII had removed the obstacles in his way. His first meetings with the Pope, with the Cardinal Nephew and with the powers on the Roman scene must have soon convinced him that the hopes of achieving positive results thanks to the new climate in the Eternal City had been a naïve illusion.

Cesi tried to hearten Galileo, urging him once again not to be hasty and to accept the rules that governed the ceremonial of the papal as well as of any other court. He agreed with Galileo that: 'the court is a source of infinite cares and troubles [...].'[37] Still, if he wanted to attain the goal he had set himself, there were no short-cuts: 'I would like you to proceed as gradually as you can, observing caution above all else'.[38]

Galileo took Cesi's friendly advice. He extended his stay in Rome for a few weeks and engaged in an intense series of meetings, in the course of which he grew convinced that, if he had had infinite time at his disposal, he would have been able to shift at least by a few metres the large boulders that blocked his reform project. As he wrote to Cesi, he could accept the 'very protracted negotiations' of the court only on one condition: 'that nature agrees to transform my few remaining days into as many years or months'.[39] If he could obtain that privilege, he was confident that he would be able to accomplish 'some of those intentions that we discussed together'. But unlimited time would not suffice; it would also require an infinite amount of 'phlegm and patience'.

By now Galileo had realised that he would not obtain the hoped for results. He found himself at sea in a world that, because of the rules that governed it, remained completely indifferent to the causes for which he was fighting. He also suffered because the continuous and vain courtly ceremonies absorbed precious time from the 'desire [...] to complete some of my speculations'. Hence his determination to return 'as soon as possible to my quiet and freedom'.[40]

---

36 OG XIII, p. 177.

37 Ibid.

38 Ibid.

39 Ibid., pp. 178–9, letter of Galileo to Cesi, 15 May 1624.

40 Ibid., p. 179.

His pessimistic reading of the situation did not, however, induce him to abandon the struggle. Although during his meeting with the Cardinal of Santa Susanna he had not managed to bring the discussion around to questions dear to his heart,[41] the results of his visit to Cardinal Georg Zollern had appeared less disappointing.[42] On this occasion he had been able to explain the reasons behind his appeal to the Church to retract, or at least revise, the decree of 1616 and to recognise the freedom to maintain the truth of the Copernican doctrine.[43] Zollern had promised to discuss the matter with the Pope and to inform Galileo of the outcome. This was the only feeble ray of hope in a scenario that appeared hostile and indifferent to his expectations.

Cesi took note of Galileo's decision to return to Florence. Indeed, the tone of his letter to the Tuscan scientist suggests that he was relieved by the news.[44] The Prince had formed a clear idea of the insuperable difficulties that Galileo's ambitious plans would run up against on the Roman scene long before the Tuscan scientist encountered them.

Galileo waited impatiently for the outcome of Cardinal Zollern's meeting with the Pope. It was not until the beginning of June that the news arrived. That it was not what he had hoped to hear is evident from the letter he sent to Cesi on 8 June, before setting out on the return journey to Florence. Drawing up a balance of his visit to the Eternal City, Galileo chose to represent its result as a glass that was half-full rather than half-empty. He had received many attestations of consideration and affection from the supreme authorities of the Church, with the Pope at their head. Besides plying him with gifts, Urban VIII had promised to grant a pension to his son. He had been affably received by many Cardinals, including the most authoritative ones on the

41 Ibid.: 'Three days ago I dined with the Cardinal of Santa Susanna [Scipione Cobelluzzi] but without discussing any of our most important questions [...]'.

42 Ibid., p. 181, letter of Faber to Cesi, 24 May 1624: 'Galileo [...] expects to leave Rome six days from now. I hope that Cardinal Zollern will do him some service with the Pope regarding the Copernican system'. It was thanks to the mediation of the German colony in Rome that Galileo could meet with Cardinal Zollern to ask him to intervene with the Pope to convince him to revoke, or at least attenuate, the anti-Copernican decree. The meeting of Galileo with the Cardinal was attended by Faber as well as by Caspar Schopp, who was on extremely familiar terms with Zollern, and by the Dominican Father Niccolò Riccardi (cf. De Renzi 1996, p. 442 and Brevaglieri 2009, p. 114). It was probably on that occasion that Galileo suggested that Zollern point out to the Pope that the revocation of the decree would deprive the heretics, the majority of whom were in favour of the Copernican system, of a solid argument against the Church of Rome.

43 OG XIII, p. 179.

44 Ibid., p. 180, letter of 18 May 1624.

Roman scene: Scipione Cobelluzzi, Francesco Buoncompagni and Zollern. The latter had given him a detailed report of his discussion with the Pope on the questions that were of importance to the Tuscan scientist. The German Cardinal had put before Urban VIII the arguments that Galileo himself had suggested:

> He said that he had spoken with His Holiness regarding Copernicus, explaining that the heretics are all of his opinion, and that they hold his system to be absolutely certain; for this reason extreme circumspection was called for before taking any determination [...].[45]

In informing Galileo of the Pope's reply, Zollern had taken care not to completely dash his hopes. The Cardinal reported that Urban VIII had told him that the Church 'had not condemned the Copernican conception, nor intended condemning it as heretical, but only as "rash"'.[46] The conclusion of the Pope's reasoning must have appeared to Galileo as a sentence without the right of appeal: His Holiness had expressed the conviction 'that there was no fear that anyone would ever demonstrate it [the Copernican system] to be necessarily true'.[47] Upon hearing this part of the report of the German Cardinal's meeting with the Pope, Galileo must have been reminded of the words with which Cardinal Bellarmine had invited Father Foscarini and himself to consider the Copernican conception as a simple hypothesis: the authoritative Jesuit had expressed the same conviction that nobody would ever provide conclusive proof of the motions of the Earth.

Galileo felt that ten years had been spent in vain. The mask of the 'marvellous conjuncture' had fallen. Moreover, the opinions expressed by persons by no means hostile to him indicated that to continue along his chosen route

45 Ibid., p. 182. One of the arguments to which Galileo and the Catholic supporters of the Copernican doctrine continuously referred to stress the damage caused by the decree of 1616 was that, in revoking it, the Church would be better able to convert the heretics, who all adhered to the heliocentric system. Galileo was to use this argument in his reply to Ingoli's *Disputation* (Finocchiaro M.A. 1989, pp. 156–7: OG VI, pp. 511–2) and in the letter to the reader of the *Dialogue* (OG VII, p. 30). As we learn from a letter to Galileo of Benedetto Castelli of 16 March 1630, Campanella used the same argument when he met with Urban VIII: 'Father Campanella, speaking a few days ago with His Holiness, told him that he had had certain German gentlemen in hand to convert them to the Catholic faith and they were well enough disposed, but once they heard of the prohibition of Copernicus etc., they had been so scandalised that he had not been able to get any further' (OG XIV, pp. 87–8).

46 OG XIII, p. 182.

47 Ibid.

would lead to disaster. Father Monster and the learned Caspar Schopp, who was very close to Faber and the Lincean circle, had declared themselves certain 'that this [the Copernican system] is not a matter of faith, nor should Holy Scripture be brought in at all'. It was an explicit invitation to retire to the calm harbour of astronomy as a purely hypothetical science. After meeting him in Rome, Galileo sent Cesi a gently ironical portrait of Father Monster, who was soon to return to the scene in a major role:

> He follows neither Ptolemy nor Copernicus, but is content in his rather hasty way to posit angels who, without any difficulty or complication, move the celestial bodies as they proceed, and he believes that we should be satisfied with this explanation.[48]

Had he been given the chance to read Federico's *De caeli unitate*, Galileo would certainly have refrained from making that remark: Father Monster's conception of the angelic agents that moved the stars and planets was the same as that expounded by Cesi in his epistolary tract to Cardinal Bellarmine!

In the middle of June, Galileo set out on his return journey with a heavy heart, this time without accepting Cesi's invitation to stop in Acquasparta. Before departing, he had received tangible signs of goodwill from the Pope and the Cardinal Nephew. They had both promptly complied with his request for formal attestations of appreciation of his Roman visit, entrusting Giovanni Ciampoli with the task of writing them. Their letters would prove useful to Galileo to show first of all to his natural Princes the consideration in which the Pope and Cardinal Francesco Barberini held him. More importantly, they would serve to silence his enemies in Florence who awaited his return to celebrate the failure, once more, of his Roman mission and to rekindle their manoeuvres to denigrate and denounce him.

Ciampoli did an excellent job. The letter that he wrote as *Segretario dei Brevi* on behalf of Urban VIII was sent to Grand Duke Ferdinand II on 8 June.

48 Ibid., p. 183. On Father Riccardi, the reasons for his nickname (Father Monster), his singular cosmological conception and his hostility to mixing research on natural philosophy with the interpretation of Holy Scripture, see Camerota 2004, p. 407. Among his contemporaries, Cesi, Patrizi, the young Kepler and Lagalla had all assigned to angelic intelligences or to the individual souls of the celestial bodies the task of keeping them in motion along their respective orbits. For almost all of them, as for Cesi, the idea of the movements of heavenly bodies guided by an intrinsic intelligent principle was closely linked with the denial of the solid spheres and the affirmation of the fluidity of the celestial spaces.

It praised the Medicean dynasty for its protection of Galileo, its 'beloved son', extolling his extraordinary merits as discoverer of new celestial phenomena. The Pope also lauded his 'devotion to piety' and manifested admiration for the skills with which he was endowed. He declared that he had been highly pleased with the lively Florentine language with which Galileo had taken part in 'learned disputations' in his presence. He therefore regretted the day of his return to Florence, where he wanted him to arrive accompanied by the affection and appreciation that the Pope felt for his sincere religious sentiments: 'we wished to pay honorific testimony to his virtue and piety'.[49] It was an important parting gift that helped to relieve the disappointment after a journey that the Tuscan scientist was bound to consider a failure.

However, Galileo had no intention of abandoning the cause. After all, Urban VIII had only declared the Copernican conception 'rash', while expressing his conviction that no one would ever be able to prove it. What might have appeared as a definitive sentence was interpreted by Galileo as an invitation to play his last card: to provide that conclusive and incontrovertible demonstration that would bring about full recognition of the truth of the Copernican system.

As he had done during his discussions with Cesi and Stelluti in Acquasparta, Galileo had placed at the centre of the meetings in Rome the illustration of his interpretation of the causes of the tides, based on his new science of motion. His correspondence with his Roman friends after his return to Florence contain many echoes of the interest aroused by those discussions. Galileo

49 Ibid., pp. 183–4. A letter in a similar vein was sent the same day by the Cardinal Nephew to the Grand Duchess Christine of Lorraine (ibid., p. 184) and to the Grand Duke. Francesco Barberini wanted Galileo 'to be highly honoured with all the graces and demonstrations that with the aid of my intercession he may expect to receive from the benignity of Your Highness' (ibid., p. 185). At the conclusion of his Roman visit of 1630 to obtain the *imprimatur* of the *Dialogue*, Galileo again took pains to make sure that his Florentine patrons were convinced of the successful outcome of his mission. He was also keen to dash the hopes of his Florentine enemies, who were ready to rejoice at the failure of yet another of his initiatives. Thus he asked the Tuscan Resident, Francesco Niccolini, to send the court a reassuring message regarding the success of his visit. As the ambassador and his wife had both formed a deep friendship with Galileo, they promptly complied with his request. The letter that Niccolini wrote to the Secretary of State, Andrea Cioli, on 20 June 1630 had the tone of a bulletin announcing an important victory: 'Galileo left here last Wednesday entirely satisfied with the outcome of his delicate business, as deserved by his virtue and most genteel manners. The Pope has been very pleased to see him and has paid him very many courtesies, as has Cardinal Barberini, who has invited the Tuscan scientist to dine with him; he has been appreciated and honoured also by the whole court as is his due' (OG XIV, p. 121).

immediately set to work to put down in writing the physical arguments that imposed acceptance of the heliocentric system. Many years after its conception, the process that would lead to the publication of the *Dialogue* was finally under way. It was an ambitious project that called for long and steady application. But Galileo was in a hurry, above all to ascertain how an attempt to show that Urban VIII was mistaken in his conviction that astronomical research could only produce hypothetical conclusions would be received. The project had been taking shape in his mind before his departure from Rome. At the end of May he had communicated his intention to Johann Faber, who had hastened to inform Cesi, underlining that Father Monster advised against 'digging up again' the anti-Copernican decree. Faber was, however, convinced that Galileo would not persist in following this dangerous route.[50] But from the apparently reassuring words of the German Lincean, it transpired that Galileo had not given up the defence of his own ideas: 'Galileo will print some small work in which he will express, although indirectly, his own opinions to avoid giving his enemies a point on which to attack him'.[51]

That 'small work' soon began to assume precise contours. Galileo conceived it as a reply to the *De situ et motu Terrae disputatio* that Francesco Ingoli had addressed to him eight years earlier. It was a curious decision, inspired by the design of demolishing the arguments of the opponents of Copernicus, thus underscoring, though without explicitly subscribing to them, the much more solid reasons of the author of *De revolutionibus*.

That he intended to put his mind to that project is confirmed by the request that he sent to Rome to Mario Guiducci to obtain for him at the earliest opportunity a copy of Ingoli's *Disputation*, as he was unable to find it among his own papers. Guiducci replied promptly, enthusiastically approving Galileo's decision to 'rid himself of people like that'.[52] From a letter of Ciampoli to Galileo of the following day we learn that the *Segretario dei Brevi* and, consequently, all the Roman Linceans knew about Galileo's decision to respond to Ingoli.[53]

Guiducci obtained a copy of the *Disputation* from Cesare Marsili and sent it to Galileo, together with Kepler's reply to Ingoli. He was no less diligent in satisfying, with the aid of Tommaso Rinuccini, Galileo's request for information about whether Grassi seriously intended to reply to the *Assayer*. The news sent

50 OG XIII, p. 181, letter of 1 June 1624.

51 Ibid.

52 Guiducci told Galileo that he would ask Cesare Marsili for a copy of the *Disputation*: 'And if Marsili does not have the text of Ingoli, I will ask Bartolino to give me his exemplar [...] and then send it on to you' (ibid., p. 186, letter of Guiducci to Galileo, 21 June 1624).

53 Ibid., pp. 187–8.

to Florence by Guiducci, who continued to maintain cordial relations with the Jesuit, appeared encouraging: Grassi had read and appreciated the *Discorso del flusso e reflusso del mare* (*Discourse on the ebb and flow of the sea*) that Galileo had dedicated to Orsini in 1616 and was impatient to consult the new work which Guiducci had told him the Tuscan scientist was preparing. Explicitly referring to the verdict that Bellarmine had given in 1615, Grassi had told Guiducci that, if the Galilean interpretation of the tides proved to be fully convincing, 'it would be necessary to interpret Holy Scripture in a different way from what has been done in those passages which mention the stability of the Earth or the motion of the heavens, and this according to the opinion of Cardinal Bellarmine [*ex sententia Card.lis Bellarminii*]'.[54] However, Guiducci was unable to satisfy Galileo's curiosity regarding Grassi's intentions to reply to the *Assayer* because the Jesuit had not made the slightest reference to cometary questions ('we did not go into things of the past').[55]

In his following letter to Galileo of 13 September, Guiducci reported a new meeting with Grassi in the Collegio Romano. The Father had asked him to explain some aspects of the Galilean conception of motion. He wanted to know in particular 'whether a body dropped perpendicularly from a ship's mast fell very close to the base of the mast, no matter whether the ship was motionless or moving'.[56] Guiducci had explained Galileo's arguments to him, based on the principle of the conservation of motion, but Grassi 'remained doubtful'. However, when, after a break to listen to a lecture, their discussion was resumed, 'he told me that he had understood that what I had told him was true'.[57] Guiducci's letter concluded with a triumphal affirmation: 'I think that he is not prejudiced against the motion of the Earth, once there are good reasons for that motion and the objections to it are removed'. Knowing that Galileo was working on his reply to Ingoli, Guiducci asked him whether, once completed, he might show it to Grassi: he was convinced that it would help the Jesuit to take the final step towards embracing the Copernican conception.

Perhaps it was the enthusiasm generated by his impression that the wind was changing and his hope of having made a breach in the certainties of Galileo's most hardened and dangerous enemy, that drove Guiducci to alleviate the tension of those weeks by including in his correspondence with Galileo on the evolution of Grassi's attitude the account of a curious conversation with the

54 Ibid., p. 203, letter from Guiducci to Galileo, 6 September 1624.

55 Ibid., p. 202.

56 Ibid., p. 205, letter of 13 September 1624.

57 Ibid., p. 206.

Cardinal Nephew. On 15 October 1624, having mentioned his own perplexities regarding the argument of the sieve used in Galileo's reply to Ingoli,[58] Guiducci informed him that in a meeting that he had had with the Pope's nephew a few days earlier the discussion had focused on carriage architecture. Francesco Barberini desired a new, but obviously not ordinary, carriage. It had to intuitively communicate the image of the eminent position of its owner. Guiducci, who had every interest in satisfying his whims, mentioned to the Cardinal Nephew that, years earlier, he had heard Galileo illustrate the project for a carriage made in such a way that it seemed to be supported by a single arc. As Francesco Barberini was fascinated by the idea, Guiducci asked Galileo for more details on that project as well as a drawing that would show the design of that innovative vehicle. The rest of the letter helps to understand why Galileo's friend attached so much importance to the project of the new carriage for Francesco Barberini, to the point of calling on the assistance of the Tuscan scientist, who was occupied at the time with controversies of a different momentum:

> The Cardinal had thought of featuring the Sun in the middle of the heavens [i.e. the canopy] of the carriage, surrounded by a snake to symbolise the year, with a motto taken from Horace, *Aliusque et idem*, and that the supporting rods should be like rays of the Sun, and the surrounding frame decorated with the twelve signs of the zodiac.[59]

Guiducci emphasised that such a design did not match the Ptolemaic system of the world, while it was perfectly coherent with the heliocentric system. He had deliberately avoided mentioning those reflections to the Cardinal, but, if the project were realised, he would find an appropriate occasion to jokingly remark to Francesco Barberini that 'the Congregation of the Index will prohibit that carriage and Ingoli will denounce you'.

This trifle ('frascheria'), as Guiducci called it, provides an eloquent picture of the inextricable tangle in the pontifical court between the severe examination of perilous theories, the demand for status symbols to publicly demonstrate a superior social position, the courtiers' careful attention to opportunities of attracting the sympathy and protection of the powerful, and, as in this case, the surreptitious circulation of ideas that were opposed by the ecclesiastical hierarchies.

---

58 Finocchiaro M.A. 1989, pp. 180–2 (OG VI, pp. 540–2).

59 OG XIII, p. 217. The words *Aliusque et idem* ('another and the same') are taken from the Horatian *Carmen Saeculare*, where they refer to the Sun. On Francesco Barberini's Copernican carriage, see Lattanzi 2014, pp. 545–7.

During Guiducci's frequent meetings with Grassi between the summer and the autumn of 1624, the question of the Jesuit's reply to the *Assayer* always remained in the background. Guiducci did not want to ask him directly whether he was working on it, and Grassi took care not to disclose his plans. It seems plausible, as Redondi maintained, that Grassi was trying to take advantage of Guiducci's ingenuousness to obtain precious information.[60] His aim, however, was not to acquire elements on which to base the denunciation of the heretical implications of Galileo's atomistic interpretation of sense perceptions that the Jesuit was to formulate in the *Ratio*. As confirmed by his curiosity about the Galilean interpretation of the tides and his new science of motion, he was trying to induce Guiducci to explicitly admit that Galileo persisted in his attempt to demonstrate the physical truth of the Copernican system.

During the same weeks, pressure from the Roman Linceans for Galileo to complete his treatise on the tides was growing. Faber, also on behalf of Father Monster, asked him to send that work to Rome as soon as possible, taking advantage of the opportunity to inform the Tuscan scientist that another author who had recently confronted the question of the ebb and flow of the sea had joined the company of the dead: he referred to Marco Antonio De Dominis, the Archbishop of Split, whose corpse, after his condemnation as a relapsed heretic by the Tribunal of the Inquisition, would be burnt in public three months after his death.[61]

---

60 Redondi 1983, pp. 227ff.

61 OG XIII, p. 207, letter of 14 September 1624. De Dominis' treatise on tides was published in Rome (De Dominis 1624). The macabre ceremony took place on 21 December 1624. Faber was perfectly informed about the affair because as physician to the Congregation of the Index and the Holy Office, he had been summoned, together with the famous physician Giulio Mancini, to conduct an autopsy on the corpse of De Dominis to remove any suspicion that he had been killed in prison. For the suspicions regarding the death of De Dominis and the participation of Faber in the autopsy, see Redondi 1983, pp. 135–48; see also De Renzi 1996, pp. 433–4 and 2008, p. 78. Faber provided Galileo with many details on the carrying out of his duty. It should be stressed that the account of the death and autopsy of De Dominis was inserted by Faber in the letter in which he sent Galileo the request from Father Monster for information about the treatise on the tides: 'The Archbishop of Split [De Dominis] [...] *ivit ad plures*. On the order of the Pope, Giulio Mancino came to my house and took me to the castle [Castel Sant'Angelo], asking me to assist when the corpse of this Archbishop was opened in the presence of a notary of the Holy Office. I believe it was done so that the world could not say that he had been poisoned'. Faber told Galileo that, although De Dominis had been dead for some time, he would be tried and sentenced. On Faber's relations with Giulio Mancini see De Renzi 2006.

In his letter of the end of September, Galileo informed Cesi that he had completed his response to Ingoli, promising to let him have it without delay. He added information on the project in which he was by now completely absorbed:

> I have now returned to the tides, and have arrived at this proposition: if the Earth is at rest, it is impossible for ebb and flood to follow; while if it moves with the motions that I assign to it, it is necessary that they follow with all the accidents that can be observed in them.[62]

His tone revealed that Galileo had once again fallen prey to enthusiasm. His excitement was produced by the conviction of having finally found the decisive proof that would oblige Urban VIII, Grassi and all adversaries of the freedom of thought in natural philosophy to recognise that it was no longer possible to oppose Holy Scripture to the conclusions obtained by mathematical demonstration and sense experience. The prophecy of Bellarmine, repeated, as Zollern had reported, by Urban VIII, that no one would ever provide incontrovertible physical proof of the motions of the Earth, was close to being proved wrong.

Galileo planned to achieve this ambitious goal through two texts, closely coordinated but staggered in time. The first, conceived as a reply to the objections to the motions of the Earth raised in the *Disputation* of Francesco Ingoli, would demonstrate that not only the world system of Ptolemy but also that of Tycho were erroneous, thus pointing out the risks facing the Holy Roman Church in condemning the only doctrine – that of Copernicus – based on plausible reasons and convincing demonstrations. The hostility of the ecclesiastical authorities to the heliocentric cosmology would be dealt a mortal blow – at least this is how Galileo imagined it – by a second text, in which he would present that physical proof of the motions of the Earth, based on the ebb and flow of the tides, which he had illustrated in Rome to authoritative exponents of the Church.[63]

That Galileo was working on both texts simultaneously is shown by the fact that many pages of the first, the reply to Ingoli's *Disputation*, would be proposed again without modification in the *Dialogue* of 1632.

62 OG XIII, p. 209.

63 In his *Relazione* of the trial of 1633, Giovan Francesco Buonamici, a diplomat related to Galileo who was in Rome at the moment of his condemnation, maintained that the Tuscan scientist had decided to publish the *Dialogue* as a consequence of the encouraging words of Cardinal Zollern after the latter's meeting with Urban VIII (ibid., XIX, p. 408).

Calamitous Novelties

Galileo completed the reply to Francesco Ingoli in the month of September. By the beginning of October the text was in the hands of Guiducci, who passed it on to Ciampoli. The latter asked Guiducci to invite Galileo to reformulate the excessively polemical tone of some passages. Ciampoli would suggest the emendations himself, for which Guiducci would have to wait before transmitting the reply to Ingoli.[64] At the end of October Cesi, who was in Acquasparta, had not yet received a copy of the work. While waiting to read it (he would only be able to do so at the end of January 1625), he expressed to Galileo his surprise at an initiative that he considered tardy and inopportune: 'I shall await with infinite desire the reply that you mention, wishing to know the reason for it, and how it has come up at this time'.[65] Clearly Cesi was by now out of the game. His surprise shows that, while Guiducci, Ciampoli and Faber knew about the decision to reply to Ingoli, he was in the dark about Galileo's strategy after the disappointing results of his visit to Rome in 1624. From his letter to Galileo of 26 October we learn that Guiducci was preparing the geometrical diagrams that were missing from the copy of the text sent to him by Galileo.[66] He asked Galileo whether he should allow his reply to 'fall into the hands of Ingoli himself'. While waiting to hear from him on this point, Guiducci was spreading the news that Galileo had replied to Ingoli among learned circles in Rome.[67] In the meantime he had received from Ciampoli indication of the passages requiring moderation:[68] emendations of form rather than substance, as he hastened to reassure Galileo. In the opening pages of the text sent to Rome, Galileo had probably stressed that, in maintaining the prohibition of the heliocentric system, the Church risked the scorn of the heretics, who were all Copernicans. It was the same argument that the Tuscan scientist had recommended that Cardinal Zollern should use in his meeting with the Pope. In Ciampoli's opinion, that argument would provoke negative reactions on the part of the ecclesiastical authorities, including the Pope. He therefore suggested not 'to change the sense, but to convey it with somewhat more specious wording that cannot awaken suspicion in persons of ill will'.[69] Galileo

64 Ibid., p. 215, letter of 15 October 1624.

65 Ibid., pp. 219–20, letter of 26 October 1624.

66 Ibid., p. 221.

67 Ibid., p. 223, letter to Galileo, 2 November 1624.

68 Ibid., p. 226.

69 Ibid.

accepted the suggestion and rewrote the paragraph,[70] avoiding mention of the Protestant derision of the Catholic Church's hostility towards Copernicus.[71]

Before consigning the work to Ingoli, Guiducci considered it wise to wait until Ciampoli had informed the most authoritative members of the clergy of the reasons that lay behind Galileo's decision to reply, so that 'if Ingoli wanted to stir up any trouble, he would get stuck'.[72]

A month later, the news of Galileo's text had gone the rounds in Rome. Ingoli too had heard about it. As Guiducci wrote to the Tuscan scientist, 'Ingoli knows that I have this text and has asked me to show him it'.[73] He had refused. Ingoli would receive it from the author 'in due time'. Guiducci added an interesting detail: it was rumoured in Rome that the reply to Ingoli was not being distributed because Ciampoli was not convinced by Galileo's arguments. The insinuation had been spread maliciously: Ciampoli did not miss any opportunity to denounce to the Pope Ingoli's presumption 'in venturing to write what he does not understand, to the detriment of the reputation of these Congregations'.[74]

During those months Guiducci continued to frequent the Collegio Romano assiduously, where he often met Father Grassi. In his letter of 30 November, Guiducci informed Galileo that he had learnt from the Jesuit that he was preparing his rejoinder to the *Assayer*. To justify his decision, Grassi had said that 'he was forced to write and had done so against his will'.[75] For Guiducci, convinced of having contributed to establish cordial relations between Grassi and Galileo, the news came like a cold shower. He was profoundly embarrassed at having been duped by the seemingly friendly attitude of the Jesuit and for having repeatedly assured Galileo that peace had been made between them. To mitigate his regret and sense of guilt, he would take it upon himself to teach Father Grassi the lesson he deserved. Galileo asked him to restrain himself for the time being; in the event, after reading Grassi's work, he himself would respond.[76] On discovering Grassi's deception, Guiducci changed his mind: he would not show the Father the 'reply to Ingoli because I no longer want to have dealings with him'.[77]

---

70 Ibid., p. 242, letter of Guiducci to Galileo, 21 December 1624.

71 Finocchiaro M.A. 1989, pp. 155–6 (OG VI, p. 511).

72 OG XIII, p. 226.

73 Ibid., p. 230, letter of 22 November 1624.

74 Ibid.

75 Ibid., p. 232.

76 Ibid., p. 241, letter of Guiducci to Galileo, 21 December 1624.

77 Ibid., p. 234, letter to Galileo, 30 November 1624.

In the last months of 1624, Cesi interrupted his retreat in Acquasparta and returned to Rome, where he was to stay until the beginning of 1626 in order to participate in the Holy Year proclaimed by Urban VIII. His return to the Eternal City enabled the Prince to discover at first hand how very delicate the situation had become. The Jesuits were emboldened, as shown by Grassi's decision, with the approval of the hierarchies of the Order, to reply to the *Assayer*, as well as by the firm position adopted against philosophical novelties in the opening lecture delivered by Grassi's fellow Jesuit, Father Spinola, at the Collegio Romano at the beginning of November 1624. Spinola had vigorously condemned the audacity of those who wanted to evade the supreme authority of Holy Scripture or claimed to accommodate its meaning to their own opinions.[78]

Spinola's lecture is known through the detailed refutation of his arguments that Guiducci sent to Galileo. He transcribed the Jesuit's entire text, deconstructing it in 82 paragraphs, to each of which he added his personal remarks.[79] The incisive and caustic style of numerous passages of these severe comments leads one to suspect that Galileo had a hand in writing them.[80] Spinola's lecture left no doubt regarding the possibility of any reform of traditional knowledge being tolerated. That of the Jesuit was a harsh and menacing condemnation of the presumption and vanity of those who wanted to introduce new methods and ideas. Even if Spinola failed to mention the names of the wicked contemporary writers who dared to challenge the authority of tradition, Guiducci, who attended the lecture, could not fail to catch the pointed reference to the execrable positions adopted by Galileo: 'Whoever disagrees with common consent in philosophy – Spinola denounced – easily lapses from the shared opinion of the Fathers in

78 Guiducci hastened to inform Galileo of Spinola's violent denunciation: 'Last Tuesday, as every year, an opening lecture, or rather an extremely vehement and violent invective, against the followers of new opinions that are contrary to the Peripatetics was delivered' (ibid., pp. 226–7, letter of Guiducci to Galileo, 8 November 1626). Galileo had at first thought of replying to Father Spinola's insinuations: 'having seen the opening lecture given this year by the Professor of Philosophy in the Collegio, I thought of introducing it opportunely in a dialogue, to settle accounts with him in detail; but considering its insipidity and the enormous errors with which it is riddled, I don't know what I will do [...]' (ibid., p. 260, letter of Galileo to Cesi of 17 March 1625).

79 Father Fabio Spinola's speech was not published, nor is any manuscript version of it extant. Favaro discovered a copy of the Jesuit's text (Favaro 1983, III, pp. 1411–74), densely annotated by Mario Guiducci, among the Galilean Manuscripts in the Biblioteca Nazionale Centrale in Florence (Ms. Gal. 85, cc. 3r-38r). He published the manuscript with Guiducci's comments as an appendix to his biography of the Florentine advocate in the series 'Amici e corrispondenti di Galileo' (ibid., pp. 1451–74).

80 Favaro expressed the same suspicion (ibid., p. 1434).

theology too'.[81] It was this reckless search for novelties, the Jesuit continued, that had produced the tragic division of religious consciences responsible for the bloodshed throughout Europe: 'Would that so many seeds of heresy, which have sprung from one single fount and in their bestial instigations have plagued the Christian republic, had not caused the shedding of so much blood and the loss of so many lives'.[82] Spinola spared no pains in listing the dramatic consequences that the irresponsible departure from tradition had caused: the destruction of flourishing cities, a climate of hatred between nations, rivers of innocent blood, the devastation of monasteries and the unspeakable acts of violence perpetrated against consecrated virgins and monks, the brazen rebellion against the magisterial functions of pastors and the supreme authority of the Pope. The war that raged among Christian Princes and that was tainting Europe with blood was the direct consequence – the Jesuit concluded – of the rebellion against the principle of authority.[83]

Guiducci's comment on this 'long tirade [...] full of more ill-will than truth' seems to evoke the biting sarcasm of Galileo and his unrivalled capacity to ridicule the pronouncements of his adversaries:

> If it were true that heresies owed their origin solely to dissenting from the common consensus of the writers on questions of natural philosophy (a common consensus which, as a matter of fact, coincides with the dogmas of the Peripatetics), the vigilant pastors of the human flock with their supreme authority would have foreseen and prohibited any contradiction of that doctrine long ago. But we see that everyone is free to follow Plato and the other philosophers who are contrary to Aristotle. Therefore, your pathetic lamentations are not based on truth, and it is evident that, lacking strong arguments against the new opinions in philosophy, you have combined the opinions and interests of Aristotle with those of theology and faith.[84]

It has until now escaped notice that, in writing these passages brimming with severe denunciations, Spinola had on his desk the text of another staunch defender of the theory of the irreparable damage caused by the promoters of novelties: the *Replicationes* of Francesco Ingoli to the *Responsio* in which

81 Ibid., p. 1473.
82 Ibid.
83 Cf. Redondi 1983, pp. 162–5.
84 Favaro 1983, III, p. 1473.

Kepler had refuted the theses contained in the prelate's *De situ et motu Terrae disputatio* of 1616.[85] Galileo had regarded the *Replicationes*, although formally conceived as a reply to the censures of the Imperial astronomer, as an explicit attack on his own person. The sections of the *Replicationes* from which Spinola had drawn inspiration were the final ones ('On the first theological response against the motion of the Earth').[86] Having defined as 'most dangerous' the freedom to interpret 'against the consensus of the Fathers' those scriptural passages in which arguments of faith were not at stake, Ingoli had denounced the dramatic consequences of the irresponsible conduct of those who had rebelled against the absolute truth of the sacred texts taken literally:

> Would that the European provinces of the most illustrious Germany had rejected this proposition at the start of the outburst of the Lutheran heresy, for it would not have abandoned its Mother, the Holy Roman Church, causing such a loss of human lives, nor would it have led astray and corrupted the faith and the apostolic doctrine.[87]

It was above all Ingoli's passionate denunciation of the dramatic consequences of the rebellion against the supreme authority of the Roman Church that had struck Father Spinola, leading him to take it as his model:

> I can scarcely hold back the tears – Ingoli had written – when I recall and consider the most holy men of the German Church, famous for their martyrdom and commitment, their pious deeds, the writings of such learned men, and the unity with the Roman faith of Saint Peter that has been lost.

The striking analogies also in literary style and terminology[88] between the texts of Ingoli and Spinola were not fortuitous. They were signs of the formation of a compact front of resolute opponents of the theses, not only of Kepler (a soul by now lost but safe from the anger of Rome), but especially of Galileo, above whose head hung the sword of Damocles of the 1616 decree and the dire consequences that would result from obstinately continuing to ignore its precepts.

85 See above, pp. 239ff.

86 Cf. Bucciantini 1995, pp. 203–5.

87 Ibid., p. 204.

88 Both Spinola and Ingoli introduced their denunciation of the calamitous effects of the quest for novelties with the exhortative preposition *utinam*. Moreover, Ingoli's expression *cum tanta hominum pernicie* (Bucciantini 1995, p. 204) was reproduced almost literally by Spinola: *tanta cum animorum pernicie* (Favaro 1983, III, p. 1473).

That Ingoli's *Replicationes* lay on the desk not only of Spinola, but also on that of Guiducci, and that the Florentine advocate read the texts of the Monsignor and of the Jesuit in parallel, is shown by the reference to Bellarmine in his (or Galileo's?) gloss on the paragraph in which the Jesuit blamed the drama of the bloody religious wars in Europe on the partisans of novelties:

> I am all the more confirmed in this opinion [the gratuity of the accusation of having provoked the wars of religion] since, having glanced at the controversies of Cardinal Bellarmine, they do not contain any passage stating that today's heresies against the Catholic faith are derived from new opinions in philosophy.[89]

Bellarmine was not mentioned in the text of Spinola to which this gloss was appended, while Ingoli had inserted an explicit reference to the authority of the recently deceased Cardinal in the conclusion of his violent attack on those who claimed the freedom of thought in natural philosophy and, above all, in the interpretation of Holy Scripture:

> [...] as copiously illustrated – Ingoli had affirmed – in the writings of the Fathers and as Cardinal Bellarmine stated in his book *De verbo Dei*, in book 4 of *De Romano Pontifice* at the beginning, and in the book *De Concilijs*.[90]

The new dynamics of the Roman context were bound to seriously alarm Prince Cesi on his return to the Eternal City after a lengthy absence. In addition to the interventions of Ingoli and Father Spinola, the general situation caused him concern and called for the utmost caution. The internal struggle between the hierarchies of the Church had reached fever pitch. The pro-French party of the Barberini faced the dogged opposition of the pro-Spanish Cardinals led by Cardinal Borgia. And Europe was inflamed by violent conflicts to which no end appeared in sight.

This climate of violent opposition between the partisans of traditional and of new knowledge is confirmed by the letter to Galileo in which Guiducci sought to uplift the downcast spirits of the Tuscan scientist after the attack of Father Spinola. He wrote that Giuliano Fabbrizi, nicknamed the 'poet of the moment' (*poeta del caso*) for his remarkable virtuosity at improvisation, had

---

89 Ibid.

90 Bucciantini 1995, p. 204.

launched an assault against the defenders of tradition and the supporters of Aristotle during a meeting of the Accademia dei Desiosi held in the palace of the influential Cardinal Maurice of Savoy, in the presence of the Cardinal Nephew and other authoritative representatives of the papal court.[91] In his speech, entitled *Dell'ambitione del letterato*, Fabbrizi had condemned those who 'defer so much to the authority of the writers and put more trust in what they have read than in what they have always seen with their own eyes'.[92] In the printed version of his lecture, Fabbrizi included an explicit condemnation of the hostility shown by the guardians of tradition towards the celestial discoveries of Galileo (though without naming him) and his telescope.[93] The 'poet of the moment' also aimed his darts at the philosopher who had refused to look through the lens of the telescope (an obvious reference to Cesare Cremonini):

> And I know that when the telescope discovered phenomena on the Moon that were unknown to antiquity, there was a famous philosopher who, refusing to put his eye to the lens, quoted a text of Plutarch, preferring to dream of the stars in books rather than in the sky.[94]

The Turncoat

As mentioned, Cesi was unable to read Galileo's reply to Ingoli until the beginning of January 1625. On reading that text his initial surprise at Galileo's late decision gave way to profound concern about its consequences. After carefully analysing the situation with Guiducci and Ciampoli, he arrived at the conclusion that to give Galileo's reply to Ingoli and to circulate it would provoke dangerous recriminations. Nonetheless, reluctant to play once again the role of warning censor and fearing that Galileo would pay no heed to his words, he hesitated in manifesting his perplexities. Thus, rather than addressing his reservations directly to the Tuscan scientist, he entrusted Guiducci, recently nominated for induction into the Academy, with the task of passing on to Florence the decision not to deliver to Ingoli Galileo's reply to the *Disputation*. Guiducci should remind Galileo that the rules of conduct of the Linceans prescribed

91 OG XIII, p. 253, letter of 8 February 1625.

92 On this episode and on the Accademia dei Desiosi, see Favaro 1983, III, pp. 1431–40; see also Redondi 1983, pp. 87–98 and Favino 2009.

93 Mascardi 1630, pp. 71–9.

94 Cited in Favaro 1983, III, p. 1433.

caution and patience, especially in this instance – Cesi added – given that the *Assayer* had been denounced by a 'pious person' on the suspicion of adhesion of its author to the Copernican doctrine prohibited by the decree of 1616.[95]

As we have seen, thanks to the initiative of an unnamed Cardinal, the evaluation of the denunciation was assigned to Father Guevara.[96] His verdict was that the work did not contain elements to justify the opening of an inquisitorial proceeding against its author. In Guiducci's opinion, the absence from Rome at that moment of an influential protector, the Cardinal Nephew, who was engaged in a delicate political mission in France, constituted a further argument against delivering Galileo's reply to Ingoli. There was reason to fear that, in the event of a new denunciation, without the support of Francesco Barberini the outcome might not be so favourable.

Evidently Cesi had strong reasons to consider a new move by the 'pious person' not unlikely:

> The opinion of Copernicus is expressly defended in the reply to Ingoli, and although you say openly that it is demonstrated by a superior light to be false, nevertheless the malicious will not believe it and will stir up trouble again.

Both Cesi and Guiducci regarded the situation as all the more delicate due to the opposition of 'another prominent person', who – as Guiducci took care to underline – had adhered to the heliocentric hypothesis in the past:

> For all these reasons it has seemed best [...] to wait and leave the question undisturbed for a while, rather than to stir it up with attacks on the adversaries and then have to defend yourself from those who are able to deliver harsh blows.[97]

Cesi had understood that the insinuations against Galileo of Grassi, Spinola, Ingoli, the 'pious person' and their powerful patrons were not isolated episodes, but the consequence of a concerted strategy. A few days after the

95 OG XIII, pp. 265–6, letter of 18 April 1625. On the identity of the 'pious person' see above, pp. 289ff.

96 Giovanni Guevara was on cordial terms with Cesi. He would be one of the few to receive a copy of the *Apiarium* from Federico: 'I saw Prince Cesi in Perugia', Guevara wrote to Galileo on 1 November 1626, 'who gave me his *Apiarium*' (ibid., p. 342). On Guevara and his role in the Galileo affair, see Favino 2010.

97 Ibid., pp. 265–6.

dispatch of Guiducci's letter to Florence, Cesi wrote in person to Galileo. He had heard from Ciampoli that the Tuscan scientist intended to return to Rome for the Holy Year. Federico was delighted at the prospect of seeing him again, but their friendship obliged him to warn against that project because of the 'unfavourable present conjuncture'. Hardly a year had passed: the 'marvellous conjuncture' after having rapidly declined to 'good',[98] had now become 'unfavourable'.

The identity of that 'prominent person' who caused the anxiety of Guiducci and Cesi naturally arouses curiosity. It is reasonable to suppose that he was a Cardinal close to the Collegio Romano, probably allied with the pro-Spanish party opposed to the pro-French policy of the Barberini. This seems to be confirmed by Guiducci's reference to Urban VIII's annoyance at the situation. These characteristics would fit Cardinal Ludovico Ludovisi, well-known to Cesi, nephew of Gregory XV and, from 1621, Prefect of the Congregation *De propaganda fide*.[99] As a result of his open siding with the pro-Spanish faction, he would be banished from Rome in 1632 by Urban VIII. Cardinal Ludovisi was also closely connected with the Society of Jesus, and in particular with Father Grassi. In 1626 he would invest an enormous sum of money in the construction of the Church of St Ignatius in Rome, a triumph of the power and artistic conception of the Society of Jesus, entrusting to Father Grassi the task of designing that grandiose building.[100] It was certainly Ludovisi, in his role as Prefect *de propaganda fide*, who invited Ingoli to denounce the close link between the search for novelties at any price and the tragic explosion of a bloody rebellion against the universal magisterium of the Pope.

While identification of the 'prominent person' as the nephew of Pope Gregory XV has many convincing aspects, there is a detail that clashes with the portrait sketched by Guiducci: there is no evidence that Cardinal Ludovisi had ever defended the Copernican doctrine, even if we cannot exclude the possibility that he did so in his discussions with Ciampoli, with whom he was on familiar terms.[101]

An even more plausible candidate for the role of the 'prominent person' is Monsignor Giovan Battista Agucchi.[102] Like Ludovisi, he was a supporter of the

98 See above, p. 280.

99 See Pastor 1931, XIII (1931), pp. 1009–15.

100 Cf. Redondi 1983, pp. 148–51.

101 Cf. Pagano 2009, p. XXIV, n. 30: 'After Cardinal Alessandro Ludovisi had been elected Pope taking the name of Gregory XV, the career of Ciampoli, who knew his nephew Ludovico Ludovisi well, made easy progress […]'.

102 On Agucchi see Favaro 1983, I, pp. 373–92.

Spanish party in the Vatican, close to the Jesuits, and a longstanding friend of Ingoli. During the brief papacy of Gregory XV, Ludovico Ludovisi and Agucchi both held prestigious offices, the first as Cardinal Nephew, the second as Segretario di Stato e dei Brevi. They shared also a passion for art and antiquities. Unlike Ludovisi, however, Agucchi had studied astronomy, at least until he assumed positions of high responsibility in the Curia. He had corresponded with Galileo after their meeting during the Tuscan scientist's visit to Rome in 1611. Agucchi could certainly appear as a 'turncoat', as Guiducci had defined him in his letter to Galileo. He had initially shown a great interest in the celestial discoveries and particular appreciation of the writings on hydrostatics of Galileo, with whom he had collaborated in the observations of the sunspots. It will be recalled that in July 1613 Agucchi had dissuaded Galileo from embarking on the interpretation of Holy Scripture against the common consent of the Church Fathers.[103] When scrutinised more closely, both that letter and the earlier one, dated 1 December 1612, seem to suggest that the 'prominent person' mentioned by Guiducci might be identified with Agucchi. In the first of his two letters, after reading the manuscript of Galileo's third letter on the sunspots, Agucchi had expressed great enthusiasm, confessing that he was now inclined to take into serious consideration the Copernican system. He went so far as to declare that he found the hypothesis of the continuous physical interactions between the Sun and the planets that revolved around it by no means improbable:

> As for the possible cause of the sunspots [...], it seems to me that you open the eyes of the mind to speculate on them when you suggest that the planets, *which all revolve around the Sun* above its tropics, could draw their matter from its body. And certainly, if we observe what happens in our tropics, we see that, contrary to the opinion of the ancients, where the Sun passes perpendicularly there are perpetual clouds and rains, and the winter prevails there because of the great strength of the Sun to attract vapours from the sea and the earth; and this is the only reason for the flooding of the Nile and of the Niger in Africa, as Fracastoro demonstrates with great learning. In the same way, although the planets do not have the same properties and strength as the Sun, nevertheless they are greater in number and some of them are closer to the Sun than the Sun is to the Earth. Moreover, they attract a smaller quantity of matter in comparison with the size of the Sun, than the Sun does from the Earth, which in that area struck by the rays of the Sun is almost constantly covered

103 See above, pp. 158–60.

> with clouds. From which one may conjecture that the planets raise from the surface of the Sun the matter of the sunspots, which is then dissolved by its rays, rather than that the Sun itself attracts that matter from the planets and from the Earth, which are so distant from it; because during such a long journey it would be completely destroyed by the Sun's brightness before reaching it.[104]

Seven months later, in July 1613, Agucchi admitted that Galileo's Copernican arguments had shaken his certainty about the centrality and immobility of the Earth:

> I have never gone into it [the Copernican system] thoroughly, not having put too much faith in its truthfulness – he wrote to the Tuscan scientist. Your authority alone, ever since I have known that you shared a similar opinion, prompted me to incline in that direction [...].[105]

However, his trust in the extraordinary talent of Galileo had not been sufficient to induce him to abandon the traditional system of the world:

> I remain extremely dubious – he stated – about it [the Copernican system] and tend rather towards the opposite conclusion [...].

This precious autobiographical testimony, and above all the confession that he had been tempted by the Copernican hypothesis, were certainly known to Cesi. It is therefore reasonable to suppose that the Prince was referring to Agucchi when he informed Guiducci of the hostility of a 'prominent person' who had once given credit to the Copernican cosmology.

The confession of a dangerous temptation, rejected in the name of obedience to the supreme authority of the Church, marked the end of the epistolary exchange between Agucchi and Galileo. The same was to happen a few years later with Luca Valerio,[106] who broke all communication with the Tuscan scientist after the publication of the anti-Copernican decree.

---

104 OG XI, pp. 442–3, emphasis added. For the work of Fracastoro on the flooding of the Nile, see Fracastoro 1563, cc. 244–8.

105 Ibid., pp. 532–3, letter of 13 July 1613.

106 It was Valerio who presented Agucchi to Galileo during the Tuscan scientist's visit to Rome in 1611: 'Once only I happened to meet you – Agucchi wrote to Galileo – in your residence on Trinità de' Monti when I was introduced by Luca Valerio' (ibid., p. 265, letter from Rome of 9 September 1611).

When Guiducci informed Galileo of the hostility of the 'prominent person', Agucchi was no longer in Rome. He had moved to Venice at the end of 1624 to take up the position of Papal Legate to the Venetian Republic. He had been sent there by Urban VIII, who was intent on clearing the field of Gregory XV's protégés. The position of *Segretario dei Brevi* had passed to Agucchi's assistant, Giovanni Ciampoli. Leaving the position in the Curia for that of Legate in Venice meant a considerable decline in prestige and influence.

However, Agucchi was still perfectly capable of making himself heard in Rome and of exploiting the influence of his authoritative friends and the anti-Barberini circles to which he belonged.

Elephants and Mites

Whoever the elusive 'prominent person' may have been, it was somebody who was troubled by the growth of reputation and influence of Galileo and the Linceans in the new climate of Barberini Rome. The Copernican question constituted an important element in the complex political equilibriums of that scenario. Cesi knew that the game had become dangerous: Galileo and the Academy had to keep out of it at all costs. The Tuscan scientist would have to shelve his reply to Ingoli for the time being. That document appeared explosive to Cesi. It was, above all, explicitly Copernican. It is enough to cite a single passage from Galileo's reply to Ingoli to realize that the Prince was perfectly right on this score:

> If any place in the world is to be called its centre, that is the centre of celestial revolutions [...]; everyone who is competent in this subject knows that it is the Sun rather than the Earth which is found therein.[107]

The repeated declarations of submission to the supreme authority of Holy Scripture were bound to appear purely instrumental to the reader. Cesi was not wrong on this score either. It was evident that the author wanted his deliberately ostentatious expressions of obedience to the precepts of the Church to appear as instrumental simulation. Notwithstanding the blows that he had

107 Finocchiaro M.A. 1989, p. 179 (OG VI, p. 539). Galileo's reply to Ingoli was republished with introduction, notes and commentary by Bucciantini and Camerota in Galilei 2009, pp. 189–246.

received, he was still confident that he would eventually succeed in convincing the ecclesiastical authorities to revise their positions on questions bearing on the real structure of the universe and on the relation between the Book of Scripture and the Book of Nature. He had not budged an inch from the positions assumed before the emission of the 1616 decree.

Prince Cesi, on the other hand, continued to view that decree as a magisterial decision that could be neither changed nor revoked. Besides, because of his social position, the education he had received and the role that he played on the Roman scene, he had not understood during the clashes that had ended with the prohibition of *De revolutionibus,* and failed to understand now, why the affirmation of the physical truth of Copernicanism was of such decisive importance for Galileo.

It was not only Galileo's elusive attitude towards an official injunction of the Church that led Cesi to view the reply to Ingoli as extremely perilous. In that text, the Tuscan scientist took to their extreme consequences the implications of the new system of the world, calling into question the traditional paradigms in which the relations between man, nature and divinity had been solidly framed. Anyone would be bound to clearly perceive those implications hammered home by Galileo with obsessive insistence.

The decision to found on the Copernican vision not only a new cosmology but also a new anthropology had already surfaced, at least to attentive readers, in his work on sunspots, as well as in the *Copernican letters* and the *Assayer.* It had now become almost a programmatic manifesto: the necessary precondition for a full understanding of the nature of man, for denouncing the pathologies of his instinctive ways of thinking, and for demonstrating, once and for all, the method to be followed in the study of nature. Galileo had realised long before that to favour the affirmation of the Copernican vision it did not suffice to replace the dynamics of Aristotle with a science of motion totally new in its principles and laws. To achieve that goal, it was equally important to remove the deep-rooted conviction that the world had been created by God to the measure of man's capacity to understand it and to reserve a privileged position for human beings. This was the most serious obstacle to the design that he had conceived. That it was a particularly delicate and difficult task was brought home to him by the perplexity of a person who held him in the greatest esteem and considered him a fundamental ally in the battle for the renewal of traditional knowledge: Federico Cesi, who had never concealed his hesitation to follow Galileo along that route. Nor was Federico alone; his closest associates were of the same opinion. In fact, among the Linceans, Fabio Colonna was the only one who adhered to Galileo's radical conception of the indifference of nature towards man's needs and expectations.[108]

We have deliberately defined the reply to Ingoli, as well as the pages of the *Dialogue* in which Galileo proposed many of its passages again, as a manifesto of a new anthropology. The pages in that text which Galileo regarded as the most important were not those in which he refuted the fragile astronomy of Ingoli, but those in which he demonstrated that the arguments used by Ingoli to condemn the Copernican system were founded on the preconception of the physical, intellectual and moral centrality of man in the universe. Galileo rejected that view in no uncertain terms. There is not – he wrote – a single Sun, our Sun, in the universe, on which we depend for light and heat, essential elements for the life of all creatures: all fixed stars are Suns.[109] The refusal of Copernicus on the grounds of the lack of proportion that would result from the enormous expansion of the universe in his system was branded by Galileo as a paralogism born once again of the presumption of the privileged position and role of mankind in the universe: the existence of whatever does not correspond to our instinctive idea of proportion must be denied. Galileo replied that we do not know whether the universe is finite or infinite. If it were infinite, it would be impossible to derive rules of proportion from it:

> If the universe were in fact finite and bounded, what reason would you have for saying that the stellar sphere was disproportionate to the Earth's annual orbit [...]? If this is a valid reason, disproportionate are all those things of the same kind which are such that one is this many or more times greater than another. Thus, since there are fish so small that a whale can hold a greater number of them, and since an elephant can hold a greater number of mites, therefore whales and mites would be disproportionate animals; hence in your opinion they do not exist because such disproportions are not allowed by nature.[110]

On the other hand, he continued, we cannot absolutely say whether the world is large or small based on our sense perceptions alone:

108 Cf. above, pp. 209–10. Olmi has rightly emphasised that among the Linceans Colonna was the closest to the interests and methods of Galileo, for whom he constantly manifested his sincere, although never passive, veneration. See Olmi 1987, pp. 53–4.

109 'It is most reasonable to say each fixed star is a Sun' (Finocchiaro M.A. 1989, p. 171: OG VI, p. 530).

110 Finocchiaro M.A. 1989, p. 171 (OG VI, pp. 529–30).

> It is very large compared to the world of earth-worms and the like, which have only the sense of touch to measure it and so cannot regard it as greater than the space they occupy; and I do not find it repugnant that the world perceived by our senses, compared to the universe, may be as small as the world of worms compared to ours.[111]

The need to transform the absolute concept of proportion used by Ingoli into the simple relation of magnitude between different entities led Galileo to conclude peremptorily that the errors made when pronouncing judgements on nature and the cosmos depend on our expectations:

> To consider such an immensity excessive is an effect of our imagination, not a flaw in nature.[112]

The Tuscan scientist continuously mortified his interlocutor, demonstrating that his paralogisms were a consequence of replacing reason with imagination:

> These discussions being based on an idle imagination, they should not, therefore, be put forth in the study of serious and important questions [...].[113]

Proportion, commensurability, elegance; all were mental constructs that impose no obligations on nature, which could have adopted very different criteria from those that seem more perfect to us:

> Who can guarantee to us that planetary motions are not all mutually incommensurable and hence liable, indeed bound, to eternal emendations since we can only handle them as commensurable?[114]

In the arguments against Ingoli's claim that in her works, nature respects our criteria of elegance, economy and proportionality, we perceive the echo of the reflections that Galileo had deployed many years earlier in his long letter to

111 Finocchiaro M.A. 1989, p. 172 (OG VI, p. 530).

112 Ibid.

113 Finocchiaro M.A. 1989, p. 173 (OG VI, p. 531).

114 Finocchiaro M.A. 1989, p. 175 (OG VI, p. 534).

Gallanzone Gallanzoni,[115] as well as in his reply to Cesi's questions about the Copernican system.

As he developed his reasonings, Galileo's tone grew increasingly direct and peremptory:

> Here, before I go any further, I must tell you that in natural phenomena human authority is worthless. Like a lawyer, you seem to capitalize on it; but nature, Dear Sir, makes fun of constitutions and decrees of Princes, Emperors, and Monarchs, and at their request it would not change one iota of its laws and statutes.[116]

Nature operates according to the laws that she judges most opportune and practical. The scrupulous observation of natural phenomena and mathematical reasoning are the only way to understand the operations of nature, even those that appear to us inconvenient, disproportionate or arbitrary.

Ingoli's conviction that Ptolemy and Tycho describe the true structure of the cosmos because their systems perfectly match our criteria of proportion, elegance and symmetry must not be considered an element of proof. The method had to be reversed: the works of nature are always proportionate, elegant and symmetrical, even when they do not appear to be such to our intellect and imagination. Man – Galileo concluded – must not 'adjust the architecture to the edifice,' as Ingoli did, but rather build 'in accordance with the precepts of architecture'.[117] In the case in question – the evaluation of the plausibility of the Copernican vision – this meant making every effort to understand without preconceptions how the world is made, and only after arriving at a concept that satisfies all observational data, concluding that it is the most proportionate, elegant and symmetrical of constructions:

> You say the following: Copernicus attributes motion to all the luminous parts of the sky, namely to the planets, and he denies it to the Sun, which is the most luminous of all, in order to attribute it to the Earth, which is an opaque and dark body; but nature, which is judicious in all her works, does not do these things. You should rearrange this, Signor Ingoli, and

115 OG XI, p. 149, letter to Gallanzone Gallanzoni, 16 July 1611: 'I judge serious and frequent the error of the many who wish to make their own knowledge and understanding the measure of the understanding and knowledge of God, taking only what they consider as perfect to be perfect'.

116 Finocchiaro M.A. 1989, p. 178 (OG VI, p. 538).

117 Finocchiaro M.A. 1989, p. 195 (OG VI, p. 559).

> say: Copernicus attributes rest to all the luminous parts of the world, which are the fixed stars and the Sun, and he gives motion to all the opaque and dark ones, which are the planets and the Earth, the latter being like the planets; this is what one would expect from nature, which is very judicious in all her works.[118]

The letter to Ingoli marked a turning-point in the Galileo affair. It concluded a long series of battles undertaken by the Tuscan scientist to get recognition for the freedom of thought not only in natural philosophy. By the end of 1625 he had become convinced of the impracticality of his project, but his awareness of the futility of his attempts to convince the Church to revoke the anti-Copernican decree of 1616 did not lead him to withdraw from the field. In the following years he was to put all of his energy into the composition of the *Dialogue concerning the two chief systems of the world*, a long cherished work of synthesis of his research on mechanics, natural philosophy and cosmology. By now he had abandoned his project of composing that work in the form of an organic treatise which, as suggested by the title initially conceived for it, would have presented the incontrovertible physical proof of the motions of the Earth provided by the ebb and flow of the sea.[119] It is no coincidence that Galileo abandoned the treatise format at the moment when he realized that the 'marvellous conjuncture' had been an illusion. That sad realisation imposed a change of strategy: instead of highlighting in discursive style the strength of the arguments in favour of the Copernican system, he chose to stage a dialogue between advocates of the new and of the traditional cosmology, with a third actor in the role of impartial participant. Cesi and the Linceans closest to him had proposed the form of a dialogue between supporters of different opinions on several occasions, including for the composition of the *Assayer*. But for a long time Galileo had rejected that suggestion, fearing that it might appear as his retreat from the battle for the recognition of the truth of the Copernican vision.

That the year 1625 marked an important turning-point not only for Galileo but also for Cesi and the Linceans is confirmed by the sudden decline in the correspondence of Galileo with Cesi, Ciampoli, Faber and Stelluti. In the five

118 Finocchiaro M.A. 1989, p. 197 (OG VI, pp. 560–1).

119 Galileo had announced on many occasions his project to write a work on the systems of the world labelling it as *Discourse on the ebb and flow of the sea*. Cf., for instance, the letter of Galileo to Cesare Marsili, 7 December 1624: 'I am making progress with my Dialogue on the ebb and flow […]' (OG XIII, p. 236). On the process that led Galileo to decide to compose his work on the main world systems in the form of a dialogue, see the introduction by Ottavio Besomi and Mario Helbing in Galilei 1998, I, pp. 3–42.

remaining years of his life, Cesi sent Galileo a total of only five letters. Ciampoli's negligence was even greater – a single letter – while Faber sent none at all. The most intense contacts with Galileo were maintained by Stelluti – eight letters, the last of which, dated 2 August 1630, informed the Tuscan scientist of the death of the Prince on the previous day.[120]

Galileo and Cesi would not meet one another again. The Tuscan scientist was unable to visit the dying Prince in the spring of 1630, when he returned to Rome in an attempt to obtain the license to print the *Dialogue*.

It has also to be stressed that between 1626 and 1630, practically no official meeting of the Academy was held, while only three new members, none of them skilled in mathematics, natural philosophy or cosmology, were inducted.[121]

It might indeed be said that the history of Cesi's Academy drew to a close between the end of 1625 and the beginning of 1626. Paradoxically, however, the years between 1625 and 1630 witnessed the most intense publication activity of Federico and the Linceans close to him. During those years, works conceived by the Prince and his most faithful collaborators ever since the foundation of the Academy finally saw the light. Cesi's first published work,

---

120 OG XIV, pp. 126–7. After the death of Cesi, Stelluti sent only a few letters to Galileo (see CL, index).

121 The only meeting (the last in the brief life of the First Academy) was held on 27 January 1629. The minutes do not indicate who took part, simply listing the names and brief biographical profiles of the last three associates: Sforza Pallavicino, Pietro della Valle, and Lucas Holstenius (Contributi I, pp. 547–8). As Gabrieli noted (CL, pp. 1000–1), the choice concentrated on individuals with little or no connection with the scientific discussions. This tendency finds further confirmation in the failure to enrol candidates with skills in mathematics, medicine and natural philosophy proposed above all by Faber and Cassiano dal Pozzo: Giovanni Camillo Gloriosi, Marco Aurelio Severino, Mario Schipani, Johann Bayer, Stephan Rodriguez de Castro, and even Francis Bacon. During Cardinal Francesco Barberini's delicate mission in France, his personal secretary, the Lincean Cassiano dal Pozzo, wrote to Faber from Paris about the works of Bacon that had made a great impression on him, to the point of wishing to induct the British philosopher into the Academy: 'A book has come into my hands by an author whom, if he were not in England, I would like us to make every effort to recruit; it is the author who has published *Essays* and *De sapientia veterum*, Francis Bacon, who less recently also published *De dignitate et augmentis scientiarum*, a splendid work from which great benefit can be derived for the advancement of speculations in all the sciences [...]' (ibid., p. 1060). The first two works referred to by Cassiano had appeared in Venice in Italian translation by the Secretary of State of the Grand Duchy of Tuscany, Andrea Cioli (Bacon 1621). As an authoritative Medicean functionary, Cioli was to personally follow all the delicate stages of the trial of Galileo in 1633.

the *Apiarium*, appeared in 1625.[122] Three or four years later a sizeable segment of the *Tabulae Phytosophicae*, the work that had entirely absorbed Federico's energy during the final years of his life, was printed. The Prince even managed to publish the first section[123] of the most ambitious project conceived by the Academy of the lynx-eyed – the so-called *Thesaurus Mexicanus* – the only endeavour based on the programmatic collaboration between its associates. Finally, on the eve of its dissolution, in 1630, the last work of the Academy written during Cesi's lifetime was published: Francesco Stelluti's annotated translation of the moralising fables of Persius Flaccus.[124] The frontispiece of that work was a sort of manifesto of Federico's conception of the intrinsic connection between knowledge of nature and rules of conduct inspired by elevated ethical values. Within the same classical architectural frame already used in the frontispiece of the *Assayer* (it would also be re-proposed in the *Thesaurus Mexicanus*) the images of poetry and of moral philosophy appear juxtaposed (fig. 10). The latter has a lion on the leash, symbol of the mastery of the passions, while in the other hand she holds a set square to underline the close connection between moral conduct and the mathematical study of nature.[125]

To resolve the apparent contradiction between a substantially dormant phase of the academic life and the sudden explosion of its publications, it is necessary to go back in time. We left the research project designed by the Prince in the *Indicatio* in the first half of the 1610s to concentrate on the battle waged by Galileo and the other Linceans after the anti-Copernican decree of 1616 until the composition of the reply to Ingoli eight years later. It is now time to change the point of observation and to focus on the themes on which the Prince was working after abandoning the *Coelispicium*, on their significance, and above all, on Federico's perception of the relation between the research he was passionately conducting and the questions at the centre of Galileo's agenda.

---

122 Cesi 2005–2006.

123 It contained Faber's illustration of the animals of New Spain.

124 Persius Flaccus 1630. The ecclesiastical authorities hesitated for some time before granting permission to print Stelluti's work.

125 Cf. Ardissino 2007, pp. 169–71.

CHAPTER 10

From the Heavens to the Bowels of the Earth

The Merging of the Two Projects

The Prince had not abandoned his longstanding project of submitting Aristotelian natural philosophy to a radical reform. As we have seen, the first outline of that project had been sketched in the complex Frontispieces of the *Indicatio* during the first half of the 1610s. Also the Frontispiece devoted to the *Coelispicium*, traditionally considered as the most eloquent proof of Cesi's firm adhesion to the Copernican agenda of Galileo, represented the logical development of the original vision of the founders of the Academy and, in a very particular way, of its Princeps.

That the entry of Galileo on the scene had not led to an abrupt deviation from the Academy's original ideals and expectations is clearly shown by an examination of the Lincean correspondence in its entirety, instead of privileging – as has usually been the case – the correspondence between Galileo and his fellow Linceans. The letters exchanged between Cesi, Faber and Stelluti as well as with other correspondents present a significantly different scenario to the one that emerges from their correspondence with the Tuscan scientist.

The encounter with Galileo did not cause Cesi to change direction. The motivations and programmes at the centre of the Prince's research programme outlined in the *Zibaldone* appear as the linear development of the goals defined in the first precarious years of the Academy.

The Lincean correspondence yields precious information in this respect. The letters sent by the Prince to Galileo from the start of their relation until his death hardly ever contain any clues regarding the activities in which Federico was engaged. In those rare cases in which he mentioned to the Tuscan scientist his immersion in researches that fascinated him and consumed his energy, he refrained from specifying their nature or giving away the conclusions that he had reached or hoped to reach. Likewise, in the few extant letters from Galileo to Cesi, there is no sign of any curiosity regarding the Prince's studies.

On the other hand, the epistolary exchanges of Cesi, Stelluti and Faber with one another, with the other Linceans, and with a restricted number of correspondents outside the Academy provide precious details on the development of the Prince's studies, which grew more intense after 1620. Cesi systematically avoided mentioning to Galileo the personal researches to which he himself attributed the greatest importance (fossils, miraculous rains, intermediate

 | DOI 10.1163/9789004342323_011

natures, systems of classification of the vegetal world, etc.). Nor did he send him the texts in which he was bringing into focus his own conception of the causes of natural phenomena (*Oposcopia, De caeli unitate, Coelispicium, Tabulae Phytosophicae*). This suggests that he was aware that the questions with which he was concerned were of marginal interest to Galileo, or suspected that they might even appear idle or misguided to the Tuscan scientist.

The picture of Cesi that emerges from the correspondence with Galileo is that of an unconditional admirer, ready to exert his influence to enable the Tuscan scientist to achieve his ambitious plan, assiduous in moderating his enthusiasm and recommending prudence from fear of the serious consequences that might otherwise ensue. It is an attitude that could almost be characterised as paternal, had Cesi not been much younger than Galileo.

If we focus on the rest of Cesi's correspondence, leaving aside his exchanges with Galileo, we obtain a very different picture. As though by magic, precise clues emerge that attest his constant and impassioned dedication, in spite of continuous economic and domestic emergencies,[1] to extremely complex fields of research, driven by the ambition of throwing new light on vast areas of the study of nature that boasted illustrious and consolidated traditions.

In order to understand this apparently schizophrenic behaviour, it is necessary to closely examine how Cesi viewed the relation between his personal research programme and that of the Academy, and the battle that Galileo was waging. The two men were united by their strong motivation to affirm the principle of freedom of thought in natural philosophy and this was undoubtedly their most important common ground. All the same, significant differences can be pinpointed in their individual interpretations of that principle. To reach that goal, Cesi set up a reserved and protected space, the Lincean Academy, where, sheltered from the suspicions or censure of political and religious authorities, debates and research could be freely conducted. During the lengthy period of his relations with Galileo, Cesi never failed to invite his admired companion to avoid presenting the principle of the freedom of thought as a demand that called for public recognition. He should be content with exercising that freedom within the protected space of the Academy, while publicly dissimulating the purely hypothetical nature of his own conclusions

1 His father, Federico Cesi senior, transferred all his property to his son on 18 April 1618, including the onerous obligations that he had contracted, amounting to around 43,000 scudi. To this enormous sum were to be added the burdens arising from Federico junior's obligation to guarantee an annual allowance to his two brothers, Angelo and Giovanni Federico, as well as to fulfil payment of the dowry of his sister Caterina; cf. Schettini Piazza 2007, p. 472.

and continually emphasising his submission to the superior authority of the Church and the powers that be.

Galileo paid no heed to those well-meant recommendations. It was not enough for him to give free expression to his new ideas among trusted friends receptive to his vision. He would not bow down to the tyranny of the presumptuous and the incompetent, as he considered his resolute opponents, whether pedantic followers of Aristotle or ceremonious theologians, who, under threat of severe punishment, imposed the unquestionable authority of Holy Scripture and their exclusive right to interpret its true meaning. The tensions and the divergences on the public or private arena in which to fight for the affirmation of the principle – essential to both of them – of the freedom of thought *in naturalibus* can be perceived throughout the course of the relations between Galileo and Cesi and they would never be resolved.

What is more, Galileo was even reluctant to accept the limitation *in naturalibus*. In his Copernican writings, in the letter to Ingoli, and in many other epistolary documents he manifested his determination to overstep that limit, extending that principle to other fields, such as that of the question of the position of man in the universe and of the relations between man and nature.

In spite of his different interpretation of the meaning and scope of the freedom of thought *in naturalibus*, Cesi regarded this principle as the ground on which the motivations that had prompted him to found the Academy and the innovative programme of Galileo reached their highest level of convergence. But could that shared principle, considering their different interpretations of its scope, be sufficient to justify, in the eyes of Cesi, making Galileo the emblem and leading light of the Academy? And would that move not have ended up by transforming the programmes that had inspired and continued to inspire Federico and his closest associates? Cesi had to find some other connection, more direct and less abstract than the principle of the freedom of thought *in naturalibus*, to convincingly suggest the organic convergence between the horizons of Galileo's researches and the objectives pursued by the Academy and its Princeps.

The *Indicatio* and, in particular, the little that we know of the *Coelispicium* provide important clues as to which of the implications of Galileo's discoveries and his reform of cosmology seemed to harmonise perfectly with Federico's personal agenda: the fluidity, corruptibility and unity of the heavens. We have already seen how the research programme outlined in the *Indicatio* and the Paduan manuscript containing the text on the *laserpitium* and the *Oposcopia* clearly indicate the Prince's intention of taking the new Galilean sky as the basis for an ambitious programme of innovative research on the nature of the most extraordinary phenomena to be observed on and below the surface of the Earth and in its atmosphere. In Cesi's view, only the hypothesis of a continuous communication between the heavens and the Earth, based on the principle of their complete homogeneity could account for the phenomena. Federico had conceived this project before Galileo arrived on the scene as a consequence of his reading of Telesio, Patrizi, Persio and even Bruno. The celestial discoveries – the irregular surface of the Moon, the sunspots, and later the Galilean interpretation of the comets – had provided fundamental confirmation of a principle which had essential importance for his project to reform natural philosophy.

In the eyes of Federico, it was not only in the demand for the freedom of thought *in naturalibus* that Galileo's trajectory converged with the programmes of the Academy; nor only in the determination with which the Tuscan scientist demolished the false premises on which Aristotelian natural philosophy rested. Cesi saw the objective of a radical reform of natural philosophy conceived by the Academy and the extraordinary innovative researches of Galileo as fully complementary: if the Tuscan scientist was unveiling the true nature of celestial bodies, he and his fellow Linceans would shoulder the equally important and perfectly integrated task of reforming the traditional interpretation of the phenomena observed on, above and below the surface of the Earth.

It is probable that at some point Cesi realised that Galileo would not support that ambiguous coupling. It did not take him long to grasp that the motivations and way of thinking of Galileo, as well as the instruments on which he relied in his researches on natural phenomena, were remote from the methods and expectations of the Lincean endeavour. Still, that awareness did not weaken his conviction of the objective convergence of the two projects.

If Galileo could not be involved in the programmes on which the Linceans centred their vision, nothing stood in the way of emphasising the direct continuity between the celestial novelties and the equally sensational discoveries that the Linceans were making through innovative investigations in a different, but no less important, dimension of the universe: while Galileo had radically reformed Aristotle's *De caelo*, Cesi and his companions set themselves the task of demolishing the principles on which the Stagirite's work on meteors was based.

That this was Cesi's framework for the relation between his personal research programme and the horizon explored by Galileo is confirmed by numerous documents. Let us begin with the laborious editing of the preface to the *Letters on sunspots* by Angelo De Filiis. The Lincean began by stressing the direct relation between 'the historians of lower nature' (a clear reference to Cesi) who set about unveiling its secrets, and Galileo, who had demonstrated new marvels in the heavens. That De Filiis intended to celebrate the Princeps of the Academy and his innovative researches is shown by the mention of 'zoophytes', the specialty of Federico. Emphasis on the programmatic complementarity between the fields of research of Federico and of the Tuscan scientist was even stronger in the first redaction of the preface, in which De Filiis had offered a more detailed description of Cesi's dedication to the reform of Aristotle's interpretation of meteors:

> If we take such delight in and praise their discoveries so much when investigators of that lower [realm of] nature that lies under our feet reveal the birth of something never before seen – be it plant, animal or deformed zoophyte – how much more should we not rejoice when we are presented with new lights into that higher realm of the highest heavens and the faces of the noblest discoveries, once veiled, are now revealed to us?[2]

The interpretation of the intrinsically complementary character of the Galilean and Cesian programmes appears no less explicit in Faber's introduction to the section on the animals of New Spain in the *Thesaurus Mexicanus*, in which he emphatically praised Galileo's celestial discoveries. This introduction, revised and approved by Cesi, was first published between 1628 and 1630. The German Lincean presented a summary of the leading astronomers and natural philosophers from antiquity to Galileo, singling out the contributions of Plato and above all of Socrates, one of the preferred philosophers of Cesi and the other Linceans,[3] of whom Faber underlined the identification of *admiratio* as a fundamental stimulus to the disinterested study of nature:

> *Admiratio* has produced so many schools and families of philosophers. For he who admires doubts, he who doubts inquires, he who inquires does not know, he who does not know wants to know: his mind is burning with desire, he investigates the causes of things, and if fortunate, by relying on the senses here and reasoning there, he tracks them both down in his hunt in the dense forest of the miracles of nature [...].[4]

It was *admiratio* that had led Endymion to turn his gaze to the Moon and to remain fascinated by it. Incomparably greater was the *admiratio* that spurred Galileo – Faber took great care to stress the Tuscan scientist's profound

2 Galilei-Scheiner 2010, p. 374 (OG V, pp. 78–80). On the central role played by zoophytes in Cesi's research see below, pp. 355ff.

3 Cf. Favino 2007.

4 TM, p. 462.

Christian faith[5] – to point the telescope, the new and more powerful 'eye of the lynx', to the skies and to discover there unprecedented novelties that had shaken time-hallowed certainties based on false arguments:

> Great and marvellous news that renders this world of the traditional philosophers exposed to interminable disputes, even more uncertain and dubious.[6]

After listing Galileo's celestial discoveries one by one, Faber went on to emphasise that the radical reform of the conception of the cosmos initiated by the Tuscan scientist had triggered processes that were changing the whole of natural philosophy:

> How can we doubt that also in our own age the philosophers are presented with fresh stimuli and a very great quantity of new material?[7]

The celestial novelties – continued Faber – had led some admiring followers of the Tuscan scientist, i.e. the Linceans, to extend the horizon of his research to include the phenomena situated above and below the surface of the Earth. The German Lincean highlighted above all the role of Prince Cesi, to whom he assigned the merit of having transferred the Galilean method of inquiry to new areas of natural philosophy, those analysed by Aristotle in his treatise on meteors. The catalogue of the most remarkable meteorological phenomena provided by Faber is bound to strike the contemporary reader:

> It would be a lengthy task to enumerate the remarkable effects produced by air and the phenomena of evaporations in the upper regions of the sky: rains of milk, blood, flesh, iron, wool and stones, not only in the era of the ancient Romans, but also in our time, as Prince Federico Cesi the Lincean [...] will disclose in his *Thaumatombria*, a work in which he explains the causes and effects of the marvellous rains.[8]

Not content with rains of milk and blood, Faber went on to present other astounding manifestations of the forces of the air and the heavens: lightning

5 'Motivated as a Christian by love of the true, triune God, Creator and Author of all the stars'.
6 Ibid.
7 Ibid., p. 463.
8 Ibid.

that struck casks full of wine, destroying the contents while leaving the casks unharmed; lightning that liquefied iron swords without damaging the sheath; lightning that killed a child by striking through the eye, leaving a body that, personally dissected by Faber, exhaled a sulphurous odour, while yellowish spots were observed in the cerebral membrane and the lungs. And, in addition, veins of metal or precious stones growing within the bowels of the Earth; water ascending to the tops of mountains; perpetual fires springing from the subsoil.[9]

Like a new Prometheus, Cesi had had the courage to shift the penetrating gaze of Galileo from the heavens to phenomena located above, on and below the surface of the Earth, thus making a fundamental contribution to a new and more complete reading of the Book of Nature. As Faber emphasised, that book encompassed not only the heavens but also the Earth ('Librum Naturae Coelum nimirum Terramque'). This was the essential mission of the Academy: to integrate the results achieved in the heavens thanks to the magisterial work of Galileo with equally innovative research on atmospheric changes, the productions of the surface of the Earth and subterranean phenomena.

The research project of the Linceans was not confined to meteorology. Faber's systematic treatment of the fauna of New Spain also reflected the spirit and aims of the Academy and its Princeps. It was Cesi who had prompted the German Lincean to take up that challenge.[10] The phenomena of animal life – the German Lincean stressed – were even more important than the celestial discoveries and the perfect knowledge of plants and minerals. The animal kingdom incited the natural desire for knowledge to its highest degree:

> How can the study of the phenomena of the upper regions and of the heavens, being so remote from our senses, possibly lead to philosophical certainty? And how can all the phenomena of the lower regions – meteors, plants, minerals – even taken all together, be compared with the lowliest creature possible that lives, breathes and moves?[11]

This remarkable operation to weld the Galilean programme to the Lincean project (with the final subordination of the former to the latter) was bound to conclude with a deferential token of homage and submission to the supreme authority of the Pope, Urban VIII. To win his favour, Faber praised the natural, moral and political virtues of the trigon of bees on the Barberini coat of arms:

9 Ibid., p. 464.
10 'Opella haec mea [...] Federici Caesij Lynceorum Principis iussu edolavi', ibid.
11 Ibid.

> Who will ever praise enough the ingenuity of bees, the image of a perfect government, and their production of honey that is so beneficial for human health?[12]

In closing with this fawning eulogy of the virtues of bees, Faber echoed the style and models with which Cesi, in those same months, was exalting the Barberini papacy in his *Apiarium*, to which we shall soon turn.

Galileo probably never had the opportunity to read these pages by Faber. If he had, he would certainly not have welcomed the emphasis on the direct continuity and complementarity between his own researches and the wonderful workings of nature at the heart of the agenda of Prince Cesi and his Academy.

That subtle operation was to enjoy a remarkable and long-lasting success. Traditional historiography on the relations between Cesi and Galileo has substantially endorsed the image of the programmatic convergence between the two projects. At least until a few decennia ago, this was the frame of reference within which research on the First Academy was conducted. Not even such a meticulous scholar as Gabrieli – without whose fundamental contributions it would be impossible today to conduct serious research on Cesi and the institution that he created – remained immune from this ambiguous paradigm. The great scholar showed a clear inclination to minimise those aspects of the thought and activities of Federico and his associates that were in striking contrast with the research methods and conception of nature of Galileo.

Up and Down the Ladder of Nature

In spite of the many studies on the First Academy published in the last thirty years, we still lack an exhaustive intellectual biography of Federico Cesi that, independently of his personal relations with Galileo, brings into focus his programme, his expectations and the results that he achieved.

It must be said that it is by no means easy to follow the development of Federico's researches, underlining the elements of continuity and the shifts vis-à-vis the original programme traced in the documents of the first eight years of the Academy. We face particularly difficult problems when we concentrate our attention on the last decennium of the Prince's life. This was the most productive period, during which he completed some of the most ambitious projects that he had conceived from the beginning of his intellectual activity.

12 Ibid.

Regrettably, many of the notebooks in which he jotted down the results of his own observations and speculations are lost. We therefore have to fall back on information contained in the correspondence and on the evidence to be found in works published during or after his lifetime by his closest collaborators.

The research programme outlined in the *Indicatio* is a fundamental resource. The Paduan manuscript containing the *Oposcopia* and the treatise on the *laserpitium*, as well as the ample evidence of his constant involvement in the project of the *Coelispicium* between 1615 and 1620, all show that Cesi closely followed the programme outlined in the *Zibaldone*, addressing one after another the main issues dealt with in that document.

The treatise on the *laserpitium* offers a foretaste of the exhaustive investigation that the Prince planned to devote to the entire vegetal world and its interactions with celestial phenomena. This ambitious project dated from the very foundation of the Academy. The title 'Phytosophical lecture' (i.e. on the science of plants)[13] of the inaugural speech that Cesi delivered to the other three founding members on 15 October 1603 shows that his researches at that time were already focused on questions on which he would continue to work for the rest of his life: the attempt to define the true nature of plants and the conviction of the need to abolish, or at least attenuate, the traditional clear-cut division between the animal and plant kingdoms.

The expression 'Phytosophical lecture' was to be echoed a quarter of a century later in the title that Cesi was to assign to one of the most demanding of his intellectual endeavours: the composition and printing of the *Tabulae Phytosophicae*. It is important to underline that, in his inaugural lecture, after tackling a problem of a theoretical nature – the definition of a plant – Cesi emphasized the essential function of chemical analysis to isolate the medicinal properties of plants in proposing to subject the hemp plant to distillation.[14] This too was a work method that he would follow throughout his life.

The centrality of the botanical programme of the young Federico also emerges from the division of labour established by the four associates at the very first meeting of the Academy, held on 12 October 1603, when Cesi was defined reader of Phytosophy.[15] The emphasis on the continuity in the 'ladder'

13 'Wanderer of the heavens then opened his phytosophical lecture with the definition of a plant, which he showed to be probably a vegetative animal', Contributi I, p. 510.

14 'Karabisque liquationem et naturalibus experimentis perutilem esse tentandam censebat', ibid.

15 Ibid., p. 507.

of all natural beings revealed by the title of Cesi's first lecture was to recur constantly in his research agenda of the years ahead. It reflected the idea that he had formed early on of the model of organisation of nature: a chain whose links succeeded one another without any break in continuity; a unitary organism, hierarchically articulated, without leaps or gaps, animated in all of its settings, as well as in all of its operations and productions. For Cesi, life was present everywhere. There was therefore no point in sharply separating the organic from the inorganic world.

The early emergence of this unitary and homogeneous vision of nature (paralleling the equally early vision of a unitary and animated cosmos) helps us to understand why plants always occupied such a privileged position in Cesi's research agenda. According to the traditional conception, vegetals were situated at the lowest level in the chain of being endowed with vital activity. Characterised by a vegetative soul, plants were subordinate in the hierarchy to man, endowed with a rational soul, and to animals, who possessed a sensitive soul. Below the vegetals came brute matter, which was inanimate. Cesi concentrated on plants because of their borderline position in the traditional ladder of being: the lowest rung above the inanimate, though clearly separate from the higher rungs occupied by animals. Federico was convinced that there were no landings that interrupted or broke up the ladder of nature. He therefore dedicated himself to attenuating the sharp division imposed by tradition between plants and animals, and between vegetals and the inorganic world. It would thus become evident that nature, the principle of life, operates organically by emanation, permeating, though to varying degrees, every corner, however remote, of the universe.[16]

This was not an original vision. In the concept of an animated nature, both celestial and terrestrial, resonated fundamental motifs of Neo-Platonism and of the Renaissance philosophies of nature formulated by authors whom Cesi had read and pondered attentively. It was precisely because they occupied the liminal zone between the animal and the inorganic world that plants appeared to the Prince as the most suitable terrain to demonstrate the unitary character of nature, as well as the ubiquitous presence of the vital principles.

Cesi did not conceive his passionate study of plants as a specialised research in a particular field. It was the privileged point of observation from which he hoped to decipher the logic and structure of the entire theatre of nature. For Cesi, botany was a comparative science based not only on the comparison

16 In the *Tabulae* he was to use the Plotinian term 'hypostasis' to describe the universal flow of the generative power of nature, cf. below, p. 466.

between different sorts of plants, but also on that between plants and animals, between plants, stones and metals, and between plants and fossils. At a certain point, he was even tempted to draw a bold comparison between vegetals and man. Although Federico did not follow up with this, he vigorously affirmed that plants provide man with exemplary lessons of moral conduct.

The particular configuration of the Prince's botanical studies and their relational character help us to understand why, from the inaugural lecture of October 1603 on, his reflections on the vegetal world appear programmatically interwoven with researches on the animal world and on natural productions traditionally considered to be inanimate. This programmatic hybridisation makes it particularly difficult to reconstruct the stages through which Cesi's ambitious project of a reform of the traditional conception and above all classification of nature outlined in tabular form ever since the *Indicatio* developed over the course of time.

One has the impression of an organic design that proceeded in a fragmentary and chaotic manner, due both to its intrinsically interdisciplinary nature and to its continuous interruptions because of practical concerns. Early on, Cesi isolated the problematic question that called for primary attention: the 'mezzane nature' (intermediate or middle natures)[17] that could be found in the animal, plant and mineral worlds. As liminal creatures, they demonstrated the absence of any break in continuity in the chain of being. These singular objects could not be dealt with using the tools and methods of traditional botany. Competence in cosmology, meteorology, geology, zoology, chemistry and medicine should be deployed at the same time. Cesi faced up to this daunting challenge with determination. He set out to reconceive the order of nature in its entirety, stressing its organic and unitary character, demonstrating the integration of celestial, terrestrial and subterranean phenomena, and arranging them all along continuous chains. This explains why it is impossible to identify stages of his application in which his interest in the classification of the vegetal world was an exclusive concern. On the contrary, other problems continually burst in: celestial and meteorological phenomena; the nature and origin of minerals and fossils; the physical, moral and literary classification of plants and animals.

For Cesi, the entities of nature can, indeed must, be distributed in categories, but these categories must not be conceived as distinct and isolated from one another. The vital force flows through the entire organisation of nature.

17 'Two-faced' (*ancipiti*) and 'ambiguous' (*ambigue*) were other expressions that he applied to these natures.

As Anaxagoras, the Presocratic philosopher whom Federico and the other Linceans held in high esteem, had posited, the same elements are found in any production of nature: 'all is in all'.

If Cesi had conceived a refined strategy to highlight the intrinsic and direct connection between the callings of the original Lincean project and the novelties introduced by Galileo, the Tuscan scientist was not interested in aligning with that perspective. Cesi's intellectual horizon appeared elusive to him. Plants, animals, fossils, marvellous rains and secrets of nature did not pertain to the world that he set out to explore. Thus it is not surprising that he paid little attention to the Prince's research priorities.

The only traces of Galileo's interest in Cesi's researches are confined to the short-lived exchange of letters on the system of the world in 1612 and to the mention of the drawings of American plants shown him by Cesi during his stay in Rome in 1611.[18] After these early signs of curiosity, no manifestations of interest on the part of the Tuscan scientist in the scientific activities in which Federico was involved emerge from the correspondence between the two men. Nor do we find evidence of a desire on the part of Cesi to keep Galileo informed about his own projects or to ask for his opinion on the results that he was obtaining.

The extant documentation indicates that Galileo knew the *Discourse on the natural desire for knowledge*, as well as the *Melissographia* and the *Apiarium*, which he received from Rome (although Cesi had never given him the slightest hint of his work on bees).[19] On the other hand, in his letters to Galileo Federico often mentioned the extremely laborious gestation of the *Thesaurus Mexicanus*, though without ever providing any detailed information on the content and goals of that work.

Cesi will certainly have illustrated his research programme to Galileo during his meetings with the Tuscan scientist in 1611, 1616 and 1624, though

18 OG XI, p. 107, letter to Piero Dini, 21 May 1611. Marco Guardo has demolished the prevalent but erroneous interpretation of the mention of the drawings of the American plants in this letter as an indication of Galileo's lack of interest in the *Thesaurus Mexicanus* (Guardo 2011). In this essay Guardo also analyzes the pages of Faber's section on animals in the *Thesaurus* in which Galileo is mentioned.

19 Stelluti informed Galileo about 'the matters pertaining to bees' that Cesi was having printed only on 10 January 1626 (OG XIII, pp. 299–300), by which date the work had already been published and presented to Urban VIII. In a subsequent letter of 28 February, Stelluti anticipated the dispatch to the Tuscan scientist of 'some sheets of those Bees observed by me [the *Melissographia*] together with other sheets printed by the Princeps on bees' (ibid., p. 308).

apparently failing to arouse his curiosity. In his letters to the Prince (most of them lost), Galileo never asked for information nor offered comments or expressed personal opinions on Federico's activities. As already mentioned, it is surprising that Galileo was kept in the dark about the letter *De caeli unitate* that Cesi sent to Cardinal Bellarmine.[20] Nor was he informed later of the Prince's decision to authorise Christoph Scheiner to publish his letter together with the rejoinder of Bellarmine.[21] No different is the scenario revealed by the work that cost Cesi the most time and trouble, the *Tabulae Phytosophicae*. Although a good number of Tables were printed between 1628 and 1630, they were not sent by Cesi to Galileo, nor were they ever mentioned explicitly in the Prince's letters to the Tuscan scientist during the many years spent on their compilation. The only information on the *Tabulae* to reach Galileo came not from Cesi but from Francesco Stelluti, who informed the Tuscan scientist that the Prince was concluding the composition of an extremely useful and beautiful work that would be published in the *Thesaurus Mexicanus*. However, Stelluti did not mention the title or describe the contents of that work.[22]

Galileo also remained in the dark regarding the progress of the intense researches that Cesi was conducting on intermediate natures, particularly fossils. Federico's correspondence with Faber and Stelluti between 1621 and 1625 shows that during those years – the years of the polemic with Father Grassi provoked by the *Discourse on comets* of Galileo-Guiducci, of the publication of the *Assayer*, and of the anonymous denunciation to the Holy Office – the

20 Galileo could read the *De caeli unitate* only after he managed to secure a copy of Scheiner's *Rosa Ursina* (Scheiner 1626–1630) in September 1631, by which time the Prince had been dead for more than a year. Father Scheiner's book was sent him by Paolo Giordano Orsiní, to whom the work was dedicated (OG XIV, pp. 294–5, letter from Naples, 9 September 1631).

21 The letter of Cesi to Bellarmine *De caeli unitate* and the Cardinal's reply were passed on to Scheiner by Faber. The German Jesuit received them on 3 July 1628 (Scheiner 1626–1630, p. 774). In July 1630 Scheiner showed the letters of Cesi and Bellarmine to three fellow Jesuits, who guaranteed their authenticity, kissing that of the deceased Cardinal as a token of devotion. On the transmission and publication of the epistolary exchange Cesi-Bellarmine on the fluidity of the heavens and on the motivations that led Scheiner to insert those documents in the *Rosa Ursina*, see Dollo 1995, pp. 146–7.

22 OG XIII, p. 113, letter of Stelluti to Galileo, 8 April 1623: 'The Prince [...] only dedicates to his studies those hours that he can steal [from practical obligations] and wishes to finish an extremely useful and beautiful work that will be added to the Mexican book that is being printed at the moment [...]'.

Prince spent a large part of the time that he managed to wrest from his administrative responsibilities[23] to explore the fossil deposits near Acquasparta and to attempt to decipher their origin and nature.

Yet the correspondence with Galileo bears not a trace of these impassioned researches. The only information that the Tuscan scientist received on Federico's study of fossils was imparted to him by Stelluti:

> He [Cesi] does not cease from his studies and most beautiful observations that he is carrying out on that mineral wood, having found very large pieces up to eleven palms in diameter, and others with iron fibres in them, or pieces similar to metal inside such woods, and others which secrete a kind of rubber like incense, even having a similar odour; moreover he has found a very large quantity of woods which have turned into stone and iron and in the most bizarre shapes.[24]

Those on fossils were not new researches for the Prince, as witnessed by the annotation to one of the drawings that he commissioned of the finds in Acquasparta indicating that it was done and taken to Rome in 1611.[25]

Stelluti's letter to Galileo contains a telling detail. Less than four months had passed since the Tuscan scientist's two-week stay in Acquasparta. Cesi had organised for him excursions to the Marmore Falls, the high hills of the surrounding territory, and the Piediluco Lake, but not to the fossil deposits.[26] Had it been otherwise, Stelluti would not have expressed his keen regret for the fact that, in declining to stop over in Acquasparta again on his return journey from Rome to Florence, Galileo had missed the opportunity to visit them:

> If you had passed through here on your return journey to Florence, you would have seen all these woods and where they originate, and some fires that burst forth from the ground, with great amazement and delight.[27]

---

23 The demands of the family property in an irreversible process of financial decline obliged the Prince to spend ever more time in Acquasparta.

24 Ibid., pp. 200–1, letter of 23 August 1624.

25 The drawing, published in Scott-Freedberg 2000 (p. 148), contains the annotation: '[…] anno 1611 a nobis invent et Romam allat […]'.

26 Gabrieli too was surprised at this missed excursion, especially as the fossil deposits between Roseto and Rosaro were not far from the Prince's residence: 'We know for certain that this visit did not take place, but we cannot understand why, given the interest and importance that Cesi attached to these *officinae naturae*' (Contributi I, pp. 214–5).

27 OG XIII, p. 201, letter of 23 August 1624; Scott-Freedberg 2000, p. 44.

In this connection, it is worth recalling that, during Galileo's visit to Rome in 1616, Cesi had organised an excursion for his guest on his estates in Tivoli to show him a curious location which he had indicated as *Opificium Tiburtinum*[28] in the list of the workshops of nature (*officinae naturae*) visited personally and recorded in the *Indicatio*. In that place there were ponds rich in small islands formed of tartar or 'spongy stones', produced by the deposits of sulphurous water on the marshy vegetation. For Cesi those spongy stones were emblematic of the transition from vegetal to mineral. He recalled that visit in his letter to Faber of 14 January 1619. The Prince, now in Acquasparta, conveyed his satisfaction that his friends and associates in Rome were discussing the relation between the tension of the strings and the sound produced, certainly stimulated by a fresh reading of the *Sambuca Lincea* (1618)[29] by the Lincean Fabio Colonna:

> How splendid and in harmony with my inclinations are the philosophical debates you are conducting. We discussed at length with our Galileo in Tivoli on musical proportions and their relation with the strings of the instruments that you mention, and it would truly have been a good thing if they were put down in writing.[30]

This document is interesting, not only because it illustrates yet another project conceived by the inexhaustible mind of Cesi, but above all because it shows that, instead of inviting the Prince to explain his reasons for attributing to the spongy stones an intermediate nature, Galileo had shifted the conversation to questions of musical harmony, in which, as the son of a talented musician, he showed a very keen interest.

Flowing Natures

While Cesi and Galileo would not mention spongy stones and fossils again in their correspondence with one another, the Prince's letters to Faber, other Linceans and illustrious figures on the Roman scene contain a continuous series of references to his feverish engagement in this field of research, particularly between 1620 and 1622, after abandoning the *Coelispicium*.

---

28 Contributi I, p. 45.

29 Colonna 1618.

30 CL, p. 678, letter of 14 January 1619.

Cesi's letters from Acquasparta during these years show that his energies were concentrated on the close examination of the fossil deposits in the vicinity of his Umbrian residence, on trying to establish the position to assign to those findings in his *Theatre of Nature*, and on endeavouring to discover their causes. Cesi got down to serious work. With the assistance of artists contracted for the purpose, he had faithful images made of the most extraordinary findings and sites.[31] At the same time, with the help of Faber – a highly competent bibliophile and passionate collector of art and other items[32] – he was building up an exhaustive bibliography on these problems, concentrating on those authors who had attempted to classify fossils, particularly the metallic and semi-mineral ones.[33]

Cesi's letter to Faber of 20 January 1621 indicates that he was trying to establish 'conjunctions, differences, divisions and affinities' between natural productions, focusing in particular on the relations between plants and minerals.[34] His letters to the German Lincean of the successive months express his enthusiasm for those studies and outline the theoretical project that he had conceived: to compose that organic table of the relations and hierarchies of all natural beings that he had envisaged ever since embarking on his studies.[35]

31 On the drawings of fossils commissioned by Cesi, see Scott-Freedberg 2000.

32 Faber was a major connoisseur and art dealer and was on familiar terms with Rubens and Elsheimer, with whom he shared Neo-Stoic leanings. From the inventory of his collection of art works after his death in 1629, published by Irene Baldriga, it emerges that Faber had amassed a considerable number of art works to embellish the rooms and corridors of his Roman residence (Baldriga 2002, pp. 285–90). The German Lincean was also a passionate book collector and occasionally traded in publications printed in Germany or the Low Countries. After his death, the executors of the inventory of his library found many books there that were on the Index of prohibited books (Brevaglieri 2009, p. 115), though he had obtained regular permits to read them. They included works by Telesio, Patrizi, Libavius and Paracelsus. Faber had also created a museum in his home, furnished with scientific instruments (telescopes, microscopes, stills, etc.) as well as mineral substances and medicaments. The section of his museum housing the anatomical preparations must have been remarkable. Besides the 'skeleton of a human body, that is all the bones of a man assembled from head to foot' (Baldriga 2002, p. 282), an entire room contained dozens of preserved terrestrial and marine animals and birds (ibid., pp. 282–4).

33 'I would like to know whether there is anyone who has orderly distinguished and listed the fossils in their classes, particularly the metallic and semi-mineral ones; and likewise whether anyone has synoptically classified the sciences in trees and tables', CL, p. 732, Cesi to Faber, letter of 20 January 1621.

34 Ibid., p. 733.

35 'I return – he wrote to Faber on 28 August 1621 – to my synoptic tables', ibid., p. 749.

At the same time he extracted singular specimens from the fossil deposits to examine their structure. During Lent, he sent Faber two specimens of fossil wood that he had had shaped in the form of a cross, inviting him to keep the smaller for himself and to give the larger to the Cardinal of Santa Susanna as a sign of deferential homage.[36]

His research on fossils and their position in the chain of being was interwoven with his efforts to introduce a new classification of the vegetal world. Two weeks after Galileo left Acquasparta, Winther – who had been recommended to Cesi as his personal physician and assistant in his studies of nature and classification – described to Faber a visit to the fossil deposits in the company of the Prince.[37] A few months later, Cesi wrote to Cassiano dal Pozzo that he was discovering 'daily marvellous things in the workshops of nature'.[38] After a few days, in the letter mentioned above, Stelluti informed Galileo that the Prince was spending a great deal of his time 'on that mineral wood'.[39] Not long after, Winther brought Faber up to date: Cesi had asked him to list all the authors who had written about 'the exotic kinds of metals'.[40] On 29 September Cesi sent a small table made of fossil wood, polished like ebony, to Francesco Barberini, whom the Prince had chosen as his personal patron, hoping that the Cardinal Nephew would help him out of his dire financial straits.[41] The situation was so serious, according to Winther, that Cesi paid the salaries of his employees very late, if at all.[42] In fact, this was what would soon lead Winther to leave his service.

36 Ibid., Cesi to Faber, 21 March 1622. Cesi suggested that Faber should hang the cross in his study 'to represent both the sacred mystery and the marvel of nature's works'.

37 Ibid., p. 880, letter from Acquasparta, 18 May 1624.

38 Ibid., p. 934, letter of 18 August 1624.

39 Ibid., p. 935, letter of 23 August 1624.

40 Ibid., p. 939, letter of 7 September 1624: 'The Prince gave me some authors to comb who have written about the exotic kinds of metals'. Winther was not happy with the commission: 'The hours of the day and the month are fully absorbed in this task [...]'.

41 Ibid., p. 947, Cesi to Cassiano dal Pozzo, 29 September 1624.

42 See the indignant letter that Winther sent to Faber on 28 July 1626, when he was on the point of abandoning Cesi's employ in Acquasparta: 'As for my provision, it has been impossible to get anything so far, so that I think that he will give me as he thinks fit for my departure and the time spent there, and will treat me like Riquius. But I do not know how he will manage to settle with all the literati: they are not all gullible and could make a fool of him one day' (ibid., p. 1126). For the reference to Cesi's similar treatment of Josse de Rycke, see ibid., n. 8 and p. 1122.

These testimonies are all eloquent proof of the fact that the innovative system of classification of vegetals that the Prince was working on would pay particular attention to the intermediate natures of metallophytes. However, the numerous references contained in the letters do not furnish clear indications of the conclusions at which Federico had arrived on the origin and causes of the singular phenomena observed in the fossil deposits of Acquasparta.

In the section of the *Thesaurus Mexicanus* on the animals of New Spain, on which Faber was working in those same months of 1624, he announced the imminent publication of a treatise by Cesi on those curious phenomena.[43] After the death of the Prince, Stelluti was to confirm that Federico had written numerous books on those singular substances.[44] If he ever completed them, none is extant. Nor can the almost two-hundred surviving drawings that Cesi commissioned of the fossil deposits and of his most important finds help to bridge that gap.[45] To try to understand his conception of their nature and origin, we have therefore to rely on the only document written by Federico that betrays his thoughts on the nature of metallophytes: the letter sent from Acquasparta to Cardinal Francesco Barberini on 1 December 1624, a few days before the Prince set out for Rome, after more than a year's absence, to take part in the celebrations of the Holy Year.[46]

Although important, this letter does not completely satisfy our curiosity about the idea that Cesi had formed regarding those strange finds. Knowing of the Cardinal Nephew's keen interest in the most remarkable natural phenomena, the Prince gave him a detailed account of the 'new form of middle nature between plants and metals, on the veined and varied mineralised wood discovered by me some years ago in this region'. He went on to illustrate their appearance:

> I have found a great variety of petrified objects, that is some completely petrified, of the consistency of stone, or brick; or mixed – part wood, part stone – or similar condition, with internal and external combinations of many forms.[47]

---

43 TM, pp. 502–3. See below, p. 359.

44 See below, pp. 364ff.

45 For an English translation of Cesi's letter to the Cardinal Nephew, see Scott-Freedberg 2000, pp. 389–90.

46 CL, pp. 965–7. I use the English translation of Cesi's letter to the Cardinal Nephew in Scott-Freedberg 2000, pp. 389–90.

47 Scott-Freedberg 2000, p. 389 (CL, p. 965).

He declared that he had been even more astonished and puzzled by the discovery of 'metallized woods, and those which were similarly either completely or partially transformed into metal'.[48] The Prince stressed that these were new phenomena that could not be explained by the principles of traditional natural philosophy. They called for the closest scrutiny and unprejudiced reflection. What was the most extraordinary discovery in his eyes was 'the communication between these terrestrial natures', confirming that there were no clear-cut boundaries between the vegetal and the mineral worlds. It was therefore necessary to revise radically the traditional systems of classification of the natural world:

> I have found many ambiguities among them, which display mixtures in various ways: of wood and soil, of wood and stone, of wood, stone and metal together, and wood and mineral fluids, brought together and participating equally in the substance, in an almost neutral body, or rather these things are reduced to quite apparent unity; from which the very obscure nature of those terrestrial liquids we call bitumens may receive quite some light.[49]

Such 'ambiguous', 'neutral' bodies were composed of different natures. The Prince was fascinated by these singular hybridisations. He proposed to 'skim over them in writing, not only to carefully describe them but also to search for their causes and to examine what can be found about these subjects or concluded from them'.[50]

Unfortunately, neither the letter to Francesco Barberini, nor the letters written by the Prince down to the moment of his death offer the slightest clue as to how the Prince interpreted the causes of these phenomena; nor do they disclose his 'theories, made over many years', that now found confirmation in the discovery of these 'furnaces kindled by nature' and in the 'essence and causes of these [subterranean] fires which, although very little known and hidden to human intellects, always seemed to me for the most part to be full of light'.[51]

The letter to the Cardinal Nephew probably represents the most advanced point reached by Cesi's researches in this domain. He announced to Francesco Barberini his imminent return to Rome to take part in the Holy Year celebrations, where he was to spend almost two uninterrupted years. Far from

48 Scott-Freedberg 2000, p. 389 (CL, p. 966).
49 Ibid.
50 Ibid.
51 Ibid., pp. 389–90 (CL, p. 996).

Acquasparta and his beloved fossil deposits, his interest in metallophytes faded. New and urgent commitments and needs obliged him to turn his gaze elsewhere.

Besides the silence on fossils, there are other indications in the Prince's post-1624 correspondence that suggest that the most intense phase of study of those singular subterranean finds had come to a close by that date. The lengthy section of the *Thesaurus Mexicanus* on the animals of New Spain had been practically brought to conclusion by Faber during the same weeks of 1624 in which Cesi was preparing his return to Rome. Among the numerous references to the Prince's activities that Faber inserted in that work, he did not fail to mention the researches on the fossil woods of Acquasparta:

> Our most illustrious Prince Cesi [...] recently sent skilfully fashioned tables[52] made of fossil wood that stupefied us and the whole city as a miracle. He was the first to discover this fossil wood sprinkled with the most beautiful flecks and waves and to have it extracted not far from his city of Acquasparta [...]. The interested reader will learn all these things more clearly from Cesi's writings themselves, which are very soon to be published, in which many new things discovered and observed by him will be presented.[53]

'Very soon' to be published was overoptimistic. Cesi's impassioned researches on fossils never saw the light, leaving in the dark the conclusions that he had reached on their causes. Nor do we have any outlines, drafts or preparatory works. Only a large number, 199, of drawings of the fossil finds and deposits that had impressed him most have come down to us. These drawings contain concise annotations that unfortunately do not help to throw light on the Prince's interpretation of the process through which these intermediate natures were formed.

It is, however, certain that, long before the letter to Francesco Barberini, the Prince had already determined the position to assign to his studies of fossils within the ambitious design of the *Theatre of Nature* and of the *Mirror of Reason*. In the *Indicatio* he had proposed to dedicate to the ambiguous and 'neutral' fossils one of the Frontispieces of his *Theatre* entitled 'On the intermediate natures of the universe'.[54] We find in the outline of that Frontispiece the same

---

52 Faber makes reference here to the tables made of fossil wood that Cesi had sent to him and to Francesco Barberini from Acquasparta (see above, p. 356).

53 TM, pp. 502–3.

54 Contributi I, p. 40.

terms as those contained in the letter to the Cardinal Nephew: 'On two-faced nature [...], On ambiguous nature [...], Species sharing in two natures'.[55] The programme of study delineated in that document indicates that Cesi intended to focus his attention on the most singular hybrid phenomena of the subterranean world: 'Zoophytes, Lithophytes or Metallophytes, Zoolithophytes'.[56] Cesi also assimilated to this category of dual or triple nature the prodigious meteors, such as the one that fell in Tuscany in 1614, of which he asked Galileo for a detailed description and a drawing.[57]

Thus the idea of dedicating a treatise to the fossil woods of Acquasparta came from long before, at least from the years of composition of the *Indicatio*, between 1612 and 1615. That it went back earlier still is attested by the only dated drawing among Cesi's collection of fossil drawings in the Royal Collection in Windsor, which bears the date 1611 and the note that it was taken to Rome.[58]

Mother Earth

After the letter to Francesco Barberini of December 1624, Cesi was to make brief mention of his fossil studies once again in the first of the *Tabulae Phytosophicae*.[59] This was the table of tables, as Fabio Colonna called it,[60] because it offered a panoramic view of the entire natural kingdom, although Cesi also referred to the supernatural one. At the top of the classification, immediately below God and the angels, came man, the rational being, followed by viviparous animals endowed with sensoriality and movement. Below them Cesi placed oviparous animals, considered as transitional elements between animals and plants because of the analogy that he drew between the *vis vegetandi* latent in the egg and that of the seed. Plants were to be considered almost animals (*proanimal*) since they displayed a vital force that was not exclusively vegetative. Below plants came those objects which had assumed the aspect of stones ('quae Petrefacta sunt'), immediately followed by those which had become metals (*Metallifacta*), which he claimed to have been the first

55 'De ancipiti natura [...], De ambigua natura [...], Species duarum naturarum particeps'.
56 'Zoophita, Litophita an Metallophita, Zoolitophita', ibid., p. 41.
57 See above, p. 216.
58 See above, n. 25.
59 TM, p. 905.
60 CL, p. 1188, letter of Colonna to Stelluti, 10 November 1628.

to discover.[61] These were ambiguous entities, gradually transformed from one into the other by the workshop of nature. He further subdivided the category of animals (*bruta*) into more and less perfect. Then, having recourse to a creative nomenclature,[62] he listed the categories into which he grouped the various animate creatures that he defined as imperfect because characterised by a dual or triple nature:

- Phytozoa, animals retaining some characteristics of plants;
- Zoophytes, partaking of the nature of both plants and animals;
- Zoolithophytes, stone-plant-animals, a term we have devised to indicate organisms endowed with a triple nature.[63]

The vast population of the vegetal world appeared even more articulated. It was distributed in two main categories (once again, more and less perfect). As in the case of the *bruta*, Cesi only provided further specification for imperfect plants:

- Metallophytes, ambiguous between plants and metals; a term we have devised and explained in several books;
- Lithophytes, plants tending to stone, first introduced in natural history by us;

---

61 TM, p. 905.

62 Cesi's peculiar nomenclature to describe the intermediate natures was not completely original. It was the extension of the method of unification of two different lemmata used by Cesalpino in his *De plantis* of 1583 (Cesalpino 1583). In his classification of the plant kingdom, based on the observation of the reproductive organs of plants, Cesalpino had reserved a space between the inorganic world and the vegetal substances for natural species such as fungi or truffles, in which he did not recognise the presence of seeds: 'zoophytes, between animals and plants'. He had called these plants 'imperfect' (an expression largely used by Cesi). Furthermore, he had defined their nature as intermediate 'between plants and inanimate matter', an expression that is once again echoed in Cesi's '*mezzane nature*': 'Only one type of plants remains, which does not produce any seed, and thus does not bear a true stalk or flower, but either lacks them altogether, or produces certain things that correspond proportionately to stalks and seeds, such as the horsetail [...]. Certain plants bear no seed at all, which, since they are extremely imperfect, are generated through putrefaction; this enables them to feed and to increase, but they are unable to reproduce themselves; these are intermediate between plants and inanimate matter, as zoophytes are situated between animals and plants; examples of these are fungi, duckweeds, lichens, and many marine plants' (ibid., p. 28, see Ottaviani 2000, p. 349).

63 'Phytozoum animal plantae nonnihil retinens; Zoophytum Plantanimal duabus naturis instructum; Zoolithophytum, lapiplantanimal, confusis tribus naturis a nobis inventum'.

– Lapisurgens, a stone on the verge of planthood, or considered by us to be plant-like but closer to the lower category.[64]

The reader will have no difficulty in recognising in this classification the same scenario that Cesi had outlined to the Cardinal Nephew in his letter of December 1624. The *Tabula Prima* was sent to Fabio Colonna in October 1628,[65] almost four years after the letter to Francesco Barberini. It shows the progress that the Prince had made in incorporating the ambiguous finds in the deposits of Acquasparta within his synoptic vision of the whole of nature. Metallophytes, Lapisurgens, Zoolithophytes: each represented a rung in the grand Ladder or Chain of being. Those intermediate or even triple natures showed that there were no breaks in continuity between the kingdoms of nature. The first of the *Tabulae* provided precious indications regarding Cesi's conception of the Chain or Ladder of being: it was hierarchically organised but neither fixed nor rigidly stable. It was possible to ascend or descend the ladder: stones could be transformed into metals, plants into stones, just as metals could place one foot on the vegetal rung and the other on that of stones; and the same was true of stones and vegetals. It was a clearly delineated hierarchy with a granular articulation in constant vibration: entire categories of beings manifested instability. In Cesi's words, 'they were transformed step by step from one condition to another through the artifice of Nature'.

Federico was perfectly right to stress that these were new observations and new conclusions that called into question traditional systems of classification. It is hardly surprising that he considered this contribution as important as the radical transformation of the image and structure of the universe brought about by Galileo. He felt that he was extending to a new dimension the profound renewal of knowledge produced by the Tuscan scientist's astronomical discoveries. There is no need to hypothesise an improbable presentiment of the theory of evolution[66] to appreciate the originality of his classification of natural beings, his precise observations on intermediate or middle natures, and the large number of accurate drawings that he commissioned of the most singular findings discovered in the bowels of the Earth.

64 'Metallophytum, anceps inter stirpes et metalla a nobis inventum quod pluribus libris explicavimus; Lithophytum, planta ad lapidem accedens a nobis primo in historiam physicam adductum; Lapisurgens, ad plantam vergens lapis, sive plantalis a nobis consideratus cui in inferioribus proximior'.

65 CL, p. 1185, letter of Colonna to Stelluti, 14 October 1628.

66 Cf. Freedberg 2002, pp. 375–86.

However, not even the *Tabula Prima* helps to understand the causes the Prince held to be responsible for the processes leading to the generation of ambiguous, dual or triple natures. He made no explicit pronouncements on this score, contenting himself with attributing to the artifice of nature responsibility for those productions. He had announced in the letter to Francesco Barberini the imminent publication of the results of his researches on the causes of these phenomena, traditionally interpreted as *monstra* produced by a nature that freely exercised its own plastic energy. Cesi did not consider them as 'tricks' of an exhibitionist nature. If they had not belonged to the regular course of nature, there would have been no point in trying to classify them.

The question of how Cesi interpreted the causes of those phenomena is of crucial importance for an understanding of his general conception of nature. Should we suppose that he really did write the many books on the intermediate natures of which he spoke to the Cardinal Nephew as a *fait accompli* and whose imminent publication Faber announced in the same months?

An important clue as to their existence is offered by a small block of text printed in the right-hand margin of the first of the *Tabulae Phytosophicae*, next to the paragraph on imperfect plants. The block, highlighted by a pointing finger to draw the reader's attention,[67] states that the category of beings to which the intermediate natures belong is simply mentioned in the first of the *Tabulae*, but would be treated more extensively elsewhere and reduced to a syntactical table. However, the Prince does not specify when or where.[68] At any rate, not in the *Tabulae*, as Cesi was not to return to these ambiguous natures in any of the successive ones.

The project of an in-depth treatment of intermediate natures was therefore still in his mind at the time of composing the first *Tabula*, but we cannot establish whether he had already written the *syntaxis* of intermediate natures or whether he intended to do so later. Nor does the Prince inform us whether that *syntaxis* would be limited to the description of these curious finds, or whether he intended disclosing his personal interpretation of the causes and processes of their generation.

There is reason to doubt that Federico's reflections on the dynamics of the continuous transformations of categories attested by the intermediate natures were ever finalized. It is hard to imagine that, if Cesi had in actual fact

67 This is the only instance in the twenty *Tabulae* in which Cesi uses the *manicula* to call the reader's attention to a block of text.

68 'Quae nunc sumenda & in Syntaxin redigenda & contemplanda', TM, p. 905, see below, fig. 17.

developed an exhaustive interpretation of the causes of those phenomena, Francesco Stelluti would not have illustrated it in his treatise on fossil wood of 1637, in which he was to provide an accurate report on the Prince's researches on the fossil deposits of Acquasparta.

Stelluti was constantly involved in Cesi's studies of lithophytes and metallophytes, and it was he who kept those Linceans whose fields of interest were closest to those of the Prince, above all Fabio Colonna, informed about Federico's progress in this area. Stelluti accompanied the Prince on his deathbed and, appealing to Cassiano dal Pozzo for help, took pains to secure the Prince's literary legacy and the large corpus of drawings from nature that he had collected.[69] It was also Stelluti who completed and published the manuscript drafts left by the Prince, of the *Tabulae Phytosophicae,* as well as the long cherished project of the *Thesaurus Mexicanus*. And finally, after Federico's death, the task fell to him – as the last surviving member of the founders of the Academy in 1603 – to furnish the first information about the Prince's discoveries in the deposits of Acquasparta in a manuscript text, the *Treatise on Fossil Mineral Wood,* that enjoyed a wide circulation on both sides of the Alps.[70]

Before examining Stelluti's *Treatise,* we should consider the testimony that he provided in his heavily annotated translation of the *Satires* of the Latin poet Persius Flaccus.[71] That testimony is absolutely reliable, since Cesi had read and approved the work prior to its publication. In the annotation to the text in which Persius had emphasised the extremely high quality and extraordinary beauty of cedar wood, Stelluti compared it to the wood extracted by Cesi from the fossil deposits of Acquasparta. He described its 'most beautiful veins in a hundred diverse forms of undulating waves, that are like camlet and other forms of shot silk that are in use today', while its lustre was like the sheen that ebony assumes after thorough polishing.[72] Stelluti was at pains to point out that this material was not timber:

> This is not the wood of tree trunks, but grows underground where there are other mineral deposits and is formed together with them, gradually taking on the form and nature of wood. For this reason he [Cesi] wanted to call it Metallophyte, finding it to be of an intermediate nature between plants and minerals, first discovered and identified by him.

69 Contributi I, p. 929; see also Solinas 1989a and 2000. On Stelluti's role as faithful interpreter of Federico's conception of fossil wood, cf. also Gould 1997–1998.

70 Stelluti 1637. See also Scott-Freedberg 2000, pp. 400–3.

71 Persius Flaccus 1630.

72 Ibid., p. 21.

Stelluti further informed the reader that his affirmations were based on direct experience:

> On many occasions I accompanied him [Cesi] to that place where this wood originates when he went there to observe it, and not without amazement I saw that earth transformed into wood [...].[73]

Stelluti's account adds further information to that contained in the letter to the Cardinal Nephew. Federico did not regard those fossil finds as having anything in common with the wood of trees; therefore they could not be classified as vegetals. He was convinced that they were the product of a combined action of soil and subterranean juices and heat.

In reply to the requests of the many who asked him for information about the Acquasparta deposits and begged him for a copy of the manuscript treatise in which he had described them, Stelluti decided to publish his *Treatise* on mineral wood seven years later. He added many particulars on those strange natural products, as well as some engravings (after the drawings commissioned by Cesi) that showed their appearance and the sites in which the Prince had discovered them.[74] Stelluti repeated that Federico had called the fossil wood of Acquasparta metallophyte 'being part plant and part metal'. He added immediately afterwards that, upon closer inspection, these finds 'partake more of the plant than of metal'.[75] He described their appearance, stressing once again their resemblance to cedar wood, the beauty of the forms and the variety of the waves, as will be seen, he added, 'in the following figures drawn from nature and engraved'. He went on to illustrate the place where the fossil wood deposits were located. It was at this point that Stelluti put forward the idea that he had formed of how this curious kind of wood was produced:

> From what I have been able to see and observe, the wood is not generated by the seed or root of any plant whatsoever, but only from a type of earth containing much clay, which is slowly transformed into wood, nature thus operating until all the earth is converted into the said wood. This I believe to take place with the assistance of the heat of subterranean fires which exist in these areas, and which, snaking underground, continuously emit very thick smoke, and sometimes flames, and especially

73 Ibid.

74 Cf. Freedberg 2002, pp. 305–6.

75 Scott-Freedberg 2000, p. 384 (Stelluti 1637, p. 5). On Stelluti's treatise see Solinas 1989b and Scott-Freedberg 2000 (which contains an English translation of the entire text).

> when the weather is wet, and with the further assistance of sulphurous and mineral waters.[76]

According to Stelluti, the multiple variety of forms that the fossil woods displayed depended on the diverse intensity of these subterranean fires:

> And if the heat is sufficient, the wood is browned, or lightly scorched, and remains similar to coal. If then the earthen material is not yet converted into wood, this fire bakes it and it ends up similar to vases of terracotta baked in a kiln, or to bricks.[77]

So he stressed the same connection between the presence in the deposits of Acquasparta of subterranean fires and fossil woods that Cesi had reported in his letter to the Cardinal Nephew of December 1624. Stelluti proposed a series of considerations to convince the reader that

> The essence of this wood is nothing other than earth, because I myself have seen pieces, one part of which was hard clay, the other wood and the rest similar to coal.

That those fossils were not born 'of any seed or root or branch, as are other plants' was also demonstrated by the fact that they did not have roots or branches, but only trunks which, moreover, are 'not erect, as all plants are born, but recumbent, or, as it were, lying on the earth'. Stelluti then provided final proof of the non-vegetal generation of these woods:

> The waves and veins it possesses do not extend in the same form through the whole of the wood but are variously fashioned, now long and straight, now narrow, now spiralling, now undulating, and in a thousand different guises; while in other kinds of wood, the waves [...] and their veins are extended along the length and height of the tree, which receives its nutriment from its roots but the nerves and veins grow along the whole length of the plant: but this mineral wood takes its nutriment from every part of the earth and still has such varying waves.

---

76 Scott-Freedberg 2000, p. 384 (Stelluti 1637, p. 6).

77 Ibid.

Stelluti showed his awareness that his interpretation was only one of the possible explanations for the nature and generation of metallophytes. He confessed that he himself had conceived a different hypothesis at first, which he had then abandoned as the result of a series of stringent considerations based on the comparison between the fossil finds and the trunks of trees that grow on the surface of the earth:

> Neither can these pieces of wood be considered trunks, nor [are they] fragments of other trees buried in this area, or which have fallen and been covered by earth and then given wave-like form by mineral waters that have their springs there [...] or by subterranean fires, *as was my initial persuasion.*[78]

The size of the trunks, disproportionately larger than those of the trees that grow on the surface of the earth, and their oval rather than circular shape, lacking a counterpart in nature, had convinced him to abandon the hypothesis of an organic origin, thus reinforcing his conviction that the metallophytes were generated beneath the surface of the earth by the action exerted on clay soil by endogenous heat. At this point he formulated his definitive conclusion:

> We may safely affirm that the raw material of this wood is nothing other than clay-rich earth.[79]

In his opinion, this was further confirmed by the high specific weight of these finds: while natural wood floated in water, the fossil woods sank.

Stelluti recorded another experiment that he had conducted on the metallophytes of Acquasparta, this time in the presence of numerous witnesses, including Prince Cesi himself, who had probably devised it:

> A quantity of damp earth, which was around a piece of this wood, having been removed and placed in a room in the Palace of Acquasparta, belonging to the Lord Duke Cesi, it was found after some months to be wholly converted into wood, to the great wonder of the aforementioned Lord and all the others who saw it.[80]

---

78 Scott-Freedberg 2000, p. 384 (Stelluti 1637, p. 6); emphasis added.
79 Scott-Freedberg 2000, p. 384 (Stelluti 1637, p. 7).
80 Scott-Freedberg 2000, pp. 385–6 (Stelluti 1637, p. 8).

The result of this experiment removed all doubt from the minds of Cesi and Stelluti:

> Hence, he has no doubt whatsoever that the earth itself is the seed and mother of this wood, the earth of this parts being most suitable for its generation.

The Earth is the mother and its womb is full of vital seeds that lie inert until subterranean heat causes them to germinate. The particular composition of the soil and the quality and intensity of the subterranean heat determine the appearance and nature of the fossil woods.

In another passage of the *Treatise*, Stelluti attributed the infinite variety of fossil woods found in Acquasparta to the moulding powers of nature: 'nature thus joking perhaps – he stated – to make us marvel at her rarities'.[81] The allusion to the *lusus naturae* should not lead us to suppose that Cesi and Stelluti had recourse to that popular theory to explain the generation of metallophytes. The affirmations in Stelluti's *Treatise* leave no doubt as to the fact that he and the Prince attributed to the Earth the capacity to generate in its womb productions that looked very similar to those that grow from seed on its surface. Faber's comments in the *Thesaurus Mexicanus* on Cesi's studies of metallophytes point in the same direction.[82] These were not playful tricks conceived by an exuberant nature, but the products of a combination of factors that favoured the activation of the vital forces latent in the Earth. The characteristics of the soil, the seeds scattered in it ready to be activated, and an appropriate level of heat led the earth to generate not only fossil wood but also an infinite variety

81 Scott-Freedberg 2000, p. 385 (Stelluti 1637, p. 8).

82 In a polemic with Liceti, who had excluded the possibility that the transformation and growth of minerals in veins could be attributed to the presence of a vegetative soul, Faber had referred to Cesi's thoughts on the question, according to whom 'it will not be so incongruent to classify amber between minerals and plants, for if you examine its substance, it is bitumen, but like plants it grows from the earth' (TM, p. 573). Faber explained immediately afterwards that, in his researches on intermediate natures, Cesi had reached the conclusion that life extended from the inorganic to the organic world without any break in continuity. Therefore a vegetative soul was present in minerals too: 'Our most illustrious and excellent Prince Cesi was the first to discover and reveal this intermediate nature between plants and minerals, which he treats specifically in his books on metallophytes […] as he has named them' (ibid.; cf. Ottaviani 2000, pp. 229–31 and 327ff.; 1997, pp. 65ff.). In confirmation of the Prince's views on metallophytes and intermediate natures, Faber referred also to Cesi's letter to Francesco Barberini on the fossil deposits in Acquasparta.

of combinations of wood and stone, of wood and metals, of stones, wood and metals, or, as in the most fascinating cases to the Prince, of plants, animals and stone. Without the favourable combination of those physical factors, nature could not produce any *lusus*. The fossil woods of Acquasparta were thus ordinary productions of nature. Their exceptionality and their prodigiousness disappeared once the causes of their generation were identified.

The image of a prolific Mother Earth laden with seeds perfectly expressed the idea dear to Cesi of a vital force that permeated the universe: in the heavens, on the surface and in the depths of the Earth. For Cesi, life existed everywhere in a latent state, ready to manifest itself in the most unpredictably diverse forms under favourable environmental conditions.

The concept of a nature permeated by vital principles shared and echoed by Stelluti in his *Treatise* drew on themes that were characteristic of Neoplatonism and the Renaissance philosophy of nature. The vision of the entire universe as a living organism was a central paradigm in the mind of the Prince and his first companions from the foundation of the Academy. Federico had peremptorily expressed that concept, underlining its crucial function and universality, in the *scholium* of the *Indicatio* which concluded the schematic outline of the articulation of the Frontispiece of the *Theatre of Nature* dedicated to the *Coelispicium*:

> Every body in nature, above and below the surface of the Earth, is permeated by a force which is not bound or circumscribed by any limits, and nothing remains outside its jurisdiction.[83]

It was a living universe: in the heavens (where angelic intelligences moved the stars and planets, as Cesi stated in his *De caeli unitate*), on the surface of the Earth, and within its bowels. It was the vital energy that pervaded the entire universe that determined generations and mutations. And it was thanks to its action that wood could be produced from clay to assume a hybrid nature between stone and metal.

The principle of plenitude elaborated by Cesi implied the image of nature as the mother of an infinite variety of creatures, arranged along the Chain of being without any discontinuity. It could be said that the intermediate natures appeared so to Cesi not *ex parte obiecti* (i.e., from the point of view of nature), but *ex parte subiecti* (as a consequence of man's incapacity to grasp the complex articulation of the Ladder or Chain of being). Throughout his life Cesi was to dedicate his energies, by throwing light on the imperceptible intermediate

83 Contributi I, p. 38.

transitions, to bridging the breaks between the realms of nature established by traditional systems of classification.

Botany for Metaphysicians?

The testimony of Stelluti enables us to reconstruct the idea that the Prince had formed of the nature and the process of generation of the fossil wood near Acquasparta. That singular interpretation seems at odds with the image of the convinced adhesion of the Prince to the rigorous research method theorised and practised by Galileo after his induction into the Academy.

How had the direct, close and repeated observation of the finds in Acquasparta led Cesi to believe that they were not the felled trunks of trees buried for ages and transformed over time into fossils? And how, after taking it into consideration (as Stelluti tells us), could he have rejected that interpretation in favour of the generation of fossils from juices activated in the soil by the heat of subterranean fires? It should be recalled in this connection that Winther, who had visited the deposits of fossil woods with Cesi, was also extremely puzzled by the Prince's interpretation, though he had refrained from expressing his perplexity to avoid irritating his patron.[84]

84 Winther had shown Cesi the section of the *De metallicis* (Cesalpino 1596, pp. 126–7) in which Cesalpino had wondered whether ebony should be assigned a vegetal or mineral nature: 'Having read the chapter on ebony in Cesalpino, I found a certain point in which, when shown to the Prince, he saw his opinion totally confirmed, even though there is no lack of valid arguments to the contrary [...]. But I make a show of following in the footsteps of the Prince' (CL, p. 881, letter of Winther to Faber, 18 May 1624). Winther's perplexities were well founded. Cesalpino had put forward a double explanation of the nature of ebony in his *De metallicis*. Referring to the opinion of Pausanias and Theophrastus, he had declared that ebony 'should not be called a root or tree that has been turned into stone, but a stone'. However, immediately afterwards, he stated that he inclined towards the interpretation of ebony as the effect of a slow process of petrifaction of a tree that has ended up underground: 'nor is it absurd to suppose that ebony is a wood that has been buried for a long time and has turned to stone, which is known to have happened to other kinds of wood' (Cesalpino 1596, p. 127). Cesalpino attenuated the borderline between organic and inorganic nature not only in the *De metallicis*, but also in the *De plantis* (Cesalpino 1583). He attributed to metals ends that, although different from those of plants and animals, nevertheless showed their importance in the general economy of nature. As Winther's testimony shows, Cesi was familiar with the *De metallicis*. There is no systematic study of the inspiration that Federico drew from the treatises of Cesalpino. For a first useful step in this direction see Ottaviani 2000, pp. 347ff.

The Prince's interpretation appears all the more surprising when one considers that he knew the botanical work published by Fabio Colonna in 1616, in which the Neapolitan had included the well-known dissertation on *glossopetrae*.[85] Federico had even called for its publication.[86] Colonna's concise but revolutionary contribution demonstrated that the images imprinted in stones, called *glossopetrae* because they were shaped like a tongue, were not 'tricks' that showed the free and infinite plastic force of nature, but the teeth of sharks that had ended up beneath the surface of the earth as the result of natural catastrophes like earthquakes, or of supernatural events such as the Flood. In the ground, hardened after the waters had receded, a faithful image of those teeth had remained impressed.

By means of some brilliant experiments and drawing also on chemical analysis, Colonna had demonstrated that *glossopetrae* were composed of a calcareous substance – such as that of sharks' teeth – which was different from the earth, hardened to the point of turning to stone, in which they were buried. Even though Colonna spoke of sharks' teeth, not of wood or trees, it is difficult to believe that the Prince was not led to think that the fossil wood in the Acquasparta deposits could have been produced through a similar process: trees grown from seed, uprooted by some natural catastrophe, progressively carbonised or petrified, leaving in the earth the imprint of their cortical and internal structure.[87]

---

85 Cesi sent Galileo a copy of the *Minus cognitarum stirpium ekphrasis* (Colonna 1616), containing *De purpura* and the dissertation on the *glossopetrae*, immediately after publication (OG XIII, p. 285, letter of Cesi to Galileo, 8 October 1616).

86 In the preface to his dissertations on *purpura* and *glossopetrae*, Colonna declared that he had decided to publish his opinions on such complex and controversial questions at the insistence of Cesi: 'He convinced me [...]. I owe so much to the most excellent Prince of Sant' Angelo Federico Cesi, Princeps of the Lincean Academy [...]' (*Auctor ad Lectorem*). Colonna went on to praise Cesi for having invited a qualified 'assembly of erudite men' to collaborate in the disinterested study of nature in the Academy of which he was proud to be a part. Repeating expressions that seem citations from the *Lynceographum* and the *Discourse on the natural desire for knowledge*, Colonna exalted the study of nature based not only on observations made with the eyes of the forehead, but enriched and ennobled by the systematic recourse to the eyes of the mind.

87 As Alessandro Ottaviani emphasised, Giovan Battista Brocchi in his *Conchiologia fossile subappennina* (Brocchi 1843) showed his great surprise that neither Cesi nor Stelluti had managed to free themselves from the 'ancient errors' on the origin of marine fossils, even though they both knew the pioneering researches of Fabio Colonna: 'All the efforts of Fabio Colonna did not avail to make them forget the ancient errors; he was not even able to convert his colleagues of the Lincean Academy either, nor the founder himself of that famous institution [...]' (cited by Ottaviani 2000, pp. 316–69 and republished in Ottaviani-Trabucco 2007, p. 14).

Colonna was probably not aware of the conclusions at which Cesi had arrived. If he had been, he would certainly have suggested applying to the Acquasparta fossils the same interpretive model that had enabled him to discover the true nature of *glossopetrae*. That model reduced the process of fossilisation to purely mechanical actions. It is unlikely that the Prince would have accepted such a suggestion because it called into question the concept dear to him of a living nature, fertile mother and breeding ground for an infinite variety of beings.

Some verses of Stelluti's ode to his patron on the occasion of Federico's marriage to Isabella Salviati in 1617 show that neither the Prince nor his faithful assistant had subscribed to Colonna's mechanical interpretation of the origin of *glossopetrae* or recognized it as an exemplary methodological model for investigating natural phenomena.[88] Nor were they the only persons in the Lincean circle to be sceptical about the Neapolitan's interpretation of marine fossils.[89]

If he did not inform Colonna of the ideas that he had formed on the generation of the intermediate natures of the fossils of Acquasparta, Cesi sent him, by way of Francesco Stelluti, some of his *Tabulae Phytosophicae* in the early days of October 1628.[90] The *Tabulae* were destined to be published in the *Thesaurus*

88 Though the interminable epithalamium that Stelluti wrote for that occasion is of modest poetic quality, it is interesting for the emphatic praise of the moral virtues of his patron, of whom Stelluti composed a biographical profile in verse. He emphasised Federico's interest in astronomy, manifested long before he came into contact with Galileo, whose sensational celestial discoveries were also lavishly praised (Stelluti 1617, p. 22). Writing barely a year after the anti-Copernican decree, he took special care, however, to avoid mention of their cosmological implications. He stressed the Prince's inclination towards botanical studies and their medicinal applications as well as his researches on meteorology and fossils. In this section of the poem, Stelluti still credited the interpretation of marine fossils as *lusus* that bore witness to the boundless plastic power of nature: 'Observe a hundred lifeless objects of various forms on those briny sands: Some of them have the appearance of hard stone, whence admire Nature since here Medusa with the head encircled by serpents for hair has no power' (ibid., p. 25).

89 Francesco Imperato also refused to believe that those singular finds were animal remains, in spite of his father's declaration in his *Historia naturalis* (Imperato Fe. 1599, p. 666) that they were the imprints of skeletons of living beings (cf. Stendardo 2001, p. 44). Francesco regarded fossils, *glossopetrae* and bucardia (a marine fossil resembling an ox heart) as products of the vegetative power (*vegetabilis facultas*) of nature. Only many years later, in the *Discorsi intorno a diverse cose naturali* (Imperato Fr. 1628), would Francesco, certainly under the influence of Colonna's dissertation on *glossopetrae*, accept the organic origin of fossils (Ottaviani 2000, pp. 326–7).

90 CL, p. 1185, letter of Colonna to Stelluti, 12 October 1628.

Mexicanus, to which Colonna was to make a fundamental contribution.[91] Cesi needed to submit his work to a competent evaluator before proceeding with publication. Who better than Colonna could review a work that proposed to design a new system of classification of the entire vegetal world and its connections with the other domains of nature?

As we have stressed, in the first of the *Tabulae* Federico mentioned the intermediate natures, making implicit reference to his researches on the creatures and substances that populate the subterranean world. After receiving the *Tabulae* and taking a quick look at them, Colonna made haste to thank Stelluti, also for the repeated eulogistic comments regarding his person that the Prince had made in them.[92] He confessed that he was astonished by the 'great care with which natural things have been examined and their relations established'. Evidently he had been struck by the complex layout of the *Tables*.[93] However, Colonna requested time before expressing his opinion. A few days later, on 20 October, Colonna transmitted to Stelluti his impressions. As was his custom, he did not hesitate to frankly manifest his perplexities:

> I have seen the subtlety of his Excellence and how much he has sublimated the plants, which he has not only placed above the earth in these *Tables*, but has also made them heavenly and ideal. These *Tables* will not be understood by herbalists, but only by metaphysicians.[94]

As he felt he did not have the competence to follow the Prince at this sublime level of analysis, he avoided formulating judgements, confining himself to expressing his admiration at the extraordinary construction: 'Passing then to the lowest level', i.e. to the analysis of specific cases and the proposed classification of all vegetals, Colonna nevertheless promised to 'see whether there is anything that can be added; although I do not think so, having observed how carefully categories have been distinguished'.[95]

The impression of a cool reaction, albeit hidden behind expressions of appreciation, is confirmed in the letter that Colonna sent to Stelluti twenty days later. He had by then had time to examine the first of the *Tabulae* with greater

91 On Colonna's contribution to the composition of the *Thesaurus Mexicanus* see Ottaviani 2007 and Mottana 2013.

92 Fabio Colonna is repeatedly mentioned in the *Tabulae*, always with expressions of praise.

93 CL, p. 1185.

94 Ibid., p. 1186.

95 Ibid.

attention. He was struck by it, as every reader will be 'once such things will be printed'. But the sting was in the tail:

> To tell the truth I do not know what the stony plant-animal called *zoolithophytum* is. I would like to know about it, since it is something I haven't seen [...]. The same with the plant-metal – unless it is the branchy silver described by Imperato.[96]

Unable to identify in nature entities that could be assigned to those intermediate entities, Colonna asked the Prince to illuminate him on this point. In mid-December he decided to address the author of the *Tabulae* directly. After reaffirming his 'great delight' at the 'composition of the whole', he returned to those tricky intermediate natures. He had attempted to identify natural species that could be assigned to the, for him, mysterious category of *lapiplantanimal* (*zoolithophyton*):

> I have pondered for a while on whether I had anything that could be classified in such a category, but can find nothing that gives me total satisfaction.

However, to demonstrate his efforts, he tried to come up with a concrete proposal taken from the testacean order, with which he was obviously very familiar:

> If we take the testacean bone to be stone, we could assign to that category that zoophyte that clings to stones and from one source produces other slightly smaller zoophytes in larger numbers, while the first functions as a trunk.[97]

But he was not satisfied with that identification either. This is hardly surprising, for the categories including dual or triple natures to which Cesi referred in the first *Tabula* were located, not on land or in the sea, but in the bowels of the Earth. Their study required accurate observation of the slow and subtle processes of transformation in the workshop of nature. That between the Prince and Colonna was a dialogue of the deaf. Had the Neapolitan understood

---

96 Ibid., pp. 1187–8, letter of 10 November 1628. I have used Freedberg's translation of this passage (Freedberg 2002, p. 382).

97 Ibid., p. 1190, letter of 16 December 1628.

Federico's ideas correctly, he would nonetheless have been extremely puzzled by his interpretation of the generation of fossil wood.[98] Unlike Cesi, Colonna believed in the discontinuity between the organic and the inorganic world. He could not countenance the existence of intermediate natures that filled the gaps in the great Chain of being. Nor could he imagine that minerals and stones were endowed with a vegetative soul that allowed them to grow.[99]

The Neapolitan Lincean recognised the diversity of the paradigms on which Cesi's research was based and beat a retreat, no longer asking for explanations of the ambiguous categories into which Federico had articulated the plant kingdom. In his successive letters to Stelluti, he was to express his warm appreciation of the *Tabulae*, going so far as to call Cesi a new Theophrastus. All the same, he would continue to manifest his perplexities at the excessively abstract character of the Prince's researches,[100] as well as the choice of an organisation of the content that seemed conceived to discourage the interested reader:

> I have received [...] the proofs of the tables of Your Excellency, which I consider a grand ornament to the volume and of extreme usefulness for those scholars who will be able to understand them, i.e. who are skilled in the study of plants and their natures. Others will fear not being able to draw any benefit from them rather than be stimulated to gaze on them, as they are so profound, or rather, so subtly constructed and arranged.[101]

'Subtly constructed and arranged': it would be difficult to devise a more synthetic and effective description of the terrific conceptual effort made by the Prince in organising his classificatory scheme of the entire vegetal universe as

98 In spite of the declarations of great respect that accompanied them, Colonna's critical remarks clearly reveal the diversity of his method. As Alessandro Ottaviani stressed, Colonna never shared the Prince's encyclopaedic vision nourished by metaphysical expectations. The Neapolitan preferred a 'detailed recognition of vegetal and animal species per geographical region in order to privilege the unprecedented, the precision of the description, and the richness of details of the illustrations. It was difficult for Colonna to identify with the scenarios designed by Cesi' (Ottaviani 2000, pp. 336–7).

99 Ibid., p. 325.

100 On Colonna's progressive transition from initial hesitation to the sincere expression of his profound admiration for the *Tabulae*, see Ottaviani 1997, pp. 42–6. Ottaviani highlighted other aspects of the contents of the *Tabulae* that could not have appeared convincing to the Neapolitan naturalist.

101 CL, p. 1198.

well as in orderly disposing the different categories within the space of a single page.[102]

There is another aspect to be considered in the correspondence between Colonna, Stelluti and Cesi. In his letter to Stelluti of 10 November 1628, the Neapolitan stated that he had appreciated the section of *Tabula I* in which, while listing them separately, Cesi had treated meteors, the Earth and the sky as unitary and homogeneous entities.[103] Immediately afterwards, Colonna asked Stelluti to recommend the Prince to proceed with the greatest caution in tackling:

> corporeal things, on which you need to curtail your mind, because if the scholastic theologians are of one opinion today, they will have another much harsher tomorrow, and to the wise man the motive is enough ['sapienti basta il motivo'].[104]

What potential danger had Colonna spotted in the first of the *Tabulae*? It is possible that in the manuscript or proof sent him by Stelluti, the Prince had cited and interpreted some biblical passage. This is what Colonna's warning to beware of the inconstancy of theologians would seem to suggest. Such an eventuality would also explain the difficulties that Faber had encountered a few months earlier in obtaining permission to publish the *Thesaurus Mexicanus*.[105] This interpretation also seems to be confirmed by the fact that, immediately after begging Cesi to stay away from the minefield of biblical exegesis, Colonna asked him to pass the same advice on to Galileo, who was engaged during those months in writing the *Dialogue on the two chief world systems*:

> As I wrote on another occasion to you, I think it would be a good idea to warn Galileo not to expatiate too much on scriptural matters, particularly with regard to the miracle of the furnace in which the three [holy] children are seen to walk and to praise God – in case these men [the theologians] find the occasion to prohibit his work, in order to make themselves the prior inventors of all his inventions, as they were unable to do with

102 Cf. Ottaviani 2000, pp. 337–9.

103 TM, p. 905.

104 Freedberg 2002, p. 343 (CL, p. 1198).

105 Ibid., pp. 1163–4, letter of Faber to Cesi, 25 April 1628: 'I have been at the Palace all morning and have done more than Charles in France to obtain the *publicetur* […]. One has to do as they wish and be grateful to obey them'. The letter is revealing of the obstacles raised by the censors.

> regard to the movement of the Earth and the Moon not shining in and of itself.[106]

Following Colonna's suggestion, Cesi removed any reference to Scripture from the final version of the first *Tabula*. However, it still contained many features that an eagle-eyed censor, armed with the patience to examine the labyrinthic Baroque architecture of the first of the Tables, might have considered ambiguous and potentially dangerous. The very notion of a continuous Ladder of being was treading on delicate ground, as it did away with the clear traditional

---

106 Freedberg 2002, p. 343 (CL, p. 1188). Colonna is referring to the debate triggered by Orazio Grassi's mention of the miracle of the fiery furnace (*Daniel* 3) in the *Astronomical balance* (Drake-O'Malley 1960, pp. 128–9: OG VI, pp. 174–5). According to the prophet Daniel, the Babylonian King Nebuchadnezzar had summoned the people to the inauguration of a large gold statue that he had decided to have erected in the plain of Dura. The King ordered the crowd to prostrate themselves in adoration of the statue. Anyone who disobeyed would be cast into a fiery furnace. Three Jews refused to worship the statue and were thrown into the furnace, which had been heated in the meantime to such a temperature that the men ordered to cast them into the flames were scorched to death. The King was astonished to see the three Jews walking unscathed in the flames, accompanied by a fourth 'who resembled the Son of God'. The miracle convinced the King that the Jews were protected by the true God and he was converted to their faith. Based on the indisputable authority of Holy Scripture, Grassi cited the episode as incontrovertible proof of the transparency of fire against the argument of Galileo-Guiducci, who had maintained in the *Discourse on comets* that if a comet consisted of burning matter, it would not be possible to observe the stars in the sky behind it. Grassi assigned considerable importance to this argument to refute the interpretation of comets as optical illusions caused by the reflection of solar light on terrestrial exhalations. In the *Assayer* (OG VI, p. 366), Galileo refuted Grassi's interpretation of the passage from the *Book of Daniel* by introducing a strained distinction between opaque flame and transparent fire. According to Galileo, the angel had saved the three Jews from being burnt alive by miraculously eliminating the flames surrounding their bodies (which prevented the King from seeing them), while leaving the fire through which the King's gaze could pass. While accompanied by a declaration of Submission to the superior authority of the theologians, the only authorised interpreters of Holy Scripture ('I leave such determinations to the care of the masters of divinity'), Galileo's interpretation recalled the hermeneutic exercises conducted on other passages of Scripture in the *Copernican letters*. Grassi was to issue his rejoinder in the *Ratio*, adducing a long series of testimonies from the Church Fathers and most respected theologians contrary to Galileo's interpretation (ibid., pp. 496–7). Moreover, he declared that to establish the genuine significance of Holy Scripture was not the task of Galileo. On the interpretation of this episode see the essay of Eileen Reeves, who provided detailed information on the lively theological debate on the miracle of the fiery furnace (Reeves 1991, pp. 1–27). I doubt, however, that Galileo was aware of those learned discussions. See also Freedberg 2002, pp. 342–5.

boundary between the organic and the inorganic world, and also attenuated, if not completely cancelled, the distinction between man and animals. Besides the traditional reason for placing man at the apex of the Ladder of Nature because endowed with reason and intellect, Cesi added a new, and in his opinion, more effective motivation: man must be distinguished from animals because he is *vestibile*, that is 'he can cover his nudity and his miserable condition with the practice of virtues and the arts'. This implied a vision of mankind characterised by an active life rather than by contemplation of the divine. In the first of the Tables the smooth transition from *homo* to *brutum* appeared no less delicate. The need to demonstrate the continuity of the great Chain of being led Cesi to bring the classes of man and animals closer together and to populate the traditional divide with numerous intermediate creatures arranged in a hierarchical order according to their moral conduct:

> from wise to ignorant, from civilised to savage, arriving finally at the lowest category which includes monkeys, parrots and Socratic dogs.[107]

Furthermore, the *Tabula Prima* revealed the perilously thin boundary established by Cesi between God and Nature. The solemn hymn at the foot of the *Tabula* in which he praised God, creator of the Ladder of Nature, and exalted the spiritual elevation produced by its study was pervaded by a strong religious air in which the distinction between Nature and its creator seemed to vanish:

> From God and to God who is Alpha and Omega. Raise your mind on high just as you raise your eyes to heaven and to the light: seek light from the Father of Lights. Go up via the steps that the many easy ladders reveal on every side. Let this pyramid bring you through the creatures of this admirable factory of the Universe to that peak and lead you to love of the Supreme Creator, so that you may be enriched with that blessed wisdom, whose works you should always observe and admire to His glory.[108]

Even the image of the ladder or pyramid with God at its apex seemed to echo pantheistic conceptions, an impression that was reinforced by the layout of the first *Tabula*, deliberately conceived to suggest this effect (fig. 17): at a glance, the texts displayed the characteristic form of a ladder or pyramid with the Supreme Being at the top.

107 TM, p. 905.

108 Ibid.

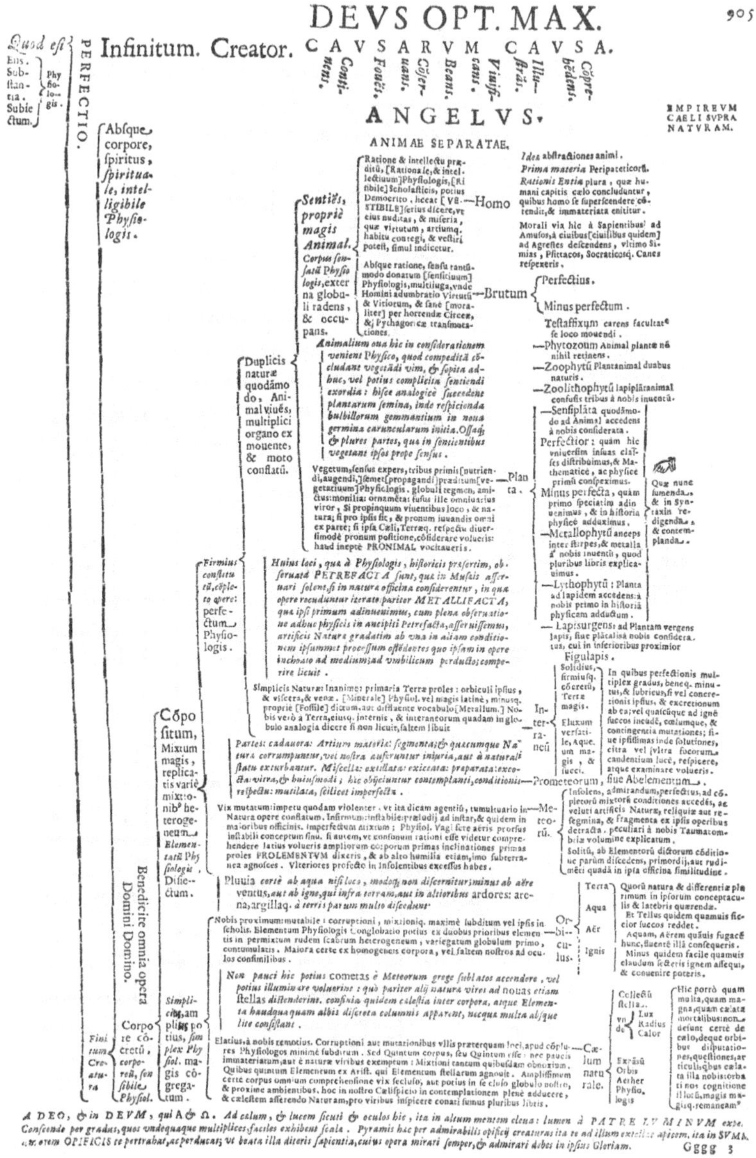

FIGURE 17 *The layout of the first* Tabula Phytosophica *in the form of a ladder or pyramid* (THESAURUS MEXICANUS, p. 905)

The Fate of Cesi's Fossil Wood Researches

In the years immediately subsequent to the death of the Prince, his letter to the Cardinal Nephew and the pre-publication manuscript of Stelluti's treatise on fossil wood were read and commented on, and not only in Italy. Those texts made a strong impression. Odescalchi mentioned the keen interest of the Lincean *manqué* Nicolas-Claude Fabri de Peiresc and of Gabriel Naudé in the fossil finds and deposits described by Cesi and Stelluti. Odescalchi was the first to publish the account of the *in situ* observations carried out in 1635 at the request of Peiresc by Jacques de la Ferrière, the personal physician of Cardinal Alphonse Richelieu.[109]

De la Ferrière provided a detailed account of the deposits near Acquasparta, which he had visited accompanied by a local guide who was familiar with the site explored by Cesi as well as with his studies of those finds. The French physician had also interviewed the local residents, who had expressed the opinion that those singular items were trees covered by earth after the Flood or some other cataclysm. De la Ferrière had become acquainted with Cesi's interpretation of the generation of those fossils through one of the numerous manuscript copies of Stelluti's as yet unpublished treatise. And his judgement of it was severe:

> As for what Prince Cesis [sic] says to prove this generation, that he had seen pieces that were like hard chalk at one end and wood for the rest, I believe that he was mistaken, taking the start of petrifaction for the start of the generation of this wood [...]. But it is easier for nature to transform wood into baked clay and stone than to turn earth into wood [...].[110]

In spite of his sincere admiration for the First Academy and the scientific work of its founder, Odescalchi took note, although with some embarrassment, of Cesi's misunderstandings regarding the nature and generation of fossil wood. However, he invited those tempted to draw negative conclusions from them regarding the intellectual and scientific stature of Cesi to take into account the context in which those theories had been formulated and the condition of natural history at the time.[111]

109 Odescalchi 1806, pp. 173–8.

110 Ibid., p. 177.

111 Ibid., p. 173.

De la Ferrière's harsh criticism of Cesi's interpretation of fossil wood, followed by the even more severe judgement of Gabriel Naudé, who had visited the fossil deposits of Acquasparta shortly after the French physician, provoked the vigorous reaction of another scholar, Ada Alessandrini, author of important contributions on the history and protagonists of the First Academy.[112] The pages in which Alessandrini heatedly contested Naudé's negative verdict provide striking evidence of the risks facing historical reconstructions with programmatically celebratory intentions. Alessandrini considered Naudé's harsh judgement an unpardonable offence. She portrayed that highly refined and learned French intellectual as a perfidious unbeliever: 'he was even considered a free-thinking atheist' and follower of an 'arrogant, unscrupulous and cynical Machiavellianism'.[113]

This embarrassment about Cesi's interpretation of fossil woods and the criticisms directed at it explain why marginal attention was paid to this area of the Prince's researches until recently, when interest in those singular subterranean finds dear to Federico was revived by David Freedberg. In collaboration with other scholars, he published an important volume in which almost two-hundred drawings of fossils commissioned by Cesi and kept today in Windsor were reproduced.[114] Freedberg contributed to that volume with two important essays, which he later developed in an exhaustive monograph on the history of the First Academy and Galileo's role in it.[115] The analyses

112 Alessandrini 1978, pp. 128–35.

113 Ibid., pp. 129–30. Alessandrini's resentment at these critical remarks on the founder of the Lincean Academy reached its peak in her comment on a passage from the letter to Peiresc of 29 March 1636, in which Naudé confessed that for a long time he had credited Cesi's interpretation of the generation of fossil wood from the earth because of the high esteem in which he regarded the Prince's capacities as a scrupulous observer and experimenter. His confidence in Cesi had changed, he added, after Jean-Louis de Bonnaire, who had known the Princeps of the Linceans personally, had informed him that Federico was the most credulous man in the world, ready to accept the most improbable notions. Alessandrini indignantly replied: 'This accusation is completely unfounded. Nothing corresponds less to reality than to judge a Lincean (and Federico Cesi in particular) as an inquirer who paid almost no heed to his own judgement, deferring excessively to that of others. It is certainly not with this uncritical method that the Linceans would have managed to attract the original and superb genius of Galileo into their zone of influence, to welcome him in their community, and even, on some occasions, to control and discipline him [...]!' (ibid., pp. 133–4).

114 Scott-Freedberg 2000, pp. 91–353. The catalogue of the drawings was edited by Andrew C. Scott.

115 Ibid., pp. 19–77; Freedberg 2002.

proposed by Freedberg in his panoramic contribution appear innovative and fully convincing. In my view, however, his reconstruction of the reasons that induced Cesi not to disclose his interpretation of the process of generation of fossil woods remains puzzling. In his opinion, out of respect for the late Prince's wishes, Stelluti deliberately concealed Cesi's true view of the generation of the metallophytes in Acquasparta from concern for the serious consequences that its publication would entail. What was so dangerous in Federico's interpretation of the fossil wood and, more generally, of the subterranean transformations on which he had focused his attention? According to Freedberg, the Prince understood that the phenomena of the subterranean world demonstrate that nature was not created once and for all, but is in a continual process of transformation. He realised that the introduction of the temporal dimension, that is, the idea that the Earth had a history, would run up against the narrative of the Creation as a perfect and thus immutable divine miracle described in the *Book of Genesis*. Fear of hostile reactions from ecclesiastical authorities drove him to censor his own ideas.[116]

I confess to finding this reconstruction problematic. There is too much explicit and convergent evidence suggesting that Cesi's ideas on fossil wood substantially coincided with those reported by Stelluti in his treatise. Further confirmation is offered by Cesi's handwritten notes on some of the fossil drawings that he commissioned.[117] In this case, also in Freedberg's important contribution, one perceives the distorting paradigm of Cesi's full endorsement of Galileo's methods and goals at work.

If instead of fitting Cesi into Galilean clothes we focus on his authentic personality and pinpoint the questions that most aroused his curiosity, we easily understand the reason that induced him to adhere to the interpretation of fossil woods as the product of the plastic energy that pervades nature. Instead of anachronistically labelling that interpretation as erroneous, we should recognise in it the natural outcome of the Prince's overall conception of nature, of the priorities of his agenda, and of the method used to accomplish them.

To repeat, there never was any Galilean conversion of either the Prince or the Academy. If we follow the course of Cesi's intellectual motivations and activity step by step, avoiding their conflation with Galileo's visions and methods,

116 Freedberg 2002, esp. pp. 342–5.

117 For example, no. 53 (ibid., p. 172) includes a note that clearly indicates the transformation of clay into a metallophyte: 'branch of clay not yet lignified turning into the branch of a tree'. No. 57 (p. 177): 'section through a piece of *cretilignum* [wood made from clay] view from the compacted side'. No. 66 (p. 188): 'dried *cretilignum* almost turned into wood'.

the fact that Federico was convinced that wood could be generated by clay soil through the action of endogenous heat ceases to surprise. This should not lead us, however, to pass negative judgements on the Prince's work.

Clearly distancing himself from the severe verdict on Cesi's credulity expressed by Bonnaire to Naudé,[118] Freedberg stated that 'everything we know about Cesi, if not necessarily about his colleagues, suggests that he was not credulous. On the contrary'.[119] If we follow Odescalchi's wise advice to place Cesi's views within the context in which he lived, belief in the transmutation of metals into gold, in the Philosopher's Stone, or in the possibility of the alchemical production of bread[120] was not taken to be proof of credulity, at least not by everyone. As incontrovertible evidence shows,[121] Cesi believed

---

118 Cf. above, n. 113.

119 Ibid., p. 69.

120 One of the most extraordinary secrets that fascinated Faber, Stelluti and Cesi was the alchemy of bread: that is, the alchemical production of bread with a nutritional content that Faber had enthusiastically communicated to Welser. Welser was not impressed: 'The alchemy of bread seems to me a great miracle if it not only fills you up, which would be useless in itself, but also produces an appreciable feeling of satiety; you know that I am very slow to believe in this kind of novelties' (CL, p. 373, letter of 2 August 1613). A week earlier Welser had displayed similar scepticism regarding information about other miraculous secrets sent him by Faber in a letter that is not extant: 'I had to laugh at the strangeness of that Neapolitan who promised to deduce from a person's handwriting not only his temperament and habits, but also a prediction for the following one or two years' (ibid., p. 171, letter of 25 July 1613). The Neapolitan was probably Prospero Aldorisio (see ibid., n. 2), who had also practised his art on the handwriting of Galileo after Franciotto Orsini had shown him an autograph letter by the Tuscan scientist (cf. OG XI, p. 557; letter of Orsini to Galileo, 24 August 1613, which includes Aldorisio's prediction). From Welser's letter to Faber we learn that the graphologist was in contact with Scipione Cobelluzzi. That strange person reminded Welser of another character – he had heard of him again from Faber – 'who had promised to record the human voice and words and transported them from one place to another'. In concluding his letter Welser expressed his own wry verdict: 'I think that they want to play a joke and promise similar items for that reason, not because they are convinced that they are feasible and practical' (CL, pp. 371–2).

121 Confirmation of Cesi's frequent contacts with alchemists can be found in the undated letter (probably from 1613) in which the learned Roman prelate Giovan Battista Lauro, who knew that Cesi was extremely curious about the most singular secrets, asked him to investigate the arcane powers of 'lightning gold', a highly explosive powder (ibid., p. 403; on the nature of this chemical compound, see ibid., pp. 403–4, n. 1). To bring into focus Cesi's attitude towards alchemy, it is useful to take into account the testimony of the German physician Winther, who was at Acquasparta in the service of Cesi. From there he wrote to Faber with disappointment, if not disgust, at the Prince's refusal to hear

that those transmutations were attainable by man; but this did not make the majority of his contemporaries regard him as credulous. To consult renowned astrologers to know whether the post-partum deaths of male heirs depended on negative astral conjunctures may be considered a sign of credulity today;[122] but in doing so Cesi was convinced that he would obtain precious information to solve a problem that concerned him; and most of his peers would have thought the same. It is evident from Cesi's writings and correspondence that he was profoundly convinced that the sky influences the Earth and its creatures; this belief – which is taken as a sign of credulity today – constituted the fundamental presupposition of his innovative and strong claim to the homogeneity of the universe. When it comes to his belief that the transmutation of metals was attainable by man, we are bound to admit that it was based on the conviction that nature herself is responsible for the imperceptible transition from one rung to another of the Ladder of Being, thanks to a vital force that is latent everywhere, activated by light and heat, as in the metallophyte deposits of Acquasparta; and that those operations of nature could be emulated by a natural magician – such as Giovan Battista Della Porta, putative father of the Academy – with the arts and techniques of chemistry, isolating the latent vital principles that are present also in bodies that seem inanimate.

If we take all those documents and testimonies into account and make a wise use of terms like 'mistake' or 'credulity', we find ourselves in a better position to grasp, without in the slightest diminishing their remarkable intellectual value, not only the originality of Cesi's researches on fossil woods, but also the entirety of his experience of life and research.

---

of chemical medicaments: 'the Prince will not hear of metals and transmutations, but only of ordinary waters which old ladies still know how to make in Germany' (ibid., p. 872). Given the copious evidence for Cesi's interest in transmutations, perhaps Winther's outburst should be understood as a reaction to the Prince's lack of esteem and sympathy for him, about which the German physician was to complain constantly to Faber.

122 As mentioned earlier, via Fabio Colonna Cesi repeatedly asked the renowned physician, astrologer and Lincean *manqué* Mario Schipani to give his opinion on the conjunctures responsible for the continual loss of his wife's male offspring. Cf. CL, pp. 1101, 1109, 1117 (and n. 3) and 1123.

CHAPTER 11

The Immaculate Conception of the Barberini Bees

Honey as a Gift from the Heavens and the Earth

In his important letter on the fossil deposits of Acquasparta to Francesco Barberini on 1 December 1624, Cesi announced his imminent return to Rome. The Prince wanted to take advantage of the extraordinary occasion of the Holy Year to strengthen his friendly relations with the Pope and the Cardinal Nephew, recently inducted into the Lincean Academy. The circumstance prompted him to write a learned and highly ostentatious tribute to the bees that adorned the stemma of the Barberini. That deferential homage was to be materialised in two distinct but coordinated formats designed to exalt the inimitable virtues of bees, who emblematically adumbrated the exceptional qualities and the supreme destiny of the Pope and his family: a plate, entitled *Melissographia,* with the three bees of the Barberini stemma at the centre, represented as they appeared when seen through a microscope; and a large broadsheet, the *Apiarium*, crammed with text.

The project of the *Apiarium* was conceived during the weeks immediately after the election of Maffeo Barberini as Pope Urban VIII. The proclamation of the Holy Year and the recent appearance of the microscope on the Lincean horizon encouraged the Prince to throw himself into the new enterprise, which was to occupy him for the whole of 1625, spent uninterruptedly in Rome.

If Cesi's motives for temporarily abandoning his drawings and writings on fossils and the intricate classificatory schemes of vegetals are clear,[1] little is known about how he came to conceive of the new project or of the stages through which it passed. We do know, however, that the decision to publish a folio with an engraving showing the bees of the Barberini trigon under microscopic enlargement from three different angles was rapidly put into practice: Cesi sent Francesco Barberini the result of his project at the beginning of February 1625. From Federico's letter to the Cardinal, which opened with a courtly salutation ('to the virtuous, nay most virtuous bees'), it appears clear

1 The temporary suspension of these researches is attested by Cesi's letter to Faber of 25 July 1626, in which he announced that, now that the large folio on the bees had been printed, he could return to the studies that he had abandoned: 'I have now resumed my dormant books' (CL, p. 1124).

 | DOI 10.1163/9789004342316_012

that Federico had enclosed the plate of the *Melissographia* (fig. 18) splendidly engraved by Matthias Greuter:

> May the accompanying sheet, remarkable for its content, although accompanied by a succinct text ('tenue per la penna'), represent to Your Excellency that which is imprinted perpetually in my mind, and may it be at the same time a portrait, albeit feebly sketched, of the greatness of the object and of the reverence of he who contemplates it.[2]

In spite of what the expressions used by Cesi in his letter seem to suggest, not one of the very few extant exemplars of the *Melissographia* bears a dedication to Francesco Barberini, while that to Urban VIII is clearly evident in all of them. It is therefore likely that Cesi had a personalised exemplar printed for the Cardinal Nephew to wish him success in the delicate legation that he was about to undertake in France. The Prince might also have thought that the dedication to Urban VIII included the entire Barberini family.

The *Melissographia* is well known not only for its remarkable beauty but also because the engravings of the bees are the very first printed representation of an insect observed through a microscope as well as the first public appearance of the term 'microscope' – a word coined by another Lincean, Johannes Faber, shortly before.[3] As the brief inscription states, the scrupulous observations had been carried out not by Cesi himself, but by his trusted friend and assistant Francesco Stelluti: 'observed by Francesco Stelluti, Lincean of Fabriano, by means of a microscope'.

---

2 Ibid., p. 1022, letter of 1 February 1625. According to Gabrieli (ibid., n. 5), the possibility that it was a first draft of the *Apiarium*, perhaps still in manuscript form, cannot be excluded. Guerrini, on the other hand, suggests that Cesi might have sent the Cardinal Nephew only a section of the *Apiarium* (Cesi 2005–2006, I, p. 85). However, the expression 'tenue per la penna' unequivocally points to the *Melissographia*.

3 On the image in the *Melissographia* of the bee seen through a microscope, see Belloni 1972, Freedberg 2002 (pp. 160–3) and Cesi 2005–2006 (pp. 89–93). Faber first mentioned the microscope, calling it a 'bellissimo occhialino', in a letter to Cesi of 11 May 1624: 'I spent yesterday evening with Galileo [...]. He has given a very fine *occhialino* to the Cardinal of Zollern for the Duke of Bavaria. I observed a fly [...]; I was astonished, and said to Galileo that he is another creator' (CL, p. 875). Later Faber sent the Prince for approval the draft of the passages of the *Thesaurus Mexicanus* on the discoveries of Galileo in which he proposed for the 'occhialino' the name 'microscope' by analogy with 'telescope': 'And I also mention this new *ochiale* to see small things that I call a microscope [...]' (ibid., p. 1038, letter of 13 April 1625). After Cesi's approval, the text was published in the *Thesaurus Mexicanus* (pp. 473–4).

FIGURE 18 *Engraving by Matthias Greuter of the Barberini bees viewed through a microscope in the* Melissographia

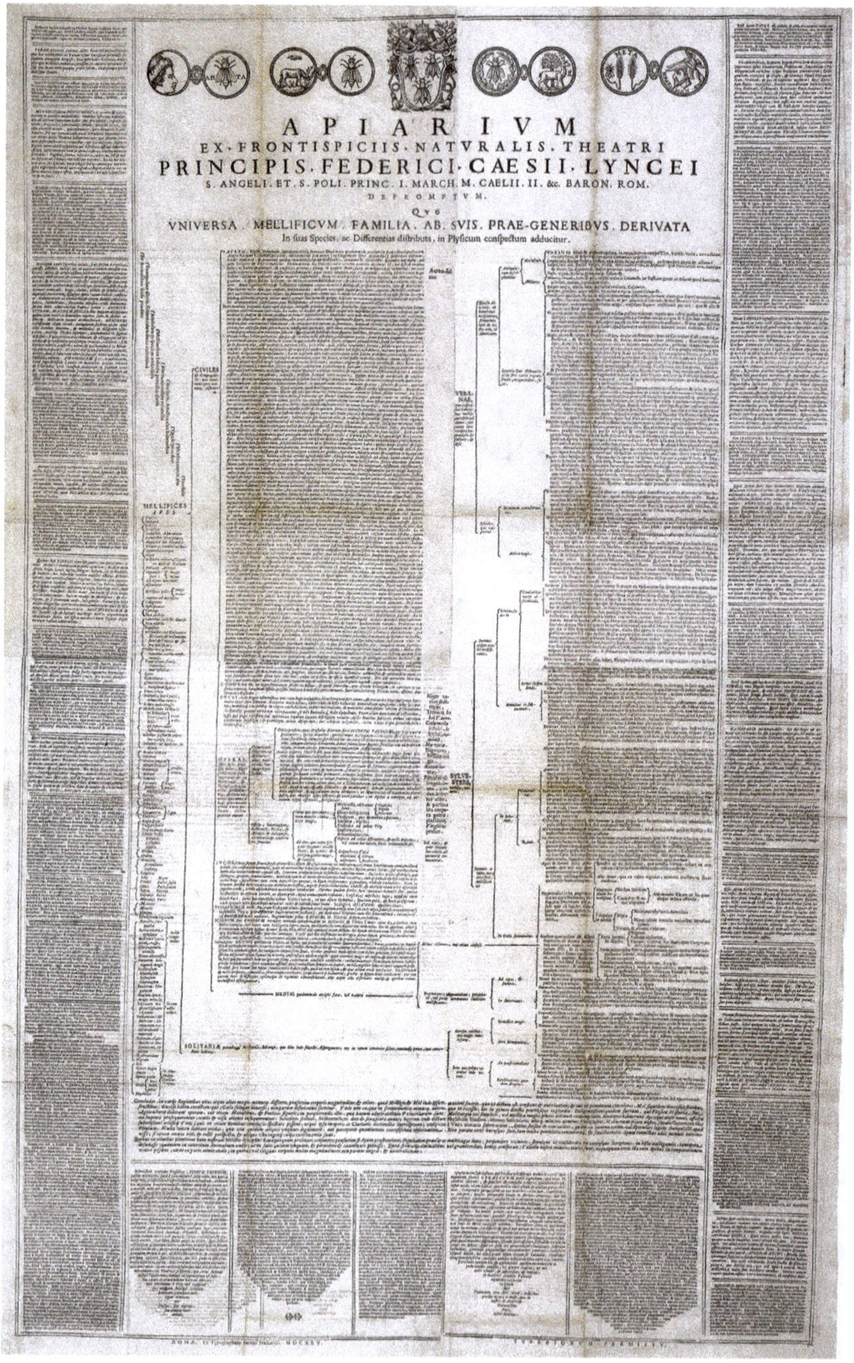
APIARIVM
EX·FRONTISPICIIS·NATVRALIS·THEATRI
PRINCIPIS·FEDERICI·CAESII·LYNCEI
S. ANGELI. ET. S. POLI. PRINC. I. MARCH. M. CAELII. II. &c. BARON. ROM.
DEPROMPTVM.
QVO
VNIVERSA. MELLIFICVM. FAMILIA. AB. SVIS. PRAE-GENERIBVS. DERIVATA
In suas Species, ac Differentias distributa, in Physicum conspectum adducitur.

FIGURE 19 *The large folio of the* Apiarium (CESI 1625)

In the cartouche at the foot of the sheet, the work of another Lincean, Josse de Rycke, the eulogistic homage to the Barberini *genus* was followed by praise of Cesi for having devoted himself to studying the bee, a creature 'than which nature has nothing more remarkable', taking advantage in this delicate and almost sacred task of the new optical instrument: 'the greatest miracles emerge with the polished glass and the eye learns to have greater faith'.[4] Note the omission of the fact that the microscope had first been shown to the Linceans by Galileo.

The second and more relevant component of the celebratory project conceived by Cesi was also the more complex. This was the large folio of the *Apiarium* (fig. 19), the first of the many Frontispieces of the *Theatre of Nature* conceived by Federico to be published.[5] This document, which is fundamental for an understanding of the intellectual identity of the Prince, is remarkable also for its intricate lay-out.[6] The gigantic dimensions of the folio (107 x 69.5 cm), its extremely complex structure, articulated in 94 small blocks of text, the use of eight different founts (roman, italic, small capital and capital) obliged Cesi, and to an even greater extent, the Roman typographer and printer Iacopo Mascardi, to engage in a laborious exercise of calibration of the texts. This implied the constant presence of the author in the typographical workshop to expand or contract the texts so that all the frames and boxes of the broad sheet were completely filled. The difficulties faced in the composition and layout of the *Apiarium* are clear evidence of the extreme value that Cesi attributed to making it possible to take in at a single glance a synoptic view of all the information, focusing on every aspect connected with the apian world.

This innovative and arduous typographic solution was imposed by the very nature of the project and corresponded to the objective pursued by Cesi. As the examination of the *Tabulae phytosophicae* in the following chapter will confirm, each of the Frontispieces of the *Theatre of Nature* and *Mirror of Reason* would be composed following the model of the *Apiarium*.

An excellent edition of the *Apiarium* was produced by Luigi Guerrini with an Italian translation of Cesi's extremely challenging Latin by Marco Guardo and a one-to-one facsimile of one of the four extant exemplars of the large sheet. The facsimile makes it possible to appreciate the panoramic vision of the

4 'Natura qua nil praestantius omni [...] maxima dum tereti surgunt miracula vitro maioremque oculus discit habere fidem' (Stelluti-Rycke 1625).

5 Cesi 1625. For the critical edition of the *Apiarium* by Luigi Guerrini, see Cesi 2005–2006, I.

6 See ibid., pp. 59–87 for the process of composition of the *Apiarium*; for the complex articulation of its blocks of text, cf. Guardo 2004. Cf. also Kidwell 1970.

finely articulated organization of contents to which Cesi attributed extreme importance.[7] In his introduction Guerrini reconstructed the genesis of the project, analysing the preliminary charts of the text sent to the printer by the Prince and identifying his sources. Guerrini and Guardo are also responsible for the digital edition of the *Melissographia* and the *Apiarium* available on the website of the Museo Galileo.[8] The website permits navigating those documents, accessing the individual blocks of text of the *Apiarium* in any order, passing from the original Latin to the transcription and Italian translation, searching the texts by form and lemma, as well as consulting the bibliography. Besides enabling detailed comparisons of the four extant exemplars of the work, the web edition also provides access to the preliminary drafts of the *Apiarium* contained in Ms. Archivio Linceo IV, the proofs of the individual blocks of text with Federico's autograph corrections, and the observations of the Linceans to whom the Prince submitted his work.

Like Federico's other works, the *Apiarium* displays the variety of visual angles from which he observes the world of bees. The celebratory aim appears explicit: the entire folio is dominated by the Barberini stemma with the trigon of bees. Moreover, the exaltation of the quality of the different varieties of bees reaches its climax in the description of the *apes urbanae* (town bees, an unheard of species, invented to praise Urban VIII) to which Cesi attributed the most noble and unselfish virtues that awaken admiration and astonishment.

Following his usual procedure, the physical, symbolic, moral and social virtues of bees are illustrated by means of an exhaustive list of etymologies, testimonies, myths, images and symbols conceived by classical authors and renowned naturalists from antiquity onwards.[9] The almost divine nature of bees

---

7 Cesi 2005–6, I and II.

8 http://brunelleschi.imss.fi.it/apiarium/index.asp (accessed on 10 July 2017). This digital edition is the result of collaboration between the Museo Galileo and the Accademia Nazionale dei Lincei.

9 Massimo Baldini (Baldini M. 1990, esp. pp. 235–41) reconstructed some aspects of the discussion of the nature and social organization of bees in the works of numerous writers on agrarian matters between the sixteenth and the eighteenth centuries. For example, the popular treatise *Le venti giornate dell'agricoltura* by Agostino Gallo (Gallo 1573) exalted the divine nature of bees and emphasised the essential function of dew for the production of honey. Other highly influential writers, such as Piero de' Crescenzi (Crescenzi 1605), had insisted on the perfect moral conduct of bees and their resolute aversion to sexual practices. Massimo Baldini also stressed the widespread inclination until the mid-eighteenth century to anthropologise the model of apian society. Although Cesi's *Apiarium* was not mentioned by Baldini, his contribution is useful for placing it within the context of the lively debates on bees with which the Princeps of the Linceans was certainly familiar.

extolled by literary tradition was again stressed in those boxes of the broadsheet in which Cesi emphasised how the exceptional qualities of these small but impressive creatures transpired also from their external and internal physical constitution, from their ability to produce a substance as divine as honey, and, above all, from their reproductive process.

Although the *Apiarium* had been occasioned by a particular circumstance, it did not deviate from the research programme illustrated in the *Indicatio*. The statement that honey, like sugar, wax and manna, was to be considered an intermediate nature because it was produced on Earth but thanks to essential substances that fell from the sky linked the 1625 broadsheet[10] with the *Zibaldone*.

Moreover, Cesi drew the reader's attention to the imminent publication of his *Taumatombria*, one of the Frontispieces described with greater prominence and in more detail in the *Indicatio*:

> [...] in which [the *Taumatombria*], among several other miraculous rains, we have contemplated not only rains of honey, manna and sugar, but also of waxes, gums, and storax; so that you can embrace both the floreal and celestial gifts of honey.[11]

The *Apiarium* also reveals the crucial importance that Cesi continued to assign to Della Porta's doctrine of chemical transmutations. Bees are presented as minuscule and highly refined distillation laboratories, a specular image of the operations carried out in the great chemical workshop of nature. In his account of the process by which bees produce honey, the Prince stressed that they perform a function complementary to that of the Sun. He went so far as to call them daughters of the Sun.[12] The Sun draws fluids from the bowels of the

---

10 Cesi 2005–2006, I, pp. 21–69.

11 Ibid., p. 140 and n. 74.

12 The stress on the solar origin of bees served to exalt their capacity to produce a substance, wax, that gave (intellectual) light to humanity. According to Guerrini, Urban VIII was to draw inspiration from this block of text in the *Apiarium* to compose the distich in praise of bees inserted in his *Poemata* of 1628:

Luce fugat tenebras, cera dum vestit ab igne
Accensas faculas, aemula Solis Apis.
(In coating with wax the lighted torches,
the bee, like the Sun, dispels the shadows with light),

Barberini 1628, p. 48; Guerrini 2007a, p. 126. Campanella wrote lengthy commentaries on Maffeo Barberini's poems (cf. Bolzoni 1989 and 2001). In the comment on this distich

Earth and transforms them into drops of dew that fall back and provide plants with precious nourishment.[13]

At this point the bees suck dew from the flowers and, by means of a new transmutation carried out in their organic chemical laboratory, they transform it into honey. Honey, Cesi concluded, was a gift from Heaven and from the Earth at the same time and was thus to be considered an intermediate substance.

The section of the *Apiarium* in which Federico elaborated these reflections on the double transmutation worked by the Sun and by the bees contains an allusion to Galileo, though he is not mentioned by name. In drawing exhalations from the Earth, the Sun is marginally contaminated by them as shown by the dark spots on its surface imperceptible to the naked eye.[14]

Two of the boxes of the broadsheet were devoted to illustrating the operations performed by the bees in producing wax and honey:

> They compose, mix and work, until they become particularly waxy, those substances which often spring up amidst the flowers or seep along the bark beyond the surface of the flowers, and which the bees themselves tear into small pieces and remove.[15]

In confirmation of the continuity with the research project sketched in the *Indicatio*, a few lines further on the Prince cited the *Oposcopia* and described the elementary components of wax as irregular particles that are thicker than clots and finer than grains of sand and which dissolve completely when heated. The mention of the operation of fire is an obvious allusion to distillation, which Federico defined as the most precise scales of nature, which makes it possible to break down honey, manna and sugar into their ultimate constituents:

> Take [wax] and examine it with the scales of nature. If you want to know its nature and composition, divide it, dissolve it, coagulate it [...]. In honey you will find the characteristics of water and a salty taste; in sugar, in a milder and subtler way, those of oil, while in wax, also those

(Formichetti 1983, pp. 77–9), he used the Lincean broadsheet on the bees, seizing the opportunity to praise Cesi, the Lincean Academy, and the wise use made of the microscope by its members (Guerrini 2007a, pp. 128–46).

13 Cesi 2005–2006, I, p. 140.

14 Ibid.: 'Chymici haud difficile omnes Apollini, qui fumos elevet, qui fumis minime tingatur ipsi Furvi undequaque primas relinquent'. Obviously Cesi refers here to the Tuscan scientist's *Letters on sunspots*.

15 Ibid., p. 142.

> of sulphur and sugar. These substances melt in fire, but pure wax is fully consumed by fire. Honey, which is similar to gum, when mixed with sugar contracts; the situation of manna is different [...]; manna is halfway between wax and honey, so that it has a double nature. And if you consider its flavour and consistency, it appears as an intermediate nature between honey and sugar.[16]

For Cesi chemical analysis demonstrated that the differences between honey, manna, wax and sugar were barely perceptible. His definition of manna as an intermediate substance evokes his discussions of the metallophytes as well as the title of one of the Frontispieces of his *Zibaldone*: *De mediis naturis*.

If the scrupulous chemical analysis of honey and related substances provoked astonishment, the close microscopic examination of the bee and its organs, arranged in perfect harmony, produced even greater stupefaction. The unprecedented magnified view made it possible to admire the remarkable instruments that nature had bestowed on bees. Microscopic observation provided tangible proof that the finest demonstration of the perfect operations of nature was to be found in the tiniest creatures.[17] Once again, Cesi was echoing a passage from the *Zibaldone* – compiled, it should be emphasised, before Cesi knew of the microscope – in which he had praised the *minimorum magnitudo*:

> Our dictum: pay attention to the smallest if you want to know the greatest. For everything seems to consist of the smallest elements.[18]

The exaltation of the new vision of nature provided by the microscope did not, however, induce Cesi to conclude that it led to the discovery of the ultimate principles of reality. Even if the instrument made it possible to observe creatures and objects that were invisible to the naked eye, many others of even smaller dimensions continued to escape notice. Just as the telescope had made it possible to observe numerous celestial bodies invisible to the unaided eye but did not allow observation of an infinite number of others situated in the most remote parts of the heavens, the finest elementary structures remained beyond the reach of the microscope. Federico emphasised the limits of microscopic observation in a section in which, to refute the thesis that bees were deaf, he maintained that they did possess auditory organs, but so small that they could not be

16 Ibid., pp. 142–3.
17 Ibid., pp. 146–8.
18 Contributi I, p. 47.

seen even with the microscope.[19] The role of the natural philosopher therefore remained fundamental in throwing light on the elementary components and their operations not through observation, but by deducing them from the careful inspection of the principles and methods employed by nature, combined with the evidence furnished by the results obtained through distillation.

These affirmations help us to grasp the relation that Cesi established between the results attainable by observation of natural phenomena with the aid of the microscope and those acquired by breaking substances down into their elementary components in the alembic. Microscopic observation helped to decipher the constitution, functions and modes of operation of organisms, while chemical analysis made it possible to isolate the ultimate components of substances that could not be revealed by the optical instrument. Though performing integrated functions, the microscope and the alembic thus produced information belonging to different levels of reality.

The allusion to Galileo's discovery of sunspots was not the only reference to the Tuscan scientist in the *Apiarium*. In the block of text in which he underlined the imperceptible differences between manna, wax, honey and sugar, the Prince launched a frontal attack on the Peripatetic doctrine of qualities and the principle of a clear-cut separation between the substances associated with heat and those considered cold. He declared that Aristotle and his followers had considered heat to be a quality opposite to that of cold. This was a grave error, resulting from the Greek philosopher's failure to recognise that heat and cold both originate from the same cause, penetration:

> Penetration is the only source of heat and cold; of cold if the penetration is followed by a slow process of densification, of heat if followed by a rapid movement. Fire is the conflict of particles in solution; flame emerges when the solution is almost complete; light is born from the expansion and complete dissolution [of the particles]. We have considered these matters in detail in our *Physica Mathesis*.[20]

Once again we find a reference to the research programme outlined in the *Indicatio*, in which the *Physica Mathesis* was presented as one of the fundamental Frontispieces of the *Mirror of Reason* and the *Theatre of Nature*.

The reduction of the phenomena of heat and cold to the principle of penetration has a distinctly Galilean ring to it. Also the interpretation of light as

19 Cesi 2005–2006, I, pp. 151–2.

20 Ibid., p. 144.

the product of the most extreme dissolution of the material particles echoes the analyses conducted in texts of the Tuscan scientist familiar to the Prince, above all the letter to Piero Dini of 1615.

When Cesi composed the *Apiarium*, he came fresh from his reading of the *Assayer*, the work in which Galileo had demonstrated the subjective nature of the secondary qualities, insisting particularly on those of heat and cold. For the Tuscan scientist, it was the impact of material corpuscles on our sense organs that, depending on their structure and their various velocities, produced the sensation of hot and cold, hard and soft, and so on. Cesi was struck by that innovative analysis and grasped its revolutionary implications.[21] This was his reason for using it in his careful study of bees and in the exaltation of their virtues.

However, Cesi's assimilation of the Galilean theory of sensations was anything but passive. The penetration to which the Prince referred as the principle from which heat and cold are generated did not operate through mechanical actions. The term most often used in Federico's text on penetration cited above is *solutio* (and its opposite, *dissolutio*), which belongs to the technical language of alchemy. Cesi adapted the purely mechanical interpretation of the origin of sensations in the *Assayer* to the framework of his own general vision of nature. In the concluding part of the block of text describing the action of penetration, he declared that, although deriving from the same causes, honey, wax, sugar and manna displayed different qualities depending on the disposition of and contacts between the elementary particles. Those contacts activated processes that presented the characteristics of chemical reactions, from which various 'mixtures' were produced which our sense organs perceive as distinct qualities:

> Such substances [manna, wax, sugar and honey] flow variously, depending on the type of particles of which they are composed. Roughness emerges from the viscosity and the consequent bondings; edges from the salty taste; density and dilation from greasiness; openings, partitions and the presence of small inner spaces from the penetration of water, oil and boiling substances. All these transformations take place through subtraction and repeated combinations, through the configuration and composition of the particles.[22]

21 Guerrini (Cesi 2005–2006, I, p. 164, n. 91) expressed his reservations, which I share, on Freedberg's thesis of the linear dependence of Cesi's corpuscular theory, as evidenced in the *Apiarium*, 'on Galileo's own corpuscular view of matter' (Freedberg 2002, p. 175).

22 Ibid., p. 145.

Cesi transformed Galileo's rigorously mechanistic corpuscularism into the conception of chemical reactions between particles of various substances and between particles and sensory organs. It is a telling detail that the velocity of the impact of the corpuscles – an essential factor for Galileo's corpuscular theory – is completely lacking from Cesi's description of the interaction between particles and sensory organs.

The Prince had already expressed his intention of developing an in-depth analysis of the nature and origin of qualities in the Frontispiece of the *Physica Mathesis*. In the outline of that section of the *Theatre of Nature* he had described the transformations of the structure of matter produced by processes of coagulation or separation of corpuscles and governed by relations of sympathy and antipathy between the elements.[23] One of the sections of the *Physica Mathesis*, divided into four chapters, was entitled *de qualitatibus*. Although that extremely concise outline makes it impossible to form a precise idea of how the Prince intended to proceed, it seems likely that he would have concentrated on the chemical processes that dissolve the bonds on which cohesion of the corpuscles of the various natural substances depend.[24]

Galileo's interpretation of the subjective nature of qualities had made a deep impression among the Linceans even before the publication of the *Assayer*. The attention of Cesi and his closest associates had been drawn to Galileo's new interpretation of sensations from the time of the publication in 1619 of the *Discourse on the comets* in which these innovative conceptions were first presented. This is borne out by the lecture on fire and metals that Faber gave at the Sapienza University in Rome on 20 November 1622.[25] By then

23 Contributi I, p. 34.

24 'Luce apertione; motione, calore, frigore; tenui, humido, sicco; nexus, laxactione et resistentia quoad partes secundarias' (ibid., p. 34).

25 That lecture reflected the lively debates on the nature of vitriol fuelled by the professional rivalry between Faber and Pietro Castelli, a physician held in the highest esteem in Rome. For Castelli, vitriol was an acid and cold substance, even if it could cause burns. Faber, who had joined the debate in defence of the opinions of the German Lincean manqué Raymund Minderer, who had come under attack from Castelli, declared resolutely that vitriol should be classified among the hot substances. An echo of those polemics is also to be found in Faber's letter to Schreck of April 1622. Faber, who must have let off steam with the Jesuit and former Lincean for the absurd positions assumed by Castelli, had received an obliging reply: 'Castelli seems to have risen to such a high level of abstraction that he now denies that hot objects cause a sensation of heat. For to heat to the highest degree is to burn, and I can hardly see how this differs from being caustic' (CL, p. 756). For a detailed analysis of Faber's lecture, see De Renzi 2007, pp. 271–313. For a reconstruction of the animated debates on chemical medicine in Rome in the 1620s and 1630s, see Clericuzio 2009, esp. pp. 286–302, and Clericuzio-De Renzi 1995.

the manuscript of the *Assayer* had already been in Rome for a month, but Faber would not be able to read it until March 1623. Though without explicitly mentioning Galileo, Faber referred to him twice. In reviewing the opinions of the ancient philosophers on the nature of fire, in particular those of Plato, who, like Telesio, attributed a fiery nature to the heavens, and of Aristotle, who considered it the element closest to the sky of the Moon, Faber declared that a modern emulator of Prometheus had cleared the field of those erroneous theories:

> Flying in secret to the concavity of the Moon, he snatched that fourth element of fire from the Aristotelians and, acting as a new Prometheus, he also robbed the Platonists of their celestial fire.[26]

Faber had apparently appreciated the innovative character of the interpretation of heat generated not by motion, as Aristotle had affirmed, but by the friction of bodies illustrated in the *Discourse on the comets*.[27] In confirmation of that principle, Galileo-Guiducci had proposed the example of a file that wears iron away by rubbing, causing its temperature to rise sharply because of the emission of 'very fine parts' which, moving, 'penetrate the pores of our flesh'.[28] As evidence that he had read that page carefully, Faber repeated the same example:

26 Cited by De Renzi 2007, p. 291. The image of Galileo as a second Prometheus would be proposed again by Cesi in the seventh of the *Tabulae Phytosophicae*, cf. below, p. 442. Cesi and his closest associates wanted to exalt the Tuscan scientist's fundamental contribution to the recognition of the homogeneous and unitary nature of the heavens: as Prometheus had broken the barrier between gods and mortals, so Galileo had brought the heavens into direct communication with the Earth. In their writings and correspondence the Linceans often compared Galileo to famous mathematicians and philosophers of antiquity and the Renaissance, as well as to the mythical protagonists of great feats. Stelliola compared him to Atlas; Ciampoli drew an analogy with Socrates (cf. Favino 2007), while Faber, in the *Thesaurus Mexicanus*, compared him with both Socrates and Endymion. In the dedication of the *Assayer* to Urban VIII, Galileo's celestial discoveries were assimilated to those of Columbus and Vespucci, while Colonna evoked the voyage of the Argonauts (CL, p. 262); to Heckius Galileo's discoveries recalled the important intuitions of Anaxagoras and Archimedes (ibid., p. 473), while Faber saw in him the reincarnation of Epicurus and Pythagoras (ibid., p. 835). In his letter to the Dupuy brothers of June 1633, Bouchard compared Galileo with the ancient philosophers (ibid., p. 1242).

27 Drake-O'Malley 1960, pp. 30–2 (OG VI, pp. 53–6).

28 Ibid., p. 32 (OG VI, p. 56).

> Fire will therefore only be present in a certain air, or in a fine exhalation, and, while it is being produced, it is constantly driven further and, like a file, continuously wears and wears away, opens and penetrates, reduces and heats, and whatever has that power can rightly be called fire.[29]

To indicate the particles by which the sensation of heat was produced, Faber referred not only to '*atomi*' but also to '*ignicoli*' (atoms of fire), a term widely used by Galileo in his writings on hydrostatics: when flint was struck by iron – he wrote – it emitted *ignicoli*. Antonio Persio had also used the example of flint struck by iron in his *De natura ignis*, with which both Faber and Cesi were thoroughly familiar.[30] It is not surprising that, as Guerrini pointed out, the section of the *Apiarium* in which Cesi described the ambiguous and intermediate nature of manna, wax, sugar and honey contains elements that go back not just to the *Assayer*, but also to Telesio's conception of matter – further evidence of the influence of both Telesio and his disciple Antonio Persio on the Lincean circle.[31]

The reference in the *Apiarium* to the Frontispiece of the *Physica Mathesis*, a project outlined many years before the Prince was able to use the microscope and tackle Galileo's interpretation of secondary qualities, indicates that he regarded the analysis contained in the *Discourse* and the *Assayer* as perfectly consistent and convergent with the programme of research outlined in his *Theatre of Nature*, as well as with the motivations that had induced him to create the Academy.

The largest block of text, situated in the middle of the broadsheet, leads to similar conclusions. The manuscript Archivio Linceo IV, which contains its preliminary drafts, shows that this section cost Cesi the most effort.[32] Aware of its fundamental importance, he submitted it for perusal to Faber, Stelluti and Colonna before having it printed.[33]

The block contains a series of not particularly perspicuous reflections – and not only due to the Prince's tortuous Latin – on the process governing apian reproduction. Cesi unwaveringly denied that bees were born from eggs fertilised

29 De Renzi 2007, pp. 293–4.

30 Ibid., p. 293, n. 43.

31 Cesi 2005–2006, I, p. 144, n. 91.

32 Ibid., pp. 59–75.

33 In the critical edition of Ms. Archivio Linceo IV (ibid., II) the corrections proposed by the other Linceans are highlighted and, where possible, their individual interventions identified.

in the mother's womb. Bees were formed by the direct imprint of the father's vital spirits on the 'generative matter', which Cesi identified with honey. Into this substance the king bee introduced vital spirits which conferred what we would today call the genetic imprint of the species. The spirits imparted life to the honey-like matter, predisposed by nature to generation, gradually imprinting the forms of the organs and limbs of the bee on it:

> The spirits themselves that are imprinted in this generative substance are sufficient for the act of formation. They are released from the whole body of the father, very distinct and bearing the seal, as it were, of multiple breath, marked by figures more firmly impressed, internally or externally they establish on every part of the surface of the substance the position from which the link with the aspect and configuration of the bee fully emerges.[34]

Cesi's model of apian reproduction can be regarded as a case of immaculate conception. Although he was convinced of the immaculate nature of that reproduction – similar to the process by which the Earth produced metallophytes – it also served his purpose in praising chastity, purity of mind, nobility of character and the incomparable ingenuity of bees that emblematically adorned the Barberini Pope's stemma. However, fearing that the imaginative reproductive sequence might not appear convincing, he invited the reader to refrain from scepticism because the figures of the limbs impressed by the king's spirit on the honey and their progressive transformations through a process of maturation to become living organisms endowed with harmoniously disposed limbs could not be seen with the naked eye, nor with the aid of the microscope. Did not the fact that the invisible particles emitted by bodies were distinctly perceived by our sense of smell as pleasant or unpleasant odours show that nature worked through agents invisible to man? And who could contest the claim that the perception of different scents depended on the varied composition of the elementary particles composing natural substances as well as the fact that each of them carried out its action according to its specific original imprint?

> All particles bear the maternal characteristics – the Prince stated – although they are invisible to our eyes.[35]

34 Ibid., I, p. 167.

35 Ibid.

To further support his argument, Cesi stressed that man's sense of smell clearly perceives a multiplicity of odours that are not discernible through sight or touch. It was in this context that he inserted another unequivocal reference to the Galilean conception of the subjective nature of secondary qualities and the corpuscular interpretation of the genesis of sensations. This time the allusion to those famous passages in the *Assayer* was more explicit, even in its terminology. The previously illustrated theory of the penetration of corpuscles, characterized by a transparent chemical approach, was now enriched by the more genuinely Galilean concept of the collisions and the diffusion in space of extremely minute particles:

> So you distinguish flavours and odours by the various collisions of particles; with your nostrils you perceive vapours emitted from things as different, even though they have the same appearance; and you even perceive those extremely minute and inconspicuous particles that emanate from musk and are spread through the entire bedroom.[36]

However, even in this passage, which displays his closest adhesion to Galilean corpuscular mechanics, Cesi made no mention of the determinant function attributed by the Tuscan scientist to the various shapes and configurations of the particles and, above all, to their variable velocity.

'This Work Has Been Done for the Sole Purpose of Pleasing Patrons'

Even when he came closest to Galileo's revolutionary theory of the subjective nature of sensations and their production, Cesi put that conception to structurally different uses from those that had led Galileo to give them pride of place in pages which he regarded as of fundamental importance for gaining acceptance of his own mechanical-corpuscular conception of the structure of matter. Though the Prince had read the analysis of the genesis of sensations in the *Assayer* very closely, he had not passively subscribed to their conclusions, nor to the theoretical framework on which they depended. For Cesi, the particles emitted by the body were not figured neutral matter in motion at different velocities, but corpuscles that stably maintained the constitutive characters of the living organism from which they emanated. The highly developed sense

36 Ibid.

of smell of dogs was eloquent proof of the processes activated by nature in its workshop:

> And if you consider the power of a dog's sense of smell, that is, its highly sensitive nostrils, you will learn to perceive the different scents given off by the bodies of individual animals, which leave their traces in the streets even after a long time and even after rains. For unless particles remained in the figures of the maternal constitution, they would be barely able to transmit those scents and our senses could not perceive and distinguish those particles, which all derive from their maternal constitution and stamp themselves on the senses on which they act.[37]

The section on apian reproduction contains many other clues that help us reconstruct the Prince's mental laboratory. His fascination with conceptions that go back to the Paracelsian and Neo-Platonic tradition appears evident in his terminology. The particles present in matter contain rudimentary figures of the organisms that they are destined to compose. Those particles are set in motion by heat, which in the course of time is converted into fervour. The boiling process triggered by that fervour produces a series of transformations that amalgamate or dissolve the corpuscles. Nature follows a lucid plan in governing the process that causes the gradual emergence from these chaotic processes of bonds that tend to stabilise between similar particles. Some elements coalesce to form fluids, while others harden and become more compact. At a certain point, these processes reach the equilibrium wisely programmed by nature:[38]

> And to prevent excesses in either direction, nature has tempered movement with rest and freedom with gentle restraints. Thus the exhalation produced by vital heat is impeded beyond fever point, and is permitted except below the threshold of dissolution, maintaining a balance between agitation and torpor, allowing that gentle heat to settle as though it had taken root.[39]

---

37 Ibid., pp. 167–8.

38 Cesi deliberately uses the word 'providdamente' (providently) to underline that nature never operates randomly but is guided and predefined by the wisest design. The Prince's corpuscular theory is not only substantially chemical, but it is also structurally providential. It thus clearly differs from Democritus' atomism (a point of reference for Galileo), in which chance, that is, the absence of a preordained plan, plays a fundamental role.

39 Ibid., p. 165.

Besides the echo of the discussion of secondary qualities in the *Assayer*, the section on bee reproduction contains analyses and themes characteristic of chemical philosophy, integrated with concepts and terms taken from the Aristotelian-Galenic doctrine of the temperaments. For Cesi, life and health consist of an equilibrium of substances, movements and gradations of heat. The excessive heat of the initial stages of the generative process is progressively tempered, allowing the 'more subtle and less excited' particles to act as a 'spirit' that governs in an orderly manner the organs that have now been formed.

For the Prince nature is a prolific mother. Creatures emerge

> from the womb [of nature]; nor should it appear strange that very many living organisms can be produced from a single operation of hers, and that their parents are contained in her forces.[40]

Aware of how close his interpretation of these processes was to the doctrine of generation from putrefaction,[41] Cesi felt the need to specify that the generative power of nature could not be expressed with a term like putrefaction, which was not only vulgar, but also inadequate to illustrate her wise workings:

> This prolific fertility belongs to the Great Mother, who has unjustly been given the name of putrefaction in the Schools.[42]

The terminology changed, but not the forces, nor the faculties that came into play in the generative processes. For Cesi, matter in its ordinary state is predisposed to generation. The heat and the imprint of the parents are sufficient to

40 Ibid.

41 Paracelsus had assigned an essential function to putrefaction in the generative processes; while the heaven was father, the soil of the Earth and the elements were the wombs containing the seeds. Like Cesi, he had underlined the strong analogy between seeds and eggs, i.e. between plants and animals. Generation was a germinative process that began with putrefaction and developed under the heat of the Sun; the various vegetals with their specific configuration implicit in the seed were generated by putrefaction (Pagel 1989, pp. 96–7). Paracelsus and Cesi also shared the conception of vital spirits that emanated from the stars and permeated the whole of nature, both organic and inorganic (ibid., p. 98). Pagel maintained that the theory of generation elaborated by Paracelsus is 'simply the separation of objects preformed in an Anaxagoran chaos' (ibid., pp. 250–1). Our previous analysis of Cesi's *Indicatio* as well as that of the *Tabulae Phytosophicae* below confirm the strong fascination that the Presocratic philosopher's conception of nature held for the Prince.

42 Cesi 2005–2006, I, p. 165.

unleash its latent creative energy. As in his interpretation of the production of fossils, Cesi advanced a modified version of the traditional doctrine of generation from putrefaction in which the spontaneous workings of nature continued to play a fundamental role, though attaining perfection only through the intervention of the parents:

> Thus in this wonderful procreation of these extraordinary and minuscule creatures, nature seems to collude with the parents, or rather to cooperate in generating them [...]. From this convergence figures and connections are produced, as well as the first and successive heats of the unions and incubations; and you suddenly find nutrients suitable for the growth of the body.[43]

This remarkable alliance between nature and parents achieved the most extraordinary results in the procreation of the king bee. In this case, the generation was not the result of a gradual process, as for other bees. The king was already perfectly formed at the moment of procreation:

> There is indeed a remarkable force both of nature and the father in the procreation of new kings through a prodigious synthesis. The choice of the most excellent matter (that condensed matter of golden honey) and the activity of an enormous amount of vital spirits cause them to attain perfection very quickly; they do not delay a moment in the larval stage or in sleeping in the nymphal one, but immediately display the greatest splendour through the size of their body and their intelligence as in a flower, as Pliny put it,[44] culled from the abundance of everything: through a regal generation and a to some extent leonine nature.[45]

The peculiar theory of the generation of bees through the imposition by the king on the predisposed matter of the spirits that contained the preformed limbs of bees puzzled Fabio Colonna, one of the Linceans whom Cesi had asked to read the section on reproduction and comment on it before the *Apiarium* went to press. The Neapolitan had difficulty in accepting that the king

43 Ibid., p. 168.

44 'The king has from the start the colour of honey, as if made from a special blossom chosen out of the whole supply, and is not a maggot but has wings from the start. The remaining throng when they begin to take shape are called nymphs, while the sham ones are called sirens or drones', Pliny, *Naturalis Historia* XI, 16.

45 Cesi 2005–2006, I, pp. 168–9.

could generate bees simply by imparting to the honey his own heat and the spiritual seal containing the image and corporeal structure of a bee. He could not imagine the generation of any natural creature, including the bee and the king bee, without seed or egg. Besides, he wondered, if the formative principle was conferred by the king, why could bees only be created from honey and not from any other substance?

> I say only this: I do not think that the [generation of bees] can depend on the king imprinting matter with his heat; for if this were sufficient cause, it would always be present. But we see the eggs of doves and ducks hatched by chickens and vice versa. For there is one cause of one effect, and if many are adduced, none of them is true. If the king imprinted his heat on the matter like a seal of the image and physical configuration of a bee, it would be possible to generate bees not only from honey but also from other suitable matter, which is clearly not the case. A seal does not imprint the same image equally in wax and on paper or any other material.[46]

Colonna had reservations regarding not only some of the statements made in the broadsheet on bees, but also Stelluti's engravings of the magnified Barberini bees in the *Melissographia.* Falling back on the oath to assist one another that members of the Academy swore at their induction, Colonna passed a severe verdict on the engraved images of the bees. He chose his words carefully to avoid offending the Prince and Stelluti:

> I have sat down to draw the rostrum of the Neapolitan bee, which is perhaps different from the Roman one, as I have indicated to Stelluti, inviting him to observe it better; but he has not replied to me at all. Now that I have seen that the copper engravings differ on several points from mine, I have decided to send you my drawing.[47]

He urged Cesi to observe the bee's rostrum more carefully through his own microscope to see whether it was different from what he had observed in Neapolitan bees, though he added 'I do not believe this to be the case'. Transmitting his drawing of the rostrum, which is regrettably not extant, Colonna reiterated

---

46 Ibid., II, pp. 130–1.

47 CL, p. 1085, letter of 9 January 1626. Note how Colonna, to avoid ruffling the Prince's feathers, pretends to believe that Neapolitan and Roman bees are not the same.

that his critical remarks were motivated not by emulation, but 'because I am bound to act in this way by my academic obligation'.[48]

Colonna's reservations on the *Melissographia* were not confined to the bee's rostrum. A few months later, he sent Cesi the drawing of the foot of the bee 'very different from the one printed'. He must have suspected some resentment on the part of Stelluti after his comments on the bee's rostrum, which prompted him to repeat that he was acting in the spirit of the Lincean oath:

> And this [drawing of the foot of the bee] is done by me to serve Your Excellency; and also to pay honour to Stelluti, for I desire nothing but to honour my patrons and friends, according with the Lincean precepts, and would not like him to be hurt, for I am more ready to serve him with facts, if I could, than with words.[49]

The message was received. Cesi and Stelluti resumed more diligent microscopic observations and reached the conclusion that Colonna was perfectly right. Stelluti was to recognise the fact in his *Persio*, in which he inserted an image of the bee seen under the lens of the microscope and significantly modified vis-à-vis the version printed in the *Melissographia* (fig. 20). Although with some ambiguity, he credited the correction to the Neapolitan Lincean:

> And for greater clarity and understanding of the reader and viewer, I shall describe its members one by one, in agreement with the Lincean Fabio Colonna, who, having conducted (at my instigation) the same observation of this tiny creature […], found the members the same as they had been observed and drawn by me […]. But afterwards, having examined the bee more closely, I obtained a better impression of the shape of its whole body.[50]

Transmitting to Cesi his impressions after reading the *Apiarium*, Colonna once again displayed his rigour in respecting the Lincean oath. Above all, he did not appreciate the decision to compress an enormous quantity of texts using tiny fonts on such a large sheet, making the reading of it 'extremely uncomfortable'. His criticism was not confined to the layout of the large broadsheet, a project

48 Ibid.

49 Ibid., p. 1123, letter to Cesi, 17 July 1626.

50 Persius Flaccus 1630, p. 47: on the discussions of the bee's rostrum between Colonna and Stelluti see Cesi 2005–2006, I, pp. 147–8, n. 99.

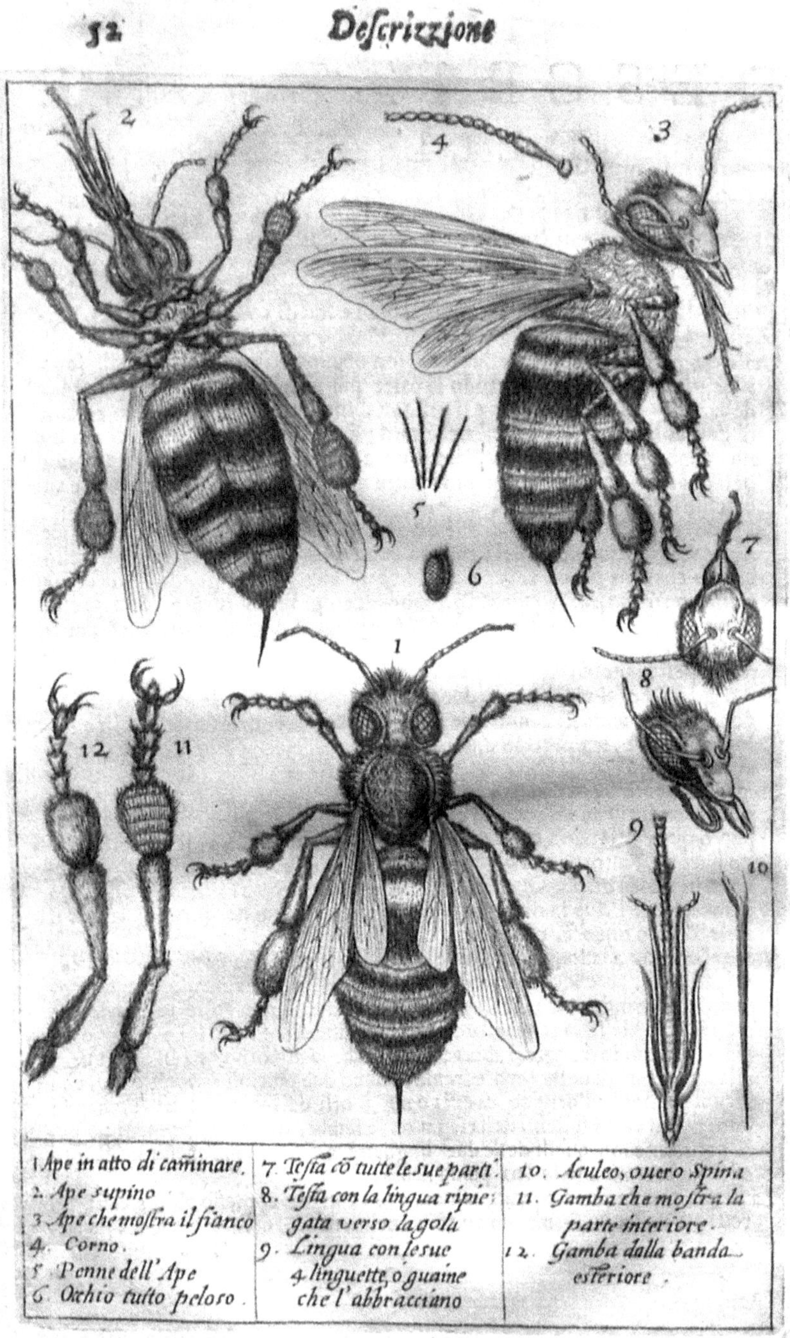

FIGURE 20 *The revised image of the Barberini bees observed under the microscope in the* Persio *of Stelluti* (PERSIUS FLACCUS 1630, p. 47)

that Cesi must have conceived as a work permanently in progress, given that he requested Colonna to propose corrections even though the *Apiarium* had already been presented to the Pope. The Neapolitan bluntly turned down Federico's invitation. He considered it useless to spend his time on revising those texts since 'this work has been done for the sole purpose of pleasing patrons'. Nor did he conceal his disapproval of the fact that the first publication of the Prince represented only

> [...] a fragment of your efforts, and such a small one compared with the grand enterprise of the *Theatro* [*Theatre of Nature*], although the subject is worthy [...] and that as first outcome of such an ambitious title as that of *Theatro* such a small thing is exhibited [...]. I would prefer that the first publication of such a Prince should not be such a brief work [...]. And I say this because I am your sincere servant.[51]

These words are eloquent testimony of how puzzled Colonna was by the penchant of the Prince and his closest collaborators to devote enormous energy to pleasing their Roman patrons. Nor did the Neapolitan Lincean approve of the gullibility of Federico, Faber and Stelluti before promises of secrets, accounts of occult phenomena, or notions like that of generation without either seed or egg. His scepticism about the most sensational transmutations that fascinated his fellow Linceans emerges in his reply to the letter in which Stelluti had informed him that Faber had been eye-witness to the transmutation of a base metal into gold:

> As for the gold, in spite of the credibility of the witness [Faber], yet I suspect a sleight of hand, even more in the case of one part being transformed into gold and the other remaining metal. This is impossible, because anything impure and terrestrial cannot be entirely purified and transformed without fusion of the rest, as no part of it could remain unchanged.[52]

Colonna explained that his doubts derived from his firm conviction that it was impossible to create or modify a natural substance. He regarded this principle as a strict law of nature:

51 CL, pp. 1100–1, Colonna to Cesi, 13 February 1626.

52 Ibid., pp. 1181–2, letter of 15 September 1628.

> I have always maintained this axiom, that no living creature can produce any natural product, no matter how minute and imperfect.[53]

From Colonna's embarrassment at the Prince's first publication (and later at that of the *Tabulae Phytosophicae*)[54] we can imagine how Galileo would have reacted after reading the *Apiarium*. Unfortunately, we can only rely on speculation, as he took the greatest care not to make the slightest comment on the activities and publications of the Prince and his closest collaborators. Galileo received the *Apiarium* and the *Melissographia* from Stelluti towards the end of March 1626, more than a month after those publications had reached Colonna.[55] Stelluti enclosed in the letter to Galileo a short note explaining the genesis and purpose of the broadsheet, evidently with the intention of justifying on grounds of expediency the emphatic dedication to the Pope, the singular format, and the only marginally scientific content of the texts:

> [...] composed [the *Apiarium*] by His Excellency to please Our Lord, dealing in detail with bees, but in a concise manner to fit the limited space of the sheet.[56]

We are left wondering whether Galileo ever read that impervious text; at any rate, he did not betray any hint of his own impression. Unlike Colonna, he would probably have appreciated the celebratory intention frankly admitted by Cesi. The Prince's objective – praise of the Pope to make him well-disposed toward the Academy – promised to benefit his personal battle too. If Galileo ever did unravel that dense mass of characters, in particular those sections in which the Prince had made use of the pages on secondary qualities from the *Assayer* for his own ends and in the context of his own particular conception of the workings of nature, it is unlikely that he would have recognised himself there.

Cesi, on the other hand, had by now formed a precise idea of Galileo's indifference to the comprehensive project that he was pursuing. Thus, it is not surprising that, though he submitted the draft of the *Apiarium* to Colonna and the Linceans in Rome for observations and suggestions, he avoided asking Galileo for his opinion.

---

53 Ibid.

54 Cf. above, p. 404.

55 Stelluti informed Galileo of the imminent dispatch of an exemplar of the *Apiarium* and six copies of the *Melissographia* in his letter of 14 March 1626 (OG XIII, p. 311).

56 Ibid.

CHAPTER 12

Plants as Compendium of Nature

Laying Out the Pages of the Book of Nature

If the *Apiarium* appeared enigmatic to the handful of persons who read it, the *Tabulae* were bound to produce bewilderment and disconcertment.

As Federico's letter to Faber of 19 November 1622 shows, a significant number of *Tabulae* had been completed before being put aside due to his intensive application to the composition and printing of the *Apiarium*:

> I find myself in that great chaos of the methodical distribution of plants and believe that I have almost totally overcome it, which will form a good part of my *Mirror of Reason* and *Theatre of Nature*. I have done what I could and hope to arrive at the completion, which I am approaching by the grace of God.[1]

The reference to the *Mirror of Reason* and *Theatre of Nature* suggests that Cesi regarded the *Tabulae* as an important section of the organic research programme outlined in the *Indicatio*.[2]

Although never completely suspended, preparation of the *Tabulae* proceeded slowly during the Prince's uninterrupted stay in Rome, from the end of 1624 to the early summer of 1626. He spent most of those months working not only on the *Apiarium* and on accelerating preparation of the *Thesaurus Mexicanus*,[3] but also on coping with his critical family finances.[4]

---

1 CL, p. 778.

2 Guerrini finds a clear break in continuity between the *Oposcopia* (and, more generally, the research programme outlined in the *Indicatio*) and the *Tabulae*. In his opinion this is demonstrated by the fact that in the *Indicatio* Cesi does not emphasise the project of an exhaustive treatment of the vegetal world (Guerrini 2006, pp. 158–60).

3 See Faber's letter to the Cardinal Nephew of 12 March 1626, in which he stated that 100 folios (660 pages) of the *Thesaurus* had already been printed (CL, p. 1107 and n. 8). Writing to Cassiano dal Pozzo four weeks later, Faber gave a more modest estimate of the progress of the publication: 'My book is going ahead at a good pace and we are now almost at 50 folios [...]' (ibid., p. 1115, letter of 1 April 1626).

4 The financial situation had worsened as a consequence of Federico's dispute with his father (see ibid., p. 1093, n. 6).

 | DOI 10.1163/9789004342316_013

From a letter sent to Faber at the end of July 1626, after his return to Acquasparta, we learn that Federico had just resumed the *Tabulae* project, to which he was to apply himself continuously in the following months: 'I have re-awakened my dormant books'.[5]

The first reference to the decision to publish the *Tabulae* as an appendix to the *Thesaurus Mexicanus* is contained in a letter from Stelluti to Galileo of 14 August 1627.[6] A few months later a number of *Tabulae* had reached an advanced stage of composition.[7] This is confirmed by a letter from Colonna to Stelluti of 14 October 1628, in which the Neapolitan acknowledges receipt of the first ten *Tabulae*,[8] as well as by Stelluti's letter to Galileo of 2 December:

> The Prince's tables on plants are now being printed; they will be appended to the Mexican book and printing will continue without delay.[9]

The absence of documentation makes it impossible to reconstruct in detail the process of compilation and publication of the *Tabulae Phytosophicae*. The first eight were probably printed in 1628.[10] The letters from Colonna to Francesco Stelluti analysed in Chapter 11 seem to suggest that the publication of the first ten *Tabulae* was completed before the end of 1629.[11] In any case, thirteen of them had been printed before the Prince's death on 1 August 1630.[12]

5 Ibid., p. 1124, Cesi to Faber, 25 July 1626.

6 OG XIII, p. 374: 'and there [in the *Thesaurus Mexicanus*] there will also be something by the Prince on the same subject, which will be pleasing and will be a start of what should come afterwards'.

7 For a detailed reconstruction of the process of publication of the *Tabulae* see Guerrini 2006, pp. 161ff.

8 CL, p. 1185: 'I have received the ten *Tabulae Phytosophicae*'.

9 OG XIII, p. 459.

10 They are preserved on loose sheets in the Biblioteca Apostolica Vaticana (Stampati Barberini, M.IV, 28), within a volume also containing Colonna's annotations and additions to the text by Recchi, and the dedication of the Linceans to Cardinal Francesco Barberini (see Accademia dei Lincei 1991, pp. 122–3 and Guerrini 2006, p. 163).

11 See above, pp. 372–3.

12 Cf. Marini Bettòlo 1992, pp. 23–8 and Guerrini 2006, pp. 166–7. On the complex definition of the project and its implementation in stages down to publication see Antetomaso 2013. On the content and the intricate editorial vicissitudes of this monumental work, see Freedberg 2002, Brevaglieri-Guerrini-Solinas 2007, Guerrini 2007 and 2008, Brevaglieri 2007, and Cadeddu-Guardo 2013 (particularly the contributions by Guardo 2013 and Pignatti 2013).

A glance at the intricate layout of the *Tabulae* explains why it took more than four years of intense editorial and typographical work (1626–1630) to arrive at their partial publication. Here, to an even greater extent than in the *Apiarium*, Cesi showed the great importance that he attached to the layout of the page and to a conception of print publication as a work in progress, permanently open to modifications and additions. His project to provide a panoramic view of the extremely articulated organisation of the whole of nature obliged him to follow a process of exhausting revision. The primary objective of offering a bird's-eye view of the various aspects of the vegetal universe entailed not only an intense conceptual effort to render the connections and filiations of individual plants intuitively evident, but also the challenging task of compressing their descriptions in formulations so concise as to make them often exceedingly difficult to decipher.

A further restraint was imposed by his obsessive determination to fill every blank space with text so as not to give the impression of any gaps in the panoramic representation of the scenario presented by each *Tabula*. Cesi's efforts to avoid limping columns or paragraphs and empty spaces in the layout is almost maniacal. Every page is saturated with typographic characters, with elegantly distributed blocks of text and with orderly and symmetrical classificatory schemes.

The decision to cram every page with text obliged him to continuously intervene in the proofs and to compose the blocks of text in the most diverse shapes. The most eloquent example of the graphic and typographic virtuosity demanded by the Prince to ensure completeness, homogeneity and proportion to the layout of individual pages is offered by the seventh of the *Tabulae*, to be examined in more detail below. Here blocks of text describing individual flavours are of significantly different lengths. To preserve the balance and the symmetry of the page the Prince organized them in a variety of different forms reminiscent of the figures of plane geometry. The entries on individual flavours thus assumed the appearance of the elegant flower-beds of an Italianate garden.

Although the format of the *Thesaurus Mexicanus* was large, it did not allow the Prince to implement the design that he had conceived and put into practice for the *Apiarium*: to enclose the central topic analysed in each *Tabula* in a typographically unified space that could be taken in at a glance. Only the first four *Tabulae* respect this rule. *Tabula* V occupies three pages; VI four; VII no less than eight; VIII six; IX one; X three (of which the third is closed by the Prince's family stemma to mark the conclusion of the first phase of publication between 1626 and 1629); XII two; and XIII (only partly completed by Cesi) two.

This Cyclopean effort, which appreciably raised the costs of typesetting, was imposed by the Prince's desire to suggest the continuous and granular articulation of the Great Chain of Being. The layout of the *Tabulae* displayed his conceptual, and consequently typographic, *horror vacui.* Empty spaces and any lack of proportion or density in the texts were signs of incompleteness or disorder: embarrassing evidence of a negligent omission of rungs on the continuous Ladder of Nature, of its complex structure and of its vertical and horizontal articulations.

After the laborious experiment of the *Apiarium*, the publication of the *Tabulae* tested Cesi's editorial capacities and the professional skill and patience of the typographer even more severely. The variety of fonts and characters that Cesi deployed in the *Tabulae* is impressive: six different capital roman fonts (also in bold); four roman fonts in two different heights; and three capital italic fonts in two different heights.

The constant alternation of texts in roman and italic satisfied different demands. In many cases, where the name of the family or class analysed is indicated in roman characters, the sequence of its ramifications is set in italics, and vice versa. Sometimes Cesi uses italics for reflections of a methodological nature or for formulating general principles to distinguish this more abstract level of reflection from the hierarchical classification of the vegetal kingdom set in roman. A text in italic is frequently inserted in a block of text in roman or vice versa to highlight key words or expressions. In the discursive scholia in which he enhances the general principles underlying his classificatory schemes, he alternates blocks in roman with blocks in italic to render the page densely saturated with homogeneous typefaces more appealing.

A graphic project of this complexity aimed at offering an exhaustive visual representation of Cesi's *Theatre of Nature* without the need for images, which are completely absent from the *Tabulae*. At the same time, it reproduced the logical order, the *Mirror of Reason*, on which the classificatory schemes of the vegetal world were modelled.

In spite of the penetrating remarks of Yvette Conry and Sandro Pignatti on how to approach the dense tangle of curly brackets that, from left to right, delineate the genealogy of the classes and families of plants, as well as their derivations and dependences, we still lack a systematic study of the typographic project conceived by the Prince to intuitively visualise his innovative synopsis of the vegetal world and its rational foundations.[13]

13 Conry 1997; Pignatti 2013.

The extraordinary layout of the *Tabulae* is by no means a marginal aspect of Cesi's project. It is a great pity that, unlike the case of the *Apiarium*, the certainly numerous manuscript drafts of the *Tabulae* have not come down to us. As in his preparation of the broadsheet on bees, Cesi must have patiently compiled an infinite number of cards, on which he recorded the distinctive characteristics of the various genera. He then arranged them in orderly sequences to bring out the affinities with the other classes into which he was organising the entire vegetal universe. We have evidence of this method in his letter of November 1622, in which he informed Faber that he was immersed in 'the chaos of the methodical distribution of plants'. 'Chaos' suggests the presence on his desk of a vast number of cards and images that he was arranging in hierarchical sequences in order to progressively refine the analysis down to the level of the individual plant.[14]

The request in the same letter to Faber of the two volumes of the *Icones stirpium* of Matthias de l'Obel indicates that, to dominate panoramically the intricate forest of the 'plantamentation'[15] that he was delineating, Cesi intended to detach the numerous images of plants contained in that work, arranging them in sequences corresponding to his schemes of classification to verify their plausibility and degree of precision.[16]

To accomplish this highly complex operation, he needed not only lengthy periods of continuous application but also the adequate research facilities with which he had furnished his Umbrian residence: a vast collection of books, a botanical garden, and a chemical laboratory. Descriptions of these resources are supplied in the letters from Cesi and Winther to Faber.[17] It is therefore

14 CL, p. 778, Cesi to Faber, 19 November 1622.

15 The word, coined by Cesi, was modelled on *castramentatio*, the laying out of a military camp.

16 Ibid.

17 See for example the letter to Faber in which Cesi described the conditions that Winther would enjoy if he agreed to become his assistant: 'In simples and distillation and books, if he wishes, he will become a distinguished and respected student, and the same in anatomy [...]; moreover in the distillery, gardens and countryside here he will be able to learn everything and I will teach him the names [...]. You know that I have many exquisite books on medicine [...]' (ibid., p. 859, letter of March-April 1624). Winther confirmed to Faber the presence in Acquasparta of the necessary research facilities to profitably devote himself to the study of the remarkable natural finds in the surroundings: 'The Prince summoned me to the tower in the garden of the house, a delightful place fitted with shelves that he calls his refuge (*Perfugiolum*) [...]. The Prince took me with him to a large garden outside the gates, where are many simples planted by him and exotic trees' (ibid., p. 928, letter of 27 July 1624).

hardly surprising that it was above all during his stays in Acquasparta that the Prince made progress with the *Tabulae* project.

His method consisted of card-indexing many books and manuscripts and then classifying this large mass of data for easy consultation. Light is thrown on this method by some passages of the *Zibaldone*, in which Federico explained how to produce a table containing a well-organized synopsis of philosophical sentences:

> A table is an orderly grid of authors and subjects which places the sentences of philosophers before your eyes [...].[18]

Such a grid permitted a hierarchical exploration as well as the immediate visualisation of the horizontal connections between genera and families:

> Rows or extended series of sentences culled from various authors, flanked simultaneously by different parallel sequences, to form the appearance of an ordered list of subjects and places.[19]

With the *Tabulae Phytosophicae* the project of the Prince to exploit the potential of typography to visualize the tight texture of connections that permits recognition, behind the infinite variability of the natural world, of a unitary and finely structured system reached its apex. It was an objective to which Cesi had assigned extreme importance from the beginning of his intellectual career. In the *Indicatio* he had already provided an orderly tabulation of the key issues of his research programme aimed at defining the connections and dependences of the productions of nature.

---

18 'Tabella est craticulato ordine ab authoribus et materiis disposita quascumque philosophorum sententias ob oculos ponens', Contributi I, p. 57.

19 'Ordines seu protractae series digestarum materiarum diversorum scriptorum diversae distinctis simul excurrentibus parallelis productae in aspectum ordinatarum materiarum et locorum', ibid. Freedberg made penetrating comments on the relation that Cesi perceived between the chaos of the collection of preparatory materials and the orderly organization of the highly structured pages of the *Tabulae* (CL, p. 778, letter to Faber of 19 November 1622). As he pointed out (Freedberg 2002, p. 377), the term *chaos* refers to cosmogony, while the order of classification is conceived by Cesi as the formation of 'an almost uncountable army of plants' (ibid.). On Cesi's use of military metaphors, cf. his repeated use of the term *militia* (military discipline) to indicate the organisation of the Academy; see also nn. 15 above and 136 below.

Federico was confident that this solution would render intuitively evident the unitary architecture of nature, as well as the rules of philosophical reasoning and even the principles and models of exemplary moral conduct. He was convinced that, without this systematic conceptual and visual cornice, the study of natural phenomena was inevitably condemned to failure.

Syntax, Painting, Theatre, Garden

Cesi did not like signing dedications of Lincean works to illustrious individuals, not even when he had written them himself. It is therefore not surprising that the dedication to Francesco Barberini of the first print-run of the *Tabulae*, although most probably written by the Prince, was signed by the Linceans as an Academy.

The dedication underlines the significance of the *Tabulae*, emphasising their main novelties by comparison with traditional botanical studies of which the Prince condemned the approximate nature and above all the lack of systematicity. In his opinion, those who consulted that vast literature, wanting in coherent principles of organization, ran the risk of getting lost like explorers in a dense forest.[20] The fundamental contribution offered by the *Tabulae* was that of establishing a rational order fixing rigorous genealogies between plants. This work was conceived as a syntax that reproduced the rules of nature and the processes that govern the correct use of reason.[21] It was the necessary premise for the production of a systematic inventory of the entire vegetal world because it helped to establish a rigorous method of inquiry. The Prince calls the *Tabulae* a painting (*diagraphen*) of the complex articulation of a garden in which plants are arranged in orderly flowerbeds that make it possible to take in the various families and their relations and dependences at a glance.[22]

Syntax, painting, theatre, garden: these are terms carefully chosen to metaphorically underline the connections and resonances between blocks of text, geometric and pictorial representations of forms, theatrical scenographies, and the rational distribution of plants by family and genus in vegetable and flower gardens.

20 '[...] velut in Sylvia invia, aut potius inextricabili quodam labyrintho [...] sine physicorum principiorum et naturae filo', TM, p. 903.

21 '[...] omnem Syntaxim plantariam [...] non ad Polycleti sed naturae normam ex Philosophiae principiis, imo e matris naturae gremio deducere conatus est', ibid., p. 904.

22 'Amplissimi horti theatrales primos conspectus', ibid.

Placed at the conclusion of the *Thesaurus Mexicanus*, the *Tabulae* were bound to appear as an extraneous appendix to the rest of the work. While the treatise on the animals and plants of New Spain was a copiously illustrated narrative, the *Tabulae* were densely filled with schemes of an unprecedented complexity, but completely lacking in illustrations or diagrams.[23] Having perused the sections of the *Thesaurus* edited – under the watchful eye of the Prince[24] – by Schreck, Faber and Colonna, the reader could not help perceiving a strong break in continuity when arriving at the *Tabulae*.

We do not know why the Prince decided to create such an apparent hybrid instead of publishing the *Tabulae* as a separate work. This choice appears especially puzzling if one bears in mind that the pilot edition of the *Thesaurus*, printed between 1626 and 1630, did not include the full series of *Tabulae*, nor the sections on the flora of New Spain by Schreck, nor the *Additiones* by Fabio Colonna.[25] In spite of his youth, the Prince might have had a presentiment of death and been anxious to publish the results of his work even though incomplete. However, it is more likely that he felt the need to offer Francesco Barberini an ostentatious token of his personal devotion and of that of his fellow

23 The insertion of the *Tabulae* at the conclusion of the *Thesaurus Mexicanus* was the result of a carefully made choice. In their dedication, the Linceans stressed the importance of a systematic method of inquiry without which it would have been impossible to faithfully represent the organisation of nature: 'We [Linceans] felt that it was very appropriate to add it [the *Tabulae*] to this Mexican Dioscorides lest such a pleasing and rewarding discipline should later seem to be lacking in rigorous method, which should be placed as the culmination of the entire plant world to establish that study on firm ground' (ibid., p. 904).

24 Though he did not write any of the sections of the *Thesaurus Mexicanus* himself, Cesi collaborated with Stelluti in revising, correcting and integrating the contributions of Schreck, Faber and Colonna, as can be seen from the letters that he sent to Faber between 1620 and 1622. For example, on 18 October 1622, Cesi wrote: 'Stelluti and I are expending much effort in arranging, comparing, correcting and modifying the sheets of the *Mexicanus* for printing'. He also wrote that they would prepare 'a short preface' (*prefationcella*) for the section on plants by Schreck (CL, p. 776; as Gabrieli indicates, ibid., note 6, the *prefationcella* is probably the text contained on p. 346 of the *Thesaurus Mexicanus*). The Prince had previously intervened in the section on animals authored by Faber (ibid., p. 702, letter of 14 December 1619). On that occasion he stated that he was trying to extend to the animal world the systematic rational classification of plants: 'in a certain digression I have become immersed and almost drowned in contemplation of the differences between all the animals and in establishing their classes [...]' (ibid., p. 704, letter to Faber of 14 January 1620).

25 The section authored by Schreck and Colonna was published in 1649.

Linceans in order to receive authoritative protection for the Academy and effective support to overcome his disastrous financial situation.[26]

The Frontispiece and dedication of the first print-run of the *Tabulae* are evidence that the work was actually presented to the Cardinal Nephew, whose curiosity regarding the most singular natural phenomena was well known to Cesi.[27]

A Galilean Syntaxis?

While the *Tabulae* are mentioned in many scholarly essays on the history of the First Academy we still lack a close analysis of its content. In this connection, it is still useful to consult, in spite of its limitations, the classic work of Romualdo Pirotta, who produced a complete edition of the *Tabulae*.[28]

Other important contributions have been made by Yvette Conry, David Freedberg and Luigi Guerrini,[29] who have underlined the Prince's effort to

26 Cesi had followed a similar strategy in the case of the publication of the *Apiarium*, see above, pp. 385–6.

27 To win the Cardinal Nephew's favour and to assuage his own curiosity, Cesi compiled and sent him a list of Mexican plants, together with notes on their characteristics, properties and place of origin, so that he could order living specimens to take to Rome that would confer an exclusive character on the garden of the powerful Cardinal. Cesi was confident that on the Cardinal's request the missions in New Spain would collect and send them to Rome: 'I believe that they [exotic plants] can be very easily sent at a simple sign from your Excellency, not only via the Jesuit Fathers or Dominicans in Mexico and in their own countries, but also from Seville, where the Spanish ministers have organised the unloading and trade' (ibid., p. 826; Cesi to Francesco Barberini, 25 November 1623).

28 Cesi 1904. The most serious limitation of Pirotta's edition is the failure to reproduce the layout of the original printing of the *Tabulae* as well as the variation and alternation of typefaces used by Cesi. For example, Pirotta spread *Tabula I* over two pages instead of the single one in the *Thesaurus*, making it impossible to distinguish the pyramid-like design deliberately adopted by the Prince to underline the continuous Ladder of Being and its hierarchical structure. Pirotta also disrupted the original layout of *Tabula VII*, with the consequent loss of the highly integrated character of the treatment of the 'flowerbeds of flavours', as well as of the schemes of quantity-quality and motion-quality that Cesi had organised to be taken in at a glance.

29 Freedberg 2002, Conry 1997 and Pignatti 2013. Guerrini 2006 analyzed each of the twenty tables published in the definitive edition of the *Thesaurus Mexicanus* between 1649 and 1651, though concentrating on the first thirteen, the only ones edited by Cesi.

define an effective method to systematically represent from multiple angles the entire organisation of the plant world, its highly refined articulations, filiations, derivations, as well as the relations between the genera and species that densely populate it.

Almost all the scholars who have taken into account the *Tabulae* share a central concern: to bring to the fore the Galilean principles on which Cesi modelled his own syntax of the plant kingdom.[30] According to the prevailing opinion, the *Tabulae* offer the most explicit evidence of the 'Galilean conversion' of Cesi and the whole Lincean Academy after 1611. Distancing himself from these reconstructions, although without denying the presence of Galilean elements in the *Tabulae*, Luigi Guerrini has stressed the continuity of the Prince's final endeavour with the research horizon that he had delineated at the start of his intellectual career – a horizon whose principal points of reference lay in the *Magia Naturalis* of Giovan Battista Della Porta, Renaissance naturalism and chemical philosophy.

Given the crucial importance of this question for the present work, I will concentrate in particular on the three elements on which the supporters of the thesis of the Galilean matrix of the *Tabulae* have insisted: Cesi's research method, based on the balanced integration of observation and mathematical speculation; his analysis of secondary qualities, inspired by the innovative model proposed by Galileo in his *Discourse on the comets* and the *Assayer*; and the corpuscular conception of the structure of matter modelled on the Tuscan scientist's writings on hydrostatics and on his texts bearing on the cometary debate.

Before proceeding in this analysis, I must make it clear that I do not intend to question the fact that Cesi was fascinated by Galileo's revolutionary ideas. My purpose here is to understand *how* the Prince assimilated those ideas and, above all, included them in his own research horizon, in which, from the outset, the *Tabulae* occupied a pre-eminent position.

The Multiple Gaze of the Botanist

Tabula II presents the various angles from which the world of plants can be observed. The starting point (that Federico defines as *centrum*) is the general definition of a plant, that is, the fundamental characteristics shared by the entire vegetal kingdom. At the immediately lower level of abstraction (*circa*

30 David Freedberg shares this view (cfr. Freedberg 2002, esp. pp. 367ff.).

centrum) comes the substance of which plants are composed. The area furthest from the centre (*a centro*) is reserved for the enumeration and illustration of the diverse species.

By means of a series of dichotomous filiations, the Prince distinguished six different ways[31] of approaching the plant universe, presented in descending order of generality: *Phytichnographia*, that is the representation of plants for their full cognition;[32] *Phytoscopica contemplatio*, the aspects of plants to be considered and understood;[33] *Phytognosticae regulae*, the method for studying plants;[34] *Triphytologia*, the analysis of the three main parts of plants;[35] *Loganatomia*, the reduction of plants to their constituent parts;[36] and *Phytomia*, the enumeration of those parts.[37]

It is important to point out that Cesi assigned here a purely metaphorical meaning to the geometric terms (centre and circumference), which should not be intended as evidence of his design to submit the universe of plants to rigorous mathematical analysis.[38] He used these terms to bring out the hierarchical sequence of the points of observation: from the centre (general view) to the periphery (particular observations).

Tabula III presents an analytical scheme of the characteristics of the vegetal world that the botanist must consider to form a coherent syntax, that is, an orderly classification of all the information that can be extracted from the careful examination of plants. Before proceeding to examine the characteristics that can be subjected to quantitative analysis, Cesi's botanist must analyse the composition and structure of plants, their vitality, medicinal properties, moral message, titles of nobility and beauty, the adequacy of their names, and, finally, the opinions of earlier writers. He should then tabulate the elements of similarity and affinity between different species ('similitudinis quibus plantae conveniant') and record their differences ('quibus ab invicem recedant ac disiungantur'). It is only through careful inspection of affinities (with their derivations and dependences) and diversities that the botanist can arrive at a syntactic representation of the entire vegetal universe.

31 Cf. ibid., p. 387.
32 'Representatio plantae ad plenam cognitionem', TM, p. 906.
33 'Quae consideranda in plantis et cognoscenda', ibid.
34 'Quibus plantae considerandae et cognoscendae', ibid.
35 'Summi explicati plantarii generis tripartitio', ibid.
36 'Priorum membra et quae a summis interdividuntur cognita', ibid.
37 'Quae individuorum complentur numero', ibid.
38 For a different opinion, cf. Freedberg 2002, pp. 386–90.

In *Tabulae III* and *IV* Cesi defines the method that he intends to follow in his own rational classification of plants.[39] The progressive transition from the general to the particular on which it is based is expressed by the schemes that start on the left margin and proceed towards the right through continuous series of dichotomies. Those sequences represent not an abstract exercise of deduction, but a model of rational organisation of information obtained through the direct observation of nature.

The fundamental role of observation is reaffirmed in the opening of *Tabula IV*, although Cesi emphasises the need to integrate close observation of plants with theoretical reflection in order to isolate their common features and to distribute individual plants in classes:

> Their [of plants] nature is first to be examined individually by subjecting it to our gaze [...]. At a later stage our reflections and deductive reason are brought to bear on it.[40]

Setting out from this awareness, Cesi stresses that the botanist can observe the world of plants from three points of view: mathematical, physical and medical.

The mathematical botanist focuses on the external appearance of plants and analyses those aspects, such as figure and form, that bear on quantity.[41] The physical botanist carries out careful observation of vital phenomena: the motions of plants and their organs. Unlike the mathematical botanist, he does not stop at external appearance and the quantitative dimension; he directs his attention to the qualities hidden within the deepest recesses of the plant.[42] As for the medical botanist, although his methods and aims are different from those of the physical botanist, he too investigates the qualitative dimension of plants. The clearly more articulated and detailed description reserved for the task of the medical botanist reflects the prime importance that Federico attributes to knowledge of the curative properties of plants:

39 TM, pp. 907–9.

40 'Priori nostro intuitui, ipsarum subijciente Natura, singillatim displicienda [...]. Posteriori reflectione nostra ex obiectis colligente et deducente ratione', ibid., p. 908.

41 'In fronte magis, quae prius apparent et sunt QUANTITATIS, cuius modus FIGURA, quae pluribus hoc in genere FORMA', ibid.

42 'In recessu magis, quae exterioribus pensitatis innotescunt et sunt QUALITATIS, quae praecipue requiri in plantis solet', ibid.

> ENERGY, or in others MOTIONS [i.e. the effects produced on diverse organisms by medicinal simples]: multiple and highly diverse USES when applied to our body as we test them and derive benefits from them.[43]

Federico takes care to point out that the analysis of medicinal properties cannot be deduced from abstract principles: it must be based on the attentive observation of the effects produced. The medical botanist can examine the properties of plants in two different conditions: when they are extracted from the soil and partly or completely crushed, or when they vegetate intact in their own habitat. The latter is the better condition for detecting the plant's energy, that is, the presence of those vital and sensory functions that lead to a blurring of the traditional sharp boundary between plants and animals.

The mathematical, physical and medical approaches do not exhaust the professional skills required of the botanist by Cesi. The moral message of plants ('plantae morale praeceptum seu virtus') is also a fundamental field of investigation. Analysis of the moral precepts that can be derived from plants marks the transition from an inquiry based on observation and experiment to a purely speculative activity.[44]

While the control, certification and decoding of the symbolic significance of the names of plants ('Plantae titulus') is the preserve of the philological botanist, the highest level of abstraction is reached by the metaphysical botanist who aims at defining the universal concept of plant and isolating those characteristics that are common to the entire plant world.[45]

It is important to stress that Cesi is not proposing here a division of labour: the naturalist who sets himself the task of grasping the significance of the vast and articulated vegetal kingdom must consider it from all six points of view.

This challenging task reflects Federico's critical evaluation of the contemporary state of research, not only on plants, but on the whole world of nature. He wished to underline the fact that only a panoramic vision would permit grasping the architectural criteria employed by nature. Thus he had not confined himself to filing a dense mass of information accompanied by indices and mnemotechnic tools for ease of retrieval. His objective was to represent, as in a theatre, the operation and organisation of nature, its instruments, the

43 'VIRES, sive in alijs MOTIONES: USUS nobis corporis applicatione, qui et multiplex et maxime varius nil non tentantibus, et extorquentibus ad propria commoda hominibus', ibid.

44 'Interius adhuc insumpta necnon conspectu ex ipso', ibid.

45 'Planta inconstituta, quae universim concludantur', ibid.

order and relations between beings and the dynamics that govern vital activity. In his dedication to the Cardinal Nephew, he praised the *Tabulae* not so much for the quantity of information as for the faithful representation that they provided of the Ladder or Chain of Being, highlighting its affinities, differences, dependences and internal motions. This vision corresponded to the conception of nature which had found its first schematic formulation in the *Indicatio*.

Among the many ways of observing plants that Cesi demanded of the botanist, that which appeared most attractive to him was the metaphysical gaze, in which he recognised the full accomplishment of the mission that he had set himself. In commenting that the *Tabulae* seemed to be written for metaphysicians rather than for herbalists, Fabio Colonna showed that he had perfectly understood the spirit of the work.[46] The Prince was convinced that without a lucid general vision of the organisation of nature and its processes it was impossible even to exploit the properties of plants for the practical needs of human beings.

Like the *Apiarium*, the *Tabulae* were presented by Cesi as one of the Frontispieces of his *Theatre of Nature*. The centrality of the idea of the *Theatre* had already been highlighted in the *Discourse on the natural desire for knowledge*, in which the Prince had drawn a clear distinction between the vast production of compendia and encyclopaedias, on the one hand, and his own project, on the other. He had warned against promoting learning as a process based exclusively on the assimilation of a vast mass of material, or on the aimless reading of an infinite number of volumes. It was a clear way of expressing his scepticism regarding a conception of the study of nature based on the accumulation and memorisation of large quantities of information. Why waste one's energies on the compilation of compendia packed with details – Federico stressed – when those instruments were already available?

> There are herbals, chronologies, theatres, polyanthuses, gardens and a variety of books of medicinal recipes; there are libraries that offer us all the books together with judgements on their value and arranged by author or by subject matter [...].[47]

What was lacking was a comprehensive vision of the system of knowledge and of its articulations. As a consequence he had dedicated himself to the creation

46 CL, p. 1186, Colonna to Stelluti, 20 October 1628. See above, p. 373.

47 Altieri Biagi 1969, p. 65.

of instruments that would make it possible to intuitively grasp the foundational structures of the edifice of nature and their perfect correspondence with the categories of rational thought. In the *Discourse* of 1616, he had contrasted the chaotic assimilation of a vast quantity of information with

> the synoptic art itself, with its types, which represents to us all together all things and their dependences, conjunctions, divisions and unions, and all conditions, as we have sought to present to the eye of the viewer in our *Mirror of Reason*, in order that the keenest memory and most alert intellect which is able to master the principles proposed in it can proceed straightforwardly ahead in invention and composition.

Being the most mature product of Federico's synoptic vision of nature, as well as of his conviction of the full correspondence between the *Theatre of Nature* and the *Mirror of Reason,* it is not surprising that he put so much effort into trying to enclose the *Tabulae* within unitary and compact graphic structures. Only in this way could the organization and the articulations of the entire system of nature be comprehended.

Having established in the first four *Tabulae* the architectural rules and the conceptual structures supporting his synoptic vision, Cesi could now define the articulation of the various panoramas opened up by the multiple specialist points of view from which the botanist observes the vegetal universe, its organisation, and the relations between the entities that populate it.

The Elusive Geometry of Plants

Tabula V analyses the field covered by the mathematical botanist and his methods. It is particularly important for present purposes because it outlines the new prospect of a quantitative treatment of botanical science. This explains why those scholars who have stressed the presence in the *Tabulae* of a strong Galilean influence have insisted on the three dense pages in which Cesi set out to demonstrate that the plant world can be analysed in terms of number, measure and proportion and reduced to the perfect figures of geometry. The unprecedented claim that the botanist requires mathematical competence seems in itself to attest the influence of the Tuscan scientist and to legitimise the definition of Cesi as a pioneer of quantitative botany.

As usual, the schematic illustration of the mathematical botanist's methods and his areas of competence is organised in dichotomous series from left to right, i.e. from the general to the particular, to bring out their hierarchical articulation and reciprocal relations. Cesi begins by stating that the mathematical botanist can consider the quantitative dimension of plants from a double perspective: absolute ('simpliciter consideratae'),[48] or in relation to the position they occupy vis-à-vis other plants, the environment, or the disposition of their constituent parts ('in planta sunt quantitatis sitae').[49] With regard to the former, Cesi distinguishes discrete quantity (the field of arithmetic) from continuous quantity (that of geometry). The quantitative characteristics of plants considered in absolute terms are then divided into three categories: their general constitution; their organs; their fibres, tissues and substances.

Each of these categories is then further analysed via a dense sequence of dichotomies. In the column reserved for the general constitution, Cesi proposes dichotomies of the type: *constituentium/complentium; internarum/externarum; formalium/accidentalium; genuinarum/adscititiarum; certarum/ambiguarum* [...]; *similarium/dissimilarium* [...]; *fluentium/solidarum*, etc. Towards the right margin of the page, the principal organs of plants are identified (roots, leaves, flowers, seeds, juices, etc.) while the adjacent column further to the right lists the substances present in plants (fibres, petioles, etc.).[50]

So far there is no trace of a quantitative treatment. It makes a timid appearance in the columns extending from the centre to the right of the opening page of the *Tabula*. The first column divides the quantitative aspects of plants (considered in absolute terms) into five categories (*habitus, excessus, paritas, defectus, privatio*). Once again the quantitative treatment appears fleeting. In the column reserved for the numerical evaluation (*numerus*) Cesi states that simplicity (*simplicitas*) can be double, triple, etc., while rarity (*raritas*) is to be considered orderly or disorderly.

The successive categories in the top half of the page refer to discrete quantity, the preserve of arithmetic. Here Cesi again has recourse to dichotomies such as *multitudo/paucitas* (referring to the *habitus*), *simplicitas/multiplicitas* and *densitas/raritas* (expressing the modes of *excessus*), *constructio/coagmentatio* and *connectio/conspiratio* (which determines the *paritas*, that is, the symmetry and proportion between the parts).

Further down come the categories of continuous quantity, the preserve of geometry, broken down into the categories of *defectus* (expressed by the

48 Ibid., p. 909.

49 Ibid., p. 916.

50 '[...] fibrarum, foliolorum, squamularum, alarum, petiolorum, ramulorum [...], pericarpiorum, ossium', etc., ibid., p. 909.

opposition *crassitudo/exhilitas; humilitas/brevitas; altitudo/longitudo*) and *privatio* (characterized by the dichotomies *plenitudo/parvitas; magnitudo/amplitudo*). The adjacent column lists the geometric figures that can be recognised in the external appearance of plant stems: *spiralis ratio* [...], *cochleata, sphaerica, quadrata, angulata, fastigiata, oblonga, collecta, expansa, diducta*.

In the second main division of the *Tabula*, Cesi analyses the quantities of relation, distinguishing between those inherent to the structure of the plant ('quoad corpus proprium') and those pertaining to the spatial distribution of its internal parts and of its relation with the environment or with other organisms ('quoad contermina corpora respectu').[51] Proceeding as usual in progressive dichotomies from left to right, the structure and spatial limits ('compositionem et terminos') of vegetal organisms are distinguished from the disposition of their parts that serve motion ('motum dispositio partium'). Through a further series of successive dichotomies the internal aspects are separated from the external ones: consistency (rarity, density, solidity, fragility, etc.), how plants feel to the touch (leathery, cartilaginous, fibrous, rough, fine, etc.) and differences in colour ('ab albo ad nigrum, a plumbeo ad caeruleum [...]'). The quantitative characteristics of the external aspect of plants are then reduced to the figures of geometry: 'anguli sinus sect. curvit, plan. inaequal. Rotund. oblong. Teret Cylind'.

The influence of the *Assayer* appears transparent here: Cesi is trying to demonstrate that the close observation of plants confirms that the Book of Nature is written in mathematical characters. However, he immediately attenuates this bold affirmation. In plants, he states,

> the figures of geometry are almost all mixed, so that they cleverly elude all our rules.[52]

This is an important proviso: in spite of his determination to stress that plants too show nature operating in terms of number, weight and measure, Cesi is conscious of the extreme difficulty of conclusively demonstrating that principle. He explains that we do not easily recognize circles, rectangles, triangles and so on in plants because those perfect figures are all mixed and combined. He reiterates this explanation in a boxed text directly linked to the list of geometric figures present in plants and their organs:

51 Ibid., pp. 910–11.

52 'Mixtae fere omnes figurae, et quae omnes nostras regulas perbelle eludant', ibid., p. 910.

> You find lines in rushes, spheres in bulbs, in fruits and in tubers, ovals in the heads of the cabbage, and the aspect of a pyramid in the pointed and florid stalk of a campanula. Why should I enumerate them one by one when all figures are in all plants and are disposed in multifarious ways? If you seek the use of the compass, go to the fungi.[53]

Passages like this leave no doubt that *Tabula v* reflects the Prince's meditations on the pages of the *Assayer* on the characters in which the Book of Nature is written. Nor can it be denied that his promotion of the competence and the methods of the mathematical botanist betrays the influence of the Galilean model.

That model is even more evident in the large section in which Cesi classifies in quantitative terms the vital movements and the disposition of the parts of plants. A plant can perform its vital functions through greater or lesser movements. Among the latter Cesi mentions the 'internae suctiones vitales' (suction of vital substances), the 'externa evaporatio' (evaporation) and the processes of growth. His analysis of the suction of vital substances then focuses on the organs that perform this function based on number, size, shape, mutual relation, as well as on inspection of the ducts through which the vital spirit and nutritional juices are assimilated. To illustrate the transmission mechanism of juices from the roots to all the parts of a plant, Cesi uses the image of the syphon, thanks to which liquids can be conveyed upwards:

> The digested, assimilated and excreted parts are the result of the processes by which the dissimilar is produced from the similar in varying degrees of rarefaction and is disposed in obedience to the figure.
>
> It is imbibed as in a mouth by the root, which takes its nourishment from the earth's juices, and distributes it to the other parts of the plant through the medulla, which operates like a mechanical syphon […].[54]

---

53 'Lineas in Iuncis habes, sphaeras in bulbis pomis ac tuberosis. Qui ambitum in Brasicae capitibus, Pyramidis aspectum infastigiato beneque florido Campanulae illius thyrso. Quid vero enumerem plures in eadem planta omnes in omnibus figuras et multifariam dispositas? si vero usum circinij quaeris Fungos pete', ibid.

54 'Digerentes, digestae, assimilatae, excretaeque partes unde similaribus dissimilares constituantur, rarius, densiusque, dispositis, et in figurae obsequium subiectis.
Ab radice, qua alimentum terrei succi, quasi ori analoga hauritur mittitur in plantae reliqua membra ampliore per medullam, veluti syphonem mechanicum scilicet Diabetem via […]', ibid.

The few scholars who have analyzed *Tabula V* agree that the image of the syphon implies the vision of the organism of the plant as a device that functions according to the rigorous laws of mechanics – further confirmation of Galileo's influence on the Prince in his study of plants.[55] It should, however, be stressed that while the definition of the syphon as a mechanical instrument is perfectly adequate as to its material composition, it is completely misleading in this context because, like all his contemporaries, Cesi did not view the working of a syphon as a mechanical operation. At least until 1644, when Torricelli demonstrated that the functioning of that device depended on the weight of air, the raising of a column of water by the syphon was explained in terms of nature's *horror vacui*. For Cesi, the working of the syphon evoked the occult powers of nature, not the rigorous laws that govern the operations of mechanical instruments. Incidentally, this passage also shows the Prince's lack of familiarity with the actual workings of a syphon, to which he assigned an impossible task: that of transporting juices from the depths of the Earth to the fibres of the plants that grow above its surface.

In this respect another element which has until now escaped the notice of scholars should be considered. Cesi had taken the model of the syphon not from Galileo's mechanical analysis of natural phenomena, but from Della Porta. In his *De aeris transmutationibus*, the Neapolitan had recurred to the syphon to explain how resinous and oily fluids rise from the roots and spread throughout the entire organism of the plant.[56] The identification of Cesi's

55 Yvette Conry underlined how Cesi's recourse to the model of the syphon is proof of his convinced adhesion to the Galilean conception of nature (Conry 1997, p. 190). On the syphon see also Guerrini 2006, pp. 189–90.

56 'Quod Sol aestate resinosos et oleoginos succos maturet atque e *radicibus* per *medullam* quasi per *syphonem* sursum *hauriat*', Della Porta 2000a, p. 112. The italicised words are literally repeated in Cesi's text. Della Porta had analysed the syphon in his *I tre libri de' spiritali* (Della Porta 1606 and 2008), in which he questioned the traditional interpretation of its functioning in terms of *horror vacui*. He contested that a vacuum – whose very existence in nature was denied by Aristotle and his followers – could overcome the force of gravity, constraining the water to rise against its nature in the syphon. His explanation was based on a different argument: the performance of the syphon was the consequence of the absolute determination with which nature attempts to maintain her unitary structure, which is her essential character. The unity of nature 'is produced by contact and by the constant reciprocal touch and embrace of its extremes with one another; they combine so reciprocally that she [nature] prefers to suffer anything rather than to allow their separation, and will prefer to create every great marvel and extraordinary thing'. The Neapolitan concluded that 'the continuous and reciprocal love of the things of nature overcomes the force of gravity because it is more natural to preserve itself than to obey

direct source indicates that, even in the passages in which he showed his greatest admiration for Galileo's mathematical and mechanical interpretation of the operations of nature, he continued to move within a horizon – that of the natural magic of Della Porta and his conception of a nature endowed with animated principles – which he does not seem to have regarded as incompatible with the Tuscan scientist's Book of Nature.

Cesi saw clear proof of the mathematical structure of nature not only in the process of internal suction, but also in the emission of substances by plants through evaporation. Referring to the scented effluvia of flowers and to the rains of dew that play an important part in the production of honey and manna, he underlines that the processes of evaporation are activated by local upward, downward and lateral movements of the internal ducts of plants. Through these movements, the lighter and finer particles of the vital spirits rise, the thicker and heavier ones sink, while the intermediate ones remain midway. The Prince qualifies as local motions of plants also the agitation of the particles of their stalks and supporting structures, as well as the oscillation of the elementary components of certain species which, when shaken, emit a sound. The boxed text below the subclass of vegetal movements that produce sound indicates that Cesi is here drawing not on personal experience but on literary sources:

> Certain poppies are called Sciachaich by the Arabs from the sound like shattered pottery and in Africa there is said to be a herb that emits the sound Huahua when touched.[57]

The last page of *Tabula v* contains the genealogies relative to the quantitative analysis of plants, no longer considered as individual, unrelated entities but as organisms formed by several parts that occupy space and maintain relations with the environment and proximate bodies. Cesi first takes into account the disposition of plant parts ('partium positio'). With the first dichotomy he distinguishes the analysis of the disposition of the parts of the plant conceived as a unitary organism from the examination of its single organs. Vegetals display a variety of bearings: 'erectio, rectitudo, declinatio, obliquitas, flexus, supinitas,

qualities. And this desire for conservation is not only present in air, water and all the other elements, but also in animals and in every creature' (Della Porta 2008, pp. 96–7; see the introduction by Oreste Trabucco).

57 'Papavera quaedam Arabes a fracti vasculi sono simili sciachaich vocant, et in Aphrica herba dicitur sonum Huahua ab homine tacta emittere quasi persentiat', TM, p. 910.

prostratio'.[58] The erect, prone and other bearings depend on the relations between the parts of the single plants and on the energy of their vital spirits. The internal disposition can be analysed by considering the parts of the plant one by one, by placing them in relation to the whole vegetal organism, to similar or to slightly different parts.

Analysing in the successive dichotomy the quantitative relations of the plant to its environment, Federico recognizes that this is treacherous terrain for the mathematical botanist because the influence of the environment can be grasped more easily by means of a qualitative analysis.[59] The environment and human operations modify nature and, consequently, the external aspect of plants. Plants may be exposed to outside influences to a greater or lesser degree. They are affected by the heat of the Sun, wind, impact of other bodies, shade, rain, human agricultural activities, by the climatic conditions of their geographical location, as well as by the influence of the sky through which the essential element of light is radiated.[60]

This eulogy of the determinant function of light concludes the *Tabula* devoted to the quantitative analysis of plants, whose innovative character has been rightly emphasised. The introduction of the mathematical botanist does indeed represent a revolutionary novelty. Nor can there be any doubt that Cesi was driven to explore this new horizon under the influence of the pages on the Book of Nature of the *Assayer* (and even earlier of the *Discourse on the comets*).

However, acknowledgement of the Galilean matrix of *Tabula* V should not lead one to conclude that in those three dense pages Cesi achieved in the science of plants the same results as Galileo in cosmology, the science of motion and the conception of the structure of matter. If read carefully, *Tabula* V appears not as the illustration of the results of a quantitative analysis of the universe of plants, but rather as an attempt to define what we might call a feasibility study to ascertain whether the conditions to proceed along that route existed.

Cesi's quantitative analysis of plants is couched in terms of more or less, big and small, heavy and light, one and many, absolute and relative. Though he often uses the term 'number', he never proposes a precise numerical relation between different plants or their parts. Moreover he often refers to the circle, the spiral and the pyramid, but the only example of forms drawn by nature-

58 Ibid., p. 911.

59 'Qui quidem si cum producendi et conservandi natura consideretur, a quantitatis modis, mensuraeque rationibus ad eos, qui qualitatis sunt vergit', ibid.

60 'Diverso vertice caelesti varie distributa lumina, radiosque infundente, imprimente permovente', ibid.

using a compass that he provides is the cap of fungi – a singular example, since the Prince classified them among the imperfect plants!

Other elements help to bring into focus the peculiar way in which Cesi made use of Galileo's paradigm of the Book of Nature in defining his new quantitative botanical science. As Federico had pointed out in *Tabula IV*,[61] the mathematical botanist's horizon was confined to the exterior appearance of plants and their parts, their reciprocal spatial proportions and relations, and the influence on their morphology and aspect of neighbouring plants and of the environment. Among the skills required of the botanist capable of radically reforming plant science, Cesi had reserved an important but circumscribed one for the mathematical botanist. His understanding of plants would be of limited use without the contributions of the physicist, the physician, the moral philosopher and the metaphysician. As we shall see, at the heart of Cesi's project of setting botany on a new foundation lay not so much the study of numerical relations and figures perceptible in the external appearance of plants, as the analysis of the internal causes that govern their development and determine their functions and properties. His vision called also for the interpretation of the enigmas, hieroglyphs, symbols, parables and fables imprinted by God in plants. That delineated in the *Tabula* V is the same integrated hermeneutic principle adopted by Cesi in his research on the animal and mineral kingdoms.

For Galileo, the quantitative interpretation of natural phenomena discounted surface appearances entirely. The mathematical structure of the Book of Nature could be read by the eyes of the mind, not those of the forehead. It was a hidden weave, impossible to perceive by observing nature only in its external aspects.

Cesi's study of nature was nourished by pressing philosophical and practical – medical, moral and religious – motivations that he attempted to harmonise with Galileo's physico-mathematical method of inquiry. Federico did not passively assimilate the Tuscan scientist's novelties, nor did they lead him to radically modify or abandon the project that he had conceived from the start of his intellectual career.

Tabula V offers further confirmation of the ambiguity, inadequacy and misleading character of those interpretations of the relation between Cesi and Galileo based on concepts like that of the 'Galilean conversion' of the Prince after 1611, from which a rigorously quantitative botany, not only inspired by but programmatically modelled on the principles and methods of the new Galilean science, would have derived.

61 See above, p. 420 and n. 41.

Glimmerings of Consciousness and Sexual Drives

Tabula VI tackles the fundamental questions of the origin of vegetal life and the causes of plant motions, the preserve of the physical botanist. It opens with the passage from *Genesis* in which the generation of plants is described as the germination in the womb of the Earth of an infinity of seeds that give life to diverse vegetal organisms.[62] Cesi does not evoke supernatural powers to explain the origin of life. He limits himself to focusing on the environmental conditions indispensable for initiating vital processes, paying particular attention to celestial influences, especially those of the Sun, whom Anaxagoras and Aristotle had called 'Father':

> Solar rays, the light of the stars, the transmission of corpuscles. Rains, dews and snow.[63]

We perceive here a clear echo of the themes outlined by the Prince in his *Taumatombria* (not surprisingly the *laserpitium* is mentioned in *Tabula* VI), centred on the fundamental principle of the unitary and fluid nature of the heavens. For Cesi the Earth too played an important role: the spirits and juices essential to plant nutrition were produced by the soil's more humid components, which was why, he added, Aristotle had rightly called the Earth 'Mother'.

Besides a Mother and a Father, other conditions are required for life to develop. Cesi locates them in warm, moist matter and in simple or composite particles endowed with motion. The movement to which Federico refers has little in common with the local motion that plays a fundamental role in Galileo's conception of nature. He alludes here to the vital process triggered by the motions of activation (*excitatio*), maturation (*eductio germinis*) and configuration (*figuram inducens vis*). These processes are perfectly programmed in the seed, where they remain latent until quickened by the moist warmth of the soil.

In the dense block of text on the origins of vegetal life at the centre of the first page of *Tabula* VI Cesi proposes once more the interpretation of the nature of life, its origin and preservation that he had presented in the *Apiarium*, as well as in his studies on meteors and fossils. Skirting the question of the origin of life, he confines himself to underlining its essential character in a

62 'And the earth brought forth grass, the herb yielding seed after his kind, and the tree yielding fruit, whose seed was in itself, after his kind' (*Genesis* I, 12).

63 'Radij, lumina syderum, immissio particularum. Imbrium, rorulentaeque aspergines et Nivium fomenta', TM, p. 912.

succinct definition: 'Life itself: the act of living: vigour: verdancy'.[64] The vital functions of plants depend on the vegetative soul (*vegetandi anima*), 'which operates through the organs suited to the various functions and in obedience to the parts'.[65] It is the soul present in all vegetals that, cooperating with the juices of the soil and endogenous heat, guarantees the health and conservation of plants, as well as the symmetry and harmony of their organs. That soul not only presides over the vegetative functions; it also permits plants to experience sensations and even to have glimmerings of consciousness and sexual drives:

> Sense, pain, pleasure, coupling, and those traces of intellect and cognition, which are asserted by Empedocles, Anaxagoras and perhaps even Plato [...].[66] By analogy, they [plants] may be attributed sexuality similar to that of animals.[67]

We have already come across these motifs in the analyses of the *Apiarium* and of *Tabula I*. For Cesi, there are no gaps in the Great Chain of Being: the transitions between the links that form it are imperceptible. Hence it is impossible to draw a line of demarcation between plants and animals. The only noticeable difference is that plants lack locomotion.

Referring again to Anaxagoras, evoked three times within the space of a few lines, Cesi declares:

> The soul by which they [plants] live is present in the whole body. It is distributed from the roots through the medulla and the heart, through visceral branches to the limbs and to the nerves and fibrous parts, and those parts enclosing the interstices, and is also present in the refuges and cavities of the bark, like those things which consist of a compressed, thick moisture that is not easily dissipated.[68]

---

64 'Vita ipsa: vivendi actus: vigor: viror [...].

65 'Quae aptis ad eas, quae infra functiones organis et partium obsequio usa', ibid.

66 Though, he adds, Theophrastus, Aristotle and Galen viewed this terminology as purely metaphorical, ibid.

67 'Sensum; dolores; voluptates, connubia interim, et ulterius mentis, ac cognitionis vestigia illa, quae ex Empedocle, Anaxagora, nec invito forsan Platone, asseruntur [...] non nisi metaphorice acceperis et eleganter. Analogice vero si libet sexum quemdam ad animalium modos', ibid.

68 'Insidet porro ea qua vivunt Anima in corpore toto, praecipue vero ab radicis fontibus per medullam illam, quasi ramosa viscera, et distensum cor, in membra diffunditur, nervosas itidem, ac fibrosas partes, interveniorumque plexus, nec parum in ipso cortice, utpote ijs, quae bene subacto lentoque nec facile dissipabile succo consistunt, in eius foculos et latibula', ibid.

The soul perfectly orchestrates the functions pre-programmed in the latent state of the seed. Cesi calls the extraction from the seed of the vital soul latent in it *educatio*. This extraction does not take place at random, but only under suitable conditions: 'the beneficial influence of the surrounding air, sunlight and rain'.

Man too can activate the vital force latent in the seed by artificially creating the right conditions as does a farmer: '[*educatio*] is an art'.

Cesi's vocabulary here recalls that of Anaxagoras, for whom the active principles of living nature were diffused everywhere. The Presocratic philosopher's conception of vital phenomena as states of aggregation of minuscule particles had drawn the interest of the Linceans ever since the foundation of the Academy. In this connection it is worth recalling that Heckius had compiled a compendium entitled *Philosophia Anaxagorae*.[69]

Federico's conception of nature as a living organism appeared perfectly symmetrical with the principle of plenitude on which his vision of a Ladder of Being without any leaps or gaps was based.

After having analytically classified the affections and motions of plants, Cesi proposes, in the long concluding scholium of Tabula VI, a series of more general reflections on the position of vegetals on the Ladder of Being. The length of this scholium is evidence of the importance that Cesi attached to this question. Here, the physical botanist leaves ample scope for the philosophical and metaphysical botanist.

The scholium is divided into two columns of text side by side (fig. 21) containing, as the general title of the page makes clear, reflections 'On the degree of life in plants, and its diversity from that in sentient beings'.[70] Although this title extends across both columns it is laid out in such a way as to signal that the first column deals with plant life, the second with the differences between plants and animals: 'On the degree of life in plants' is perfectly centred above the left column, 'its diversity from that in sentient beings' above that on the right. Once again we find an eloquent example of the almost maniacal attention that Cesi paid to the layout of the page, not for aesthetic reasons, but for conceptual needs and for ease of communication. Cesi devised this graphic solution to indicate that the scholium contrasts the account of the vital functions present in plants, on the left, with the comparative analysis of the vital processes of animals and vegetals, on the right. Both columns are set in italics,

69 Cf. Contributi I, pp. 286 and 291.

70 'Ad vitae gradum in Plantis, et eius ab eo qui in sentientibus diversitatem', TM, p. 915.

PRINC. FED. CAESII LYNC. PARS I. 915

Ad vitæ gradum in Plantis, & eius ab eo qui in sentientibus, diuersitatem.

Quæ plenissima in cultu obseruatio, & penetralia ad vsque pertingens, ea nobis stirpium, quæ de vitæ gradu sunt ostendat vlterius, vt scholio quodam hic subnotemus. Animalis eam agnouimus ab eo secto trifariam, adstipulante cum pluribus Philosophis Peripato. Sui autem peculiaris generis, & ingens à Brutis ad Plantas, perfectis, perfectas, interuallum. Viuacitas à spiritu est paulò supra Interraneorum halitus promoto, superioribus radijs. Corpus vita motum si cum ferino comparationem subeat, Vniuersi membris, scilicet terrenæ cutis parte, imò aëre, & aquis integratur. Benè nimirum Anaxagoras ventrem ipsi, seu ventriculum potius à terra substrata agnoscit, & circumfusa. Verum in conceptione dicamus, & appensione radicis quasi ab vmbilico alimentum, & deinde à perpetua quasi nutrice: sinum præterea quo semet tuto recipit aduerso cælo: matrem titulis pluribus. Hinc os est, nec vllis opus est organis, crassioribus ingerendis, concoquendis, excernendis: si purum putum chilum, cibos scilicet omnes in succum redactos, ab radicis osculis exugunt. Rectè ex Galeno Epar. quid Cardane stercora hinc expectas? potius vt Scaliger falsè à te vrinas plãtis expostulet in iudicia febriũ. Musculosa, neruosa, venosa traductio ab Epate eoq. & capite in ramos est; in vicarios inde flores, & partus. hoc vegetum animal. Excrementa, sudores, verrucæ, tubercula: vltimarũ vtpotè digestionũ: nõ crassũ quid, aut fœculentum. Quid alia quæritas mẽbra? an ne splenem quoque in his, & pulmonem? respirant ne & illa præstant, quæ nos cum reliquis sensu præditis eiusmodi organis. Diuersis indigent instrumentis functiones diuersæ: indigebunt talibus stirpes ad functiones non suas? Ramosa illa traductio à radicis capite, & cor, & cerebrum, & viscus omne est: & membrum, & membra. transferri nõ debet Animal, nec artubus vllis permoueri, nec ita sẽtire. Quid de appetitu? Imbres terra appetit, brutum escam: medium infra hoc, supra illud, loces appetitum, qui de stirpibus dicitur, nec erraueris. Idipsum de odio, fuga, amore credideris. Sensuum modo affectiones, altioris indaginis. Tactum vult Guilandinus obscuritatem excusat nostrorũ hebetudine sensuum, stirpium ruditate. non tamen facilè vtut obscurum quis coniectare poterit, nedum cõparentem vnquam. Siquidem disiungere à sentientiũ membrorum motibus æquum minimè videri potest. Tacta retrahitur Spõgia *vt superioris gradus. In planta excisa, aut percussa, quo talis innotescat conditio, motum habemus nullũ. Scimus plures ridere quã referre rusticos, arborem coram vibrata bipenne, antequã ipsum adactũ sit nutare ferrum. Cur nõ potius in nutrimenti arrhabonem affusorũ liquorum gustatus quidam, & genitalibus in operibus spirituũ conceptione, & emissione quasi venereæ lætitiæ nõnihil suspicandum, indagandum in stirpibus, pro organorum, & analogiæ modo? quod sentientibus respõdeat, citra tamen sensum. dum haudquaquam pro saporum ratione eligere, aut iucundis titillationibus in Venerem duci debent; nec voluntate, aut instinctu, sed vegetæ determinatione naturæ in semen parturiant: affusum nõ quæsitum, succum cõbibant. Cõtactus similitudo ab internis quidem, in alimenti augumenti partus ipsius, & sanitatis, necnõ excretorum modo, læuiuscula profert indicia. Potiora quæ ab exterioribus, nullum ferè vestigium referunt. Quæ tactu patiuntur animalia licet infima, nunquam membrorũ abscissionibus gaudere, ac prosperiora reddi solent. In Plantis insitio facilis, grata putatio, pãpinatio opportuna, faciliora in vitã germina à resecatis radiculis, à contortis, attritis, scissis, contusis ramulis; nedum dissectis ab omnibus; & læsionem ferè omnem à sectione excludere videntur, & quemuis huiusmodi tactus dolorem, & sensum. Si quis autem eo quo diximus modo inest, is ab vstione feruentium radiorum, aut rigentis brumæ iniurijs, intrinsecus etiam peruadentibus in spirituum, & complexus perniciem, sumendus est. Nec sexus à sterili fœmina mare fœcundo, vt perperam, & præter nascendi quæ in Animalibus leges, fit; sed à diuersa, quæ ab eodem semine prosurgit prole, & maris, ac fœmina connubijs quasi, & conuictu; vti in suis inferius ostendemus locis. Porrò diuerso in genere sumus, ea vtere analogia quã eius diuersa exigit ratio; & quæ præcipuè eam sequantur diuersitatem cape.*

Vegetantis huius Animalis vita affixa est. Terræ fouentur gremio, aëris sinu, aquæ lacte, corporis partem, mundi ipsius partem habent, conuiuunt elementis, accedit tellus ipsa in membrũ, vniuersi spiritus fouent.

Constitutio inde non vna, nec facies, nec totum simul & semper; abijcere, amittere membra potest, & recuperare renouatione quadam. Contrahere sese, & in radice; vel radice se abscondere, & bulbo, reliquis ammissis omnibus; vel in radice, & caudice reducere, frondibus nec dolenter dimissis. Exerere, vel sese Superficie dicta tota scilicet ab radice resurgentibus omnibus: [renouatis dixeris, & renatis iterum, atque iterum] vel nouas frondes promittere, quod diuersimodè; aut eodem ferè conspectu semper perseuerare, diuersis datum, vt suis patebit locis. Omnibus certè nouella erumpunt germinatione, quæ propria.

Partes in suum ita ire totũ valent; vt in id vertantur si abscissæ à parente, magnæ Telluris parentis cõmittantur sinu, siue illum demissæ contingãt, in quo ventriculũ obtineant. Quæ seminij virtus Viuiradicis dicta efficacia, satis ostendit animam ipsam vegetam ita in qualibet parte diffusam esse; vt diuidi sine vllo reliquæ præiudicio possit, & plenitudine, ac multiplicitate longe eam, quæ in Brutis est sentientibus superare, quæ inseparabili vnitate continetur.

Alijs etiam apposita membra imò membrorum particulæ conuiuunt, & coalescunt insitione, meliusque prorsus habent. Vlterius pensatur ipsa alligatio, & compositio cum mundi membris, quod abscissa etiam artubus viuere possint, diuq; illi seruari: ramuli, & fasciculi suis cum floribus diu certè vident in aqua, sed & semina proferunt, ferèq; perficiunt. Nõ rarò Lilia ab exiccatione etiam, & ab aqua semota reflorescere solent.

Admiranda in Pulegio prorsus decennium [vt ipsi obseruauimus] iam exiccato, florem bis in anno singulis solstitijs rursus aperiente, & emittente, & vita vis & syderum ratio; in nullo iam remanente succo, arenti penitus corpore. Nouit hoc forsan Cicero. causa alibi à nobis aperienda. Notius commoueri ipsissimos plantarum liquores in vasis, dum sui generis, plantæ in flores, fructusque aguntur; colore, & partibus; quinimò in commadefactis etiam, & pertinacibus maculis consentire.

Conculcata; & contusa plura in hoc agmine latius eriguntur. Hispanicum exemplo sit Chamamelum. Triticum à Septentrionali frigore percussum miram nanciscitur à niualibus gelidis abollis, à Scythicis pruinis vbertatem. Nec Taxi tantum Aquilone, & frigoribus exhilarantur; sed ea in quibusdam in seruando nexu vis adhuc tenellis à semine missis [contra quam vllis sentientium pullis permissum sit] vt gelidas etiam compressiones exuperet. Ita Tune nobis plantula vix vnciali foliolo, benèque succoso exporrecto restiterunt nocturnis brumæ iniurijs, quibus hoc anno ampliores eiusdem generis, benèque lignosæ arbores plurimum conciderunt.

Ea virtus est, vt annorum oneribus multo minus quam sentientia succumbat; siquidem longissima in multis ætas, in multis fœcunditas; vix finem, vix sterilitatem tangat. Et annosa Iliaca, Herculea, Quercus, vel vna amplum præbeat exemplum.

Vt extensio ipsa, & crescendi impetus, quæ vagandi libertati Brutorum respondere videntur; cum hæ ramis magis, magisque distendantur, ictusque telorum etiam superare contendant, surgendi spiritu indefinito. Nec in Pino tantum Abiete, ac Cupresso, Fagis, Populisq; sed & in alijs, etiam quandoq; superbientibus, vt in Aquaspartana nostra Vlmo.

Etsi ad bruta emittenda, quæ spontè prosurgere dicuntur vis particularum in connexum cõuenientium extenditur. Iam plantæ, quæ infra consistunt [nec mirum] ortum duplicem, ortum multiplicem, & propagationes sortitæ sunt; tum ab illis omniparentis compulsis opere; tum à seminali virtute per illarum corpora pluribus in locis in prolem, & fœtum cum germine præsertim effusa.

Ars præterea non ita Naturæ iura in brutis in variam multiplicandis speciem miscere potest, & temporum leges præuertere, ac in his ferè ad arbitrium inque dies magis facere videtur.

Superauerint ea propter reliquas in Natura rerum classes quascunque, varietatis, & familiarum numero. Luserit ea in his primo ferè in figuras, firmioresque eas, & benè impressas ducens, pictura in copiam luxuria simul, & licentia.

Tab.6. Hhhh 2

FIGURE 21 *The scholium of* Tabula VI *on the differences between plants and animals* (THESAURUS MEXICANUS, p. 915)

the right-hand one in a smaller type. As usual, the text completely fills the two columns, which are perfectly justified on all four sides.

At the very beginning of the left-hand column Cesi declares that a 'large gap' separates plants from animals. However, he immediately adds that this clear distinction is perceptible only when two 'perfect' exemplars are compared. In the case of intermediate natures, that distinction fades to the point of becoming imperceptible. After explaining that vitality (*vivacitas*) is generated by the spirits activated by celestial influences ('superioribus radijs') that raise subterranean exhalations ('interraneorum halitus'), the Prince emphasises that the body of plants is entirely pervaded by vital energy. Cesi evokes here once more Anaxagoras, who had defined the Earth as 'ventrem seu ventriculum'. In Federico's opinion, the function performed by the Earth in the generation of plants can be compared to that of the uterus in animals. Similarly, the action of the roots which extract juices from the soil and distribute them within the vegetal organism could be assimilated to that of the umbilical cord through which the mother feeds the foetus. Cesi defines the Earth as a perpetual nurse who nourishes plants which have no need of organs of mastication, digestion and excretion of the superfluous residue. The roots of plants draw from the soil nutrients that have already been transformed into juices. The functions of the heart and liver are performed in plants by the branches, while the flowers correspond to the creatures to which animals give birth. Warts and swellings show that plants emit substances similar to sweat and excrement. What prevents plants – Cesi concludes – from being considered like animals? Surely not lungs, since they do not breathe.

For the Prince, the transmission of the lymph through the ducts ('ramosa traductio') performs the same functions as those carried out by the heart, brain and viscera in animals. Plants are nourished not by food but by rain. They are also endowed with sensitivity, which they manifest through movement: the sponge contracts when touched, and, if one is to believe what countryfolk say, the tree backs away from the axe raised to strike it.

Cesi does not stop here. There are grounds to suppose – he adds – that plants have a sense of taste, enabling them to distinguish the qualities of the nutritional lymph, and that they even share in the joys of sex ('venerea laetitia'). These faculties should not be denied to plants on the grounds that they lack the sense organs of animals. Taste and sexual pleasure in vegetals do not depend on will and instinct as in animals. These sensations are spontaneously produced in plants by the characters of their vegetal nature imprinted in the seed.

Moreover, plants do not suffer when mutilated:

> Among plants grafting is simple, pruning welcome, defoliation opportune, and regeneration easier from damaged roots or twisted, worn, cut and bruised branches.[71]

This is shown by their capacity to emerge again from the soil after the devastation caused by winter frosts, fire or scorching by the Sun's rays. Their remarkable vitality, which is superior to that of animals, resides in the vital spirit or soul pervading the plant. Thus plants regenerate even if only the tiniest part of their root apparatus survives. Reproduction through sexual contact between a sterile mother and a fertile father confirms that plants do not lack any of the functions with which animals are endowed.

At the conclusion of this long column of text, the Prince must have realised that he had overstepped the limit in stressing the analogies between plants and animals. He therefore explains that he had not intended to deny the differences between the two kingdoms, but simply to emphasise that plants have analogous functions to animals, though performed in different ways and with different organs. Moreover he anticipates that in the opposite column he will bring out the differences between animals and plants.[72]

In fact the column of text on the right side of the page opens by underlining the different natures of animals and plants. However, the Prince stresses once again that the latter should be considered vegetal animals:

> The life of this vegetal animal is predefined. The Earth is its womb, the air its breast, water its milk. Part of its body corresponds to part of the world itself and it lives with the elements.[73]

Referring once again, although implicitly this time, to Anaxagoras, Cesi declares that the whole plant is present in every one of its parts ('partes in suum ita ire totum valent'):

> This property of the seed amply demonstrates through the efficacy of the scion that the soul of plants is diffused in every part: so that it [every part]

71 'In plantis insitio facilis, grata putatio, pampinatio opportuna, faciliora in vitam germina a resecatis radiculis, a contortis, attritis, scissis, contusis ramulis', ibid.

72 'Porro diverso genere sumus, ea utere analogia quam eius diversa exigit ratio; et quae praecipue eam sequantur diversitate cape', ibid.

73 'Vegetantis huius Animalis vita affixa est. Terrae foventur gremio, aeris sinu, acquae lacte, corporis partem mundi ipsius partem habent, convivunt elementis', ibid.

> can be separated without any prejudice to the rest, and it [the soul of vegetals] far exceeds in plenitude and multiplicity that in sentient animals, which are bound by their inseparable unity.[74]

Forgetting his promise to highlight the differences between plants and animals, in this column the Prince insists on the presence of sensitivity and even movement in plants. For him the impetus with which vegetals develop from seed can be compared with the motor faculties of animals:

> The extension itself and impetus of growth seem to correspond to the freedom of movement of animals; for plants extend their branches more and more and even strive to avoid the blows of missiles [...].[75]

Also the reproductive energy conferred on plants by the properties of the seed is comparable to that of animals. Indeed it is even more ductile, for while nature is unable to hybridise different animal species, the botanist manages to cross plants belonging to different families:

> Art cannot bend the laws of nature or subvert those of time to generate hybrid species of animals, while among these [sc. plants] it [the botanist's art] seems to do as it wishes and increasingly with the passage of time.[76]

The Prince's conclusion is that, in the variety of their species, the number of their families and the diversity of their configurations, plants are superior to every other genus. Nature has expressed its creative energy to its highest degree in the vegetal kingdom.

Thus, in spite of the message transmitted by its title, the right-hand column abundantly confirmed the opinion formed by the reader after consulting the one on the left. Cesi had not maintained his promise to attenuate his strong

74 'Quae seminij virtus Viviradicis dictae efficacia, satis ostendit animam ipsam vegetam ita in qualibet parte diffusam esse; ut dividi sine ullo reliquae praeiudicio possit, et plenitudine ac multiplicitate longe eam, quae in Brutis est et sentientibus superare, quae inseparabili unitate continetur', ibid.

75 'Ut extensio ipsa et crescendi impetus, quae vagandi libertati Brutorum respondere videntur; cum hae ramis magis magisque distendantur, ictusque telorum etiam superare contendant', ibid.

76 'Ars praeterea non ita Natura iura in Brutis in variam multiplicandis speciem miscere potest et temporum leges praevertere, ac in his fere ad arbitrium inque dies magis facere videtur', ibid.

emphasis on the analogy and symmetry between the animal and vegetal kingdoms. On the contrary, he seems bent on demonstrating the superiority of plants vis-à-vis animals. His message is clear: the faculties and functions of animals are all present in plants, which, by virtue of a vital principle diffused throughout their entire organism, present exclusive characteristics. These characteristics offer extraordinary benefits to those who are able to recognise their properties and turn them to their own ends.

The passionate eulogy of the vegetal world, a most noble terrain of philosophical and moral speculation, as well as the emphasis on the possibility of manipulating plants for the benefit of man that conclude the scholium serve as the perfect introduction to *Tabula VII*, in which Cesi dons the garb of the medical botanist.

The Bologna Stone Again

It is in the reflections and schemes of *Tabula* VII that we find the most explicit evidence of the influence of Galileo's analysis of sense perceptions as well as precious clues to the way in which the Prince integrated that lesson within the framework of his own vision of nature.

The scheme of classification follows the by now familiar layout of a dense sequence of dichotomies extending from left to right. Cesi first takes into account the most general category that includes the properties manifested by plants whose integrity is destroyed.[77] With the first dichotomy Cesi distinguishes the two fundamental ways of obtaining this result: the breakdown into confused and mixed particles without any alteration of their constitutive nexus; and the solution by heat and fire to extract the elementary components endowed with active properties. The first of these subdivisions is further broken down: the specific nexus of a given plant remains unaltered; or is modified to a greater or lesser extent.

To understand the effects on human beings of the properties of different plants, it is necessary first of all to examine their scent and taste. These qualities, however, and particularly taste, may prove deceptive, since a strong flavour does not necessarily produce beneficial effects.[78] The medicinal properties of plants depend on the particular combination and interaction of their active qualities, which Cesi defines as 'primary', specifying that these qualities are

77 '[...] vires plantae, eius compositione destructa', ibid., p. 916.

78 'Nec tamen insignes sapores insignes consequuntur plantarum vires', ibid.

identified differently by those obsequious to medical tradition and by the followers of chemical medicine. For the former the primary qualities are hot and cold, moist and dry. Chemical medicine, on the other hand, takes the primary qualities to be salt, sulphur and mercury, corresponding to ash (taste), flame (scent) and smoke (colour).[79] Below the traditional primary qualities and gradually flowing from them are the secondary (*posteriores*) qualities, that the Prince divides into those of second (softening, hardening, astringent),[80] third (revitalizing, reactivating, cicatrizing)[81] and fourth grade (poisonous, operating as antidotes, cathartic, stimulating).[82]

He then illustrates how the active principles can be artificially extracted from plants by maceration, violent pressure (as in juices, wine, oil), moderate pressure (as in beer, thick juices), or by natural processes (as in honey, sugar, manna, wax, resin, etc.). He then subdivides the faculties of vegetals into nutritional (for animals and men) and therapeutic (medicinal, cathartic, harmful and poisonous). Cesi clarifies that when plants are broken down, the nexus between their particles can be modified in two different ways: firstly, through natural processes of putrefaction, from which different types of fungi are generated, or through the fervour of the internal spirit from which small creatures (spiders, worms, caterpillars, chrysalides, beetles, etc.) are born.[83] Secondly, through the use of fire ('perquoquente igne'), which makes it possible to extract from plants watery or oily juices, salts, or substances of a different composition and consistency, such as glass.[84] Besides modifying the internal links

79 Instead of adopting Paracelsus' interpretation of flame as the quality of sulphur, Cesi followed Libavius, whose works he possessed (see Guerrini 2006, pp. 67–8 and Biagetti M.T. 2008, index). Libavius replaced the flame with bitumen (TM, p. 916). Federico had focused his attention on bitumen in his researches on the subterranean fires of Acquasparta and the fluids on which they depended.

80 'Emolliendi, indurendi, adstringendi', TM, p. 916.

81 'Sarcotica, conglutin [...], cicatr', ibid.

82 'Venena, Alexipharm, Cathart, Impell', ibid.

83 TM, p. 920.

84 The production of glass from ash dates from the Middle Ages. Among the contemporary texts on the production of glass, Cesi regarded *L'arte vetraria* by Antonio Neri as particularly authoritative. Neri was on familiar terms with Galileo and frequented in Florence the Casino di San Marco of the alchemist Prince Antonio de' Medici, a follower of Paracelsus, to whom Neri's work was dedicated (Galluzzi 1982). After an unsuccessful attempt to obtain the book in Rome, Cesi wrote on 3 January 1614 to Galileo to request a copy: 'I hear that *L'Arte Vetraria* by P. Antonio Neri has been printed in Florence, and I think there is some good in it. Please send me a copy [...]' (OG XII, p. 12). Galileo took advantage of his familiarity with Antonio de' Medici and with Neri himself to satisfy the request. Three

between the particles, the medical botanist can have recourse to two different techniques, both based on fire, to isolate the simple elementary components: by evaporation ('fumida exhalatio'), which selects the most substantial particles, or by solution ('flammea exsolutio'), effected through the action of both heat and light, by which the finer substances are extracted from the thorns, stalks and more oleaginous juices.[85]

This section of *Tabula VII* offers an opportunity to compare Cesi's conception of light and heat with that of Galileo. The Prince states that luminous substance is ubiquitous, though in different grades of intensity: from light itself to its pale reflection ('leviusculam eius adombrationem') in the surroundings or in objects that we erroneously consider shrouded in darkness. Federico demolishes the traditional theory of light as the contrary of darkness. For Cesi light permeates the entire universe in different degrees of intensity. Like cold, which does not exist as such but is an extremely low degree of heat, light is present in the deepest darkness, even if imperceptible to human sight. It was based on this principle, like that used by Galileo to refute the positive existence of weightlessness, that Federico explained the luminosity of finds like the fossil woods of Acquasparta, which, in spite of being buried, showed evident traces of the presence of the luminous substance, which therefore even penetrates to the bowels of the Earth.

The same applies to heat and cold, which should be considered simply as different gradations of the same quality, as demonstrated by the light of the Moon, which, even though cold by comparison with that of the Sun, still has a degree of heat:

> You may find the light of the Moon's rays feeble and cold by comparison with the heat of the Sun's rays, but much less so than the difference between that degree of heat and icy cold.[86]

---

weeks later Cesi had the book in his hands and had already read it: 'I thank you for the *Vetraria* book – he wrote to the Tuscan scientist – which I find very rich in experiences and fine operations' (ibid., p. 15). Neri's work has been reissued with an introduction by Ferdinando Abbri (Neri 2001, pp. 5–34). *L'arte vetraria* is registered as no. 114 in the inventory of the library of Federico Cesi (Biagetti M.T. 2008, p. 80: 'L'Arte vetraria distincta in libris sex [...] in Fiorenza 1612').

85 TM, p. 921.

86 'Lunaribus hanc [lucem] innoxiam, frigidamque compares, si igneam Solaribus radijs, sicuti enim minus multo, quam existimari soleat caloris ipsa acies ab acuto distat frigoris gelu', ibid.

This principle served to explain phenomena such as the pyrites from Acquasparta that were luminous but cold to the touch, or extraordinary effects like that displayed by the Bologna stone.

In the text in which Cesi highlights the causes of the infinite variations of the degrees of light and heat we find a new and more explicit reference to Galileo's interpretation of the subjective nature of secondary qualities and the corpuscular composition of light. Luminous intensity, for Cesi, depends on the state of agitation of the constitutive particles of light:

> The light emitted from agitated [particles] is not separated by a large interval from that when they are calm or at rest.[87]

This brief statement seems to have been inserted almost casually in a block of text on the relations between heat and light. Federico declares that he will not insist on such a pregnant principle here, having dealt with it elsewhere ('de quibus suo egimus in loco').[88]

The conception of light as substance composed of corpuscles in variable states of agitation and as the universal principle of life had been delineated by Galileo in the letter to Piero Dini of 1615. Given the strong Ficinian echoes contained in that letter and Cesi's inclination toward Neoplatonist conceptions, one wonders whether it was Galileo who provided the Prince with the vision of light as an enlivening substance that pervades the universe, or whether Federico inspired the metaphysical conception of light expounded by the Tuscan scientist in the letter to Dini.

However that may be, in the vision of light as a pervasive corpuscular substance, the fundamental agent of vital activity, the *Tabulae* display the maximal degree of convergence with Galileo's interpretation of the structure of matter and of the causes of sensations expounded in the *Assayer*.

It has escaped notice until now that in the labyrinthine composition of this *Tabula*, of which Galileo is the tutelary deity, though without ever being cited, Cesi inserted an unmistakable reference to the Bologna stone in a block of

---

87 'Quod a concitatis [particulis] mittitur lumen haudquaquam longo disterminatur intervallo ab eo qui a placide vel in ipsam quietem disiectis habetur', ibid. (block of text in centre of page).

88 No specific analysis of this problem is to be found in Cesi's surviving writings. It is possible, however, that he referred here to the blocks of text of the *Apiarium* in which he attributes the generation of sensations of heat and cold to the penetration of the corpuscles (see above, pp. 394–6).

text[89] in which he proposes many proofs of the close affinity between light and heat. Among the conquests of the human race – Federico underlines – the greatest importance should be assigned to the capacity to generate light and heat through fire. In our century, the Prince continues, there is no lack of followers of Prometheus who have shown even greater skill than the mythical demigod in stealing light from the Sun. He is obviously referring to Galileo,[90] and the light stolen from the Sun takes on here the appearance of a stone – the Bologna stone:

> A tiny part of the light is stolen from the Sun's rays (for there is no lack of descendants of Prometheus in our day), shut up in a box and carried into the darkness.[91]

This is not the only reference to the Tuscan scientist in this highlighted block of text. In the concluding lines, Cesi regrets the tendency of natural philosophers to neglect phenomena such as fire and light that are before our eyes every day and are essential for the production and preservation of life. He attributes this neglect to the inclination to send the intellect into exile ('in exilium truditur'), following the ideas of others. As an example of this passive attitude, he cites the obstinate refusal of many to recognise the presence of mountains and valleys on the surface of the Moon.

It is impossible to establish when Cesi arrived at the interpretation of the relative nature of dark and cold and of light as a substance that permeates the universe, diffusing vital heat in infinite degrees of intensity. That his concern with these issues went back many years is shown by passages of the *Zibaldone* composed between 1612 and 1615. In the section entitled *Physica Mathesis* (mentioned several times in the *Tabulae*) Federico conferred considerable importance on the question of the nature of light and heat, though his remarks are too sketchy to enable us to reconstruct his ideas with precision. The bare mention there of a treatise *De lumine* is juxtaposed with the dichotomy *De luce calida/De luce frigida*, which seems to indicate the Prince's adherence to the conception of opposite qualities. A few lines further down,[92] his plan to conduct an in-depth inquiry into the anatomy of light (*luce apertione*), motion, heat

89 Ibid., block of text close to the left hand margin of the page.

90 On Faber's assimilation of Galileo to Prometheus in his 1622 lecture on the nature of fire, see above, p. 397 and n. 26; see also De Renzi 2007, p. 290.

91 TM, p. 921.

92 Contributi I, p. 34. Note that the term used here is *lux*, not *lumen*; cf. below, pp. 454–5.

and cold (*motione, calore et frigore*) and feeble, moist and dry (*tenui, humido et sicco*) transpires.

Further clues are offered by the passages of the *Physica Mathesis* in which Cesi manifests his intention of representing the variations in the intensity of qualities as a ladder with the opposite qualities at its two extremities and between them the rungs representing intermediate transitions. He adumbrates here the solution adopted in the *Tabulae Phytosophicae*:

> A table with different columns for different conditions that proceed by gradation between opposite extremes; for example, heat at the top of the column, cold at the bottom, and the intermediate grades in between, depending on their distance from the one or the other.[93]

The use of the expression 'opposite extremes' ('estremi oppositi') suggests that at that date Cesi had not yet subscribed to the conception of the relative nature of qualities.

It should also be mentioned that, while the Bologna stone features in *Tabula VII* as proof that light is a substance diffused everywhere, in the *Indicatio* that stone was not evoked in the Frontispiece of the *Physica Mathesis*, but in the context of the analysis of meteorological phenomena that Cesi planned to develop in the *Taumatombria*.[94] Such a context is hardly surprising, since in the projected treatise on meteorology Cesi intended to analyse the nature of pyrites, which – exactly like the Bologna stone – had a fiery appearance but were cold to the touch.

The Garden of Flavours

Tabula VII has attracted the attention of scholars mainly for its remarkable classificatory schemes of flavours and the appendices in which the Prince comments on their significance.

The table opens with the text contained in the narrow right-hand column of the first page, in which Cesi informs the reader that on the facing page he

93 'Tabula in qua pro diversis conditionibus columnae diversae pariter producuntur ab extremis oppositis gradibus procurrentes, ut in eiusdem columnae capite calor, in basi frigus, intermedii in excursu gradus, prout ab illis magis recedentes vel accedentes notantur', ibid., p. 36.

94 Ibid.

will find important elements to profoundly 'savour' the nature of flavours.[95] He also declares that he will focus on the processes from which flavours derive and elaborate a quantitative classification of their varieties and of how the particles of different substances act on the taste organ. Only in this way – he states – can the nature of flavours and their properties be precisely detected:

> It is more simple and certain to deduce the nature of flavours by taking into consideration both the recipient and what enters into contact with it; the mode of reception and contact is through motion or rest; alternation of transfer and exchange; evacuation or retention; from which particles are removed or left here or there, the body remaining in one or other position.[96]

Cesi reduces the great variety of flavours to thirteen categories, which he analyses one by one in as many blocks of text of various geometric shapes: above all triangles, trapezia and rhombi (fig. 22).[97] This original graphic solution clearly expresses his intentions: to create a garden of flavours, composed of 'flowerbeds' of geometrically shaped blocks of text that recall the layout of Renaissance gardens, like those, familiar to the Prince, of Villa d'Este at Tivoli, or the Vatican Belvedere, superintended by his fellow academician Johannes Faber.[98] This choice made the task of calibrating the individual blocks of texts even more challenging than in the *Apiarium* or the other *Tabulae*, obliging the Prince to make continual use of abbreviations and contracted expressions that rendered his Latin prose more opaque than ever.

The thirteen fundamental flavours are organised in three distinct categories (active, medium and passive)[99] and the same format is used to describe

95 TM, p. 916.

96 'Haec vero quae deinde ducuntur simplicius certe potiusque ratione partium tum recipientis rei, tum in illam advenientis; pro receptionis, connectionisque modo habeamus et utrinque, vel quiete vel motu; transfundi et transmutandi vicibus; egressu aut retentione, ablatisque sive relictis hinc inde particulis, in alio atque alio statu conlocationum corpore remanente', ibid.

97 Ibid., p. 917.

98 Yvette Conry stressed that the distribution of text in this section of *Tabula VII* is clearly modelled on the architecture of vegetable and flower gardens. Late Renaissance and Baroque topiary art contrasted the garden with the labyrinth. The latter, as a metaphor of the *silva*, represented the complexity of nature that could not be unravelled by man (Conry 1997, pp. 198–200). According to Conry, Cesi adopted this graphic solution under the influence of Vitruvius, Galileo and Palladio (ibid., p. 203).

99 Active: *amarus, acris, acidus, nitrosus, salsus, dulcis, acerbus, austerus*; Medium: *oleosus, humilentus, aquinsipidus*; Passive: *ligniterreus, adoreus*.

PRINC. FED. CAESII LYNC. PARS I. 917

Sapores, quibus scilicet Plantæ tibi sapiant, tu illas plurimum sapias.

ACRIS

AMARVS

ACIDVS

Nitrosus. SALSVS. DVLCIS. ACERBVS. Austerus.

Adstringens.

OLEOSVS. HVMILENTVS. AQVINSIPIDVS.

LIGNITERREVS. ADOREVS.

Tab.7. Hhhh 3

FIGURE 22 *The table of flavours* (THESAURUS MEXICANUS, p. 917)

each of them. In the first place, the Prince focuses on the composition of the substances that produce the sensation of flavour when they come into contact with the taste organs. A bitter flavour, for example, is produced by denser earthy particles set in motion by fire; a sharp flavour by very fine earthy particles and water; an acidic one by the force of the fire that dissolves the corpuscles;

a nitrous one by particles of earth excited by the spirits. Among the middle flavours, *oleosus* (oily) is produced by particles of water that have assumed an airy consistency, while *ligniterreus* (intermediate between earthy and woody) is produced by earthy corpuscles entirely lacking in spirits.

Then he analyses the action of the elementary components of the diverse substances on the organs of taste and concludes with the classification of the vegetal substances which produce flavours.

The table of flavours put the remarkable linguistic skills of the Prince to the test. He had to contrive effective ways of conveying all the nuances of the perception of different flavours. He proposes long strings of verbs to express the operations of the material particles on the organs of taste and the effects produced by them. To appreciate Cesi's extraordinary lexical performance it is enough to read the descriptions of bitter and acidic flavours:

> [*Amarus*] All authors agree that it is not suitable for nutrition; some of them write that it is an aperient, so as to dilate the apertures of the veins and provoke haemorrhages and thirst by causing the blood to flow, as well as thinning, cutting, biting, irritating, drying, liquefying and attracting; however, it burns, dries out and consumes less fiercely [...].
> [*Acidus*] assails the organs of taste, removes nausea, stimulates appetite, penetrates and is mordicant and corrosive; it is said to equally thicken and reduce without heat, because it is cold for the physicians, opens, thins, embitters, frees from indigestion, purges [...].[100]

After listing the sensations produced by flavours (unpleasant, sharp, ancipital, astringent),[101] Cesi takes into account the particles that generate them (watery, earthy, fiery or a mixture of them) and the influence of environmental factors (heat, cold or intermediate conditions). It is evident that he intends to contrast his own corpuscular inquiry on the generation of flavours with the analyses of traditional medicine. He states that flavours are not the result of the intrinsic

100 [*Amarus*]: 'Ineptum nutritioni solum omnes fatentur, aperire plures scribunt, ita ut orificia venarum reseret, haemorrhagiam et sitim excitet, sanguinem fluere faciens, tenuare, incidere, mordere, exasperare, abstergere, liquare, attrahere; minus tamen vehementer adurere et siccare et consumere [...]'. [*Acidus*]: 'gustum ferit, naus. tollit, appetit. citat, penetrat, mordicat, rodit absque calore, pariter densare, repercutere dicitur quod frigidum est apud Medicos, nec solum, sed aperire, abstergere, exasperare, attenuare, farctu liberare, perpurgare [...]', TM, p. 917.

101 'Tristis, acutus, anceps, stipticus [...]', ibid.

qualities of plants, but of the interaction of their elementary particles with the organs of taste.

In his analysis of the first three flavours (bitter, sharp and acidic) he once again severely criticizes the conception of contrary qualities. There is no break in continuity in the transition from hot to cold or from moist to dry. This concept plays a key role in the description of all flavours, but is put particularly forcefully in that of the acidic flavour, in which Cesi denounces the 'vulgar philosophers' who are totally ignorant of the true nature of heat and cold. He reaffirms that cold is simply an extremely low degree of heat. Even if we perceive it as cold – he stresses – there are 'atoms of fire' (*ignicoli*) in water. In this statement we find new evidence of the influence of Telesio and Persio, who had both attributed a fiery nature to water.[102] Cesi's familiarity with Telesio's interpretation of flavours is also confirmed by the transparent similarities in their terminology.[103]

Federico deems it imperative to abandon the theory of contrary qualities:

> Let us join burning and freezing, drying and humidifying, or fire and water, heat and cold, so that you may easily doubt whether all contraries differ from one another.[104]

Instead of trying to ascertain whether a specific flavour is hot or cold, the medical botanist should determine the position on the scale of heat to which each flavour should be assigned. The Prince was convinced that this innovative method would make it possible to build a rigorously quantitative science of simples, a field traditionally based exclusively on qualitative analysis and therefore lacking in rigour and consensus. Federico's ambition to transfer Galileo's mechanico-corpuscular analysis of natural phenomena to the study of plants clearly transpires in this *Tabula*.

---

102 Telesio 1590 and 1981, pp. 349–78; Telesio 1965–1974, 1976, III, pp. 24–59. In one of his opuscules published by Persio (Telesio 1590, pp. 1–15) Telesio refuted the Aristotelian doctrine of flavours. He dealt with flavours again in his *De rerum natura*.

103 Compare the series of verbs in Telesio's *De saporibus*, published by Antonio Persio in Venice ('dilatari, fundique, compingi, constringi, laevigari vel exasperari, mordicari atque erodi, spiritus dispositionem immutari'), with that of Cesi illustrated in note 100 above.

104 'Colligamus et urere et perfrigerare; siccare quamvis ab humore. Vel potius ignis esse et aquae; caloris et frigoris, contrariorum omnium ut facile dubites an a se quoque differat', ibid.

Introducing the 'garden of flavours',[105] the Prince declares that there are more flavours than those he has listed. He therefore invites the reader to supplement them as long as he suggests the terms ('si et nomina suppetunt') that clearly express their characteristics– an explicit admission of having exploited every possible terminological resource to describe the thirteen flavours in the 'flowerbeds' of text. There are, he says, as many flavours as there are foods; their variations can be multiplied almost ad infinitum by the combination of the diverse vegetal substances. The only way to master such a complex universe is to translate these combinations of basic flavours into numerical relations.[106]

Before tackling that daunting prospect the Prince highlights the limits of a simply qualitative analysis of flavours as well as its potential danger to health. He warns the reader that while some flavours unpleasant to the taste contain valuable nutritional and therapeutic properties, other more agreeable ones are not always beneficial, even though it may take some time before any nefarious effects manifest themselves:

> Flavours are first distinguished by the mouth or the fauces. We must not rely upon our sensations nor believe that they provide a perfect evaluation of all things. Similarly, we should not think that those substances which do not hurt the sense of taste can be safely assimilated. The capacity of corrupting is hidden in some vegetals, while in others it does not become manifest by brief exposure to heat in the mouth, but bursts out after longer exposure in the stomach.[107]

Proof of the ambiguity of flavours is provided by sugar, which is mild to the palate but harmful to the eyes, or mustard, which is harmless to the mouth but causes obstructions in the intestine. Plants, like men, also deceive with

105 TM, *Append. ad p. 917* (unnumbered page).

106 'A mixtione foecundius, et proportionum in ea multifarijs rationibus in omnen excurrent numerum', ibid. (block of text in the lower part of the page).

107 'Ad hominem porro hi, immo ora fauces hominum, primae scilicet gustationis instrumenta notantur; nec illud nobis arrogemus sensus nostros tamquam trutinam ac mensuram se habere rerum omnium, neve existimemus quae ora non feriunt tuto interius admitti posse [...]. Occultae siquidem in nonnullis acies, tardiores in alis, quae si non oris brevi, longiori certe stomachi calore excitata erumpant in impetus', ibid. (block of text in centre of page).

false smiles.[108] Man's physical constitution is different from that of animals, who can safely ingest substances detrimental to the human organism. Animals, moreover, are protected by their more refined sense of smell, which enables them to distinguish dangerous substances.

The principle underlying his new corpuscular classification of flavours is stated in no uncertain terms: flavours are the product of the solvent action on ingested substances by the juices of the organ of taste. Their variations depend on the configuration of the elementary corpuscles of those substances:

108 In his edition of Della Porta's treatise on physiognomy, Stelluti extended to humans the concept of the dissimulation by plants of their properties, emphasizing how good men are at hiding their thoughts and emotions. Stelluti proposed this bold analysis in the dedication to Cardinal Francesco Barberini of his vernacular edition of the Neapolitan's treatise. He underlined how the ability to recognize the true character of men from the 'signs' in their faces and bodies was essential in choosing devoted servants: 'Physiognomy is what teaches me to know the nature of these signs; it is a very useful and necessary science, especially for Princes and other nobles who need, more than others, servants and ministers. If the latter are of poor quality, the number of their secret enemies will be all the greater, this is why they should make sure to choose good, faithful and neutral ones' (Della Porta 1637, Stelluti's dedication to Francesco Barberini). Stelluti had originally thought of writing a longer preface to recount Della Porta's extraordinary divinations of the destiny of various persons through a simple observation of their physiognomies. The Lincean also referred to Della Porta's exceptional divinatory skills in his correspondence with Faber and Cesi (cf. Contributi I, p. 939). The most impressive testimony of Stelluti's confidence in the Neapolitan's divinatory powers can be found in three letters sent to an unknown correspondent in 1637 (ibid., pp. 759–61), in which he gave a detailed account of a series of extraordinary cases that he and Cesi had observed at first hand in Naples in 1604. After a mere perusal of the portrait of the King, Della Porta had predicted the tragic death of Henry IV of France two years before it took place; he had also informed his master that a servant condemned to death would be pardoned; finally he had predicted that the Sienese Giovan Battista Orlandini, who had accompanied Stelluti and Cesi to Naples, would be imprisoned by the Holy Office. Stelluti mentioned countless other examples of Della Porta's prescience without betraying the slightest shadow of doubt. He did not regard Della Porta's divinatory powers as black magic. Far from it, they were based on a perfect reading of the Book of Nature: 'So Porta who had studied this Great Book of Nature, was able to give account of its operations' (ibid., p. 761). There is no need to point out the distance separating the Great Book of Nature of Stelluti and Della Porta from that delineated in the pages of the *Assayer*. It is more useful to note how the Lincean mental horizon could encompass both rigorous demonstrations of the properties of the parabola (like those of Della Porta, of which Stelluti wrote admiringly to the unknown correspondent) and accounts full of wonder at the Neapolitan's magical and divinatory capacities.

> You would translate them [flavours] more easily into figures on the basis of the disposition of the particles and would not be astonished by their changes even though contained in the same liquid or container, if we admit that different flavours depend on the diverse distribution and disposition of those particles.[109]

The Prince reaches the most advanced level in his attempt to apply a quantitative treatment to the nutritional and therapeutic properties of plants in the extremely ragged tabulation that occupies the whole of page 918 (fig. 23).

This time the analysis does not proceed exclusively from left to right. In the section that occupies the upper part of the page, where the quantitative gradations of the effects produced by the qualities of wet and dry are tabulated, the scheme develops from top to bottom. Vice versa, the classification of the degrees of intensity of the therapeutic properties of hot and cold in the lower part of the page proceeds from left to right.

On this page Cesi analyses the two pairs of qualities traditionally considered opposites (hot/cold and wet/dry), introducing two important innovations. First, he places the terms of the two pairs of qualities on scales articulated in grades that link the extreme degrees of hot/cold and wet/dry without any break in continuity. He then reduces the potentially infinite gradations of the two scales to four categories, numbering them from 1 to 4 to quantify the intensity and therapeutic efficacy of the qualities present in the vegetal substances listed there. This approach, the structure of the scheme and the theoretical analysis underpinning it are clearly motivated by the desire to found on a strictly quantitative analysis a new science of medicinal simples to effectively exploit the therapeutic properties of plants.

This innovative conception coexisted with a traditional interpretation of illness. Cesi subscribed to the ancient vision of good health as the perfect balance of the four humours and of ill health as an alteration of that equilibrium. The quantitative graduation of the intensity of the properties of medicinal simples would provide the physician with a fundamental aid in administering the most efficacious remedies and, above all, in dosing them with precision.

The layout of both graduated scales of the two pairs of qualities highlights a central zone – that of temperateness (*Temperies*) and of balanced distribution

---

109 'Proprius certe a conlocatis in figuras duceres particulis, nec inde admirari mutationibus, quamvis liquore in eodem aut vase contingeret; si illam distributione, dispositione varia alius atque alius subinde mittatur sapor', TM, *Append. ad p. 917* (block of text in the lower part of the page).

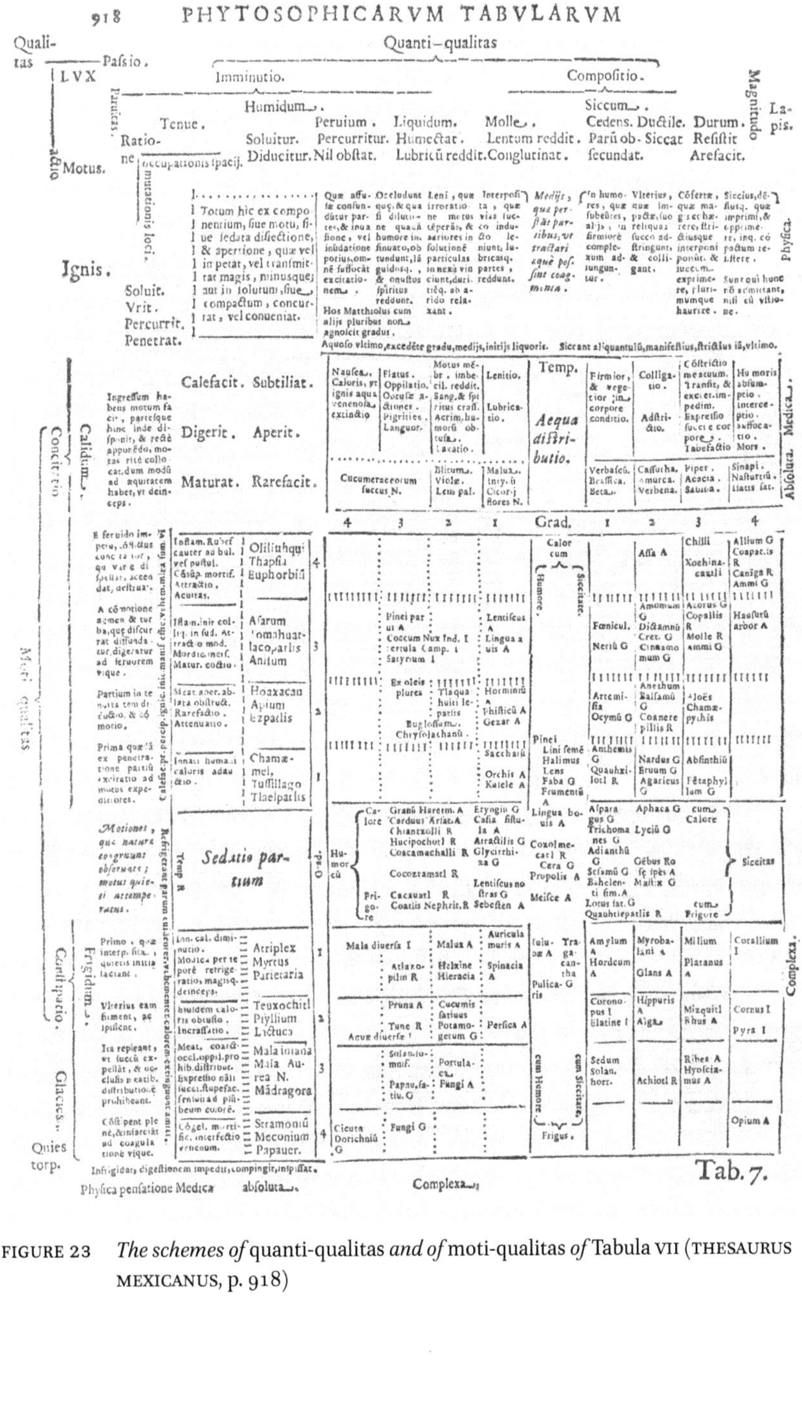

FIGURE 23 *The schemes of* quanti-qualitas *and of* moti-qualitas *of* Tabula VII (THESAURUS MEXICANUS, p. 918)

(*Aequa distributio*) – which represents the equilibrium to be restored in an ailing body by means of the appropriate administration of remedies.

Cesi entitles the section devoted to the analysis of the qualities of wet and dry *Quanti-qualitas* to underline his intention of submitting them to a quantitative treatment. After explaining that the prevalence of wetness leads to decomposition (*imminutio*), while that of dryness results in composition (*compositio*), he articulates wetness in four levels, from 4 to 1, to indicate in quantitative terms its decreasing intensity: subtle, penetrable, liquid, soft (*tenue, pervium, liquidum, molle*). For each of these four gradations, he indicates the operations of the elementary particles on which variations of intensity of the medicinal properties depend. The same distribution into four levels is applied to dryness, this time in increasing order of intensity from 1 to 4. As a result, the pair of qualities wet/dry is articulated in eight gradations plus the central zone reserved for the state of equilibrium.[110] For each of the eight gradations Cesi juxtaposes the corresponding pathologies and the vegetal substances to be taken to restore the balance of the humours.

In a block of text, framed to enhance its importance, Federico informs the reader that all the classificatory schemes on the page are based on a central concept: the degree of efficacy of the qualities depends on the state of agitation or calm of the constituent particles.[111]

The same method is followed in the lower section of the page with the scale of variations in intensity of the qualities between the extremes of hot and cold. Cesi entitles this section *Moti-qualitas* because it highlights the essential function performed by the motion of particles in determining variations in the intensity of these qualities. Heat produces movement, which in turn plays an essential part in the perception of heat or cold:

> When heat enters it produces movement, duly arranging the parts and their motions until it achieves a balance.[112]

The intensity of the sensations of heat is determined by the state of agitation of corpuscles.

110 'Mediis quae perstant partibus ut tractari aeque possint coagmenta', ibid., p. 918.

111 'Totum hic ex componentium, sive motu, sive sedata disiectione et apertione, quae vel in petat, vel transmittat magis minusque; aut in solutum, sive compactum, concurrat vel conveniat', ibid.

112 'Ingressum habens motum facit, partesque hinc inde disponit et recte apponendo motas rite collocat, dum modum ad aequitatem habet', ibid.

Mirroring the scheme of the wet/dry scale, the hot/cold scale is divided into eight gradations of intensity from 1 to 4 and from 4 to 1, with the central zone reserved for the state of equilibrium.[113] The corresponding pathologies and the vegetal substances to be taken to restore the balance of humours are juxtaposed with each of the gradations.

These two extraordinarily intricate schemes are further complicated by the introduction of a third one that acts as a hinge between the hot/cold and the wet/dry scales. This third grid is intended to enable a quantitative computation (again from 1 to 4 and from 4 to 1) of the degree of intensity of the qualities resulting from blends of the terms of the pairs hot/cold and wet/dry thus refining the classification. While the boxes of the two pairs of qualities should be read from top to bottom (dry/wet) and from left to right (hot/cold), the grid linking them makes it possible to compute (again in eight degrees of intensity over and above the central zone) the effects produced by interferences between the two scales, to pinpoint imbalances produced by specific levels of admixture and to visualise the remedies required to restore health.

Cesi's scheme of *Quanti-qualitas* and *Moti-qualitas* represents an extraordinary attempt to reform the traditional conception of qualities as well as of the properties of medicinal simples, based on theoretical principles that clearly betray the influence of Galileo's lesson (the subjective nature of sensations, the corpuscular structure of matter and the relativistic conception of qualities). Cesi, however, hybridizes the Tuscan scientist's innovative vision with instances and models drawn from Renaissance natural philosophy and the Paracelsian tradition. The influence of the latter appears particularly evident in his insistence on the extraction of the active juices from crushed plants in the chemical laboratory of the taste organ.

The importance attributed by Cesi to the function carried out in these processes by the state of agitation of the particles should not lead us to conclude that he directly transferred to medical botany the conception of the rigorously mechanical action of corpuscles with which Galileo had explained the genesis of sensations in the *Assayer*. The movement to which the Prince refers here echoes the alterative motion of Aristotle rather than the local motion of Galileo. For the Prince contact was enough to activate the particles. There was no need for the corpuscles to impact at greater or lesser speed on the taste organ. In spite of the crucial role played by the speed of impact in Galileo's corpuscular interpretation of secondary qualities, Federico had not taken into consideration the velocity of the particles in the *Apiarium,* nor did he do so now. To

113 'Sedatio partium. Quae naturae congruent observatae, motus quieti attemperatus', ibid.

describe the operations of the particles he uses here the same words as in the broadsheet on bees: *digestio, solutio, maturatio, sedatio, conglutinatio, temperatio*; terms taken from the lexicon of chemistry, not from that of mechanics.

These schemes and *Tabula VII* as a whole confirm that Galileo's interpretation of the subjective nature of sensations had made a profound impression on the Prince. But his method of research and the points of reference (Della Porta, Telesio and Paracelsus) who had inspired him right from the beginning of his studies led him, almost inadvertently, to analyse those pages through the eyes of a reader for whom all operations of nature are based on processes of transformation like those that man can emulate by means of fire and the alembic.

Cesi did not feel the need to modify the Galilean theory of secondary qualities to fit it into the horizon of his research and his vision of a living nature. The model of corpuscular actions that generate flavours proposed in the *Tabulae* must have appeared to him perfectly compatible with that expounded by Galileo in the *Assayer*. Even if the Prince were aware that his scheme represented a considerably modified version of that model, he probably thought that it was imposed by the specificity of the field to which he was applying it.

The pages devoted to the quantitative classification of the medicinal properties of plants offer other indications that help to clarify the function that Cesi assigned to the motion of particles in the perception of qualities and of their variations in intensity. Before introducing the intricate classificatory schemes commented on above, he had highlighted the fundamental principles of nature from which the qualities of plants derive. The supreme principle is light (*lux*) from which motion is produced. Motion in turn generates heat (*ignis*) which puts the elementary particles into a state of agitation. There is no motion without light and no heat without motion.

The motion to which he refers here is not local motion, unless one supposes that Cesi had not read the pages of the *Assayer* in which Galileo demolished the Aristotelian conception of motion as the cause of heat. The Prince intends to emphasise that the sensation of heat is generated by the state of agitation induced in particles by light. He was never to make any explicit pronouncement on how light produces motion. We may recall the passages in the *Apiarium* and in this *Tabula*[114] in which Federico attributes to the Sun the sole responsibility for the production of heat. That vision seems to echo Galileo's reflections on the functions of the Sun, light and heat and their relations in the letter to Piero Dini of 1615, as well as Kepler's exaltation of the Sun as the source of life in

114 'Unicus & inexhaustus caloris fons Sol', ibid., *Append. ad p. 918* (unnumbered page, block of text in the lower part of the page).

the *Conversations with the Sidereal Messenger* and other works.[115] In the letter to Dini, Galileo had distinguished *lux*, the fundamental life-giving principle, from *lumen*, produced and distributed by the Sun throughout the entire universe. Heat accompanied the universal diffusion of *lumen*, thus enlivening the whole of nature. Galileo distinguished light from heat: *lumen*, after all, could be hot or cold. The schemes of the *Physica Mathesis* confirm that Cesi shared this view. For him, it was not motion as such that produced heat, but only that of the Sun's rays. He qualifies the agitation (*fervor*) of particles caused by solar rays as *concitatio, digestio, sublimatio*; terms that once again evoke the models of action characteristic of chemical transformations rather than mechanical operations.

The special attention that Cesi dedicated to the Bologna stone for many years provides further indications of how he conceived the relations between light, motion and heat. The stone that absorbed light and restored it in darkness was not a substance in its natural state. To produce that remarkable phenomenon the stone had to be calcined. Only after the operation of fire did it absorb light and retain it for a certain period of time. Calcination played a determinant role: it transformed the bonds between the elementary particles of the natural stone. It is a telling detail that Cesi had exerted every possible pressure on Schreck, the most knowledgeable in the Lincean circle regarding the transmutations of substances by fire, to help him decipher the stone's secret and to discover the causes of the light that it emitted.

As we have stressed, for the Prince the value of the finely graduated tabulation of *quanti-qualitas* and *moti-qualitas* lay above all in the simplification and rigour that it introduced in the administration of vegetal remedies. He was convinced that his complex tabulation would offer the physician a pharmaceutical formulary that could be taken in at a glance, helping him to rapidly identify the most efficacious medicinal simples to restore the equilibrium disturbed by various pathologies. In Cesi's vision, all physicians – at least all those inclined to abandon the uncertain traditional methods – should therefore fix this folio on the wall or keep it in a prominent place on their desk.

The practical purpose of the *Tabulae*, and especially of *Tabula VII*, should be borne in mind to contrast the widespread conception that they contain exclusively theoretical analyses, of great interest for understanding Cesi's

115 In the *Conversations with Galileo's Sidereal Messenger*, Kepler had spoken of the Sun in terms that echoed the Neoplatonist motifs that Galileo was to use in the letter to Piero Dini and that Cesi would repeatedly evoke: 'In the centre of the world is the Sun, origin of life and cosmic motion', Kepler 1965, p. 45 (OG III[1], p. 123).

conception of nature (which is certainly true), but useless in the practical activities of the botanical and medical professions.[116]

The pre-eminently practical aims of the schemes of *quanti-qualitas* and *moti-qualitas* are highlighted by Cesi in a long scholium articulated in four distinct blocks of text (fig. 24), of which he underlines the direct and intrinsic relation with the quantitative tabulation of the qualities in the 'garden' of flavours.[117] In the opening text (in roman) Cesi states that he has composed that complex grid of quantitative classification of the medicinal qualities for the benefit of those physicians prepared to abandon the traditional medicine of Galen, Avicenna and their followers. Resorting once again to the metaphor of the garden, he defines the boxes of its densely interwoven scheme as 'flowerbeds'. Moreover he declares his awareness of having simply opened up the prospect of a totally new field of research. He was confident that competent readers would help to emend, enrich and refine his first attempt at an organic treatment based on a quantitative analysis. Thanks to their contributions, it would gradually become possible to implement a garden (the Prince also uses the metaphor of a chart) that reflected the order governing the relations between all the plants present in nature, thus making it possible to grasp perfectly the gradations of their therapeutic properties. At that point, the provisional garden of qualities outlined in *Tabula VII* would faithfully represent the structure and articulations of the real Book of Nature:

> A larger number of these flowerbeds (*areolae*) will not be lacking in the workshop of nature, and you will be able to fill them in with more examples [...]; if you enlarge the sheet to reach the dimensions of a chart (*mappa*), you will be able to fill in those that are still missing and are nowhere to be found on paper; and if you search in the Books of Nature you may come across a different disposition and arrangement that is preferable.[118]

116 As we recall, this was the opinion that Fabio Colonna had formed after reading the first *Tabulae*, in particular *Tabula I*: useful for metaphysicians, but incomprehensible for herbalists (see above, p. 373). But after pressing on, particularly after reading *Tabula VII*, Colonna realised that the *Tabulae* not only presented new theoretical interpretations, but also indicated the way of rendering medical practice less uncertain and most beneficial for humanity.

117 TM, *Append. ad p. 918.*

118 'Plures hic areolae non autem in Naturae vacaverint officinis, quippe ulterius repere poteris [...]; si in votis fuerit affatim folium in mappam distendere, quae vero prorsus carent, nequaquam in chartaceis erit, implere ut queas: perquirentibus autem secus evenerit in naturae libris in quibus et luculentius et alia forsan dispositione ac trutina', ibid.

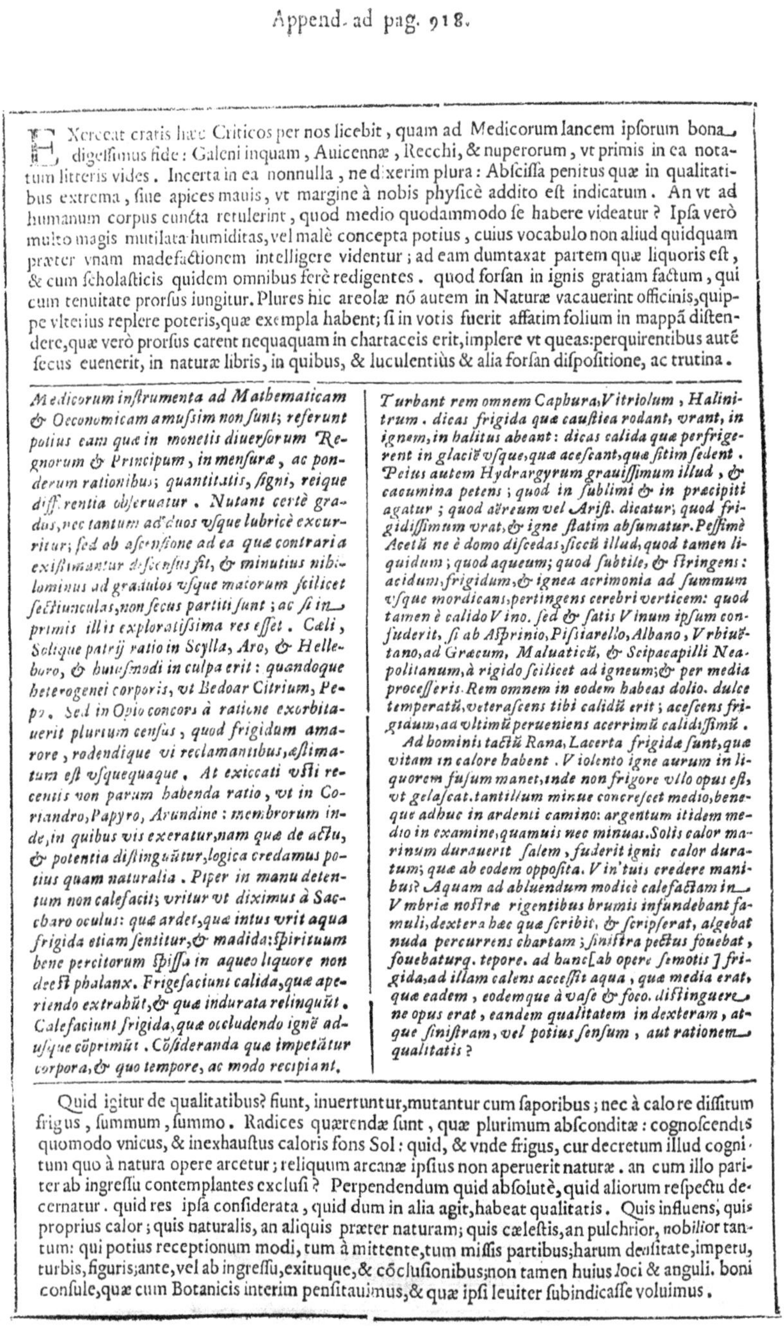

Append. ad pag. 918.

EXerceat cratis hæc Criticos per nos licebit, quam ad Medicorum lancem ipsorum bona digestimus fide: Galeni inquam, Auicennæ, Recchi, & nuperorum, vt primis in ea notatum litteris vides. Incerta in ea nonnulla, ne dixerim plura: Abscissa penitus quæ in qualitatibus extrema, siue apices mauis, vt margine à nobis physicè addito est indicatum. An vt ad humanum corpus cuncta retulerint, quod medio quodammodo se habere videatur? Ipsa verò multo magis mutilata humiditas, vel malè concepta potius, cuius vocabulo non aliud quidquam præter vnam madefactionem intelligere videntur; ad eam dumtaxat partem quæ liquoris est, & cum scholasticis quidem omnibus ferè redigentes. quod forsan in ignis gratiam factum, qui cum tenuitate prorsus iungitur. Plures hic areolæ nō autem in Naturæ vacauerint officinis, quippe vlterius replere poteris, quæ exempla habent; si in votis fuerit affatim folium in mappā distendere, quæ verò prorsus carent nequaquam in chartaceis erit, implere vt queas: perquirentibus autē secus euenerit, in naturæ libris, in quibus, & luculentiùs & alia forsan dispositione, ac trutina.

Medicorum instrumenta ad Mathematicam & Oeconomicam amussim non sunt; referunt potius eam quæ in monetis diuersorum Regnorum & Principum, in mensuræ, ac ponderum rationibus; quantitatis, signi, reique differentia obseruatur. Nutant certè gradus, nec tantum ad duos vsque lubricè excurritur; sed ab ascensione ad ea quæ contraria existimantur descensus fit, & minutius nihilominus ad gradulos vsque maiorum scilicet sectiunculas, non secus partiti sunt; ac si in primis illis exploratissima res esset. Cæli, Solique patrij ratio in Scylla, Aro, & Helleboro, & huiusmodi in culpa erit: quandoque heterogenei corporis, vt Bedoar Citrium, Pepo. Sed in Opio concors à ratione exorbitauerit plurium census, quod frigidum amarore, rodendique vi reclamantibus, æstimatum est vsquequaque. At exiccati vsti recentis non parum habenda ratio, vt in Coriandro, Papyro, Arundine: membrorum inde, in quibus vis exeratur, nam quæ de actu, & potentia distinguūtur, logica credamus potius quam naturalia. Piper in manu detentum non calefacit; vritur vt diximus à Saccharo oculus: quæ ardet, quæ intus vrit aqua frigida etiam sentitur, & madida: spirituum bene percitorum spissa in aqueo liquore non deest phalanx. Frigefaciunt calida, quæ aperiendo extrahūt, & quæ indurata relinquūt. Calefaciunt frigida, quæ occludendo ignē adusque cōprimūt. Cōsideranda quæ impetātur corpora, & quo tempore, ac modo recipiant.

Turbant rem omnem Caphura, Vitriolum, Halinitrum. dicas frigida quæ caustica rodant, vrant, in ignem, in halitus abeant: dicas calida quæ perfrigerent in glaciē vsque, quæ acescant, quæ sitim sedent. Peius autem Hydrargyrum grauissimum illud, & cacumina petens; quod in sublimi & in præcipiti agatur; quod aëreum vel Arist. dicatur; quod frigidissimum vrat, & igne statim absumatur. Pessimè Acetū ne è domo discedas, siccū illud, quod tamen liquidum; quod aqueum; quod subtile, & stringens: acidum, frigidum, & ignea acrimonia ad summum vsque mordicans, pertingens cerebri verticem: quod tamen è calido Vino. sed & satis Vinum ipsum confuderit, si ab Asprinio, Pissiarello, Albano, Vrbiuētano, ad Græcum, Maluaticū, & Scipacapilli Neapolitanum, à rigido scilicet ad igneum; & per media processeris. Rem omnem in eodem habeas dolio. dulce temperatū, veterascens tibi calidū erit; acescens frigidum, ad vltimū perueniens acerrimū calidissimū.

Ad hominis tactū Rana, Lacerta frigidæ sunt, quæ vitam in calore habent. Violento igne aurum in liquorem fusum manet, inde non frigore vllo opus est, vt gelascat. tantillum minue concrescet medio, beneque adhuc in ardenti camino: argentum itidem medio in examine, quamuis nec minuas. Solis calor marinum durauerit salem, fuderit ignis calor duratum; quæ ab eodem opposita. Vin' tuis credere manibus? Aquam ad abluendum modicè calefactam in Vmbriæ nostræ rigentibus brumis infundebant famuli, dextera hæc quæ scribit, & scripserat, algebat nuda percurrens chartam; sinistra pectus fouebat, fouebaturq. tepore. ad hanc [ab opere semotis] frigida, ad illam calens accessit aqua, quæ media erat, quæ eadem, eodemque à vase & foco. distinguere ne opus erat, eandem qualitatem in dexteram, atque sinistram, vel potius sensum, aut rationem qualitatis?

Quid igitur de qualitatibus? fiunt, inuertuntur, mutantur cum saporibus; nec à calore dissitum frigus, summum, summo. Radices quærendæ sunt, quæ plurimum absconditæ: cognoscendis quomodo vnicus, & inexhaustus caloris fons Sol: quid, & vnde frigus, cur decretum illud cognitum quo à natura opere arcetur; reliquum arcanæ ipsius non aperuerit naturæ. an cum illo pariter ab ingressu contemplantes exclusi? Perpendendum quid absolutè, quid aliorum respectu decernatur. quid res ipsa considerata, quid dum in alia agit, habeat qualitatis. Quis influens, quis proprius calor; quis naturalis, an aliquis præter naturam; quis cælestis, an pulchrior, nobilior tantum: qui potius receptionum modi, tum à mittente, tum missis partibus; harum densitate, impetu, turbis, figuris; ante, vel ab ingressu, exituque, & cōclusionibus; non tamen huius loci & anguli. boni consule, quæ cum Botanicis interim pensitauimus, & quæ ipsi leuiter subindicasse voluimus.

FIGURE 24 *The scholium on the medicinal properties of plants of* Tabula VII (THESAURUS MEXICANUS, Appendix to p. 918)

In the block of text (in italics) on the left, the Prince underlines that the intellectual and material instruments used by physicians do not present the rigour and universal objectivity of those employed by mathematicians. He compares the fluctuations, variations and uncertainties of medical methods and tools with the different mintings of the same coin. He blames this situation above all on the obstinacy of the physicians in continuing to follow the traditional doctrine of contrary qualities, refusing to recognise that they simply represent the extremities of a scale that does not display any break in continuity.

That misguided conception had made it impossible to know the true nature of many plants, in which the qualities typical of cold coexist with those characteristic of heat, such as pepper, which does not burn the hand but inflames the eyes:

> There are hot things that generate cold by opening, extracting and leaving matter hardened. There are cold things that generate heat, enclosing the fire and completely compressing it.[119]

The column of text set in smaller italics on the right contains Cesi's own critical analysis of the theory of contrary qualities. How could Aristotle define camphor, vitriol or saltpetre as cold substances, given that they are caustic, dissolve, burn and are transformed by fire into vapours? In plants as in animals we only observe heat in various degrees of intensity:

> Frogs and lizards are cold to human touch, yet have vital heat.[120]

Cesi multiplies the examples that demonstrate the absurdity of the conception of contrary qualities, with which he contrasts his own relativistic vision in the small block of text (in roman) at the foot of the page:

> So what of qualities? They are produced and are radically changed or modified together with flavours; cold is not separate from heat, but each constitutes an extremity [on the same scale].[121]

---

119 'Frigefaciunt calida quae aperiendo extrahunt et quae indurata relinquunt. Calefaciunt frigida quae occludendo ignem adusque comprimunt', ibid.

120 'Ad hominis tactum Rana lacerta frigidae sunt, quae vitam in calore habent', ibid.

121 'Quid igitur de qualitatibus? fiunt, invertuntur, mutantur cum saporibus; nec a calore dissitum frigus, summum, summo', ibid.

The Sun is the inexhaustible source of heat; cold simply represents a minimal level of heat that our senses are no longer able to perceive. This is why the quantitative measurement of the intensity of qualities can only be conducted in relative terms. It is only by operating in this way that it becomes possible to grasp the real structure of the Book of Nature, which has no gaps or oppositions:

> It is necessary to consider what is distinguished absolutely or relatively, which is the quality of a thing when considered in itself and which when it acts on others; what receives heat and what has its own heat; what is natural and what preternatural [...].[122]

Food for the Mind

The long scholium of *Tabula VII* on the medicinal properties of plants serves as the perfect introduction to *Tabula VIII* in which Cesi broadens the perspective by extending his analysis to the many fields of human activity that derive practical benefits from plants.[123] As he was to underline at the conclusion of *Tabula XVI*,

> of all the plants there is practically none that is not of some use or benefit to us, not only in virtue of one or two of its parts, but of all of them.[124]

His account begins by stressing how many substances of extraordinary utility are produced through the transformation of various parts of plants performed by minuscule creatures such as silk worms, caterpillars and spiders. A telling example of the 'mutual labour of nature and the skills of men and worms' is offered by clothing, made from weaving threads of silk produced by *vermiculi* that process vegetal material.[125] Without their precious contribution it would be impossible to avoid the immodesty and misery of nudity: 'we naked and

---

122 'Perpendendum quid absolute, quid aliorum respectu decernatur, quid res ipsa considerata, quid dum in alia agit, habeat qualitatis. Quis influens, quis proprius calor; quis naturalis, an aliquis praeter naturam [...]', ibid.

123 'Usus plantae, eius peculiaribus partibus nonnullis sumptis', ibid., p. 922.

124 'Nulla fere ex omnibus stirpibus est, quae suis non una dixerim, vel altera, sed omnibus partibus nobis aliquo modo usu venire vel prodesse non possit', ibid., p. 943 (concluding block of text in italics).

125 'Mutua et naturae et artis hominum et vermium opera', ibid.

miserable creatures seek clothing from worms'.[126] Also ropes and nets are made by twisting vegetal fibres in various ways.

But the most extraordinary product that, thanks to their ingenuity, men have drawn from vegetal material is paper. Swept away by his enthusiasm for paper, Cesi reserves no less than a page and a half of the five comprising *Tabula VIII* to emphasize its multiple benefits for mankind. Following his usual method, he illustrates in great detail, by means of a continuous series of dichotomies, the techniques for the production of paper as well as its varied uses (literary/non-literary, manuscript writings/printed works, etc.) in different parts of the globe (Europe/Asia).[127]

In composing this densely populated list, he showed off his familiarity with every aspect of the production and use of paper as well as his great admiration for its essential contribution to the communication and transmission of knowledge and to the orderly functioning of society:

> How would Princes or citizens or enterprises, whether in time of peace or war, manage without paper [...]? Who is not greatly indebted to the plants that provide paper for all the arts and sciences that spread the use of our reason far and wide? Who does not love them? Who does not cherish these most pleasant studies? From them we derive not only our herbals, books and letters, but the seeds of all virtues and disciplines. It would be ungrateful and unworthy to scorn them and to show ingratitude.[128]

Cesi displays no less enthusiasm in illustrating the practical benefits that derive from cutting and drying trunks and branches of trees. Thanks to the skills of the carpenter and the cabinet-maker, that natural material is transformed into an infinite variety of extremely useful objects, which the Prince lists in

126 'A vermiculis indumenta petimus nudi ac miselli', ibid., p. 922. As we recall, in *Tabula I*, Cesi defined man not only as a rational being, but also as the only living creature capable of hiding his nudity with clothing (see above p. 378).

127 '*Europea/Asiatica; litteraria/prophanis usibus accomodata; scriptorum/impressorum* [...]', ibid., p. 923.

128 'Quid absque charta aut principes aut cives aut negotia, sive ea belli sive pacis? [...]. Quis inquam non debere aut possit aut velit et plurimum quidem plantis per Chartam scientias artesque omnes ministrantibus, nostrae rationis usum longe lateque diffundentibus? Quis eas non amet? non haec amoenissima colat studia? Si inde praeter propriam Herbariam et libros quoscunque habemus et litteras, si virtutum disciplinarumque omnium semina? nisi ingratus prorsus et qui huismodi indignus spernat, nec agnoscat gratias', ibid.

minute detail, showing his profound mastery of the terminology of the arts and trades.[129]

Federico also insists on the practical benefits provided by plants in the following two *Tabulae*. *Tabula* IX focuses on the precious functions of plants not only when they are chopped up, distilled and reduced to fibres, but also when they vegetate: their roots render soil resistant to erosion, their leaves afford welcome shade and temper the force of winds, while their branches provided shelter to primitive man and accommodate birds' nests. They also produce abundant fruits that can be harvested without harming the plant, while their sweet scent contributes to the salubrity of the air. Meadows, flowers and woods created spontaneously by nature are a pleasing sight and stimulate admiration and peace of mind. Finally, plants offer man the resources for the creation of magnificent artificial corners of nature like the most admirable gardens of the greatest patrons of the ancient and contemporary world.[130]

Tabula X delineates the profile of the moral botanist, whose task is that of highlighting and promoting the models of noble conduct offered by plants:

> Flowers and fruits can be read like a book. Indeed they furnish you with very pleasant and useful food for the mind, a crown for the spirit and a perennial fruit for your labours.[131]

Plants communicate moral exhortations in many ways: in hieroglyphic and symbolic form (the mulberry prudence; the olive gentleness; the palm intellectual love), in the parables of Christian tradition, emblems, enigmas, myths and proverbs. They mirror the dogmas of the Church, including the ultimate dogma, that of the Trinity, emblematically adumbrated by the three-leafed clover.

Nature conveys through plants models of principled conduct and of the exercise of wisdom. These principles are shown 'not only mystically and symbolically, but through living examples easy for us to understand'.[132]

---

129 Ibid., pp. 924–5.

130 Ibid., p. 928.

131 'Ita si flores, fructusque legere volueris; Nae tu periucundum, perutilemque intellectui pabulum, coronam animo, laboribus perennem fructum tibi constitues', ibid., p. 929 (left margin in italics). Allegra has insisted on the parallels established by Cesi between the human and vegetal worlds in moral conduct. Imperfect plants such as fungi and mistletoe are examples of immoral conduct: their beauty conceals the fatal effects of their consumption, like wicked men with a smile on their lips before stabbing their victim. Parasitism, simulation and disguise are behaviours common to men and plants (Allegra 2005, esp. pp. 88ff.).

132 'Non mystice et signis tantum, sed et evidenter vivis etiam propriis exemplis', TM, p. 930.

Finally, the close study of plants illuminates the way to the supreme Creator, opening the doors of Paradise:

> Ascending to the love of the Supreme Creator, the noblest and greatest God, you will attain perfection, happiness and blessedness in perpetuity, and will enjoy in Paradise the incorruptible fruits of eternal Spring.[133]

In the universe of plants, the moral botanist discovers the logic that inspired Creation and its purpose, from which the original sin distanced mankind.

We detect in these passages once again convergences and divergences between the visions of the Book of Nature of Galileo and Cesi. They both see the perfection of the plan of Creation and the infinite greatness of its Author reflected in nature, but while Galileo does not go beyond displaying a feeling of sincere admiration for the work of the divine architect, the Prince sees in the creatures that populate the Book of Nature a powerful exhortation to the redemption of humanity from original sin.

The ability to read the moral message conveyed by the vegetal world represents a fundamental requisite for the botanist. Anyone who proposes to understand the deep significance of nature cannot confine himself to selecting one aspect and observing it from a particular angle. As the pages of the Book of Nature are closely interconnected, they should all be read with equal attention, as the exemplary lesson they communicate can be taken in only by a unitary gaze. This was the vision that had motivated Cesi's efforts to harmonise the truth of nature and of Holy Scripture – a project far removed from Galileo's claim to the full autonomy of natural philosophy from the guardianship of theology and the sacred texts. For Federico, that unitary gaze should not be confined to the external aspect of plants. It must penetrate their internal structure to understand their utility, to grasp their moral message, and to contemplate their titles of nobility. To achieve this comprehensive vision, it was also necessary to master the knowledge of the vegetal world stored in the works of the most authoritative writers ever since the golden age of antiquity.

A Preformistic Conception?

Tabula XIII, the last of the series personally edited by Cesi, delineates the professional profile of the metaphysical botanist. The structure of the *Tabula*

133 'In summi Dei Optimi Maximi amorem ascendens perfectionem, felicitatem, beatitatem ipsam virtutem perpetuo ac florentem in sempiterni viris et incorruptibilium fructuum Paradiso consequaris', ibid., pp. 931–2 (text in italics in right-hand margin).

suggests that Federico intended it to contain a general scheme of the system of plants analysed in their individual properties ('absoluta contemplatione') and in their connections and relations of dependence ('syntactica conspiratione').[134] Plants can be studied individually, to bring out the integrity and symmetry of their organs and vital functions, or by comparison with other entities, by which the Prince envisages not just other plants but the entire universe. He focuses in particular on the relations of sympathy and antipathy of individual plants with the elements (first of all with water), with animals (not surprisingly, the list opens with a reference to the *Apiarium*), and with other plants (that are distinguished into friend or foe).

The analysis gives priority to the internal relations of the vegetal universe seen from many points of view: between a single plant and its class ('planta ad classes') and conversely ('classis ad plantas'); between different classes ('classis ad classes'); between plants that display similar characteristics while belonging to different classes ('plantae a diversis classibus inter se '); and between intermediate plants and the classes on which they border ('plantae ancipites ad classes') that Federico defines as those 'which connect the classes themselves and in some way seem to be related to both' (a transparent reference to the intermediate natures dear to him).[135]

The relations of sympathy and antipathy between plants belonging to different classes are particularly interesting. The Chain or Ladder formed by the orderly sequence of vegetals is transformed here into a dense mesh formed by links that display both vertical and horizontal connections:

> [...] the chain can be considered by analogy to be composed of several columns of soldiers; it connects by hidden links things dispersed through matter, organizing them into various cohorts.[136]

Replacing the traditional image of the linear Chain of Being with this new bidimensional representation made it possible to bring out the multiplicity of relations that can be observed in the vegetal world. The organisation of nature now appeared far more complex: the ladder of continuous rungs on which

134 Ibid., p. 936.

135 'Sunt et mediae, quae internectere classes ipsas ancipiti quodammodo utrinque affinitate videntur', ibid. With the row of dots after this definition of the intermediate natures Stelluti indicated that Cesi had left it incomplete.

136 '[...] cathena, cum quadam analogia e diversis agminibus considerari potest, qua occultis vinculis licet per materias eae dispersae in manipulos alios atque alios colligentur', ibid., p. 937. Note the use of *acies* and *manipuli*, which are both military terms.

creatures were arranged now displayed branchings and transversal connections with other parallel ladders. This new vision permitted an intuitive grasp of the analogies, differences and connections between the various families and genera. Cesi projected his own orderly classification of plants in a vast amphitheatre in which the audience watched different scenes depending on where they were seated.

The rest of the text displays Cesi's attempt at linking the reticular grid of vegetals with that of the animal world. To form a correct idea of the entire architecture of nature the metaphysical botanist must master the multiple connections between plants and animals, highlighting their specificities and analogies. Based on the authority of Empedocles, Democritus and Plato, Cesi reaffirms that there are no substantial differences between plants and animals. He goes on to analyse the functions of the primary organs of vegetals, concentrating on the seed and the flower, the elements essential for the reproductive function:

> The flower is the impatient promise of fertile joy, the herald of more perfect proliferation and the messenger of spirits, the hope of abundance of fruits and of the delight of plants [...].
>
> The seed is the germ of the Great Mother, scion of the vegetal soul, the whole plant in compendium, a side shoot; foetus of the dying plant and tutelary God of its rebirth; breviary of the whole vegetal animal [...].[137]

The seed as 'the whole plant in compendium', the Earth as the 'Great Mother', 'vegetal animal': these are expressions that seem to delineate a preformistic conception of plant reproduction – the same notion that Cesi had expounded in the *Apiarium* and in the long scholium of *Tabula VII*.[138] Once again we find echoes of Anaxagoras' definition of the Earth as womb of all things: plants, animals and men. Cesi, aware that these were bold, even dangerous ideas, felt it opportune to distance himself from the Presocratic philosopher. Only plants, not animals or humans – he states – are born from the womb of Mother Earth:

---

137 'Flos qui est faecundae hilaritatis festina quaedam sponsio, perfectioris fructificationis praenuncia et premissio spirituum, spes fructiferae abundantiae et plantarum laetitia [...]. Et Semen, quod est magnae matris foetura vegetantis animae soboles, collecta in compendium planta, stirps obsiquata; morientis plantae foetus, ac renascentis genius: universi denique animalis vegetabilis breviarium [...]', ibid., p. 943.

138 See above, p. 435.

> Although Anaxagoras' notion that the womb of plants is the Earth herself, the Great Mother of all things, appears convincing, you should not believe this nor state with certainty that the organs of plants are common to animals.[139]

Nature Was Not Created Once and for All

The fundamental importance of the opinions of ancient and modern authors on plants and the vegetal world is the focus of *Tabula XII*. In the opening section Cesi delineates what might be defined as a standard format of classification for the entire vegetal universe. For each individual plant he plans to record first of all *nomen* and *cognomen* (also in double, triple, etc. form), whether it is simple or composite, referring to a single plant or to a genus, certain or uncertain, legitimate or spurious, assigned casually or based on rational analysis of the morphology and properties of the plant.[140]

Like many students of the vegetal universe of his time, Cesi attributed enormous importance to the reform of nomenclature. He regarded the adoption of a rigorous and universal nomenclature as the key to the precise identification of single exemplars: the prerequisite to compare without ambiguity the descriptions of the nature and properties of plants by ancient or modern writers and to properly place the vegetals recently discovered in the New World in the traditional classificatory schemes.[141]

That reform aimed to eliminate the many uncertainties arising from the fact that the same plant was often designated by different names, making the work not only of the taxonomist but also of the physician and the herbalist extremely problematic.

---

139 'Plantarum venter, ipsa terra, magna omnium mater facile creditur Anaxagorae; non tamen hoc, neque alia quaeque membra animantibus et plantis communia existimare poteris, nec quiquam de his determinate statuere', TM, p. 943.

140 Ibid., p. 935.

141 The Prince attached just as much importance to nomenclature in the animal world, as shown by the corrections that he made to his personal copy of the section on animals of the *Thesaurus Mexicanus* written by Johannes Faber. He added the Nahuatl name to many of the exemplars illustrated and described by Faber even though he did not know that language. He did so because he considered it important to take into account the name assigned to the animals by the inhabitants of their territory of origin. Cesi's marginal additions of the names in Nahuatl are reproduced in the facsimile edition of his personal copy of the *Thesaurus Mexicanus* (TM, *passim*; see Guerrini 2007b).

Cesi insisted on the need to identify and adopt universally a name that directly and intuitively expressed the distinctive characteristics of the exemplar, of the family, and of the genus to which the plant belonged. That name would function as an identity tag and at the same time as a 'hypostasis' of the Great Chain of the world of plants:

> To arrive at the fullest knowledge, it is necessary that there is no plant without a name, *nor a name without the hypostasis of its plant*; we will easily achieve this goal if we establish without ambiguity by whom and in what way a plant will be named, after all the clouds of antiquity have been dispelled, because all ancient knowledge must be renovated in the present age.[142]

The crucial problem of nomenclature is resumed in the *Tabulae* published by Stelluti after the death of the Prince. In *Tabula XIV*, for instance, in which the standard format of classification for the inventarisation of the vegetal universe is proposed again with some variations: 'nomen, cognomen, etymon, genus, differentiae quae peti possunt, natura in qua considerantur, quantitas, qualitas', etc.[143] To record the many nuances of the world of plants Cesi articulates each of the principal fields of his standard classification format in subsections. For instance, *genus* is broken down into *summum* (*arbor, frutex, suffrutex, herba*) and *infimum*, proceeding in specification down to the identification of the *familia*, the ultimate level of classification of the groups of plants displaying homogeneous characters.

The importance assigned to nomenclature is confirmed by the fact that Cesi insists on it in two more *Tabulae*. *Tabula XV* tackles the problem of naming plants that are extinct, new or modified by natural processes in comparison with the names and descriptions furnished by earlier writers. Cesi was aware of the lack of stability of nature; creatures are subject to transformation or even disappear (an eloquent example was *laserpitium*), while new ones emerge:

> Both the world above and that below are subject to change: gradually, over the course of centuries, something new appears among plants and

142 'Ut ad plenissimam tandem plantarum notitiam; ne scilicet planta ulla sine nomine, *neve nomen sine plantae suae hypostasi* exhibeatur, aditus facilior sit, utque intelligamus sine dubio per quem et quomodo planta nobis imprimis innotuerit, discussis omnibus vetustatis nebulis, quae plurimum studiosis offundunt tenebras, omnis antiquitas in praesentis aevi novitatem renovanda est', TM, p. 941; emphasis added.

143 Ibid., p. 938.

> animals, some elements disappear, while others display slight changes; it is not surprising that we often hear of similar changes in the appearance, faculties, origin and qualities of diverse plants.[144]

Texts like these echo the conception of the determinant influence of the celestial bodies and spaces on terrestrial phenomena and creatures illustrated in the *Taumatombria*, as well as that of the homogeneity and corruptibility of the heavens.

To grasp the significance of Federico's vision of a nature characterised by dynamics of change there is no need to hypothesize a brilliant anticipation of the theory of natural selection. It is sufficient to recall the way in which Cesi had assimilated the evidence of the presence of processes of corruption in the celestial bodies provided by Galileo's telescopic discoveries. The unity and homogeneity of the heavens and of the Earth strengthened Federico's conviction that the Theatre of Nature undergoes modifications, sometimes profound modifications, in parallel to and as a consequence of those that take place in the celestial spaces (sunspots, comets, *novae*, etc.). This dynamic view of nature is fresh evidence of the original use to which Cesi put the Galilean discoveries, transferring them to horizons far removed from those on which the Tuscan scientist fixed his gaze.

Tabula XIX deals with the field of application and the methods of *Phytonomia*, the science of exact botanical nomenclature. *Phytonomia* should be based on perfect knowledge of the affinities and differences of plants, as well as of the alterations produced in them by atmospheric and celestial influences, their location and the type of soil in which they grow:

> The factors that play an important role in the modification of plants must be taken into consideration [...]: the various configurations of the heavens and soil [...], the influence of air; the aspect and position of the soil [...].[145]

144 'Superiorum mutatione inferiora quoque mutari par est: sensim proinde ac multis saeculis in herbis et animalibus novum aliquid, aliquid amissum, multa paululum immutata detegimus; ut nihil mirum si in figuris, facultatibus, origine et qualitatibus diversarum plantarum similes plerumque audiamus mutationes', ibid., p. 940 (block of text in italics at bottom of page).

145 'Advertendum quam multum in plantis immutandis valeant [...] coeli et soli varietates [...], aeris impressionibus [...] soli facies, ac situs [...]', ibid., p. 949.

An unequivocal nomenclature, capable of recalling synthetically to the mind all the characteristics and properties of a plant, as well as its position on the Ladder of Being, constitutes a fundamental prerequisite for a rigorous analysis of the vegetal world. Federico's Book of Nature could be abridged in an *Onomasticon,* an index of names capable of faithfully representing the articulations, analogies and differences between its creatures. As he put it, [:] 'the name is the door to the knowledge of plants'.[146] The objective portrait of the essential characteristics of the plant provided by its name is even more expressive than a drawing or painting:

> Phytonomia guarantees the perfect match between a thing and its name.[147]

One wonders whether, in stating the objective value of a correct vegetal nomenclature, those passages in which Galileo drew a clear distinction between words and things, stressing that names were conventional labels bearing no intrinsic relation with the entities designated, came to the Prince's mind.[148] As he was well aware, for Galileo the book of names, however accurate and methodical, could not help us to understand the true constitution of nature and her laws. His Book of Nature was populated not by names, but by geometric figures.

Names as Shadows of Things

The creative terminology that Cesi devised to designate the intermediate natures (*metallifacti, metallophyti, ligniterreus, lapisurgens, lithophytus*, etc.) also shows that he regarded the attribution of names capable of expressing the properties of natural things as an extremely important challenge. This helps to understand why he and his associates strove to identify animals and plants that could be qualified as 'Lincean' by virtue of their symbolic connotations and/or physical or behavioural characteristics.

Naturally, they put great effort into emphasising that the name of the lynx as the symbol of the Academy evoked, both directly and metaphorically, the aims

146 'Nomen quod est ad cognitionem plantarum ansa', ibid.

147 'In Phytonomia certitudinem praescriptae copulationis nominis scilicet et rei obfirmatio', ibid., p. 948.

148 See OG IV, p. 707 (*Considerazioni appartenenti al libro del Sig. Vincenzio di Grazia*).

of the institution founded by Federico Cesi.[149] It is therefore hardly surprising that Faber's section on the animals of New Spain in the *Thesaurus Mexicanus* devoted considerable space to the lynx, in spite of the fact that it is not to be found in Central America. Faber and Cesi could not resist the temptation to vaunt the emblem of the Academy and to sing its praises in the work that was the outcome of their most ambitious and laborious project. To justify the contradiction of a long digression on the lynx in the description of the fauna of a geographical context which did not number the feline among its animal species, they devised a clever strategy: the digression was inserted in the section on the panther, a wild cat that did come from the New World (*catus pardus americanus*). In considering the image of this animal in Recchi's *Compendium*, Faber wondered whether it might be related to the lion or to the Molossian hound. He ruled out this possibility, however, on the grounds that the coat of the animal appeared to be spotted. It was at this point that Faber hypothesized that the *catus pardus americanus* might be assimilated to the European lynx or deer-like wolf (*lupus cervarius*). He was hesitant, however, because neither he nor Cesi had ever seen a live lynx, but only three coloured drawings which differed considerably from one another in terms of morphology and colour.[150] Faber had also seen lynx pelts in the shop of a Roman furrier.[151] More than twenty years after the foundation of the Academy, its Secretary did not know what a lynx looked like, and even wondered whether it really existed![152] Having made the commitment to describe it, he had to rely on information taken from those ancient, medieval and Renaissance authors who had mentioned what he defined as the most controversial of the felines.

Following the method systematically adopted in his section of the *Thesaurus Mexicanus*, Faber presented a very long list of testimonies by past authors, concluding that they were contradictory: only the direct inspection of the animal would resolve the quandaries of the literary tradition. As this proved impossible, he tried to extract reliable information from a close analysis of the names given to the animal, which, he stressed, had been designated by previous writers not only as *lynx* but also as *deer-like wolf*. He then explained why this double denomination helped to define the real physiognomy, characters and habits of the feline:

149 Cf. above, pp. 30–5.

150 TM, p. 519.

151 Ibid., pp. 526–7. It was not until 1623–24 that Cesi was able to see a live lynx. In July 1612 he manifested his disappointment to Stelluti, who had sent him drawings of a lynx based on literary sources: he would have liked at least a drawing from life (CL, p. 247).

152 TM, pp. 525–6.

> Since names are the shadows of things and their etymology often reveals some hidden thing, I am not surprised that the Lynx has been called a deer-like wolf by several recent writers.[153]

At this point, the German Lincean underlined the etymological analogies between *lynx* and the Greek name for wolf (λύκος). As many authors derived λύκος from the Greek word for light (λευκός), the etymology of the *lupus cervarius* perfectly expressed the intrinsic characteristics of the lynx reflected in the ideals of the Linceans:

> What etymology would be more appropriate for the lynx, the animal with the keenest sight of all?[154]

With a fantastic operation of patchwork, Faber provided a detailed description of the feline, including its habits and medicinal properties, based on three different drawings, on dried furs, and above all on the clues provided by etymological analysis. His exercise recalls Cesi's terminological acrobatics in the same years to deduce the sublime qualities of bees from the etymology of their names in classical languages.[155]

The distinctive qualities of the lynx are identified by Faber in the penetration of its gaze (this was the reason why – he stressed – Prince Cesi had adopted it as the emblem of the Academy) and in its voracity.[156] In going beyond superficial appearances, the feline was thus the perfect symbol of the perspicacity of the mental eye: it penetrated even greater depths than the telescope. Linceans saw 'not more but better' ('non plus, sed melius'). In combining acuteness of (mental) vision with the thirst for knowledge, the lynx expressed the noble mission of the Academy to perfection.

To have found in the animal kingdom a creature that intuitively communicated the aims of the Academy was not enough for Cesi and his fellow Linceans. The next step was to find a similarly evocative symbol in the plant world. The Mexican flora described by Recchi offered a perfect example: the flower of the *coatzontecoxochitl*, which was spotted like the lynx. Schreck, who

153 'Cur igitur nomina sint umbrae rerum et saepe vocis Ethimon habeat nescio quid reconditae rei, non miror *Lyncem* a plerisque recentioribus *Lupum cervarium* dictum fuisse', ibid., p. 526.

154 'Quod Ethymon cui convenientius quaeso quam Lynci animalium omnium perspicacissimo adaptari potuit?', ibid.

155 Cesi 2005–2006, I, pp. 129–31.

156 TM, p. 529.

FIGURE 25 *The Lincean plant with spotted flowers* (CL, p. 230)

had identified that plant in 1612, proposed calling that flower *Lincean* because of its speckled bloom (fig. 25).[157] Cesi approved enthusiastically and hastened to send the drawing of the flower to Galileo.[158] The Prince had also had a celebratory medal cast in which the Lincean flower replaced the lynx.[159]

It was not only for the superficial resemblance to the spotted fur of the lynx that Cesi had subscribed to that bold pairing. In celebrating its Lincean baptism, he emphasized that the attribution to that flower of the name *Lincean* was based on far more intrinsic analogies, which the commentary of Schreck on Recchi's description enumerated one by one:

> As this beautiful flower is considered a delightful miracle of nature by the rulers of the Indies for its elegance and grows in great profusion from various roots, so the studies of the Linceans, who strive to reveal the richest treasures concealed by nature and to place them before the keen

157 CL, p. 229, Cesi to Galileo, 2 June 1612.

158 'I will send you some Indian plants that, because of their beauty and lynx-like spots, Schreck, the commentator on that work [the *Compendium* of Recchi], has thought of calling by our name' (OG XI, p. 312, Cesi to Galileo, 2 June 1612).

159 De Toni 1908, pp. 354–5. In the *Thesaurus Mexicanus* the Lincean flower is also used to mark the end of chapters.

> minds of industrious men, deserve their gratitude as well as that of the Princes who promote and encourage literary and scientific studies. Although they [the Linceans] are few in number at present, yet their roots go deep and spread the sweetest scent of the Muses far and wide through the fertile offspring of their literary monuments, attract many students to them, and stimulate several associates to the assiduous contemplation of nature.[160]

At this point the moral ideals and the desire for knowledge that animated the Lincean enterprise resonated loudly in both the animal and vegetal kingdoms. It was an effective way to emphasise that concept of the unity of nature and of the full correspondence between its kingdoms that was so dear to the Princeps of the Academy. It also served to drive home the message that names perfectly mirror the constitutive characters of things.

In this connection it should be recalled that Federico had announced the project of publishing an in-depth analysis of the lynx and of the Lincean plant in the section of the *Zibaldone* in which he had schematically outlined the ambitious editorial programme of the *Thesaurus Mexicanus* ('pro Thesauro Rerum Medicarum novae Hispaniae'):

> A philological work on the lynx, and whatever is named and called after it, the power itself of very keen sight. Gemma Vin: Argonauta. And other philosophers. *Plant.*[161]

---

160 '[...] ut quedamodum pulcherrimus Flos hic Indorum Principibus, ob elegantiam et Naturae miraculum in delitijs habetur, et ex varia radice foecundissimus provenit; ita Lynceorum studia, quae opulentissimos clausae Naturae thesauros recludere et avidis hominum solertium ingeniis subijcere satagunt, et apud hos et apud Principes bonarum litterarum promotores et patronos gratiam mereantur, cumque pauci adhuc eorum sint numero, radices tamen altas agant, faecundaque litterariorum monumentorum sobole suavissimum Musarum odorem longe lateque spargant, plurimos studiosos sibi devinciant, pluresque ad sedulam Naturae contemplationem sibi socios excitent', ibid., p. 266. In a marginal note Cesi corrected 'various roots' (*varia radice*) to 'a single root' (*una radice*) to emphasise the unity of the Academy.

161 'Opus philologicum de Lynce animali, et quaecumque ab eo denominata ac dicta, vis ipsa acutissime videndi. Gemma Vin: Argonauta. Philosophii alique. *Planta*', Contributi I, p. 46, emphasis added. 'Gemma Vin' is presumably a reference to the Venetian edition of Lodovico Dolce's *Delle Gemme* (Dolce 1565), a book owned by Cesi (cf. above, p. 36). 'Argonauta' refers to the mythical figure of Linceus. See above, p. 31.

We detect the same motivations in Fabio Colonna's attempt in the *Thesaurus Mexicanus* to identify creatures in the animal and vegetal kingdoms that clearly reflected the characteristics of the Lincean enterprise and, above all, the vision of its founder. Colonna took on the responsibility of identifying the most suitable among the plants of New Spain to bear the name of the Princeps of the Academy. After making his choice, he wrote a description and sent it to Faber to be passed on to Federico for approval.[162] He had drawn inspiration from the colour *cyaneus* of the root of the Mexican plant *Tuxpatlis* described by Recchi (fig. 26), similar to the dark blue indicated by the adjective *caesius*, and therefore a perfect match for the name of Cesi.[163] The exclusive properties of its root and its rarity had confirmed the Neapolitan naturalist in the conviction that this was the most appropriate plant.[164]

As usual, the denomination rested not only on the physical characteristics of the flower, but also on the eloquence with which it showed its hidden powers. The root of the *Tuxpatlis* appeared suitable for that noble and exclusive name because it was endowed with signal therapeutic properties. It offered further confirmation of the Lincean paradigm of the name as omen of the intrinsic qualities of things and living creatures.

The parallel that Colonna established between the *Tuxpatlis* plant and Cesi afforded the opportunity to emphatically celebrate the Prince:

> As this plant, by virtue of a single root of this colour [*caesius*], may be called privileged above the others [...] and be considered among the first of the whole genus of plants, so the most illustrious Cesi, our Princeps, besides his ancient lineage [...], has received innumerable other and superior prerogatives [...]. He has loved the study of the liberal arts, expounded the nobler sciences and disciplines and mastered all natural and metaphysical knowledge [...]. He has set a noble example for all Princes in establishing a new kind of institution to promote knowledge

162 CL, p. 1155, Colonna to Stelluti, 28 November 1627: 'I have written a short text on the Cesian plant'.

163 'We were delighted when we saw the deep blue (*Cyaneus* or *Caesius*) colour in the root of this extremely rare plant as observed by our author [...]' (TM, p. 873). In fact, as Ettore De Toni underlined, there is no trace of blue in the root of the plant named *cesia* by Colonna. Through a banal error, that colour had been mistakenly attributed to it in Recchi's *Compendium* (De Toni 1908, pp. 355–8).

164 The properties of this plant convinced us – Colonna wrote – to call it 'after the most illustrious and excellent Prince Cesi and grace it with his surname', ibid., p. 874.

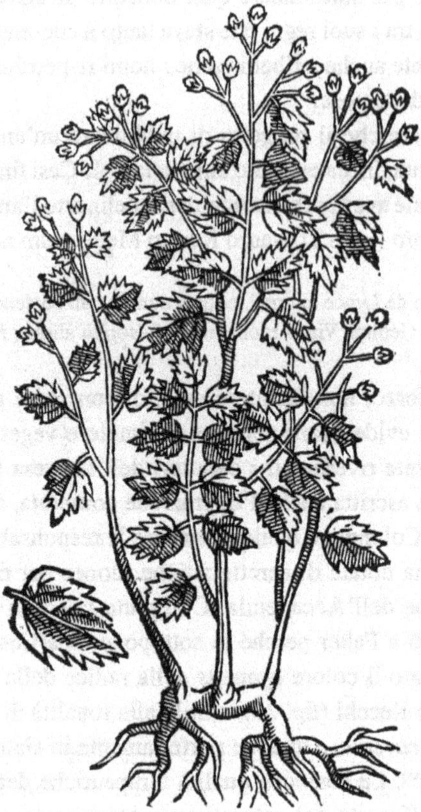

FIGURE 26
The plant cesia (THESAURUS MEXICANUS, p. 149)

[…]. He is therefore deserving of being called a second Caesius, like a son of Jupiter and brother of the blue-eyed Caesia Minerva […].

> O Cesian leader of the Linceans, the Lynx and blue-eyed Pallas retreat before you; for all is disclosed by virtue of your mind and vision.[165]

165 'Dux tibi Lynceidum Caesi Lynx, Caesi Pallas/cedunt: mente, oculis nam tibi cuncta patent', ibid. The distich at the end of the Neapolitan's long celebratory text displeased Cesi, who had the last word on the texts to be printed, including those dedicated to him. However, Colonna rejected the alternative drafts that he received from Rome and proposed in their place a solution that satisfied him: 'I have redone the distich that now flows more smoothly than the one I sent, in which although there is not that sweetness of the ones done in Rome, mine says more […] and is more allusive and celebratory, even though it is rougher' (CL, p. 1159, Colonna to Stelluti, 17 February 1628).

In token of recognition of Colonna's efforts in identifying the plant *cesia*, Cesi proposed calling a column-like tree *columnia*.[166] At this point, the list of the Lincean vegetals had to be completed with the plant *stellutia* to celebrate the only surviving one, besides the Prince, of the four founding members of the Academy.[167]

Colonna's efforts to flatter illustrious persons by naming vegetals that reflected their signal virtues after them were not confined to the *cesia* and the *stellutia*. In a letter of 20 November 1627 he informed Stelluti of his vain attempts to find among the Mexican plants described by Recchi a species suitable to exalt the virtues of Francesco Barberini, the authoritative fellow and patron of the Academy.[168] The tone of the letter betrays that he had been expressly asked to find one by Cesi. He then fell back on a plant from Virginia belonging to the *Lobelia* genus that was widespread in Southern Africa.[169] He called it 'the flower of Cardinal Barberini' (fig. 27) for the purple colour of its flowers:

> We in no way wished to delete or change that illustrious name of Cardinal, but to make it even more illustrious by naming after him a species or even an individual exemplar of this genus [...].[170]

The solemn christening of the *Lobelia Cardinalis* was naturally followed by a fulsome inventory of the extraordinary learning, gifts of nobility, humanity, prudence and piety of the Cardinal as well as of his uncle the Pope. Colonna did not hesitate to interpret the trigon of bees in the Barberini stemma as a symbolic expression of the Trinity and a presage of the responsibility of the Pastor of Christianity to which Maffeo Barberini had been destined from birth. Neither did he omit praise of Prince Cesi who, with his synoptic and symbolic tribute ('synopticis emblematicisque encomijs') to bees in the *Apiarium*, had solemnly celebrated the family of the Pope and of the Cardinal Nephew.

---

166 Cf. De Toni 1908, p. 358.

167 Ibid., p. 359.

168 CL, pp. 1155–6: 'I cannot find any among the Mexican plants that satisfies me as a *barberina*'.

169 The name *lobelia* was given to this genus in honour of the great botanist Matthias de l'Obel, whose works Cesi admired (cf. De Toni 1908, pp. 359–60). After the radical reform of nomenclature by Linnaeus, the *Flos Cardinalis* is the only plant to have maintained the name given it by the Linceans. It is still known as the *lobelia cardinalis*.

170 'Illius tamen tam praeclarum Cardinalis nomen delere aut immutare nullo pacto voluimus, imo potius clarius reddere aut sub generico nomine ad suam speciem vel proprium individuum referre', TM, p. 880.

FIGURE 27
Flos Cardinalis Barberini
(THESAURUS MEXICANUS, p. 880)

Not content to leave the acrobatic excursions in the nomenclature of the animal and vegetal worlds to his fellow Linceans, Cesi himself showed his hand at creative terminology inspired by the concept of the intrinsic correspondence between names and things: he named *Lincean* a medicine that he was convinced offered protection against contagion.[171]

171 See Winther's letters to Faber from Acquasparta of 18 August 1624 (CL, p. 934: 'I am now composing a very potent Lincean alexipharmic'), of 7 September (ibid., p. 939: 'We are now composing the Lincean antidote') and of 9 November (ibid., p. 961: 'We prepared last week the Lincean antidote, on which we have been working all summer [...])'. Winther's remarks seem to indicate that the Lincean antidote was made from a mixture of vegetal extracts.

Epilogue

'It Has Been Impossible to Persuade Him to Make a Will'

The end of the year 1625 marks an important turning point in the relations between Cesi and Galileo. The noticeable slowing down in their correspondence (this time also on the part of Cesi) suggests that both were now aware that they were following separate trajectories. Even more striking than the decline in their number is the evasive character of the letters exchanged between Florence and Rome. Cesi avoided informing the Tuscan scientist about the researches in which he was intensively engaged, while Galileo stopped turning to the Prince for information on the attitude of the ecclesiastical authorities regarding the celestial novelties and their cosmological implications. In spite of the blows received, Galileo had not given up his struggle to affirm the truth of the Copernican system, but his trusted intermediaries and informants in the Eternal City were no longer Cesi nor the Roman Lincei. Mario Guiducci and, above all, Benedetto Castelli, who was permanently established in Rome, now played that role. It was primarily through the devoted Benedictine that Galileo maintained contact with Cesi, Ciampoli and the Barberini circle.

In fact, the Accademia dei Lincei no longer existed. Stelluti was occupied in assisting Federico so that the *Thesaurus Mexicanus* could finally be published; Faber was to pass away a little less than a year before Federico;[1] and Ciampoli was involved in the demanding and delicate obligations of his position of high responsibility in a bitterly divided Curia. Federico, often in permanent voluntary exile in Acquasparta, burdened by debt and concerned about the continuous decline of his health, had abandoned the Academy. He was probably conscious of the fact that the project to which he had devoted so much energy was in irreversible decline. It suffices to read the letter in which Francesco Stelluti informed Galileo of the passing of Federico to realise that during his last years the Princeps of the Lincei had given up hope of relaunching the Academy to the point of losing interest in its future:

> I am even more saddened by the fact that he [Cesi] has left no disposition for the things of the Academy, to which he wanted to bequeath his whole library, museum, manuscripts and other precious things. I do not know

1 Faber died on 17 September 1629.

 | DOI 10.1163/9789004342323_014

> in whose hands these goods will end up [...]. Nor has it been possible to persuade him to make a will.[2]

Cesi would not read Galileo's *Dialogue concerning the two chief world systems,* even though he only just missed the opportunity. The manuscript of that magisterial work, by now in its practically definitive stage, was in the luggage that accompanied Galileo to Rome in May 1630. He had undertaken that journey to the Eternal City – the fifth, he could not imagine that it was not to be the last – in the hope of obtaining from the censors the licence to publish the *Dialogue*.

Months before facing that tiring move, Galileo had received assurances from Cesi that his work would be published at the expense and under the aegis of the Academy,[3] as had been the case for the *Letters on sunspots* and the *Assayer*.

Although he remained in Rome for almost two months, between May and June 1630, Galileo was unable to meet Cesi and to show him the results of his last and most ambitious effort. The poor health of the Prince prevented him from receiving visitors, even company as welcome to him as Galileo's undoubtedly would have been.[4] Moreover, while the Tuscan scientist was still in Rome, Federico expressed the desire to be moved to his beloved Acquasparta, where he was to spend his final days.[5]

---

2 OG XIV, pp. 126–7, letter of 2 August 1630. The Prince's mood of dejection also emerges from the letter he sent Galileo on 20 January 1630. He confessed that 'the past year has been full of afflictions for damages and losses, delays and impediments' (ibid., p. 73).

3 Galileo had expressed his willingness to have the *Dialogue* printed in Rome under the emblem of the Academy in his letter to Cesi of 24 December 1629 (ibid., p. 60). Having received no reply, he repeated the proposal in his last letter to the Prince of 13 January 1630 (ibid., p. 66). Cesi gave a positive answer on 20 January 1630 (ibid., p. 72).

4 In his letter from Florence of 6 August 1630 to Giovan Battista Baliani, Galileo recalled his visit to Rome in the spring of the same year, mentioning also Cesi: 'I was in Rome in the past months to obtain a licence for the *Dialogi* [...], and having finally overcome some difficulties, I obtained the licence for them [...]; and if circumstances had been different I would have left them in the hands of his Excellency Prince Cesi who, as he has done with other of my works, would have taken care of the printing; but His Excellency was indisposed, and, what is worse, I now hear that he is at death's door. This is why I am now trying to have them printed here [...]' (ibid., p. 130). A few months later, Galileo informed Andrea Cioli that the original idea of printing the *Dialogue* in Rome under the aegis of the Lincei had vanished due to the death of the Princeps of the Academy (ibid., p. 216; letter of 7 March 1631).

5 Cesi remained in Rome at least until the end of June, as shown by his last extant letter, addressed to the Priors of Todi, which was dispatched from the Eternal City on 29 June 1630 (CL, p. 1213).

It is natural to wonder how Federico would have reacted if a kinder fate had allowed him to peruse the pages of the *Dialogue*. He would certainly have admired the intelligence that radiated from every page of the work. He would have praised the decision to adopt the form of the dialogue between advocates of different opinions that he had recommended to the Tuscan scientist ever since the polemic with Scheiner on the sunspots. Finally, Galileo's extraordinary demonstration of the capacity of the vernacular to communicate sophisticated scientific matters and novelties that implied the revision of the traditional conception of nature with elegance and precision would probably have led the Prince to reconsider his conviction that only Latin was capable of confronting complex questions of natural philosophy.

What is certain is that a reading of the *Dialogue* would have aroused in Cesi the same preoccupations that Galileo's reply to Ingoli had provoked in him five years earlier: it would not have escaped his notice that many pages of that text that he had 'censored' had been restored to the *Dialogue*, nor that, in spite of the declarations of submission to the deliberations of the Church, Galileo's determination to fight for the recognition of the truth of the Copernican system appeared only too evident in it.

Cesi's death from a cruel illness spared him the embarrassment of having to recommend Galileo yet again to proceed with prudence and to forewarn him of the risks he ran in claiming the full autonomy of natural philosophy from theology and in resolutely defending the right to the freedom of thought in philosophy, not only *in naturalibus*.

Another question arises spontaneously: if Cesi had not departed prematurely from the scene, would he, after reading the *Dialogue*, have respected his commitment to Galileo to publish it under the aegis of the Academy? One may wonder. Federico's personal influence and that of the other Linceans on the Roman scene and in the Curia no longer had the strength it had had at the time of the publication of the *Assayer*. The promising season of the 'marvellous conjuncture' was by now a distant memory. The present horizon appeared even more menacing than it had been when the Prince had decided not to pass Galileo's response to the *Disputation* on to Francesco Ingoli.

For these reasons, it is extremely likely that, even if the Prince had had the chance to read the *Dialogo*, Galileo would have had to put the manuscript back in the bags that accompanied him on his return journey to Florence. This time he would have to face the challenge, the ultimate challenge, alone. As we know, he did not pull back.

There are parallels between the vicissitudes of Cesi and Galileo also in their passing. Cesi's demise met with a general indifference. He died in Acquasparta on 1 August 1630, far from the gaze of the powers of the Church and of the

Roman patriciate. No solemn funeral was organised for him. No public eulogy of his industrious life, of his impressive writings, or of the Academy that he had founded was recited. This silence is striking, especially bearing in mind his intellectual calibre, his important social relations and the authority of the patrician family to which he belonged.

The Prince's corpse did not receive the honourable treatment that had been reserved only a few years earlier for the mortal remains of his cousin, Duke Virginio Cesarini, Vice-princeps of the Lincei. Even his burial place in the Church of St Cecilia in Acquasparta passed unnoticed until 1872,[6] when the family vault was opened and the solemn identification of the remains of the Princeps of the Linceans was conducted.[7] However, it was not until 1920 that a celebratory inscription in his honour was appended in the Cesi family chapel. And it was only at the end of the nineteenth century that an inscription recording the Prince's name and achievements was set in the wall of his Roman residence in Via della Maschera d'oro, seat of the Academy and the setting for meetings between personalities of remarkable talent.[8]

After Cesi's death, neither he nor his writings were mentioned for almost two centuries. The *Apiarium* was extremely rare and its structure too tortuous for the reader; the *Thesaurus Mexicanus* was published only twenty years after his death, by which time scientific progress had turned that innovative work into an outdated resource. Its publication did not pay full justice to the intellectual energies and considerable financial resources that the Prince had put into that monumental work, nor to the determinant role that he personally, as well as the Academy and its associates, had played in it. Finally, Federico's *Tabulae Phytosophicae* had no place in the lively debates on systems of classification of the natural world of the following two centuries.

---

6 Odescalchi condemned the *damnatio memoriae* that extended even to Federico's mortal remains: 'his corpse was buried in the Church of St Cecilia in Acquasparta, in the family chapel and ancestral tomb, where he still lies without even a short inscription to indicate to the passer-by the place where such a learned and excellent gentleman is buried' (Odescalchi 1806, p. 199).

7 The recognition of the family tomb that led to the identification of the mortal remains of Cesi was conducted on 26 October 1872 on the initiative of the Municipality of Acquasparta and of the Accademia dei Nuovi Lincei, represented by its President, the Senator Giuseppe Ponzi. The detailed notarial deed of recognition is kept in the Archive of the Accademia Nazionale dei Lincei. I owe this reference to the kindness of Marco Guardo. A new recognition of the tomb was conducted in the spring of 2013 under the supervision of the Soprintendenza Archeologica dell'Umbria.

8 Cf. Bianchi V.E. 1892, p. 44. The commemorative stone was placed in 1872 by the Municipality of Rome.

Not only did the memory of Cesi fade into obscurity for a long time, but so did that of the brief life of his academic initiative. The only conspicuous *post mortem* homage to the Prince and to the Linceans was that provided by Galileo in the *Dialogue concerning the two chief world systems*. The Tuscan scientist qualified himself as 'Accademico Linceo' on the frontispiece of the publication printed two years after Federico's death. Moreover, in the animated discussions between Salviati, Sagredo and Simplicio, he inserted numerous asides in which he made his own voice directly heard. The person whose authoritative views were repeatedly evoked by Salviati was not presented to the readers in his official capacity as Philosopher and Mathematician of the Grand Duke of Tuscany, but as a fellow of the Lincean Academy. Galileo had devised that solution when he was still convinced that the *Dialogue* would be issued as an official publication of the Academy. He maintained it after the Prince had departed from the scene in sincere recognition of the generous and disinterested support that he had received from Cesi and from the institution that he had founded. It is not by chance that six years later, when the memory of the Academy was completely eclipsed, Galileo was to renew this gesture of gratitude in the *Two new sciences*, placing great emphasis on his Lincean affiliation. At that point, behind the voice of the Academician evoked by Salviati glimmered the disturbing image of a man condemned by the Church for being vehemently suspected of heresy.

Galileo too departed this life amid a deafening silence. No public funeral was arranged for the Philosopher and Mathematician of the Grand Duke of Tuscany, whose name was familiar not only to the most distinguished natural philosophers and astronomers throughout the world, but even to the most powerful sovereigns of Europe. Nor did anyone challenge the ecclesiastical authorities by daring to deliver a solemn funeral oration to celebrate his memory. In January 1642, the unadorned coffin in which his corpse had been laid was furtively accompanied by a few fond friends and pupils from his home in Arcetri, the gilded prison of the last twelve melancholy years of his life, to a small room in the Florentine Basilica of Santa Croce, where it remained inaccessible to the faithful and to visitors for almost a century.[9]

The posthumous fortunes, however, of the author of the *Sidereus Nuncius* and of the Princeps of the Linceans did not run parallel. Galileo was not forgotten, nor did his works disappear from circulation. Paradoxically, the

9 On the first burial of Galileo in the Basilica di Santa Croce in Florence and on the series of events that led in 1737 to the exhumation and transfer of his mortal remains to the monumental tomb erected in the same Florentine church, see Galluzzi 1993.

condemnation made him even more famous, above all in Tuscany, where he continued to play the role of mentor of Medici scientific patronage in cleverly disguised forms.[10] Thanks to the early translation of his major works into Latin, Galileo remained at the centre of discussions among the protagonists of the scientific revolution and of the new philosophy in France, England, Germany and the Low Countries: Gassendi, Huygens, Newton and many other prestigious authors did not hesitate to give him credit for having opened up and considerably advanced a radical transformation of the traditional conception of nature and of the laws governing its operations. Nor did interest in Galileo diminish in later centuries. The Tuscan scientist has remained the point of reference in debates on the laws of nature, on the position and purposes of mankind in Creation, on the relations between science and religion and on freedom of thought down to our own times.

The posthumous fate of Cesi was completely different. After a lengthy absence from the collective memory, he gradually re-emerged from oblivion in the early nineteenth century. Baldassare Odescalchi was the first to focus his attention on Federico, on his companions in adventure and on the activities of the First Academy, demonstrating that it had not been a marginal episode but rather one worthy of being brought back to light.[11] Odescalchi recovered and published the original documents that illustrate the vicissitudes of the early years of the Academy. In his remarkable essay on Cesi's intellectual vision, the image of the Prince finally assumed plausible features, revealing the objectives that had inspired the Lincei for over a quarter of a century. Odescalchi clearly emphasized the difference in the methods and views of Galileo and Cesi. Finally, he underlined the utopian nature of the prescriptions penned by Cesi in the *Lynceographum*, showing that they had had virtually no impact on the actual life of the institution or on the conduct of its associates.[12] Thanks to Odescalchi, after a long separation, the images of Galileo and Cesi were once again brought together.

However, it would have been impossible to bring into focus the project that had fired Cesi with enthusiasm and the foundations on which his admiration for Galileo rested without the intense, passionate and rigorous scholarly and editorial work carried out by Giuseppe Gabrieli over the span of forty years. Without the monumental *Carteggio Linceo*, without the innumerable series of

---

10 Cf. Galluzzi 1981.

11 Odescalchi 1806.

12 Odescalchi remarked that the *Lynceographum* had to be considered 'a novel rather than a history, an ideal picture rather than a portrait of a real, existing object' (ibid., p. 242).

extraordinary contributions that Gabrieli devoted to Cesi, to his collaborators and to the events that marked its brief existence, the renaissance of the last thirty years in studies of the history of the First Academy would have been inconceivable. Gone from memory for centuries, the writings of the Princeps of the Linceans are now the object not only of in-depth analysis, but also of critical editions. This new season of studies has contributed to a clearer picture of the commitment, energy and brilliance with which Cesi pursued his ambitious objectives.

By now it should be obvious that to restore identity and dignity to Cesi's project and to the role played in it by Galileo, the cliché of the 'Galilean' Academy must be abandoned as well as that of the Linceans' alleged 'conversion' to the new science after the induction of the Tuscan scientist in 1611. When comparing the personalities of Galileo and Cesi we should not speak of an identity of vision, but of a strategic convergence based on the shared conviction of the need to promote a radical reform of traditional knowledge. The horizons on which they had fixed their gaze were not only distant from, but in many respects irreducible to one another: the lynx and the telescope; the Theatre and the Book of Nature; the chemical interactions at corpuscular level and the mechanical actions of particles; the freedom of thought in natural philosophy submissive to the superior authority of Holy Scripture, and the conception of the full autonomy of the Book of Nature from the Book of Scripture.

The image of 'parallel worlds' portrays the nature of the relations between Cesi and Galileo more effectively than that of the alleged 'conversion' of the Prince and the Linceans to the new science. From this perspective it becomes possible to focus on the lucid plan and on the expectations that led Cesi to found the Academy, as well as on the originality of his vision and on the relevance of the results that he achieved in the course of his brief and laborious life.

Bibliography

L'Accademia dei Lincei e la cultura europea nel XVII secolo: manoscritti, libri, incisioni, strumenti scientifici: mostra storica, Fondation Dosne-Thiers-27, Place Saint-Georges, Parigi, 13 dicembre 1991–8 gennaio 1992: catalogo, eds. A.M. Capecchi, C. Forni Montagna, P. Galluzzi, A. Nicolò, G. Paoloni (Rome, 1991)

Agostino d'Ippona, *Opere di sant'Agostino* (Rome, 1967)

Aiton, E.J. and Drake, S. 'Concerning Cesi and Galileo', in *Isis* 50 (1959), 152

Alessandrini, A. *Originalità dell'Accademia dei Lincei*, in *Federico Cesi*, Convegno celebrativo del quarto anniversario della nascita (Rome, 1986), pp. 77–177

Alessandrini, A. 'Giovanni Heckius linceo e la sua controversia contro i protestanti', in *Rivista di Storia della Chiesa in Italia* 30 (1976), 364–404

Alessandrini, A. *Cimeli lincei a Montpellier* (Rome, 1978)

Alessandrini, A., Armezzani, R., Beltrame, B., Gazzini, T., Mezzanotte, E., Nicolò, A., Quagliarini, I. eds., *Francesco Stelluti linceo da Fabriano: studi e ricerche*, foreword by F. Gabrieli (Fabriano, 1986)

Allacci, L. *Leonis Allatii Apes urbanae, sive De viris illustribus, qui ab anno MDCXXX per totum MDCXXXII Romae adfuerunt, ac typis aliquid evulgarunt* (Rome, 1633)

Allegra, A. *Corpi molt'ambigui: Federico Cesi e la mezzana natura*, in Allegra-Vinti 2005, 77–90

Allegra, A. and Vinti, C. eds., *Federico Cesi e il suo tempo* (Assisi, 2005)

Altieri Biagi, M.L. ed., *Scienziati del Seicento* (Milan, 1969)

Altieri Biagi, M.L. and Basile, B. eds., *Scienziati del Seicento*, (Milan-Naples, 1980), vol. 2 of *Galileo e gli scienziati del Seicento* (Milan-Naples, 1953–1980)

Antetomaso, E. *Il censimento degli esemplari del Tesoro Messicano: nuovi esiti di un'indagine bibliografica*, in Cadeddu-Guardo 2013, 93–132

Amabile, L. *Fra Tommaso Campanella ne' castelli di Napoli, in Roma ed in Parigi: narrazione con molti documenti e 10 opuscoli del Campanella inediti* (Naples, 1887)

Apiano, P. *Astronomicum Caesareum* (Ingolstadt, 1540)

Ardissino, E. *Pietas, curiositas et poësis nell'attività dell'Accademia dei Lincei: intorno a Virginio Cesarini*, in Battistini-De Angelis-Olmi 2007, 147–73

Arrighi, G. 'P. Antonio Santini da Lucca discepolo di Galileo (1577–1662)', in *Lucca: rassegna del Comune* 9 (June 1965), 3–21

Artese, L. 'Filosofia telesiana e ramismo in un inedito di Antonio Persio', in *Giornale critico della filosofia italiana* 66 (68), ser. 6, 7 (1987), 433–58

Artigas, M. 'Un nuovo documento sul caso Galileo: EE 291', in *Acta Philosophica* X (2001), 199–214

Assemblea celeste, radunata nouam[en]te in Parnasso sopra la noua cometa, dall'academico danico il Riposato descritta in una lettera all'academico italico lo Spassionato (Milan, 1619)

Bacon, F. *Saggi morali di Francesco Bacchon, corretti e dati in luce dal sig. caualier Andrea Cioli* [...] *et un Trattato della sapienza de gl'antichi* (Venice, 1621)

Bacon, F. *Francisci baronis de Verulamio* [...] *De dignitate et augmentis scientiarum libri IX* [...], *iuxta exemplar Londini impressum* (Paris, 1624)

Baldini, M. *Le ricerche sulla nascita delle api tra '500 e '700: Swammerdam, Reaumur, Schirach, Riem*, in Conti 1990, 235–55

Baldini, U. *L'astronomia del cardinale Bellarmino*, in Galluzzi 1984, 293–305

Baldini, U. *Legem impone subactis: studi su filosofia e scienza dei Gesuiti in Italia, 1540–1632* (Rome, 1992)

Baldini, U. ed., *Christoph Clavius e l'attività scientifica dei Gesuiti nell'età di Galileo: atti del convegno internazionale: Chieti, 28–30 aprile 1993* (Rome, 1995)

Baldini, U. *The Academy of mathematics of the Collegio Romano from 1553 to 1612*, in Feingold 2003, 46–98

Baldini, U. and Coyne, G.V. eds., *The Louvain lectures (lectiones Lovanienses) of Bellarmine and the autograph copy of his 1616 declaration to Galileo*, texts in the original Latin (Italian) with English translation, introduction, commentary and notes by U. Baldini, G.V. Coyne (Vatican City, 1984)

Baldini, U. and Napolitani, P.D. 'Per una biografia di Luca Valerio: fonti edite e inedite per una ricostruzione della sua carriera scientifica', in *Bollettino di storia delle scienze matematiche* 10 (1990), 3–157

Baldini, U. and Spruit, L. 'Nuovi documenti galileiani degli archivi del Santo Ufficio e dell'Indice', in *Rivista di storia della filosofia* 56 (2001), 661–99

Baldini, U. and Spruit, L. eds., *Catholic church and modern science* (Vatican City, 2009)

Baldriga, I. *L'occhio della lince: i primi Lincei tra arte, scienza e collezionismo, 1603–1630* (Rome, 2002)

Barberini, M. *Maffaei card. Barberini nunc Urbani pp. VIII Poemata* (Codogno, 1628)

Battistini, A. *Introduzione a Galilei* (Rome, 1989)

Battistini, A. *Galileo e i Gesuiti: miti letterari e retorica della scienza* (Milan, 2000)

Battistini, A. *Galileo* (Bologna, 2011)

Battistini, A., De Angelis, G., and Olmi, G. eds., *All'origine della scienza moderna: Federico Cesi e l'Accademia dei Lincei* (Bologna, 2007)

Bellini, E. *Umanisti e lincei: letteratura e scienza a Roma nell'età di Galileo* (Padua, 1997)

Bellini, E. *Il papato dei virtuosi: i Lincei e i Barberini*, in I Primi Lincei 2005, 47–96, subsequently publ. in Bellini 2005, 109–57

Bellini, E. *Stili di pensiero nel Seicento italiano: Galileo, i Lincei, i Barberini* (Pisa, 2009)

Belloni, L. *Il microscopio applicato alla biologia da Galileo e dalla sua Scuola (1610–1661)*, in Maccagni 1972, vol. 2, 689–730

Belloni Speciale, G. *La ricerca botanica dei Lincei a Napoli: corrispondenti e luoghi*, in Lomonaco-Torrini 1987, 59–79

Beltran, A. *Cometas, ciencias y religión. La polémica Galileo-Grassi* (Madrid, 2016)

Bernardino Telesio. Atti del Convegno internazionale di studi su Bernardino Telesio: Cosenza, 12–13 maggio 1989 (Cosenza, 1990)

Berti, D. *Nuovi documenti su Tommaso Campanella tratti dal carteggio di Giovanni Fabbri* (Rome, 1881)

Berti, L. ed., *Il Pantheon di Santa Croce a Firenze* (Florence, 1993)

Besomi, O. and Camerota, M. *Galileo e il Parnaso tychonico: un capitolo inedito del dibattito sulle comete tra finzione letteraria e trattazione scientifica* (Florence, 2000)

Biagetti, A. *Federico Cesi e Galileo Galilei nel convegno di Acquasparta, 8–22 aprile 1624* (s.l., 1974)

Biagetti, M.T. *La biblioteca di Federico Cesi* (Rome, 2008)

Biagioli, M. 'Knowledge, freedom and brotherly love: homosociality and the Accademia dei Lincei', in *Configuration* 3 (1995), 139–66

Biagioli, M. *Did Galileo copy the telescope? A new letter by Paolo Sarpi*, in Van Helden-Dupré-Van Gent-Zuidervaart 2010, 203–30

Biancani, G. *Aristotelis loca mathematica ex universis ipsius operibus collecta, et explicata: Aristotelicae videlicet expositionis complementum hactenus desideratum: accessere de natura mathematicarum scientiarum tractatio atque clarorum mathematicorum cronologia, authore Iosepho Blancano* (Bologna, 1615)

Biancani, G. *Sphaera mundi, seu Cosmographia demonstratiua ac facili methodo tradita, in qua totius mundi fabrica, una cum nouis, Tychonis, Kepleri, Galilaei, aliorumque astronomorum adinuentis continetur: accessere 1. Breuis introductio ad geographiam, 2. Apparatus ad mathematicarum studium, 3. Echometria, idest geometrica traditio de echo, authore Iosepho Blancano Bononiensi e Societate Iesu* [...] (Bologna, 1620)

Bianchi, M.L. *Scheda n. 4*, in Accademia dei Lincei 1991, 53–6

Bianchi, V.E. *Le epigrafi sulle case e sui monumenti di Roma* (Turin, 1991)

Biondi, E. *Sul Trattato del legno fossile minerale di Francesco Stelluti accademico linceo da Fabriano*, with facs. repr. of the 1637 ed. (Fabriano, 1984)

Biondi, E. and Castagnani, G. 'Francesco Stelluti, un naturalista fra XVI e XVII secolo: una personalità enciclopedica', in *Natura e montagna* 26, 2 (June 1979), 45–52

Biscaretti di Ruffia, C. *Carrozzieri di ieri e di oggi*, rev. D. Jappelli (Turin, 1963)

Blumenberg, H. *The Legitimacy of the Modern Age* (Cambridge, MA, 1986)

Boaga, E. 'Annotazioni e documenti sulla vita e sulle opere di Paolo Antonio Foscarini, teologo copernicano (1562c.–1616)', in *Carmelus* 37 (1990), 173–216

Boccalini, T. *De' ragguagli di Parnaso*[...], in Venetia, appresso Giovanni Guerigli, 1614

Boccalini, T. *I ragguagli di Parnaso, or Advertisements from Parnassus* [...], put into English by Henry, Earl of Monmouth [Henry Carey] (London, 1656)

Boccalini, T. *Ragguagli di Parnaso e scritti minori*, new edition by L. Firpo (Bari, 1948)

Boehm, L. and Raimondi, E. eds., *Università, accademie e società scientifiche in Italia e in Germania dal Cinquecento al Settecento* (Bologna, 1981)

Boffito, G. *Gli strumenti della scienza e la scienza degli strumenti* (Florence, 1929)

Bolzoni, L. 'Un modo di commentare alla fine dell'Umanesimo: i Commentaria del Campanella ai Poëmata di Urbano VIII', in *Annali della Scuola normale superiore di Pisa, Classe di Lettere e Filosofia* 19 (1989), 289–312

Bolzoni, L. *Urbano VIII, Campanella e la censura dei Commentaria*, in Prosperi 2001, 265–84

Bonaccorsi, T. 'Clausos rerum aperire sinus: l'esperimento di un poeta linceo Virginio Cesarini', in *Bruniana & Campanelliana* 7, 1 (2001), 51–76

Bondì, R. *Introduzione a Telesio* (Rome, 1997)

Brahe, T. *Tychonis Brahe Dani De mundi aetherei recentioribus phaenomenis: liber secundus, qui est de illustri stella caudata ab elapso fere triente Novembris anni 1577, usq. in finem Ianuarii sequentis conspecta* (Uranienborg, 1588)

Brahe, T. *Tychonis Brahe Astronomiae instauratae mechanica* (Nuremberg, 1602)

Brevaglieri, S. *Il cantiere del Tesoro Messicano tra Roma e l'Europa: pratiche di comunicazione e strategie editoriali nell'orizzonte dell'Accademia dei Lincei*, in Brevaglieri-Guerrini-Solinas 2007, 1–68

Brevaglieri, S. *Science, books and censorship in the Academy of the Lincei: Johannes Faber as cultural mediator*, in Donato-Kraye 2009, 109–33

Brevaglieri, S. *Saperi in circolazione alla Scala di Roma: un'agenda di ricerca per il Tesoro Messicano*, in Cadeddu-Guardo 2013, 243–55

Brevaglieri, S., Guerrini, L., and Solinas, F. *Sul Tesoro Messicano e su alcuni disegni del Museo cartaceo di Cassiano dal Pozzo* (Rome, 2007)

Brocchi, G.B. *Conchiologia fossile subappennina: con osservazioni geologiche sugli Appennini e sul suolo adiacente* (Milan, 1843), 2 v.

Brückle, I. and Hahn, O. eds., *Galileo's Sidereus Nuncius: a comparison of the proof copy (New York) with other paradigmatic copies*, vol. 1 of *Galileo's O*, ed. H. Bredekamp (Berlin, 2011)

Bruni, T. *Novo planisferio, o Astrolabio universale di Theofilo Bruni veronese: istromento, che abbraccia tutti gli istromenti matematici, col quale facilissimamente s'ottengono tutte le operazioni astronomiche et geometriche, che all'uso pratico ridur si possono* [...] (Vicenza, 1625)

Brusa, G. ed., *La misura del tempo: l'antico splendore dell'orologeria italiana dal XV al XVIII secolo* (Trento, 2005)

Bucciantini, M. *Contro Galileo: alle origini dell'affaire* (Florence, 1995)

Bucciantini, M. *Teologia e nuova filosofia: Galileo, Federico Cesi, Giovambattista Agucchi e la discussione sulla fluidità e corruttibilità del cielo*, in Sciences et religions 1999, 411–42

Bucciantini, M. *Galileo e Keplero: filosofia, cosmologia e teologia nell'età della Controriforma* (Turin, 2003)

Bucciantini, M. *Reazioni alla condanna di Copernico: nuovi documenti e nuove ipotesi di ricerca*, in I Primi Lincei 2005, 301–19

Bucciantini, M. *Il copernicanesimo di Paolo Antonio Foscarini*, in Romeo-Lupi-Pupo 2008, 39–44

Bucciantini, M., Camerota, M., and Giudice, F. *Galileo's telescope. A European story* (Cambridge, MA, 2015)

Bucciantini, M. and Torrini, M. eds., *La diffusione del copernicanesimo in Italia: 1543–1610* (Florence, 1997)

Cadeddu, M.E. and Guardo, M. eds., *Il Tesoro Messicano: libri e saperi tra Europa e Nuovo Mondo* (Florence, 2013)

Camerota, M. *Galileo Galilei e la cultura scientifica nell'età della Controriforma* (Rome, 2004)

Camerota, M. 'Aristotelismo e nuova scienza nell'opera di Christoph Scheiner: intorno ad un commentario al De caelo', in *Galilæana* 2 (2005), 43–85

Campanella, T. *F. Thomae Campanellae* [...] *Philosophia, sensibus demonstrata, in octo disputationes distincta, aduersus eos, qui proprio arbitratu, non autem sensata duce natura, philosophati sunt* [...] (Naples, 1591)

Campanella, T. *F. Thomae Campanellae Calabri* [...] *Apologia pro Galileo, mathematico fiorentino, ubi disquiritur, utrum ratio philosophandi, quam Galileus celebrat, faueat sacris scripturis, an aduersetur* (Frankfurt am Main, 1622)

Campanella, T. *Lettere*, ed. G. Ernst, on unpublished preparatory materials by L. Firpo, with the collaboration of L. Salvetti Firpo and M. Salvetti (Florence, 2010)

Canone, E. ed., *Bibliothecae Selectae: da Cusano a Leopardi* (Florence, 1993)

Canone, E. ed., *Per una storia del concetto di mente* (Florence, 2005–2007), 2 v.

Capanna, E. *Observatio e admiratio: i sorprendenti animali del Nuovo Mondo*, in Cadeddu-Guardo 2013, 155–73

Capecchi, A. 'Per la ricostruzione di una biblioteca seicentesca: i libri di storia naturale di Federico Cesi Lynceorum Princeps', in *Atti dell'Accademia Nazionale dei Lincei, Rendiconti, Classe di scienze morali, storiche e filologiche ser. 8*, 41 (1986), 145–64

Caredda, B. *Aspetti e momenti del dibattito astronomico nella prima Accademia dei Lincei, 1603–1616* (PhD Diss. in History, Philosophy and Didactics of Sciences, University of Cagliari, 2008)

Caroti, S. *Un sostenitore napoletano della mobilità della Terra: il padre Paolo Antonio Foscarini*, in Lomonaco-Torrini 1987, 81–121

Carutti, D. *Breve storia della Accademia dei Lincei* (Rome, 1883)

Cavallo, S. and Gentilcore, D. eds., *Spaces, objects and identities in early modern Italian medicine* (Malden, MA, 2008)

Caverni, R. *Storia del metodo sperimentale in Italia* (Florence, 1891–1900), 6 v., repr. with an introduction by G. Tabarroni (New York, 1972)

Cerbu, Th. *Melchior Inchofer: un homme fin et rusé*, in Montesinos-Solis 2001, 587–611

Cesalpino, A. *De plantis libri XVI Andreae Cesalpini aretini, medici clarissimi, doctissimique atque philosophi celeberrimi ac subtilissimi* (Florence, 1583)

Cesalpino, A. *De metallicis libri tres, Andrea Caesalpino auctore* [...] (Rome, 1596)

Cesi, F. *Apiarium, ex frontispiciis naturalis theatri principis Federici Caesii Lyncei, quo universa mellificum familia ab suis prae-generibus deriuata in suas species, ac differentias distributa, in physicum conspectum adducitur* (Rome, 1625)

Cesi, F. *Sodales R. Lynceorum Academiae annum CCC ab eius institutione concelebrantes Federici Cesi auctoris Conlegi opus probatissimum de plantis* [...] *studio et cura cl. Conlegae Romualdi Pirotta, iterum edendum decrevere* (Rome, 1904)

Cesi, F. *Del natural desiderio di sapere, et institutione de' Lincei per adempimento di esso*, in Altieri Biagi-Basile 1980, 39–73

Cesi, F. *Lynceographum, quo norma studiosae vitae Lynceorum philosophorum exponitur*, ed. A. Nicolò (Rome, 2001)

Cesi, F. *Opere scelte*, eds. C. Vinti and A. Allegra (Perugia, 2003)

Cesi, F. *Apiarium*, ed. L. Guerrini, translated by M. Guardo (Rome, 2005–2006), 2 v.

Chambers, D.S. and Quiviger, F. eds., *Italian academies of the sixteenth century* (London, 1995)

Clavelin, M. *Galilée copernicien: le premier combat, 1610–1616* (Paris, 2004)

Clavius, Ch. *Algebra Christophori Clavii Bambergensis e Societate Iesu* (Orléans, 1609)

Clavius, Ch. *Christophori Clauii Bambergensis e Societate Iesu Opera mathematica V tomis distributa ab auctore nunc denuo correcta, et plurimis locis aucta* (Mainz, 1611–1612)

Clavius, Ch. *Corrispondenza*, eds. U. Baldini and P.D. Napolitani (Pisa, 1992), 14 v.

Clericuzio, A. *Chemical medicines in Rome: Pietro Castelli and the vitriol debate, 1616–1626*, in Donato-Kraye 2009, 281–302

Clericuzio, A. and De Renzi, S. *Medicine, alchemy and natural philosophy in the early Accademia dei Lincei*, in Chambers-Quiviger 1995, 175–94

Colonna, F. *Fabii Columnae Lyncei Minus cognitarum rariorumque nostro coelo orientium stirpium ekphrasis* [...]. *Item De aquatilibus aliisque nonnullis animalibus libellus ad ill.mum* [...] *Martium Columnam* [...]. *Omnia fideliter ad vivum delineata, atque aeneis typis expressa cum indice in calce voluminis locupletissimo* (Rome, 1616)

Colonna, F. *La sambuca lincea, ouero dell'istromento musico perfetto lib. III di Fabio Colonna linceo, nel quale oltre la descrittione et costruttione dell'istromento si tratta della diuisione del monocordo, della proporzione de' tuoni, semituoni et lor minute parti, della differenza de' tre generi di musica, de' grandi enarmonici et cromatici, et in che differiscano da quello de gli antichi l'osseruati et descritti dall'autore, con gli esempi di numeri, di musica et disegni* [...] *con l'organo hydraulico di Herone Alessandrino dichiarato dall'istesso autore* (Naples, 1618)

Conigliani, N. *Virginio Cesarini* (Piacenza, 1925)

Conry, Y. 'L'organizzazione dello spazio dei viventi nell'opera di Federico Cesi, fondatore dell'Accademia dei Lincei', in *Giornale critico della filosofia italiana* 76 (78), ser. 6, 17 (1997), 184–203

Conti, L. 'Giuseppe Neri fra Galilei e Cesi', in *Annali della Facoltà di lettere e filosofia dell'Università degli studi di Perugia, 4: Studi filosofici* 23, n.s. 9 (1985–1986), 59–98

Conti, L. *Francesco Stelluti, il copernicanesimo dei Lincei e la teoria galileiana delle maree*, Vinti 1990a, 141–236

Conti, L. *Giuseppe Neri: un matematico aristotelico all'Accademia dei Lincei* (Rome, 1990)

Conti, L. ed., *Medicina e biologia nella rivoluzione scientifica* (Assisi, 1990)

Conti, L. *Galilei, Federico Cesi e l'esperimentare*, in Allegra-Vinti 2005, 113–6

Copernicus, N. *Nicolai Copernici* [...] *De reuolutionibus orbium coelestium, libri* VI, *habes in hoc opere iam recens nato, et aedito, studiose lector, motus stellarum, tam fixarum, quàm erraticarum, cum ex ueteribus, tum etiam ex recentibus obseruationibus restitutos, et nouis isuper ac admirabilibus hypothesibus ornatos, habes etiam tabulas expeditissimas, ex quibus eosdem ad quoduis tempus quam facillime calculare poteris, igitur eme, lege, fruere* (Nuremberg, 1543)

Coresio, G. *Operetta intorno al galleggiare dei corpi solidi* [...] (Florence, 1612), subsequently publ. in OG, IV, 197–244

Cozzi, G. *Paolo Sarpi tra Venezia e l'Europa* (Turin, 1978)

Cremonini, C. *Caesaris Cremonini Centensis* [...] *Disputatio de coelo in tres partes diuisa: De natura coeli, De motu coeli, De motoribus coeli abstractis; adiecta est Apologia dictorum Aristotelis De via lactea, De facie in orbe Lunae* (Venice, 1613)

Crescenzi, P. *Trattato dell'agricoltura di Piero de' Crescenzi cittadino di Bologna, compilato da lui in latino, e diviso in dodici libri, ne' quali distintamente si tratta delle piante, e degli animali, e di tutte le villerecce utilità, già traslatato nella favella fiorentina, e di nuovo rivisto, e riscontro con testi a penna dallo 'Nferigno accademico della Crusca* (Florence, 1605)

Curtius, E.R. *European Literature and the Latin Middle Ages*, with a new introduction by C. Burrow (Princeton, 2013)

Damanti, A. Libertas philosophandi. *Teologia e filosofia nella Lettera alla Granduchessa Cristina di Lorena di Galileo Galilei* (Rome, 2010)

Danti, I. *Trattato dell'uso et della fabbrica dell'astrolabio* [...] *con l'aggiunta del planisferio del Roias* (Florence, 1569)

Danti, I. *Primo volume dell'uso et fabbrica dell'astrolabio et del planisferio, di maestro Egnatio Danti* [...], *nuouamente ristampato et accresciuto in molti luoghi, con l'aggiunta dell'uso et fabbrica di noue altri istromenti astronomici* [...] (Florence, 1578)

De Angelis, G. *Notizie inedite sulla prima giovinezza di Federico Cesi: una conferma delle fonti francescane della spiritualità cesiana*, in Battistini-De Angelis-Olmi 2007, 17–105

De Angelis, G. 'Per una storia delle conoscenze sulle orchidee della flora d'Italia', in *GIROS notizie* 51 (Sept. 2012a), 1–66

De Angelis, G. *Il Prognosticum astrologicum di fra' Tommaso Campanella per la congiunzion magna del 24 dicembre 1603 e l'inaugurazione dell'Accademia dei Lincei: contributo critico allo studio di una amicizia filosofica* (Tivoli, 2012b)

De Angelis, G. *Il ritratto filosofico di Federico Cesi conservato presso l'Accademia Nazionale dei Lincei (Roma, Palazzo Corsini): una preziosa testimonianza storica poco studiata e bisognosa di oculato restauro* (Tivoli, 2012c)

De Dominis, M.A. *Euripus, seu de fluxu et refluxu maris sententia* (Rome, 1624)

De Franco, L. *Alcune considerazioni sulla funzione del freddo nella fisica telesiana*, in Bernardino Telesio 1990, 69–82

De Franco, L. *La teoria della luce di Bernardino Telesio*, in Sirri-Torrini 1992, 53–77

De Franco, L. *Introduzione a Bernardino Telesio* (Soveria Mannelli, 1995)

Della Porta, G.B. *Phytognomonica Io. Baptistae Portae* [...] *octo libris contenta in quibus nova, facillimaque affertur methodus, qua plantarum, animalium, metallorum, rerum denique omnium ex prima extimae faciei inspectione quivis abditas vires assequatur* [...] (Naples, 1588)

Della Porta, G.B. *Io. Bapt. Portae Neapolitani Magiae naturalis libri* XX, *ab ipso authore expurgati, superaucti, in quibus scientiarum naturalium diuitiae et delitiae demonstrantur* [...] (Naples, 1589)

Della Porta, G.B. *Villae Io. Baptistae Portae, Neapolitani, libri* XII [...], *in quibus maiori ex parte, cum verus plantarum cultus, certaque insitionis ars, et prioribus seculis non visos producendi fructus via monstrantur, tum ad frugum, vini ac fructuum multiplicationem experimenta propemodum infinita exhibentur* [...] (Frankfurt am Main, 1592), subsequently publ. in Della Porta 2010–2011

Della Porta, G.B. *De occultis literarum notis seu Artis animi sensa occulte alijs significandi, aut ab alijs significata expiscandi enodandique libri* IIII, *Io. Baptista Porta Neapolitano auctore, explicantur autem inter caetera et veterum quoque scriptorum loca, hac de arte loquentia, indidemque petita* (Montbéliard, 1593a)

Della Porta, G.B. *Ioan. Baptistae Portae Neap. De refractione optices parte libri nouem* [...] (Naples, 1593b)

Della Porta, G.B. *Io. Baptis. Portae Neap. De humana physiognom[on]ia, libri VI in quibus docetur quom. animi propentes naturalibus remediis compesci possint* (Naples, 1599), subsequently publ. in Della Porta 2011

Della Porta, G.B. *Io. Bapt. Portae Neapolitani Pneumaticorum libri tres, quibus accesserunt curuilineorum elementorum libri duo* (Naples, 1601)

Della Porta, G.B. *De furtivis literarum notis vulgo De Ziferis libri quinque, altero libro superaucti, et quamplurimis in locis locupletati, Io. Baptista Porta neapolitano auctore* [...] (Naples, 1602)

Della Porta, G.B. *Coelestis physiognomoniae libri sex, Ioan. Baptistae Portae Neapolitani, unde quis facile ex humani vultus extima inspectione, poterit ex coniectura futura presagire, in quibus etiam astrologia refellitur, et inanis, et imaginaria demonstratur* (Naples, 1603), subsequently publ. in Della Porta 1996

Della Porta, G.B. *Claudii Ptolemaei Magnae constructionis, id est perfectae coelestium motuum pertractationis lib. XIII* [...]. *Io. Baptista Porta Neap. interprete* (Naples, 1605), subsequently publ. in Della Porta 2000b

Della Porta, G.B. *I tre libri de' spiritali di Giovambattista della Porta napolitano, cioè d'inalzar acque per forza dell'aria* (Naples, 1606)

Della Porta, G.B. *Io. Bap. Portae* [...] *De distillatione lib. IX, quibus certa methodo, multiplicique artificio, penitioribus naturae arcanis detectis, cuiuslibet mixti in propria elementa resolutio, perfecte docetur* (Rome, 1608)

Della Porta, G.B. *Io. Baptistae Portae Neapolitani De aeris transmutationibus libri IV, in quo opere diligenter pertractatur de ijs, quae, vel ex aere, vel in aere oriuntur* [...] (Rome, 1610a), subsequently publ. in Della Porta 2000a

Della Porta, G.B. *Della fisionomia dell'huomo del sig. Gio. Battista Della Porta napolitano, libri sei, tradotta dal latino in volgare, e dall'istesso autore accresciuta di figure et di luoghi necessarij a diuerse parte dell'opera* [...] (Naples, 1610b), subsequently publ. in Della Porta 2013a

Della Porta, G.B. *Io. Baptistae Portae Neapolitani Elementorum curvilineorum libri tres, in quibus altera geometriae parte restituta, agitur de circuli quadratura* [...] (Rome, 1610c), subsequently publ. in Della Porta 2005

Della Porta, G.B. *Della magia naturale del sig. Gio. Battista Della Porta napolitano, libri XX* [...] (Naples, 1611)

Della Porta, G.B. *Della fisonomia di tutto il corpo humano del s. Gio. Battista della Porta acc. linceo libri quattro, ne' quali si tratta di quanto intorno a questa materia n'hanno i greci, i latini, e gli arabi scritto, hora breuemente in tauole sinottiche ridotta et ordinata da Francesco Stelluti acc. linceo da Fabriano* (Rome, 1637)

Della Porta, G.B. *Coelestis physiognomonia, e in appendice Della celeste fisonomia, ed. A. Paolella, (Naples, 1996)*, vol. 8 of *Edizione nazionale delle opere di Giovan Battista della Porta* (Naples 1996-)

Della Porta, G.B. *De aeris transmutationibus*, ed. A. Paolella (Naples, 2000a), vol. 14 of *Edizione nazionale delle opere di Giovan Battista della Porta* (Naples, 1996)

Della Porta, G.B. *Claudii Ptolemaei Magnae constructionis liber primus, cum Theonis Alexandrini commentariis, Io. Baptista Porta Neapolitano interprete*, ed. R. De Vivo (Naples, 2000b), vol. 18 of *Edizione nazionale delle opere di Giovan Battista della Porta* (Naples, 1996)

Della Porta, G.B. *Elementorum curvilineorum libri tres*, eds. V. Gavagna and C. Leone (Naples, 2005), vol. 11 of *Edizione nazionale delle opere di Giovan Battista della Porta* (Naples, 1996)

Della Porta, G.B. *Pneumaticorum libri tres, e in appendice I tre libri de' Spiritali, cioè d'inalzar acque per forza dell'aria*, ed. O. Trabucco (Naples, 2008), vol. 10 of *Edizione nazionale delle opere di Giovan Battista della Porta* (Naples, 1996)

Della Porta, G.B. *Villae libri XII*, ed. L. Laserra (Naples, 2010–2011), 2 v., vol. 5 of *Edizione nazionale delle opere di Giovan Battista della Porta* (Naples, 1996)

Della Porta, G.B. *De humana physiognomonia, libri sex*, ed. A. Paolella (Naples, 2011), vol. 6.1 of *Edizione nazionale delle opere di Giovan Battista della Porta* (Naples, 1996)

Della Porta, G.B. *Della fisionomia dell'uomo libri sei*, ed. A. Paolella (Naples, 2013a), vol. 6.2 of *Edizione nazionale delle opere di Giovan Battista della Porta* (Naples, 1996-)

Della Porta, G.B. *Taumatologia e criptologia*, ed. R. Sirri (Naples, 2013b)

Delle Colombe, L. *Contro il moto della Terra*, with postils by Galileo, [1610–1611], in OG, III.1, 251–307

Del Puerto, C. *La paternidad del término telescopio*, in Montesinos-Solis 2001, 18–9

De Renzi, S. *La biblioteca di Johann Faber Linceo*, in Canone 1993, 517–24

De Renzi, S. 'Courts and conversions: intellectual battles and natural knowledge in Counter-Reformation Rome', in *Studies in history and philosophy of science* 27 (1996), 429–49

De Renzi, S. *Writing and talking of exotic animals*, in Frasca Spada-Jardine 2000, 151–67

De Renzi, S. *Un linceo alla Sapienza: la natura del fuoco e dei metalli in un'orazione di Johannes Faber*, in Battistini-De Angelis-Olmi 2007, 271–313

De Renzi, S. *Medical competence, anatomy and the polity in seventeenth-century Rome*, in Cavallo-Gentilcore 2008, 79–95

De Toni, E. 'Le piante lincea-cesia-columnia-stelluta e barbarina', in *Memorie della Pontificia Accademia dei nuovi Lincei* 18 (1908), 349–61

Dollo, C. *Tanquam nodi in tabula, tanquam pisces in aqua: le innovazioni della cosmologia nella Rosa Ursina di Christoph Scheiner*, in Baldini U. 1995, 133–58

Dollo, C. *Le ragioni del geocentrismo nel Collegio Romano: 1562–1612*, in Bucciantini-Torrini 1997, 99–166

Donahue, W.H. *The dissolution of the celestial spheres: 1595–1650* (New York, 1981)

Donato, M.P. and Kraye, J. eds., *Conflicting duties: science, medicine and religion in Rome, 1550–1750* (London, 2009)

Drake, S. 'More on Cesi and Galileo', in *Isis* 48 (1957), 65–6

Drake, S. and O'Malley, C.D. *The controversy on the comets of 1618: Galileo Galilei, Horatio Grassi, Mario Guiducci, Johann Kepler*, translated by S. Drake, C.D. O'Malley (Philadelphia, 1960)

Dupré, S. 'Galileo's telescope and celestial light,' in *Journal of the History of Astronomy*, 34 (2003), 369–99

Eastwood, B.S. 'The chaster path of Venus, orbis Veneris castior, in the astronomy of Martianus Capella', in *Archives internationales d'histoire des sciences* 32, 109 (1982), 145–58

Eastwood, B.S. 'Plato and circumsolar planetary motion in the Middle Ages', in *Archives d'histoire doctrinale et litteraire du moyen age* 68 (1993), 7–26

Fabbri, N. 'The Moon as another Earth: what Galileo owes to Plutarch', in *Galilæana* 9 (2012), 103–35

Faber, J. *Praescriptiones Lynceae Academiae, curante Ioanne Fabro* [...] (Rome, 1624)

Fantoli, A. *Galileo per il Copernicanesimo e per la Chiesa*, 3rd ed. (Vatican City, 2010)

Fattori, M. and Bianchi, L. eds., *Ordo: atti del II Colloquio internazionale [del] Lessico intellettuale europeo: Roma, 7–9 gennaio 1977* (Rome, 1979)

Favaro, A. 'Nuovi studi galileiani,' in *Memorie del R. Istituto Veneto di Scienze, Lettere ed Arti*, XXIV (1891)

Favaro, A. *Amici e corrispondenti di Galileo*, ed. with introductory note of P. Galluzzi (Florence, 1983), 3 v.

Favelleran di te sempre le stelle: Galileo, i primi Lincei e l'astronomia: Biblioteca dell' Accademia nazionale dei Lincei e Corsiniana, Roma 6 aprile-30 giugno 2009, eds. E. Antetomaso, A. Romanello and A. Trentini (Rome, 2009)

Favino, F. *Quel petardo di mia fortuna: riconsiderando la caduta di Giovan Battista Ciampoli*, in Montesinos-Solis 2001, 863–82

Favino, F. *Un caso di censura postuma: la filosofia naturale di Giovanni Ciampoli*, in I Primi Lincei 2005, 141–56

Favino, F. *Scetticismo ed empirismo: Ciampoli linceo*, in Battistini-De Angelis-Olmi 2007, 175–202

Favino, F. *Marvellous conjuncture? The academy of Maurice of Savoy in Rome between politics and the new science*, in Donato-Kraye 2009, 135–55

Favino, F. 'I chierici regolari minori fra teologia e nuova scienza: Giovanni Guevara e il caso Galileo', in *Studi medievali e moderni* 14, 1 (2010), 209–25

Favino, F. *La filosofia naturale di Giovanni Ciampoli* (Florence, 2004)

Federico Cesi e la fondazione dell'Accademia dei Lincei: mostra bibliografica e documentaria (Naples, 1988)

Feingold, M. ed., *Jesuit science and the Republic of letters* (Cambridge, MA, 2003a)

Feingold, M. ed., *The new science and Jesuit science: seventeenth century perspectives* (Dordrecht, 2003b)

Felli, M. *L'astrolabio di Galileo* (Florence, 1983)

Ferrari, G.B. *Io. Bapt. Ferrari [...] De florum cultura libri IV* (Rome, 1633)

Ferrone, V. *Galileo tra Paolo Sarpi e Federico Cesi: premesse per una ricerca*, in Galluzzi 1984, 239–53

Ficino, M. *De vita*, eds. A. Biondi and G. Pisani (Pordenone, 1991)

Figliucci, F. *Stanze sopra le stelle e macchie solari scoperte col nuovo occhiale [...] dedicate all'illustriss. [...] card. Aldobrandino [...] da Flaminio Figliucci* (Rome, 1615)

Findlen, P. *Possessing Nature: Museums, Collecting, and Scientific Culture in Early Modern Italy* (Berkeley-Los Angeles-London, 1994)

Finocchiaro, M.A. *The Galileo affair. A documentary history* (Berkeley-Los Angeles-London, 1989)

Fiorentino, F. *Bernardino Telesio, ossia Studi storici su l'idea della natura nel Risorgimento italiano* (Florence, 1872–1874), 2 v., facs. repr. (Naples, 2008)

Formichetti, G. *Campanella critico letterario: i Commentaria ai Poemata di Urbano VIII (Cod. Barb. Lat. 2037)* (Rome, 1983)

Foscarini, F. *F. Pauli Anton. Foscareni [...] Institutionum omnis generis doctrinarum tomis VII comprehensarum, syntaxis, qua methodus et ordo, in tradendis omnibus disciplinis seruandus explicatur, ut demum ad perfectam solidamque sapientiam peruenire possit [...]* (Cosenza, 1613)

Foscarini, F. *Lettera del r.p.m. Paolo Antonio Foscarini carmelitano sopra l'opinione de' pittagorici e del Copernico della mobilità della Terra e stabilità del Sole, e del nuouo pittagorico sistema del mondo [...]* (Naples, 1615)

Fosi, I. *Johannes Faber: prudente mediatore o estremo persecutore dei protestanti?*, in I Primi Lincei 2005, 189–206

Fracastoro, G. *Risposta dello eccellentissimo messer Hieronymo Fracastoro Del crescimento del Nilo a messer Gio. Battista Rhamusio*, in Ramusio 1563, 264–8

Frasca Spada, M. and Jardine, N. eds., *Books and the sciences in history* (Cambridge (UK), 2000)

Freedberg, D. *The eye of the lynx: Galileo, his friends and the beginnings of modern natural history* (Chicago, 2002)

Freedberg, D. *Heckius, Cesi and their mycological researches: the history and fate of the drawings of fungi in Paris, Montpellier and Windsor*, in Battistini-De Angelis-Olmi 2007a, 317–65

Freedberg, D. *L'occhio della lince: Galileo, i suoi amici e gli inizi della moderna storia naturale*, translated and edited by L. Guerrini (Bologna, 2007b)

Gabrieli, G. 'Verbali delle adunanze e Cronache della prima Accademia Lincea: 1603–1630', in *Atti della Reale accademia nazionale dei Lincei, Memorie della Classe di scienze morali, storiche e filologiche* ser. 6, 2 (1927), 463–512, subsequently publ. in Gabrieli 1989, vol. 1, 497–550

Gabrieli, G. 'L'archivio di S. Maria in Aquiro o degli orfani in Roma e le carte di Giov. Faber linceo', in *Archivio della Società romana di storia patria* 51 (1928a), 61–77, subsequently publ. in Gabrieli 1989, vol. 2, 1221–32

Gabrieli, G. 'Fra' Tommaso Campanella e i Lincei della prima Accademia', with an autobiographical appendix of G. Scioppio, in *Rendiconti della R. Accademia nazionale dei Lincei, Classe di scienze morali, storiche e filologiche ser. 6*, 4 (1928b), 250–67, subsequently publ. in Gabrieli 1989, vol. 1, 385–98

Gabrieli, G. 'Due prelati lincei in Roma alla corte di Urbano VIII: Virginio Cesarini e Giovanni Ciampoli', in *Atti dell'Accademia degli Arcadi e scritti dei soci* 13 (1929–1930), 171–200, subsequently publ. in Gabrieli 1989, vol. 1, 763–85

Gabrieli, G. 'Virginio Cesarini e Giovanni Ciampoli: con documenti inediti', in *Rendiconti della R. Accademia nazionale dei Lincei, Classe di scienze morali, storiche e filologiche* ser. 6, 8 (1932), 422–62, subsequently publ. in Gabrieli 1989, vol. 1, 787–817

Gabrieli, G. 'Luca Valerio linceo e un episodio memorabile della vecchia Accademia', in *Rendiconti della R. Accademia nazionale dei Lincei, Classe di scienze morali, storiche e filologiche* ser. 6, 9 (1933a), 691–728, subsequently publ. in Gabrieli 1989, vol. 1, 835–64

Gabrieli, G. 'Notizie della vita e degli scritti di Antonio Persio linceo', in *Rendiconti della R. Accademia nazionale dei Lincei, Classe di scienze morali, storiche e filologiche* ser. 6, 9 (1933b), 471–99, subsequently publ. in Gabrieli 1989, vol. 1, 865–87

Gabrieli, G. 'Scritti di Giovanni Faber linceo', in *Rendiconti della R. Accademia nazionale dei Lincei, Classe di scienze morali, storiche e filologiche* ser. 6, 9 (1933c), 276–334, subsequently publ. in Gabrieli 1989, vol. 2, 1177–220

Gabrieli, G. 'Emblematica lincea', in *Rendiconti della R. Accademia nazionale dei Lincei, Classe di scienze morali, storiche e filologiche* ser. 6, 10 (1934), 269–84, subsequently publ. in Gabrieli 1989, vol. 2, 61–70

Gabrieli, G. 'Una gara di precedenza accademica nel Seicento fra Umoristi e Lincei', in *Rendiconti della R. Accademia nazionale dei Lincei, Classe di scienze morali, storiche e filologiche* ser. 6, 11 (1935), 235–57, subsequently publ. in Gabrieli 1989, vol. 1, 479–96

Gabrieli, G. 'L'orizzonte intellettuale e morale di Federico Cesi illustrato da un suo zibaldone inedito', in *Rendiconti della R. Accademia nazionale dei Lincei, Classe di scienze morali, storiche e filologiche* ser. 6, 14 (1938a), 663–725, subsequently publ. in Gabrieli 1989, vol. 1, 27–77

Gabrieli, G. 'La prima Accademia dei Lincei (1603–1630) nella luce della recente documentazione', in *Accademie e biblioteche d'Italia* 12 (1938b), 7–25, subsequently publ. in Gabrieli 1989, vol. 1, 225–42

Gabrieli, G. 'La prima Biblioteca lincea o Libreria di Federico Cesi', in *Rendiconti della R. Accademia nazionale dei Lincei, Classe di scienze morali, storiche e filologiche* ser. 6, 14 (1938c), 606–28, subsequently publ. in Gabrieli 1989, vol. 1, 79–96

Gabrieli, G. 'Le schede fogeliane e la storiografia della prima Accademia Lincea', in *Rendiconti della R. Accademia nazionale dei Lincei, Classe di scienze morali, storiche e filologiche* ser. 6, 15 (1939), 140–67, subsequently publ. in Gabrieli 1989, vol. 1, 273–96

Gabrieli, G. 'Pratica e tecnica del telescopio e del microscopio presso i primi Lincei', in *Atti della R. Accademia d'Italia, Rendiconti della classe di scienze morali e storiche* ser. 7, 2 (1940), 18–43, subsequently publ. in Gabrieli 1989, vol. 1, 347–71

Gabrieli, G. 'Francesco Stelluti linceo fabrianese', in *Atti della R. Accademia d'Italia, Rendiconti della classe di scienze morali e storiche* ser. 7, 2 (1941), 191–233, subsequently publ. in Gabrieli 1989, vol. 1, 913–52

Gabrieli, G. 'Galileo in Acquasparta', in *Atti della R. Accademia d'Italia, Memorie della Classe di scienze morali e storiche* ser. 7, 3 (1942), 1–29, subsequently publ. in Gabrieli 1989, vol. 1, 195–221

Gabrieli, G. *Contributi alla storia della Accademia dei Lincei* (Rome, 1989), 2 v.

Galilei, G. *Discorso* [...] *intorno alle cose che stanno in sù l'acqua ò che in quella si muovono, di Galileo Galilei* [...] (Florence, 1612), subsequently publ. in OG, IV, 59–141

Galilei, G. *The controversy on the comets of 1618: Galileo Galilei, Horatio Grassi, Mario Guiducci, Johann Kepler*, translated by S. Drake and C.D. O'Malley (Philadelphia, 1960)

Galilei, G. *Dialogue concerning the two chief world systems – Ptolemaic & Copernican*, translated by S. Drake, forwarded by Albert Einstein (Los Angeles, 1967)

Galilei, G. *Sidereus Nuncius. Le messager céleste*, text, translation and notes by I. Pantin (Paris, 1992)

Galilei, G. *Dialogo sopra i due massimi sistemi del mondo tolemaico e copernicano*, critical edition and commentary by O. Besomi and M. Helbing (Rome, 1998)

Galilei, G. *Il Saggiatore*, critical edition and commentary by O. Besomi and M. Helbing (Rome, 2005)

Galilei, G. *Scienza e religione: scritti copernicani*, eds. M. Bucciantini and M. Camerota (Rome, 2009)

Galilei, G. *Lettera a Cristina di Lorena*, critical edition by O. Besomi, with the collaboration of D. Besomi, Latin trans. by E. Diodati, ed. G. Reggi (Rome, 2012)

Galilei, G. *Le Opere di Galileo Galilei: Appendice* to the National Edition, vol. II: *Carteggio* (Florence, 2015)

Galilei, G. *Sidereus Nuncius or the Sidereal Messenger*, translated with introduction, conclusions and notes by A. van Helden (Chicago and London, 2015: 2nd ed.)

Galilei, G. and Guiducci, M. *Discorso delle comete*, critical edition and commentary by O. Besomi, M. Helbing (Rome, 2002)

Galilei, G. and Scheiner, C. *On sunspots*, translated and introduced by E. Reeves, A. van Helden (Chicago, 2010)

Le vinti giornate dell'agricoltura et de' piaceri della villa, di m. Agostino Gallo (Venice, 1573)

Galluzzi, P. *Il tema dell'"ordine' in Galileo*, in Fattori-Bianchi 1979, 235–77

Galluzzi, P. 'L'Accademia del Cimento: gusti del principe, filosofia e ideologia dell'esperimento', in *Quaderni Storici* 16 (1981), 788–844

Galluzzi, P. *Motivi paracelsiani nella Toscana di Cosimo II e di Don Antonio dei Medici: alchimia, medicina chimica e riforma del sapere*, in Scienze, credenze occulte 1982, 31–62

Galluzzi, P. ed., *Novità celesti e crisi del sapere: atti del Convegno internazionale di studi galileiani* (Florence, 1984)

Galluzzi, P. *I sepolcri di Galileo: le spoglie vive di un eroe della scienza*, in Berti L. 1993, 145–82

Galluzzi, P. *Gli orologi di Galileo: l'astronomia e la misurazione del tempo*, in Strano 2008, 121–31

Galluzzi, P. 'Galileo e gli spettacoli del cielo sul Colle del Quirinale', in *Il Quirinale* 5, 8 (Mar. 2009), 82–90

Galluzzi, P. *Tra atomi e indivisibili: la materia ambigua di Galileo* (Florence, 2011)

Galluzzi, P. *Libertas philosophandi in naturalibus: I mondi paralleli di Cesi e Galileo* (Rome, 2014a)

Galluzzi, P. *'Officio veramente da Lincei.' Conferenze a Classi Riunite, Accademia Nazionale dei Lincei* (Rome, 2014b), 13–22

Galluzzi, P. 'Spotted felines and the spotted Sun. The Lincei and Galileo's *Istoria*,' in *Galilæana* 11 (2014c), 37–47

Gardair, J.M. 'I Lincei: i soggetti, i luoghi, le attività', in *Quaderni storici* 16 (dic. 1981), 763–87

Garin, E. *La cultura filosofica del Rinascimento italiano: ricerche e documenti* (Florence, 1961a), 2nd edition (Florence, 1979)

Garin, E. *Nota telesiana: Antonio Persio*, in Garin 1961b, 435–8

Garin, E. *La nuova scienza e il simbolo del libro*, in Garin 1961c, 451–65

Garin, E. 'Telesiani minori', in *Rivista critica di storia della filosofia* 26 (1971), 199–204

Garin, E. 'Fra '500 e '600: scienze nuove, metodi nuovi, nuove accademie', in *Nuncius* 1 (1986), 3–23

Gatto, R. *Tra scienza e immaginazione: le matematiche presso il Collegio gesuitico napoletano, 1552–1670 ca.* (Florence, 1994)

Gazzini, T. *La macchina crittografica circolare di Francesco Stelluti, 1625–1630 circa: Biblioteca Apostolica Vaticana, Ms. Vat. Lat. 9685, cc. 80 v.-81 r.*, with a working model of the instrument based on the drawings of F. Stelluti (Fabriano, 1986a)

Gazzini, T. *L'uomo di Saturno: Stelluti e la tradizione ermetica*, in Alessandrini *et al.* 1986b, 243–364

Ghirardelli, C. *Discorso overo consideratione sopra la notabile eclisse del sole, qual havrà a succedere alli 21 maggio del presente anno 1621. Intorno alla quale si tratta succintamente di molte eclissi solari* [...], *di Cornelio Ghirardello bolognese* [...] (Bologna, 1621)

Giard, L. ed., *Les Jésuites à la Renaissance: système éducatif et production du savoir* (Paris, 1995)

Gingerich, O. 'The censorship of Copernicus De revolutionibus', in *Annali dell'Istituto e Museo di Storia della Scienza di Firenze* 6, 2 (1981), 2, 45–61

Gingerich, O. and Westman, R.S. *The Wittich connection: conflict and priority in late sixteenth-century cosmology* (Philadelphia, 1988)

Giovio, P. *Dialogo dell'imprese militari e amorose*, ed. M.L. Doglio (Rome, 1978)

Gómez, S. 'The Bologna stone and the nature of light: the Science Academy of Bologna', in *Nuncius* 6 (1991), 3–32

Gorman, M.J. *Mathematics and modesty in the Society of Jesus: the problems of Christoph Grienberger*, in Feingold 2003, 1–120

Gottifredi, A. *In funere Virginij Cesarini oratio Alexandri Gottifredi e Societate Iesu ad S.P.Q.R. dum ei in aede Virginis Capitolinae publico sumptu parentaret* (Rome, 1624)

Gould, S.J. 'This view of life: the sharp-eyed lynx outfoxed by Nature,' in *Natural History* 106, 4 (1997), 18–28; 107, 1 (part 1), 16–23, and 5 (part 2), 23–32

Govi, G. 'Il microscopio composto inventato da Galileo', in *Atti dell'Accademia delle scienze fisiche e matematiche* ser. 2, 2 (1889), 1–33

Gozzadini, G. *Delle antiche carrozze e segnatamente di due veronesi cenni* (Bologna, 1862)

Graney, Ch. *Setting aside all authority: Giovanni Battista Riccioli and the science against Copernicus in the age of Galileo; including the first English translation of Monsignor Francesco Ingoli's essay to Galileo disputing the Copernican system and the first English translation of Riccioli's reports regarding his experiments with falling bodies and with the effect of air resistance on falling bodies* (Notre Dame, 2015)

Graniti, A. ed., *Federico Cesi: un principe naturalista: Acquasparta, 29 e 30 settembre 2003* (Rome, 2006a)

Graniti, A. *Il mondo vegetale nella visione di Federico Cesi*, in Graniti 2006b, 17–101

Grassi, O. *De tribus cometis anni MDCXVIII disputatio astronomica publice habita in Collegio Romano Societatis Iesu ab uno ex patribus eiusdem Societatis* (Rome, 1619a), subsequently publ. in OG, VI, 17–35

Grassi, O. *Libra astronomica ac philosophica qua Galilaei Galilaei opiniones de cometis a Mario Guiducio in Florentina Academia expositae, atque in lucem nuper editae, examinantur a Lothario Sarsio Sigensano* (Perugia, 1619b), subsequently publ. in OG, VI, 107–79

Grassi, O. *Ratio ponderum librae et simbellae: in qua quid e Lotharij Sarsij libra astronomica, quidque e Galilei Galilei simbellatore, de cometis statuendum sit* [...], *proponitur, auctore eodem Lothario Sarsio Sigensano* (Paris, 1626), subsequently publ. in OG, VI, 373–500

Grazzini, M.G. *Antonio Neri e l'arte vetraria: conoscenze tecniche e motivi ermetico-paracelsiani*, (Master's thesis in Philosophy, University of Florence, supervisor prof. Paolo Rossi, 1983–1984)

Grottanelli, L. 'Le carrozze nel passato e nei tempi nostri', in *La Rassegna Nazionale* 111 (1900), 288–321

Guardo, M. 'Il "ristretto" delle costituzioni lincee del 1612: fonti, stile e funzioni', in *Studia Borromaica* 19 (2005), 491–517

Guardo, M. 'Galilei e il Tesoro Messicano', in *L'ellisse* 6 (2011), 53–82

Guardo, M. *Nell'officina del Tesoro Messicano: il ruolo misconosciuto di Marco Antonio Petilio nel sodalizio linceo*, in Cadeddu-Guardo 2013, 67–92

Guardo, M. and Orioli, R. eds., *Cronache e statuti della prima Accademia dei Lincei* (Rome, 2014)

Guerrini, L. 'Le Stanze sopra le stelle e macchie solari scoperte col nuovo occhiale, di Vincenzo Figliucci: un episodio poco noto della visita di Galileo Galilei a Roma nel 1611', in *Lettere italiane* 50 (1998), 387–415

Guerrini, L. *Apis natura mira laudatur: l'universo delle api e il mondo umano in Federico Cesi*, in I Primi Lincei 2005, 157–87

Guerrini, L. *Natura, mente e verità in Galileo*, in Canone 2005–2007, vol. 2, 145–78

Guerrini, L. *I trattati naturalistici di Federico Cesi* (Rome, 2006)

Guerrini, L. *De Sole et api: Tommaso Campanella e Federico Cesi, un'amicizia filosofica nella Roma di Urbano VIII*, in Battistini-De Angelis-Olmi 2007a, 107–73

Guerrini, L. *Federico Cesi e il Tesoro Messicano*, in Brevaglieri-Guerrini-Solinas 2007b, 69–72

Guerrini, L. *Nuovi saperi e antichi primati: studi sulla cultura del primo Seicento* (Bologna, 2008)

Guerrini, L. *Galileo e la polemica anticopernicana a Firenze* (Florence, 2009)

Guiducci, M. *Discorso delle comete di Mario Guiducci fatto da lui nell'Accademia fiorentina nel suo medesimo consolato* (Florence, 1619), subsequently publ. in OG, VI, 39–105

Gunther, R.W.T. *The Astrolabes of the world* (Oxford, 1932), 2 v.

Hamel, J. and Keil, I. eds., *Der Meister und die Fernrohre: das Wechselspiel zwischen Astronomie und Optik in der Geschichte: Festschrift zum 85. Geburtstag von Rolf Riekher* (Frankfurt am Main, 2007)

Hamesse, J. and Fattori, M. eds., *Lexiques et glossaires philosophiques de la Renaissance: actes du Colloque international organisé à Rome par l'Academia Belgica* [...], *Academia Belgica, 3–4 novembre 2000* (Louvain-la-Neuve, 2003)

Hamou, P. 'La nature est inexorabile: pour une reconsidération de la contribution de Galilée au problème de la connaissance', in *Galilæana* 5 (2008), 149–77

Haskell, F. ed., *Il Museo Cartaceo di Cassiano dal Pozzo: Cassiano naturalista* (Ivrea, 1989)

Heckius, J. *De noua stella disputatio Io. Heckij I. Lyncaei Dauentriensis philosophiae, et medicinae doctoris, ad illustriss. dominum D. Federicum Caesium marchionem Monticellorum, etc.* (Rome, 1605)

Heidegger, M. *Uber den Humanismus* (Frankfurt am Main, 1939)

Heilbron, J.L. *Galileo* (London-New York, 2010) (Italian transl.: *Galileo scienziato e umanista*, ed. S. Gattei, Turin, 2013)

Horky, M. *Martini Horky* [...] *Brevissima peregrinatio contra Nuncium sidereum nuper ad omnes philosophos et mathematicos emissum a Galilaeo Galilaeo* [...] (Modena, 1610), subsequently publ. in OG, III.1, 129–45

Hoskin, M. ed., *The general history of astronomy* (Cambridge, UK, 1984)

Husserl, E. *The crisis of European sciences and the transcendental phenomenology*, translated with an introduction by D. Carr (Evanston, 1970)

Huygens, Ch. *Correspondance: 1657–1659* (La Haye, 1889), vol. 2 of *Oeuvres complètes de Christiaan Huygens* (La Haye, 1888–1950)

Imperato Fe., *Dell'historia naturale di Ferrante Imperato napolitano, libri XXVIII, nella quale ordinatamente si tratta della diuersa condition di miniere e pietre, con alcune historie di piante et animali sin'hora non date in luce* (Naples, 1599)

Imperato Fr., *Discorsi intorno a diverse cose naturali di Francesco Imperato, opera non meno curiosa che utile, e necessaria a' professori della natural filosofia* (Naples, 1628)

Kepler, J. *Astronomia nova aitiologetos, seu Physica coelestis, tradita commentariis de motibus stellae Martis, ex observationibus G.V. Tychonis Brahe, jussu et sumptibus Rudolphi II Romanorum imperatoris etc, plurium annorum pertinaci studio elaborata Pragae, a* [...] *Joanne Keplero* [...] (Heidelberg, 1609)

Kepler, J. *Ioannis Kepleri* [...] *Dissertatio cum Nuncio Sidereo nuper ad mortales misso a Galilaeo Galilaeo mathematico Patavino* (Prague, 1610), subsequently publ. in OG, III.1, 99–126

Kepler, J. *Epitome astronomiae Copernicanae usitata forma quaestionum et responsionum conscripta, in*[*que*] *VII libros digesta, quorum tres hi priores sunt de doctrina sphaerica* [...] *authore Joanne Kepplero* [...] (Linz, 1618–1622)

Kepler, J. *Tychonis Brahei* [...] *Hyperaspistes aduersus Scipionis Claramontii* [...] *Anti-Tychonem, in aciem productus a Ioanne Keplero, quo libro doctrina praestantissima de parallaxibus, deque nouorum siderum in sublimi aethere discursionibus, repetitur, confirmatur, illustratur, cum indice rerum memorabilium* (Frankfurt am Main, 1625)

Kepler, J. *Gesammelte Werke*, ed. Kepler-Kommission der Bayerischen Akademie der Wissenschaften (Munich, 1937)

Kepler, J. *Conversation with Galileo's Sidereal Messenger*, with an introduction and notes by E. Rosen (New York, 1965)

Kidwell, C.S. *The Academia dei Lincei and the Apiarium: a case study of the activities of a Seventeenth Century Scientific Society* (PhD. Diss., The University of Oklahoma, Norman, 1970)

Koestler, A. *The sleepwalkers: a history of man's changing visions of the universe*, with an introduction by H. Butterfield (London, 1968)

Lagalla, G.C. *De phoenomenis in orbe Lunae noui telescopii usu a d. Gallileo Gallileo nunc iterum suscitatis physica disputatio, ad Iulio Caesare La Galla* [...] *necnon De luce et lumine altera disputatio* (Venice, 1612), subsequently publ. in OG, III.1, 310–99

Lagalla, G.C. *Iulii Caesaris Lagallae De coelo animato disputatio* (Heidelberg, 1622)

Lattanzi, M. *Il progetto di Galilei per una carrozza di gala*, in Galluzzi 2014a, 545–7

Lattis, J.M. *Between Copernicus and Galileo: Christoph Clavius and the collapse of Ptolemaic cosmology* (Chicago, 1994)

Lerner, M.P. 'Sicut nodus in tabula: de la rotation propre du soleil au seizième siècle', in *Journal for the history of astronomy* 11 (1980), 114–29

Lerner, M.P. 'Le problème de la matière céleste après 1550: aspects de la bataille des cieux fluides', in *Revue d'histoire des sciences* 42 (1989), 255–80

Lerner, M.P. *La physique céleste de Telesio: problèmes d'interpretation*, in Bernardino Telesio 1990, 83–107

Lerner, M.P. *Tre saggi sulla cosmologia alla fine del Cinquecento* (Naples, 1992)

Lerner, M.P. *L'entrée de Tycho Brahe chez les Jésuites, ou Le chant du cygne de Clavius*, in Giard 1995, 145–85

Liceti, F. *De spontaneo viventium ortu libb. quatuor, in quibus de generatione animantium, quae vulgo ex putri exoriri dicuntur, accurate aliorum opiniones omnes primum examinantur, caussae singulae propositi deinde cum generatim, tum etiam speciatim ex rei natura deteguntur, patefacto praesertim efficiente proximo univoco eorum, quae in fungorum, plantarum, zoophytorum, et animalium genere sponte nascuntur* [...] *autor Fortunius Licetus* (Vicenza, 1618)

Liceti, F. *Litheosphorus, siue De lapide Bononiensi lucem in se conceptam ab ambiente claro mox in tenebris mire conseruante liber Fortunij Liceti Genuensis* [...] (Eutin, 1640)

L'Obel de, M. *Icones stirpium, seu plantarum tam exoticarum, quam indigenarum, in gratiam rei herbariae studiosorum in duas partes digestae, cum septem linguarum indicibus, ad diuersarum nationum usum* (Antwerp, 1581–1591), 2 v.

Lomonaco, F. and Torrini, M. eds., *Galileo e Napoli* (Naples, 1987)

Longo, O. 'Un'impresa galileiana: l'emblema della stamperia Alle stelle medicee', in *Atti e memorie dell'Accademia patavina di scienze lettere ed arti, già Accademia dei Ricovrati, pt. 3, Memorie della Classe di scienze morali lettere ed arti 399*, 110 (1997–1998), 5–15

Lüthy, C. 'Atomism, Lynceus and the fate of seventeenth-century microscopi', in *Early science and medicine* 1 (1996), 1–27

Maccagni, C. ed., *Saggi su Galileo Galilei* (Florence, 1972)

Marini Bettòlo, G.B. *Una guida alla lettura del Tesoro Messicano*, in Rerum medicarum 1992 (accompanying booklet to the facs. repr. Rome, 1992)

Martinez, R. A. 'Il manoscritto ACDF. Index. Protocolli. vol EE f. 291r-v', in *Acta Philosophica* 10 (2001), 215–42.

Mascardi, A. *Prose vulgari di monsignor Agostino Mascardi cameriere d'honore di n. sig. Urbano VIII* (Venice, 1626)

Mascardi, A. *Saggi accademici dati in Roma nell'Accademia del sereniss. Prencipe Cardinal di Savoia da diversi nobilissimi ingegni, raccolti e publicati da monsig. Agostino Mascardi* (Venice, 1630)

Miniati, M. *Museo di storia della scienza: catalogo* (Florence, 1991)

Montesinos, J. and Solis, C. eds., *Largo campo di filosofare: Eurosymposium Galileo 2001* (La Orotava, 2001)

Morello, N. *La macchina della Terra: teorie geologiche dal Seicento all'Ottocento* (Turin, 1979a)

Morello, N. *La nascita della paleontologia nel Seicento: Colonna, Stenone e Scilla* (Milan, 1979b)

Morosini, D. *Lettere del conte Domenico Morosini, nobile veneziano, al signor abate Francesco Cancellieri di Roma, e di questo a quello intorno ad alcune cifre spettanti all'Accademia de' Lincei, e per la seconda volta pubblicate* (Ceneda, 1865)

Mottana, A. *Il Tesoro Messicano: il commento di Fabio Colonna (1628) e i contributi innovativi alle conoscenze mineralogiche*, in Cadeddu-Guardo 2013, 175–241

Neri, A. *L'arte vetraria distinta in libri sette del r. p. Antonio Neri fiorentino, ne' quali si scoprono effetti marauigliosi et insegnano segreti bellissimi, del vetro nel fuoco et altre cose curiose* (Florence, 1612), subsequently publ. in Neri 2001

Neri, A. *L'arte vetraria*, ed. F. Abbri (Florence, 2001)

North, J.D. 'The astrolabe', in *Scientific American 230*, no. 1 (1974), 96–105 (Italian transl.: 'L'astrolabio', in *Le Scienze* 68 (1974), 62–73)

Novati, L. ed., *Genesi e natura: racconti di creazione e modelli scientifici* (Brescia, 2009)

Odescalchi, B. *Memorie istorico-critiche dell'Accademia de' Lincei e del principe Federico Cesi secondo duca d'Acquasparta, fondatore e principe della medesima* (Rome, 1806)

Olmi, G. *In essercitio universale di contemplatione e prattica: Federico Cesi e i Lincei*, in Boehm-Raimondi 1981, pp. 169–235, subsequently publ. in Olmi 1992, 315–79

Olmi, G. *La colonia lincea di Napoli*, in Lomonaco-Torrini 1987, 23–58

Olmi, G. *Libertà di filosofare e virtuose fatiche: l'Accademia dei Lincei nell'Italia della Controriforma*, in Accademia dei Lincei 1991, 1–14

Olmi, G. *L'inventario del mondo: catalogazione della natura e luoghi del sapere nella prima età moderna* (Bologna, 1992)

Orbaan, J.A.F. *Documenti sul Barocco in Roma* (Rome, 1920)

Ottaviani, A. 'La natura senza inventario: aspetti della ricerca naturalistica del linceo Fabio Colonna', in *Physis* 34 (1997), 31–70, partially publ. in Ottaviani-Trabucco 2007

Ottaviani, A. 'Il fascino indiscreto delle nature ancipiti: un saggio della istoria naturale nei secoli XVII e XVIII', in *Giornale critico della filosofia italiana* 79 (81), ser. 6, 20 (2000), 316–69, partially publ. in Ottaviani-Trabucco 2007

Ottaviani, A. *La filologia scomoda dei Lincei*, in I Primi Lincei 2005, 101–24

Ottaviani, A. *La parte di Fabio Colonna nel Tesoro Messicano dei Lincei*, in Battistini-De Angelis-Olmi 2007, 367–93

Ottaviani, A. and Trabucco, O. *Theatrum naturae: la ricerca naturalistica tra erudizione e nuova scienza nell'Italia del primo Seicento* (Naples, 2007)

Pagano, S. *I documenti vaticani del processo di Galileo Galilei (1611–1741)*, new edition, augmented, revised and annoted by S. Pagano (Vatican City, 2009)

Pagel, W. *Paracelso: un'introduzione alla medicina filosofica nell'età del Rinascimento*, introduction by E. Garin (Milan, 1989)

Pagnini, P. 'Di alcuni documenti galileiani recentemente ritrovati', in *Atti della Società Colombaria di Firenze*, (1933–1934), 269–312

Pannocchieschi, A. *Considerazioni sopra il Discorso del sig. Galileo Galilei intorno alle cose che stanno in sù l'acqua o che in quella si muovono [...] fatte a difesa e dichiarazione dell'opinione d'Aristotile da accademico incognito* (Pisa, 1612), subsequently publ. in OG, IV, 145–96

Panofsky, E. *Galileo as a critic of the arts* (The Hague, 1954)

Pantin, I. 'Galilée, la Lune et les Jésuites: à propos du Nuncius sidereus Collegii Romani et du problème de Mantoue', in *Galilæana* 2 (2005), 19–42

Partini, A.M. 'I primi Lincei e l'ermetismo', in *Atti della Accademia nazionale dei Lincei, Rendiconti, Classe di scienze morali, storiche e filologiche* ser. 8, 41 (1986), 59–83

Patrizi, F. *Francisci Patricii Nova de universis philosophia, in qua Aristotelica methodo non per motum sed per lucem et lumina ad primam causam ascenditur* (Ferrara, 1591)

Patrizi, F. 'Emendatio in libros suos novae philosophiae', ed. P.O. Kristeller, in *Rinascimento* (1970), 215–8

Pegler, D. and Freedberg, D. eds., *Fungi*, volume ed. M. Clayton (London 2005), 3 v., Ser. B, pt. 2 of *The Paper museum of Cassiano dal Pozzo: a catalogue raisonné: drawings and prints in the Royal library at Windsor castle, the British museum, the Institut de*

France and other collections, general eds. F. Haskell, J. Montagu, assistant general ed. H. McBurney (London, 1997)

Pelseneer, J. 'Gilbert, Bacon, Galilée, Képler, Harvey et Descartes: leurs relations', in *Isis* 17 (1932), 171–208

Persio, A. *Disputationes libri nouarum positionum Antonii Persii, triduo habitae Venetiis anno MDLXXV, mense Maio, ad eximium equitem* [...] *Ugolinum Gualterutium* (Florence, 1576a)

Persio, A. *D'Antonio Persio Trattato dell'ingegno dell'huomo* [...] (Venice, 1576b)

Persio, A. *Index capitum librorum abbatis Antonii Persii Lyncei Materani* [...] *De ratione recte philosophandi et de natura ignis et caloris* (Rome, 1613)

Persio, A. *Trattato dell'ingegno dell'huomo*, ed. L. Artese (Pisa, 1999)

Persius Flaccus, *Persio tradotto in verso sciolto e dichiarato da Francesco Stelluti accad. linceo da Fabriano all'ill.mo et r.mo sig.re il sig. cardinale Barberino* (Rome, 1630)

Petrucci Nardelli, F. 'Un illustre mostro tipografico: il Tesoro Messicano', in *Rara Volumina* 1 (1998), 33–71

Peuerbach, G. *Theoricae novae planetarum Georgii Purbachii astronomi celebratissimi*, (s.l., 1472–1473)

Pignatti, S. *Il Tesoro Messicano e la nascita del concetto di biodiversità nella Roma del XVII secolo*, in Cadeddu-Guardo 2013, 133–53

Pirro V. ed., *Federico Cesi e i primi Lincei in Umbria: atti del Convegno di studi nel IV centenario della fondazione dell'Accademia dei Lincei, Terni, 24–25 ottobre 2003* (Arrone, 2005)

Poppi, A. *Cremonini e Galilei inquisiti a Padova nel 1604: nuovi documenti d'archivio* (Padua, 1992)

Poppi, A. 'La lettera del cardinale Carlo Conti a Galileo su cosmologia aristotelica e Bibbia (7 luglio 1612): l'approdo galileiano alla nuova ermeneutica biblica', in *Atti e memorie dell'Accademia patavina di scienze lettere ed arti, già Accademia dei Ricovrati, pt. 3, Memorie della Classe di scienze morali lettere ed arti 398*, 109 (1996–1997), 131–58

Poulle, E. *Les instruments de la théorie des planètes selon Ptolémée: équatoires et horlogerie planétaire du XIIIe au XVIe siècle* (Geneva, 1980), 2 v.

Preti, C. *Lagalla Giulio Cesare*, in DBI, 63 (2004), 63–5

Price, J.D.D.S. *Strumenti di precisione fino al 1500*, in *Storia della Tecnologia*, ed. C. Singer, 7 v. (Turin, 1961–1965), vol. 3 (1963), 592–627

I Primi Lincei e il Sant'Uffizio: questioni di scienza e di fede, Roma, 12–13 giugno 2003 (Rome, 2005)

Prosperi, A. ed., *Il piacere del testo: saggi e studi per Albano Biondi* (Rome, 2001)

Ptolemy, *Claudii Ptolemaei Magnae constructionis, id est perfectae coelestium motuum pertractationis lib. XIII* [...] *Io. Baptista Porta Neap. Interprete* (Naples, 1605), subsequently publ. in Della Porta 2000b

Puliafito, A.L. *La fisica telesiana attraverso gli occhi di un contemporaneo: Francesco Patrizi da Cherso*, in Sirri-Torrini 1992, 257–70

Quattromani, S. *La philosophia di Bernardino Telesio ristretta in brevità, et scritta in lingua toscana dal Montano Academico Cosentino* (Naples, 1589)

Quietanus, J.R. *Observationes et descriptiones duorum cometarum qui anno 1618, mense novembri usque ad finem anni currenti, in aetherea regione visi sunt* [1619], Biblioteca dell'Accademia dei Lincei e Corsiniana, Rome, Ms. Archivio Linceo I, 32–9

Raimondi, E. *Anatomie secentesche* (Pisa, 1966a)

Raimondi, E. *Paesaggi e rovine nella poesia di un virtuoso*, in Raimondi 1966b, 42–72

Raimondi, E. *Scienziati e viaggiatori: l'Accademia dei Lincei*, in *Storia della letteratura italiana* (Milan, 1967), vol. 5, 270–5

Ramusio, G.B. *Primo volume et terza editione delle nauigationi et viaggi raccolto gia da m. Gio. Battista Ramusio, et con molti et vaghi discorsi da lui in molti luoghi dichiarato et illustrato, nel quale si contengono la descrittione dell'Africa* [...] *con la relatione dell'isola Giapan* [...] *et alcuni capitoli appartenenti alla geographia, estratti dell'Historia del s. Giouan di Barros portoghese, con tre tauole di geographia in disegno* [...] *con due indici* [...] (Venice, 1563)

Rati, G. *Federico Cesi e Galileo Galilei ad Acquasparta*, in Pirro 2005, 75–90

Redondi, P. 'Galilée aux prises avec les théories aristotéliciennes de la lumière, 1610–1640', in *XVIIe Siècle* 136 (1982), 267–83

Redondi, P. *Galileo eretico* (Turin, 1983)

Redondi, P. *La luce messaggio celeste*, in Galluzzi 1984, 177–86

Redondi, P. *Galileo heretic* (London, 1988) [see Redondi 1983]

Redondi, P. 'Fede lincea e teologia tridentina', in *Galilæana* 1 (2004), 117–41

Redondi, P. *Libro della natura e libro della scrittura*, in Novati 2009, 91–126

Reeves, E. 'Daniel 5 and the Assayer: Galileo reads the handwriting on the wall', in *Journal of Medieval and Renaissance Studies* 21 (1991), 1–27

Reeves, E. 'Speaking of sunspots: oral culture in an early modern scientific exchange', in *Configurations* 13, 2 (2005), 185–210

Reeves, E. 'Variable stars: a decade of historiography on the *Sidereus Nuncius*,' in *Galilaeana* 8 (2011), 36–72

Reeves, E. and Van Helden, A. *Verifying Galileo's discoveries: telescope-making at the Collegio Romano*, in Hämel-Keil 2007, 127–41

Rerum medicarum novae Hispaniae thesaurus, seu Plantarum animalium mineralium Mexicanorum historia, ex Francisci Hernandez novi orbis medici primarij relationibus in ipsa Mexicana urbe conscriptis, a Nardo Antonio Reccho [...] *collecta ac in ordinem digesta, a Ioanne Terrentio* [...] *notis illustrate* (Rome, 1649), facs. repr. (Rome, 1992)

Ricci, S. 'Accademie e vita civile nella esperienza lincea', in *Quaderni Utinensi* 11–12 (1988a), 165–9

Ricci, S. 'Federico Cesi e la Nova del 1604: la teoria della fluidità del cielo e un opuscolo dimenticato di Joannes van Heeck', in *Atti della Accademia nazionale dei Lincei, Rendiconti, Classe di scienze morali, storiche e filologiche* 385, ser. 8, 43 (1988b), 111–33, subsequently publ. in Ricci 1994, 7–31

Ricci, S. *Secolo antico molto più pubblico, secol d'hoggi molto più privato: Federico Cesi e la scienza come bonum publicum*, in Federico Cesi 1988c, 31–42

Ricci, S. 'Rivoluzione del cielo fisico, riforma del cielo morale: scienza e vita civile da Giordano Bruno ai Lincei', in *Studi filosofici* 12–13 (1989–1990), 245–96, subsequently publ. in Ricci 1994, 61–105

Ricci, S. 'I primi Lincei e l'alchimia: Paracelso superstitione nudatus', in *Atti della Accademia nazionale dei Lincei, Rendiconti, Classe di scienze morali, storiche e filologiche* ser. 9, 3 (1992), 253–73, subsequently publ. in Ricci 1994, 35–57

Ricci, S. *Una filosofica milizia: tre studi sull'Accademia dei Lincei* (Udine, 1994)

Ricci, S. 'Nicola Antonio Stigliola enciclopedista Linceo', together with an edition of the treatise *Delle apparenze celesti* edited by A. Cuna, in *Atti della Accademia nazionale dei Lincei, Memorie, Classe di scienze morali, storiche e filologiche* 393, ser. 9, 8 (1996), 3–149

Ricci, S. *I Lincei: l'invenzione della mediazione accademica: nuova scienza, religione, vita civile*, in Sciences et religions 1999, 205–34

Ricci, S. *Il caso Heckius*, in I Primi Lincei 2005, 207–34

Ricci, S. *I primi Lincei e l'enciclopedismo*, in Graniti 2006, 123–38

Ricci Riccardi, A. *Galileo Galilei e fra Tommaso Caccini: il processo del Galilei nel 1616 e l'abiura segreta rivelata dalle carte Caccini* (Florence, 1902)

Rienstra, M.H. 'Giovanni Ecchio linceo: appunti cronologici e bibliografici', in *Atti della Accademia nazionale dei Lincei, Rendiconti, Classe di scienze morali, storiche e filologiche* ser. 8, 23 (1968), 255–66

Romeo, L., Lupi, F.W., and Pupo, S. eds., *Foscarini e la cosmologia moderna: atti del Convegno internazionale di studi per il quarto centenario della fondazione dell'Accademia montaltina degli Inculti: Cosenza-Montalto Uffugo, 6–7 dicembre 2001* (Cosenza, 2008)

Rosen, E. *The naming of the telescope*, foreword by H. Shapley (New York, 1947)

Rosen, E. Review of Panofsky 1954, in *Isis* 47 (1956), 78–80

Rossi, P. *Immagini della scienza* (Rome, 1977a)

Rossi, P. *Sfere celesti e branchi di gru*, in Rossi 1977b, 109–47

Rossi, P. 'Galileo Galilei e il libro dei Salmi', in *Rivista di filosofia* 69 (1978), 45–71

Rycke, J. *De vita viri praestantissimi Virginii Caesarini Lyncei, Iuliani, civitatis novae ducis, baronis romani, f. liber auctore Iusto Riquio Belga* (Pavia, 1629)

Santagata, S. *Istoria della Compagnia di Giesù appartenente al Regno di Napoli* [...], *Parte terza-Parte quarta* (Naples, 1756–1757)

Sapori, G., Vinti, C. and Conti, L. *Il Palazzo Cesi di Acquasparta e la rivoluzione scientifica lincea* (Perugia, 1992)

Sarpi, P. *Lettere ai protestanti*, 1st critical edition edited by M.D. Busnelli (Bari, 1931), 2 v.

Scheiner, Ch. *Tres epistolae de maculis Solaribus scriptae ad Marcum Velserum [...] cum observationum iconismis* (Augsburg, 1612a), subsequently publ. in OG, V, 21–70

Scheiner, Ch. *De maculis Solarib. et stellis circa Iouem errantibus, accuratior disquisitio ad Marcum Velserum [...] perscripta, interiectis obseruationum delineationibus* (Augsburg, 1612b)

Scheiner, Ch. *Disquisitiones mathematicae de controuersiis et nouitatibus astronomicis, quas sub praesidio Christophori Scheiner [...] publice disputandas posuit, propugnauit, mense Septembri die [...] Joannes Georgius Locher [...]* (Ingolstadt, 1614)

Scheiner, Ch. *Rosa Ursina, sive Sol ex admirando facularum et macularum suarum phoenomenon varius necnon circa centrum suum et axem fixum ab occasu in ortum annua, circaque alium axem mobilem ab ortu in occasum conversione quasi menstrua, super polos proprios libris quatuor mobilis ostensus, a p. Christophoro Scheiner [...]* (Bracciano, 1626–1630)

Scheiner, Ch. *Prodromus pro Sole mobili et Terra stabili contra [...], Galilaeum a Galilaeis authore r.p. Christophoro Scheinero [...], ante annos 20 et amplius elucubratus, qui nunc primum in publicam lucem prodit [...]* (*S.l.*, 1651)

Schettini Piazza, E. *Spigolando nell'archivio Cesi*, in Battistini-De Angelis-Olmi 2007, 467–79

Schettini Piazza, E. *The Sidereus Nuncius in Federico Cesi's library*, in Brückle-Hahn 2011, 23–30

Schofield, C.J. *Tychonic and semi-Tychonic world systems* (New York, 1981)

Schofield, C.J. *Tychonic and semi-Tychonic world systems*, in Hoskin 1984, 33–44

Sciences et religions de Copernic à Galilée, 1540–1610: Rome 12–14 décembre 1996: actes du Colloque international organisé par l'École française de Rome (Rome, 1999)

Scienze, credenze occulte, livelli di cultura: Convegno internazionale di studi: Firenze, 26–30 giugno 1980 (Florence, 1982)

Scott, A.C. and Freedberg, D. 'Fossil woods and other geological specimens', with documentation provided by F. Solinas, with contributions by J. Taylor (London, 2000), ser. B, pt. 3 of *The Paper museum of Cassiano dal Pozzo: a catalogue raisonné: drawings and prints in the Royal library at Windsor castle, the British museum, the Institut de France and other collections*, general eds. F. Haskell, J. Montagu, assistant general ed. H. McBurney (London 1997)

Sirri, R. and Torrini, M. eds., *Bernardino Telesio e la cultura napoletana* (Naples, 1992)

Sirtori, G. *Hieronymi Sirturi [...] Telescopium, siue Ars perficiendi nouum illud Galilaei visorium instrumentum ad sydera: in tres partes diuisa, quarum prima exactissimam perspicillorum artem tradit, secunda telescopii Galilaei absolutam constructionem et artem aperte docet, tertia alterius telescopii faciliorem vsum et admirandi sui adinuenti arcanum patefacit [...]* (Frankfurt am Main, 1618)

Sizi, F. *Dianoia astronomica, optica, physica, qua Syderei nuncij rumor de quatuor planetis a Galilaeo Galilaeo [...] recens perspicilli cuiusdam ope conspectis, vanus redditur, auctore Francisco Sitio [...]* (Venice, 1611), subsequently publ. in OG, III.1, 205–50 (with postils by Galileo)

Snell, W. *De maculis in Sole animadversis et tamquam ab Apelle in tabula spectandum in publica luce expositis Batavi dissertatiuncula* (Leyden, 1612)

Solinas, F. ed., *Cassiano dal Pozzo: atti del Seminario internazionale di studi* (Rome, 1989a)

Solinas, F. *Il Trattato del legno fossile minerale di Francesco Stelluti e i quattro volumi della Natural history of fossils nelle raccolte della Biblioteca reale di Windsor*, in Haskell 1989b, 84–94

Solinas, F. ed., *I segreti di un collezionista: le straordinarie raccolte di Cassiano dal Pozzo (1588–1657)* (Rome, 2000)

Stabile, G. *Lo statuto di 'inesorabile' in Galileo Galilei*, in Hamesse-Fattori 2003, 269–85

Stelliola, N.A. *Il telescopio over Ispecillo celeste, di Nicolò Antonio Stelliola linceo* (Naples, 1627)

Stelluti, F. *Il pegaso, epithalamio del signor Francesco Stelluti linceo nelle nozze degl'illustrissimi et eccell.mi principi Federico Cesi et Isabella Salviati* (Rome, 1617)

Stelluti, F. *Trattato del legno fossile minerale nuovamente scoperto nel quale breuemente si accenna la varia et mutabil natura di detto legno, rappresentatoui con alcune figure che mostrano il luogo doue nasce, la diuersità dell'onde che in esso si vedono e le sue così varie e marauigliose forme, di Francesco Stelluti accad. Linceo da Fabriano [...]* (Rome, 1637)

Stelluti, F. and Rycke, J. *Urbano VIII Pont. Opt. Max. cum accuratior Melissographia à Lynceorum Academia, in perpetuae devotionis symbolum ipsi offeretur Franciscus Stellutus Lynceus Fabris microscopio observabat* (Rome, 1625)

Stendardo, E. *Ferrante Imperato: collezionismo e studio della natura a Napoli tra Cinque e Seicento* (Naples, 2001)

Storia della letteratura italiana, directors E. Cecchi, N. Sapegno, texts by A. Roncaglia [...] *et al.* (Milan, 1967)

Strano, G. *Orologi astronomici e planetari*, in Brusa 2005, 129–39

Strano, G. ed., *Il telescopio di Galileo: lo strumento che ha cambiato il mondo* (Florence, 2008)

Strano, G. *L'astrolabio linceo*, in Galluzzi 2014a, 539–44

Telesio, B. *Bernardini Telesii Consentini, Varii de naturalibus rebus libelli ab Antonio Persio editi, quorum alij nunquam antea excusi, alij meliores facti prodeunt, sunt autem hi De cometis, et lacteo circulo, De his, quae in aere fiunt, De iride, De mari, Quod animal uniuersum, De usu respirationis, De coloribus, De saporibus, De somno, Unicuique libello appositus est capitum Index* (Venice, 1590), subsequently publ. in Telesio 1981

Telesio, B. *De rerum natura iuxta propria principia*, text, foreword, translation and notes edited by L. De Franco (Cosenza, 1965–1974), 2 v.

Telesio, B. *De rerum natura: libri VII-VIII-IX*, critical text and Italian translation by L. De Franco (Florence, 1976)

Telesio, B. *Varii de naturalibus rebus libelli*, critical text by L. De Franco (Florence, 1981)

Telesio, B. *De iis quae in aere fiunt et de terraemotibus, De mari*, with an Italian translation by F. Martelli, ed. L. De Franco (Cosenza, 1990)

Tognoni, F. *Naturamque novat: simboli e significati delle medaglie di Galileo* (Pisa, 2013)

Torrini, M. 'Galileo copernicano', in *Giornale critico della filosofia italiana 72*, ser. 6, 13 (1993), 25–42

Torrini, M. 'La natura della nuova scienza', in *Nuncius* 17 (2002), 409–22

Torrini, M. *Far divenir libera la natural filosofia e le scienze tutte*, in I Primi Lincei 2005, 405–19

Torrini, M. *Le scienze e le accademie, 1600–1800*, in Vesentini-Mazzarolli 2006, 49–67

Trabucco, O. *Il corpus fisiognomico dellaportiano tra censura e autocensura*, in I Primi Lincei 2005, 234–70

Turner, G. L'E. 'The Florentine Workshop of Giovan Battista Giusti 1556-c.1575', in *Nuncius* 10 (1995), 131–72

Valleriani, M. *Galileo Engineer* (Dordrecht-Heidelberg-London-New York, 2010)

Van Helden, A. *The invention of the telescope* (Philadelphia, 1977)

Van Helden, A., Dupré, S., Van Gent, R., Zuidervaart, H. eds., *The origins of the telescope* (Amsterdam, 2010)

Vesentini, E. and Mazzarolli, L. eds., *L'esperienza delle accademie e la vita morale e civile dell'Europa* (Venice, 2006)

Vinti, C. ed., *Galileo e Copernico: alle origini del pensiero scientifico moderno* (Assisi, 1990)

Vocabolario degli Accademici della Crusca (Venice, 1612)

Von Pastor, L. *Storia dei papi nel periodo della Restaurazione cattolica e della Guerra dei trent'anni: Gregorio XV (1621–1623) e Urbano VIII (1623–1644)*, Italian version by P. Cenci (Rome, 1931), vol. 13 of *Storia dei papi dalla fine del Medio Evo: compilata col sussidio dell'Archivio segreto pontificio e di molti altri archivi* (Rome, 1908–1934)

Wackernagel, R.H. *Der Französischer Krönungswagen von 1696–1825* (Berlin, 1610)

Wedderburn, J. *Quatuor problematum quae Martinus Horky contra Nuntium sidereum de quatuor planetis nouis disputanda proposuit, confutatio per Ioannem Wodderbornium Scotobritannum* (Padua, 1610)

Westfall, R. *Galileo and the Accademia dei Lincei*, in Galluzzi 1984, 189–200

Wilding, N. *Faussaire de la Lune. Autopsie d'un imposture, Galilée et ses contrefacteurs*. Conference Léopold Delisle, Bibliothèque Nationale de France (Paris, 2015)

Ziggelaar, A. *Jesuit astronomy North of the Alps: four unpublished Jesuit letters, 1611–1620*, in Baldini U. 1995, 101–32

Index of Names*

by Laura Manetti

* There are no entries for Galileo Galilei and Federico Cesi as their names appear on almost every page. The names of mythological characters are italicized.

www.ingramcontent.com/pod-product-compliance
Lightning Source LLC
LaVergne TN
LVHW010637300926
849161LV00005B/969